OLNEY'S MATHEMATICAL SERIES.

A

GENERAL GEOMETRY

AND

CALCULUS.

INCLUDING BOOK I. OF THE GENERAL GEOMETRY, TREATING OF LOCI IN A PLANE; AND AN ELEMENTARY COURSE IN THE DIFFERENTIAL AND INTEGRAL CALCULUS.

BY

EDWARD OLNEY,

PROFESSOR OF MATHEMATICS IN THE UNIVERSITY OF MICHIGAN.

NEW YORK:
SHELDON AND COMPANY.

PREFACE.

THIS volume presents a course in the General Geometry and the Infinitesimal Calculus, which is thought to be as extended as is practicable for the general student in the regular undergraduate course in our American colleges. If we can secure a sufficiently high grade of preparation, so that students in the Freshman year can complete a respectable course in Elementary Geometry, including Plane and Spherical Trigonometry, and in Algebra, it is thought that during the Sophomore year the contents of this volume can be readily mastered. Such is the purpose in this University; and it is already well nigh realized.

As to the propriety of including the study of both these subjects in the regular undergraduate course, there can be but one opinion among those competent to judge. No man can justly claim to have a good general education, who is ignorant of the elements of the processes by which all extended operations in the exact sciences are carried forward, and which are the foundation of all the arts based upon mathematical science. The man who is ignorant of the General Geometry and the Calculus, is not only a stranger in one of the sublimest realms of human thought, but knows nothing of the instruments in most familiar use by the engineer, the astronomer, and the machinist in any of the higher walks of art. In short he is ignorant of the characteristic processes of the mathematician of his day.

Nor is it impracticable for the majority of students to become intelligent in these subjects. They do not lie beyond the reach of good common minds, nor require peculiar mental characteristics for their mastery. The difficulty hitherto has been in the methods of presentation, in the limited and totally inadequate amount of time assigned them, and more than all in the preconceived notion of their abstruseness.

The mathematician will see in the plan of the first part of

this volume, as well as in its title (*General* Geometry), a recognition of the profound views of *Comte* upon the philosophy of the science. This science is *a method of Geometrical reasoning*. Its characteristic feature is that it represents *form*, as well as magnitude, by equations, and hence makes algebra its instrument. It is consequently indirect. Its ultimate object is breadth of comprehension,—the discussion of general problems. In accordance with this conception, the first purpose is to exhibit the method of translating geometrical forms into algebraic equations, *i. e.* to show how loci are represented by equations. While the prominence is given to the Conic Sections which their importance in physical science demands, the student is not led to think that this is merely a scheme for treating these curves. He is taught to look upon it as a method of investigation—as designed to embrace the discussion of all loci. For this purpose many Higher Plane Curves are treated. After the student has become familiar with the equation as the representative of a locus, and has learned how to produce the equation of a locus from its definition, he has obtained the instrument. He is now to learn how to apply it for the purposes of Geometrical investigation. In carrying forward this part of his study the Calculus renders invaluable service. Moreover, this preparatory study of the General Geometry gives him exactly the needed means for illustrating the elementary processes of the Calculus. He has, therefore, come to a point where his further progress requires a knowledge of the Differential Calculus, and he has also the requisite preparation for its study. Hence, after having become familiar with the first three chapters of General Geometry, he reads the Differential Calculus.

By this arrangement the Calculus is seen in its true relations, as an independent abstract science, grand and beautiful in itself, and rendering most efficient service in the more immediately practical science of Geometry, as it is afterwards seen to do in Physics. Having obtained the needed acquaintance with the Differential Calculus, the student returns to pursue his Geometrical studies, with equations of loci as his instruments, and the Calculus to aid in the manipulation of them. But the peculiar features of the treatise are too numerous to be enumerated here, and can be seen in their true light only by a perusal of the work.

In the treatment of the Calculus I have used the Infinitesimal method instead of the method of Limits, on account of its greater simplicity, as well as because it is the only conception which enables us to apply the Calculus to practical problems with any degree of facility. The general use of the method of limits in our text books has done not a little to prevent the common study of this elegant and useful branch of mathematics. This method is not only exceedingly cumbrous, but it has the misfortune that its element, a differential coefficient, is a ratio. The abstract nature of a ratio, and the fact that it is a compound concept, peculiarly unfit it for elementary purposes. The beginner will never use it with satisfaction, for it does not give him simple, direct and clearly defined conceptions. But while I have adopted the infinitesimal theory, I have felt free to introduce the doctrine of limits, and to illustrate and apply it. The metaphysical objections to this method, if not rebutted by equal difficulties of a similar character encountered in the method of limits, are immensely overborne by its practical advantages; for, let it be remembered that no writer adheres to the Newtonian method throughout, but glides into the other in the Integral Calculus, and adopts it exclusively in most geometrical and physical applications.

The sources from which the material has been drawn will be readily perceived by the mathematician, and need not be enumerated here. That the treatise is sufficiently different from others of a similar purpose to justify its existence, the author feels more sure than that these differences will commend themselves to his fellow laborers in the work of mathematical training. One thing, however, is certain, nothing in matter, arrangement, or manner of treatment, has been introduced without careful reference to the capabilities and wants of such students as I have been accustomed to meet in the class room for more than twenty years; and few things will be found in the volume but what have been put to the test of class room use many times over.

A second volume, treating of Loci in Space, and affording a more extended course in the Calculus, will be published as soon as it can be prepared. The present is thought sufficient for all students except such as make mathematics a specialty; and for the latter the other volume will be designed.

In conclusion I must do myself the pleasure to acknowledge my indebtedness to my accomplished colleague and friend, Prof. J. C. Watson, Ph. D., for the original, direct, and simple method of demonstrating the rule for differentiating a logarithm, which is given on page 25, and which banishes from the Calculus the last necessity for resort to series to establish any of its fundamental operations. I am also indebted to my friend and pupil, J. B. Webb, B. S., for many valuable suggestions, and much careful labor in reading both the manuscript and proof. To his quick and accurate eye, and his good taste and logical acumen, I am indebted for the elimination of not a few defects which might otherwise have disfigured the work. That there is not much of the same sort of pruning yet needed, I have not the vanity to think. But, such as it is, I commend my work to the consideration of teacher and student, with the hope that it may contribute to aid the one in imparting, and the other in acquiring, a knowledge of the elements of two branches of science which, in their fuller developments, exhibit the profoundest and most sagacious workings of the human mind, and reach to the farthest verge of the hitherto explored realms of human thought.

EDWARD OLNEY.

ANN ARBOR, Mich., *July*, 1871.

N. B.—*A shorter course in the General Geometry, without the Calculus, may be taken from this volume by such as desire it. For this purpose, the first three chapters are to be read, and then the course completed by reading the XIV. and XV. Sections of Chapter IV. If time and purpose permit, Articles* (**194, 195**) *might be read with profit by such students. This will be found to comprise a course on Plane Co-ordinate Geometry somewhat more full than is found in our common text-books.*

CONTENTS.

INTRODUCTION.

A BRIEF SURVEY OF THE OBJECTS OF PURE MATHEMATICS AND OF THE SEVERAL BRANCHES.

GENERAL GEOMETRY.

BOOK I.

OF PLANE LOCI.

CHAPTER I.

THE CARTESIAN METHOD OF CO-ORDINATES.

SECTION I.

DEFINITIONS AND FUNDAMENTAL NOTIONS.

SECTION II.

CONSTRUCTING EQUATIONS, OR FINDING THEIR LOCI.

SECTION III.

THE POINT IN A PLANE.

SECTION IV.

THE RIGHT LINE IN A PLANE.

SECTION V.

OF PLANE ANGLES, AND THE INTERSECTION OF LINES.

SECTION VI.

OF THE CONIC SECTIONS.

SECTION VII.

EQUATIONS OF HIGHER PLANE CURVES.

CHAPTER II.

THE METHOD OF POLAR CO-ORDINATES.

SECTION I.

OF THE POINT IN A PLANE.

SECTION II.

OF THE RIGHT LINE.

SECTION III.

OF THE CIRCLE.

SECTION IV.

OF THE CONIC SECTIONS.

SECTION V.

OF HIGHER PLANE CURVES.

OF PLANE SPIRALS.

CHAPTER III.

TRANSFORMATION OF CO-ORDINATES.

SECTION I.

PASSING FROM ONE SET OF RECTILINEAR AXES TO ANOTHER.

SECTION II.

PASSING FROM RECTILINEAR TO POLAR CO-ORDINATES, AND VICE VERSA.

CHAPTER IV.

PROPERTIES OF PLANE LOCI INVESTIGATED BY MEANS OF THE EQUATIONS OF THOSE LOCI.

SECTION I.

TANGENTS TO PLANE LOCI.

(*a*) BY RECTILINEAR CO-ORDINATES.

PAGE

(*b*) TANGENTS TO POLAR CURVES.

SECTION II.

NORMALS TO PLANE LOCI.

(*a*) BY RECTANGULAR CO-ORDINATES.

(*b*) NORMALS TO POLAR CURVES.

SECTION III.

DIRECTION OF CURVATURE.

(*a*) BY RECTANGULAR CO-ORDINATES

(*b*) BY POLAR CO-ORDINATES.

SECTION IV.

SINGULAR POINTS.

SECTION V.

TRACING CURVES.

SECTION VI.

RATE OF CURVATURE.

PAGE

SECTION VII.

EVOLUTES AND INVOLUTES.

SECTION VIII.

ENVELOPES TO PLANE CURVES.

SECTION IX.

RECTIFICATION OF PLANE CURVES.

SECTION X.

QUADRATURE OF PLANE SURFACES.

PAGE

SECTION XI.

QUADRATURE OF SURFACES OF REVOLUTION.

SECTION XII.

CUBATURE OF VOLUMES OF REVOLUTION.

SECTION XIII.

EQUATIONS OF CURVES DEDUCED BY THE AID OF THE CALCULUS.

SECTION XIV.

OF TANGENTS AND NORMALS.

[WITHOUT THE AID OF THE CALCULUS.]

SECTION XV.

SPECIAL PROPERTIES OF THE CONIC SECTIONS.

THE

INFINITESIMAL CALCULUS.

INTRODUCTION.

CHAPTER I.

THE DIFFERENTIAL CALCULUS.

SECTION I.

DIFFERENTIATION OF ALGEBRAIC FUNCTIONS.

RULES FOR DIFFERENTIATING ALGEBRAIC FUNCTIONS.

SECTION II.

DIFFERENTIATION OF LOGARITHMIC AND EXPONENTIAL FUNCTIONS.

SECTION III.

DIFFERENTIATION OF TRIGONOMETRICAL AND CIRCULAR FUNCTIONS.

SECTION IV.

SUCCESSIVE DIFFERENTIATION AND DIFFERENTIAL COEFFICIENTS.

SECTION V.

FUNCTIONS OF SEVERAL VARIABLES, PARTIAL DIFFERENTIATION, AND DIFFERENTIATION OF IMPLICIT AND COMPOUND FUNCTIONS.

SECTION VI.

SUCCESSIVE DIFFERENTIATION OF FUNCTIONS OF TWO INDEPENDENT VARIABLES, AND OF IMPLICIT FUNCTIONS.

SECTION VII.

CHANGE OF INDEPENDENT VARIABLE.

CHAPTER II.

APPLICATIONS OF THE DIFFERENTIAL CALCULUS.

SECTION I.

DEVELOPMENT OF FUNCTIONS.

SECTION II.

EVALUATION OF INDETERMINATE FORMS.

SECTION III.

MAXIMA AND MINIMA OF FUNCTIONS OF ONE VARIABLE.

CHAPTER III.

THE INTEGRAL CALCULUS.

SECTION I.

DEFINITIONS AND ELEMENTARY FORMS.

SECTION II.

RATIONAL FRACTIONS.

SECTION III.

RATIONANIZATION.

SECTION IV.

INTEGRATION BY PARTS.

SECTION V.

INTEGRATION BY INFINITE SERIES.

SECTION VI.

SUCCESSIVE INTEGRATION.

SECTION VII.

DEFINITE INTEGRATION AND THE CONSTANTS OF INTEGRATION.

INTRODUCTION.

A BRIEF SURVEY OF THE OBJECT OF PURE MATHEMATICS AND OF THE SEVERAL BRANCHES.

1. *Pure Mathematics* is a general term applied to several branches of science, which have for their object the investigation of the properties and relations of quantity—comprehending number, and magnitude as the result of extension—and of form.

2. *The Several Branches* of Pure Mathematics are Arithmetic, Algebra, Calculus, and Geometry.

3. Arithmetic, Algebra, and Calculus treat of number; and Geometry treats of form and magnitude as the result of extension.

4. *Quantity* is the amount or extent of that which may be measured; it comprehends number and magnitude.

The term quantity is also conventionally applied to symbols used to represent quantity. Thus 25, m, xi, etc., are called quantities, although, strictly speaking, they are only representatives of quantities.

Sch. 1.—It is not easy to give a philosophical account of the idea or ideas represented by the word *Quantity* as used in Mathematics; and, doubtless, different persons use the word in somewhat different senses. It is obviously incorrect to say that "Quantity is anything which can be measured." Quantity may be affirmed of any such concept; nevertheless, it is not the thing itself, but rather the *amount* or *extent* of it. Thus, a load of wood, or a piece of ground, can be measured; but no one would think of the wood or the ground as being the quantity. The *quantity* (of wood or ground) is rather, the *amount* or extent of it. The word is very convenient as a general term for mathematical concepts, when we wish to speak of them without indicating whether it is number or magnitude that is meant. Thus we say, "m represents a certain quantity," and do not care to be more specific.

As applied to number, perhaps the term conveys the idea of the whole, rather than of that whole as made up of parts. It is, therefore, scarcely proper to speak of multiplying by a quantity; we should say, by a number.

On the other hand, when we apply the term quantity to magnitude, it is with the idea that magnitude may be measured, and thus expressed in number.

The distinction between quantity and number is marked by the questions, "How much?" and "How many?"

SCH. 2.—So, also, the word *Magnitude*, as used in mathematics, is not easily defined. Sometimes it has reference to quantity in the aggregate, or mass, and sometimes to the relation which one quantity bears to another. Thus, we speak of a line, a surface, or a solid, as a magnitude, simply meaning thereby that these have extent,—are extended. A circle, a triangle, a cube, are magnitudes,—*i. e.*, they have extension. Again, we speak of the magnitude of a circle, meaning its size,—area as compared with some other surface. The magnitude of a line is expressed by telling how many times it contains another line of known length. In like manner the magnitude of a surface or a volume is made known by comparing the surface with some unit of surface, and the volume with some unit of volume. In one aspect, therefore, number is an expression for the ratio of magnitudes.

5. ***Number*** is quantity conceived as made up of parts, and answers to the question, "How many?"

ILLUSTRATION.—Thus, a distance is a quantity; but if we call that distance 5, we convert the notion into number, by indicating that the distance under consideration is made up of parts. Now, the distance may be just the same, whether we consider it as a whole, or think of it as 5,—*i. e.*, as made up of 5 equal parts. Again, m may mean a value, as of a farm. We may or may not conceive it as a number (as of dollars). If we think of it simply in the aggregate, as the worth of a farm, m represents quantity; but if we think of it as made up of parts (as of dollars), it is a number.

6. Number is of two kinds, ***Discontinuous*** and ***Continuous.***

7. ***Discontinuous Number*** is number conceived as made up of finite parts; or it is number which passes from one state of value to another by the successive additions or subtractions of finite units, —*i. e.*, units of appreciable magnitude.

8. ***Continuous Number*** is number which is conceived as composed of infinitesimal parts; or it is number which passes from one state of value to another by passing through all intermediate values, or states.

ILL.—The method of conceiving number with which the pupil has become familiar in arithmetic and algebra, characterizes *discontinuous number.* Thus the number 13 is conceived as produced from 5 by the successive additions of finite units, either integral or fractional. In either case we advance by successive steps of *finite* length. If we say 5, 6, 7, etc., till we reach 13, we pass by one kind of steps; and, if we say 5.1, 5.2, 5.3, etc., till we reach 13, we pass by another sort of steps (*tenths*), but as really by *finite* ones. If, however, we call the line **AB**, *Fig.* 1, x, and **CD**, x', and conceive **AB** to slide to the position **CD**, increasing in length as it moves so as to keep its extremities in the lines **OM** and **ON**, it will pass by infinitesimal elements of growth from the value x, to the value x'; or, it will pass from one value to the other by passing through all intermediate values, and thus becomes an illustration of *continuous* number.

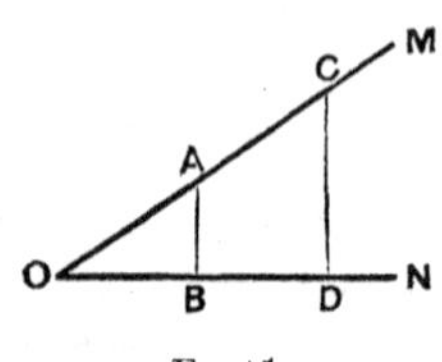

Fig. 1.

Again, if the line **AB**, *Fig.* 2, be considered as generated by a point moving from **A** to **B**, and we call the portion generated when the point has reached **C**, x, and the whole line x', x will pass to x' by receiving infinitesimal increments, or by passing through all states of value between x and x'.

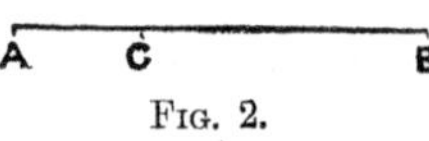

Fig. 2.

A surface may be considered as generated by the motion of a line, and thus afford another illustration of continuous number. Thus let the parallelogram **AF** be conceived as generated by the right line **AB** moving parallel to itself from **AB** to **EF**. When **AB** has reached the position **CD**, call the surface traced, namely **ABCD**, x, and the entire surface **ABEF**, x'; then will x pass to x' by receiving infinitesimal increments, or by passing through all intermediate values.

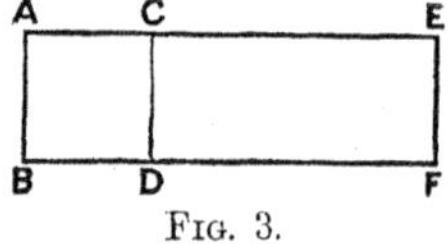

Fig. 3.

Finally, as volumes may be conceived as generated by the motion of planes, all geometrical magnitudes afford illustrations of continuous number.

We usually conceive of *time* as discontinuous number, as when we think of it as made up of hours, days, weeks, etc. But it is easy to see that such is not the way in which time actually grows. A period of one day does not grow to be a period of one week by taking on a whole day at a time, or a whole hour, or even a whole second. It grows by imperceptible increments (additions). These inconceivably small parts of which continuous number is made up are called *Infinitesimals.*

Motion and force afford other illustrations of continuous number. In fact, the conception which regards number as continuous, will be seen to be less artificial—more true to nature—than the conception of it as discontinuous.

9. *Arithmetic* treats of *Discontinuous Number*, of its nature and properties, of the various methods of combining and resolving it, and of its application to practical affairs.

For an outline of the topics of Arithmetic, see the Complete School Algebra, Art. 9.

10. *Algebra* treats of the *Equation*, and is chiefly occupied in explaining its nature, and the methods of transforming and reducing it, and in exhibiting the manner of using it as an instrument for mathematical investigation.

For a full account of the province of Algebra, see the Complete School Algebra, ART. 10.

11. *Calculus* (The Infinitesimal Calculus) treats of *Continuous* Number, and is chiefly occupied in deducing the relations of the infinitesimal elements of such number from given relations between finite values, and the converse process, and also in pointing out the nature of such infinitesimals and the methods of using them in mathematical investigation.

12. *Geometry* treats of *magnitude* and form as the result of extension and position.

SCH. 1.—The principal divisions of the science of Geometry are :

1. The Ancient, Platonic, Special, Graphic, or Direct Geometry (the common Geometry of our schools), including Trigonometry, Conic Sections, and all other geometrical inquiries conducted upon these methods.

2. The Analytical, Modern, Cartesian, General, or Indirect Geometry (the theme of this volume), and

3. Descriptive Geometry.

SCH. 2.—The first system of geometrical investigation probably took its rise *as a science* in the school of Plato (about 400 B. C.), and was brought almost to its present state of perfection, as far as its methods are concerned, by the time of Euclid (about 300 B. C.); hence it is called the Ancient or Platonic Geometry. As the argument is carried forward by a direct inspection of the forms (figures) themselves, delineated before the eye, or held in the imagination, it is called the Direct or Graphic method. Inasmuch as it discusses *particular* instead of *general* problems it is properly characterized as *Special.* With this method of geometry the student is supposed to be acquainted before commencing the study of this volume.

The fundamental notion of the Modern Geometry (a system of coordinates), was developed by Des Cartes in the earlier part of the 17th century, and hence the names *Modern* or *Cartesian.* The term *Analytical* has come to be applied in mathematics in the sense of Algebraical, all investigations carried forward chiefly by the aid of Algebra being called Analytical. This use of the term is quite unfortunate, inasmuch as the processes of Algebra are no more analytical, in the true sense of that term, than are those of the Special Geometry. Again, as a name for the General Geometry, even if used in the sense of algebraic, the term does not distinguish the *system* from any other application of algebra to geometry.

The true character of the Modern Geometry is expressed by the terms *Indirect*, and *General.* This system of geometrical reasoning proposes the solution of *general* problems, and effects its purpose by first translating geometrical forms into equations, then carrying forward the investigation by means of these equations, and finally returning to the geometrical forms by a re-translation. The *indirectness* of this method is apparent, and might seem, in itself, a serious objection; but it is found to be of great advantage, inasmuch as it makes the discussions much more comprehensive (*general*). To illustrate this *general* (comprehensive) character of its discussions, we have only to notice some of its problems. Thus the *Special Geometry* discusses the problem of the tangent to a circle, and, on an independent basis, investigates the properties of a tangent to any other curve, making a special problem with respect to each separate curve studied. On the other hand, the *General Geometry* proposes the problem in this way: *To find a formula sufficiently general to embrace the properties of tangents to* ALL *plane curves*;—in technical language, *To find the equation of the tangent to any plane curve.* Again, in the Special Geometry, the area of a *circle* is obtained (approximately). But the General Geometry proposes to investigate the problem on a broader basis, and find a formula which shall be applicable in finding the area of *any* plane curve.

13. *Descriptive Geometry* is that system of geometry which seeks the graphic solution of geometrical problems by means of projections upon auxiliary planes.

This is the ordinary definition of the Descriptive Geometry, and it would be out of place to attempt any elucidation of it here.

SCH.—From the definition of Geometry, as well as from the detailed study of its propositions, it will be seen to embrace two classes of problems; viz., *Problems relating to Position*, and *Problems relating to Magnitude.* Problems of the latter class were solved by the aid of algebra before the time of Des Cartes; but it was reserved for him to invent a method by which problems of both kinds could be so discussed. This system constitutes the foundation of the General Geometry.

14. The inquiries in the General Geometry may be divided into two classes, *viz.:*

1. Concerning Plane Loci,
2. Concerning Loci in Space.

In accordance with this division the present treatise is divided into *Two Books.**

SCH.—This division is found especially convenient when the subject is treated by the aid of the Calculus, as it corresponds to the distinction between functions of *a single* variable, and functions of *two* variables.

---

* The Second Book is reserved for another volume, which will also contain an advanced course in the Calculus.

BOOK I.

OF PLANE LOCI.

CHAPTER 1.

THE CARTESIAN METHOD OF CO-ORDINATES.

SECTION I.

Definitions and Fundamental Notions.

1. The term ***Locus*** as used in geometry is nearly synonymous with *geometrical figure*, yet having a latitude in its use which the latter term does not possess. The locus of a point is the line (geometrical figure) generated by the motion of the point according to some given law. In the same manner, a surface is conceived as the locus of a line moving in some determinate manner.

2. The General Geometry is a system of geometrical investigation in which the loci under consideration are represented by equations, and the inquiries carried forward by means of these equations, the final object being the discussion of general problems.

[NOTE.—While it is true that the only way to obtain a full comprehension of the nature of a science is by the detailed study of its parts, it is, nevertheless, important, at the outset, to comprehend as clearly as possible the *general aim* of the science, in order that the *tendency* of the several steps in our progress may be perceived, and the *symmetry* and *unity* of the whole may appear. According to our definition it will be our first purpose to exhibit a scheme by which points, lines straight and curved, the magnitude of angles, surfaces, etc., which we have characterized as "geometrical forms" (loci), may be represented by equations. This will be done in Section 1st of this chapter. Section 2nd will then exhibit a method of constructing the geometrical figure represented by any given equation. Then will follow a series of sections showing how the equations of loci are derived from the definitions of the figures. This series of sections comprises what may be termed the translation of geometrical forms into algebraic equations, and will answer such questions as: "What equations represent points? What straight lines? What circles? What ellipses? etc., etc." Section 2nd, which shows how equations are translated into geometrical forms, might, perhaps, with strict logical propriety, *follow* instead of precede this series of sections; but it is thought the present arrangement will promote clearness of conception. The first three chapters will be seen to be *preparatory*. It is not their purpose to develop geometrical truths, but simply to prepare instruments (the equations of loci) to be subsequently used in conducting geometrical inquiries. In the fourth chapter it will be our purpose to show how geometrical truth can be developed by means of these equations.]

3. A device by means of which we are enabled to represent loci by equations is called a

METHOD OF CO-ORDINATES.

4. There are two systems of co-ordinates in common use, viz.:

1. The system of Rectilinear Co-ordinates,
2. The system of Polar Co-ordinates.

5. There are two varieties of the rectilinear system of co-ordinates, the *rectangular* and the *oblique*. (In our study, the rectangular system will always be used unless otherwise specified.)

6. In order to locate a point in a plane by the method of rectilinear co-ordinates, two lines intersecting each other are assumed as fixed in position. These lines are called ***Axes of Reference,*** or, simply, ***The Axes.*** The system is called rectangular or oblique, according as these lines make a right or an oblique angle with each other.

7. One of these axes is called the ***Axis of Abscissas,*** and the other is called the ***Axis of Ordinates.***

8. ***The Origin*** is the intersection of the axes.

9. ***The Co-ordinates*** of a point are its distances from the axes, the distance to either axis being measured on a line parallel to the other, or on that other axis.

10. ***The Abscissa*** of a point is the co-ordinate which is measured parallel to or on the axis of abscissas, and is the distance of the point from the axis of ordinates measured on a line parallel to the axis of abscissas.

11. ***The Ordinate*** of a point is the co-ordinate which is measured parallel to or on the axis of ordinates, and is the distance of the point from the axis of abscissas measured on a line parallel to the axis of ordinates.

SCH. 1.—These lines, when spoken of separately, should be distinguished as abscissa and ordinate; but, when taken together, they are called co-ordinates.

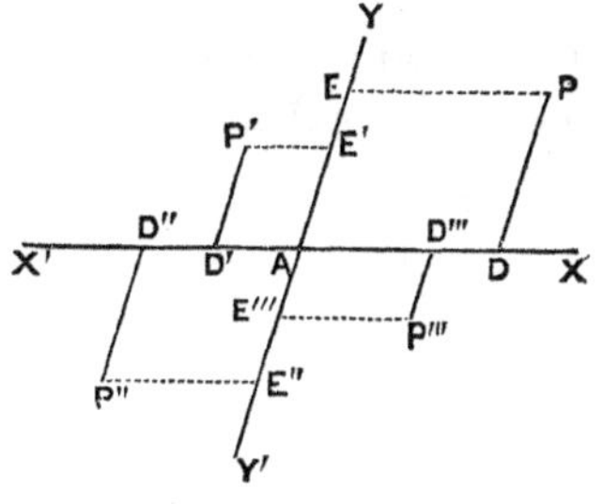

FIG. 4.

ILL.—Definitions 3 to 11 may be illustrated thus: Let the plane in which the loci are situated be represented by the surface of the paper, *Fig.* 4. In this plane assume two fixed, indefinitely extended, straight lines, as **XX'** and **YY'**, intersecting each other at **A**, and to which all points in the plane are to be referred. These lines are the *Axes*, and **A** is the *origin*, *i.e.*, the point at which the co-ordinates are conceived to originate, and from which they are reckoned. One of these lines, as **XX'** (in ordinary use the horizontal one) is called the *Axis of Abscissas*, because abscissas are reckoned on it; and the other, **YY'**, is for a like reason called the *Axis of Ordinates*. The system is called *rectangular* or oblique according as **YAX** is a right or an oblique angle. It is evident that we can now define the position of any point in this plane by giving its distances from these two fixed lines, or axes. For convenience, we measure these distances on lines parallel to the axes. (In the case of rectangular axes, the co-ordinates will become the perpendicular distances of points from the axes.) Thus the location of the point **P** is determined by giving the lengths of **PE**, the abscissa of **P**, and of **PD**, the ordinate of the point. Usually, **AD** is called the abscissa, instead of **PE**. **PD** and **AD** taken together are called the co-ordinates of the point **P**.

SCH. 2.—The pupil will see that this device for locating points is not unlike the method of locating places on the earth's surface by means of latitude and longitude.

12. Abscissas are represented in the notation by the letter x, and ordinates by y.

13. The four angles into which the plane is divided by the axes are distinguished thus: The angle *above* the axis of abscissas and at the *right* of the axis of ordinates is called the *First Angle;* and the numbering proceeds from right to left. **YAX** is the *First Angle*, **YAX'** is the *Second*, **X'AY'** is the *Third*, and **XAY'** is the *Fourth*.

14. In order to indicate in which of the four angles a point is located, the signs + and — are used on the following principles: abscissas reckoned from the origin to the right are marked +, and those reckoned to the left are —; ordinates reckoned upward from the axis of abscissas are +, and those reckoned downwards are —. Accordingly, the abscissas of points in the 1st and 4th angles, as **AD** and **AD'''** are +, while those in the 2nd and 3rd angles as, **AD'** and **AD''**, are —. Ordinates in the 1st and 2nd angles, as **PD** and **P'D'**, are +, and those in the 3rd and 4th, as **P''D''** and **P'''D'''** are —.

15. The quantities used in General Geometry are distinguished as *Constant* and *Variable.*

16. A constant quantity is one which maintains the same value throughout the same discussion, and is represented in the notation by one of the leading letters of the alphabet.

17. Variable quantities are such as may assume in the same discussion any value, within certain limits determined by the nature of the problem,* and are represented by the final letters of the alphabet.

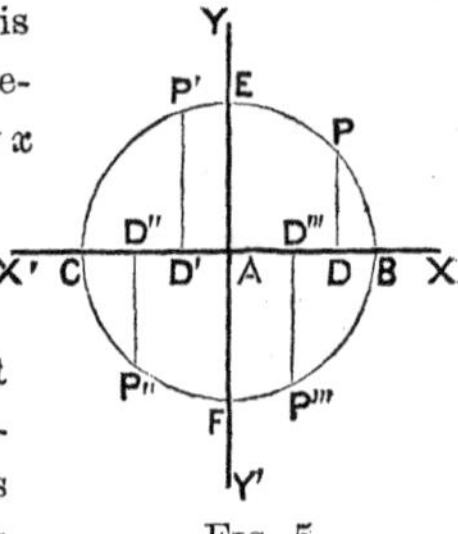

Fig. 5.

ILL.—In *Fig.* 5 let **BECF** be a circle whose radius is R, and **XX**′ and **YY**′ be the axes of reference. Represent the abscissa of *any* point in this circumference by x and the *corresponding* ordinate by y; so that when x signifies **AD**, y shall represent **PD**; when $-x$ is **AD**′, y shall be **P′D**′; when $-x$ is **AD**″, $-y$ shall be **P″D**″, etc. Now, suppose it possible to represent the relation between x, y and R by an equation so general as to be true for all points in this circumference, as **P**, **P**′, **P**″, etc. (It will subsequently appear that this equation is $x^2 + y^2 = R^2$.) In such an equation R would be *constant,* for it remains the same for all positions of the point **P**; and x and y would be *variables*, since they vary in value with every change of the position of **P**. In such a problem it is evident that x or y could not exceed R, hence these variables could have all values between the limits of $+R$ and $-R$.

SCH. 1.—Care should be taken not to confound the terms *constant* and *variable* as here used, with *known* and *unknown* as used in algebra: especially as the notation would suggest an identity which does not exist. Both the *known* and *unknown* quantities of Algebra are *constants;* moreover the *constants* in General Geometry may be either *known* or *unknown;* and the same, in a certain sense, may be said of the *variables.*

SCH. 2.—In order that the variables may retain their peculiar characteristic, we cannot have as many equations arising from a particular problem as there are variables; thus, if there are two variables involved, we have but one equation. In algebra such problems are called *indeterminate,* since the equation does not determine definite values of the unknown quantities, but can be satisfied by an infinite variety of values. From this feature of the General Geometry it is sometimes called *Indeterminate Analysis.* The *Calculus* is also embraced under the same term, as its problems involve a like feature.

---

* Our limits do not permit a discussion of the continuity of functions and the general geometrical interpretation of imaginary co-ordinates, and hence for simplicity we retain the conception of imaginaries as impossible quantities.

18. *To construct an equation*, or find its locus, is to draw the geometrical figure represented by it.

SECTION II.

Constructing Equations, or Finding their Loci.

19. A curve is ***continuous*** when its course is uninterrupted both in extent and in the character of its curvature.

ILL.—A circle, an ellipse, and the curve in *Fig.* 6 are examples of curves *continuous* in extent and curvature. They may be traced throughout by the uninterrupted movement of a point. The curve *Fig.* 7, is *discontinuous* in *extent*; and in *Fig.* 8, we have an example of a curve *discontinuous* in *curvature*. *Fig.* 9 affords an example of discontinuity both in extent and curvature.

FIG. 6.

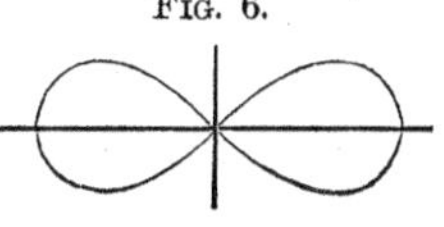

FIG. 7.

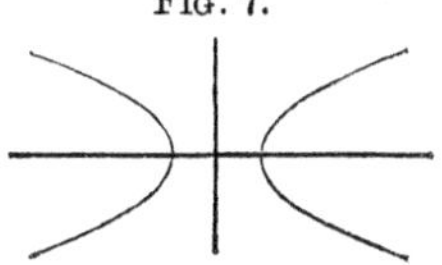

20. *A Branch* is a continuous portion of a curve. In *Figs.* 7 and 8 the curves have two branches each. In *Fig.* 9 there are *four* branches.

FIG. 8

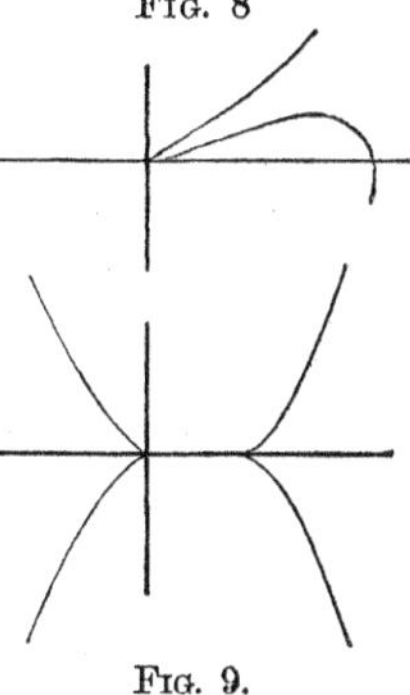

FIG. 9.

21. A curve is ***symmetrical*** with respect to either axis, or to any line, when it has the same form on both sides of the line, or when every point on one side of the line has a corresponding point on the other. The curves in *Figs.* 6, 7 and 9 are symmetrical with respect to the axis of abscissas, and the first two with respect to both axes. The curve in *Fig.* 8 is not symmetrical with respect to any line.

---

22. *Prob.* *To locate a Point whose co-ordinates are given.*

SOLUTION.—Lay off from the origin, on the axis of abscissas, a distance equal to the given abscissa, to the right if the abscissa is $+$, and to the left if it is $-$. Through the point thus found draw a line parallel to the axis of ordinates, and lay off on it a distance from the axis of abscissas equal to the given ordinate, above if the ordinate is $+$, and below if it is $-$. The point thus found will be the one required.

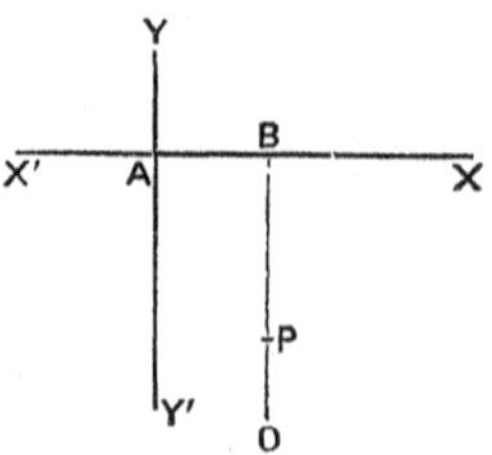

FIG. 10.

Ex. 1. Locate the point $x = 3$, $y = -5$.

SOLUTION.—Draw the axes **XX'** and **YY'**. Lay off **AB** $= 3$ to the *right*, as x is $+$, and draw **BO** parallel to **YY'**. Then take **PB** $= 5$ *below* the axis of abscissas as y is $-$, and **P** is the point required.

SCH.—Points are usually designated by mentioning simply their co-ordinates, as the point 3, —5, for the point in the last example. The abscissa is mentioned first.

Exs. 2 to 9. Locate —6, 2; —5, —7; —3, 0 ; 0, —3 ; 0, 0 ; 5, —1; 2, 0 ; 0, 4.

QUERIES.—Where are points situated whose abscissas are 0? Where are points situated whose ordinates are 0? What are the co-ordinates of the origin? In what line are 3, —2 ; 3, 5 ; 3, 0 ; and 3, 4 situated?

---

23. Prob. *To find the locus of an equation between two variables; i. e., to construct the equation.*

SOLUTION.—Solve the equation with respect to one of the variables. Then, since the equation expresses the relation between the co-ordinates of all points in the locus, substitute for the other variable any values which give *real* values for the first, and locate the points thus determined. These will be points in the locus; and, by determining a sufficient number, the locus can be sketched through them.

Ex. 1. Construct the equation $\frac{y - 2x}{3} = 1$.

SOLUTION.—Solving the equation for y, we have $y = 2x + 3$. Now attributing *arbitary* values to x, we make the following table of corresponding values :

When $x = 1$, $y = 5$, giving the point 1, 5 ;
" $x = 2$, $y = 7$, " " " 2, 7 ;
" $x = 3$, $y = 9$, " " " 3, 9 ;
etc., etc., " " " etc.

Noticing that all positive values of x give *real, positive*, and *single* values to y, we discover that the locus has but one branch which extends to the right of the axis of ordinates, extends indefinitely, and lies above the axis of abscissas.

Again, giving negative values to x, we have

When $x = -1$, $y = 1$, giving the point —1, 1 ;
" $x = -2$, $y = -1$, " " " —2, —1 ;
" $x = -3$, $y = -3$, " " " —3, —3 ;

and for all subsequent negative values of x, y has *real, negative*, and *single* values. Hence we learn that the locus has a single branch extending indefinitely in the third angle.

If we make $y = 0$, $x = -1\frac{1}{2}$; whence we see that the locus cuts the axis of *abscissas* at $-1\frac{1}{2}$, 0. If we make $x = 0$, $y = 3$; and hence the locus cuts the axis of *ordinates* at 0, 3.

Finally, locating these points, as in *Fig.* 11, we find that the line **M N** includes all the points, and hence conclude that it is the required locus.

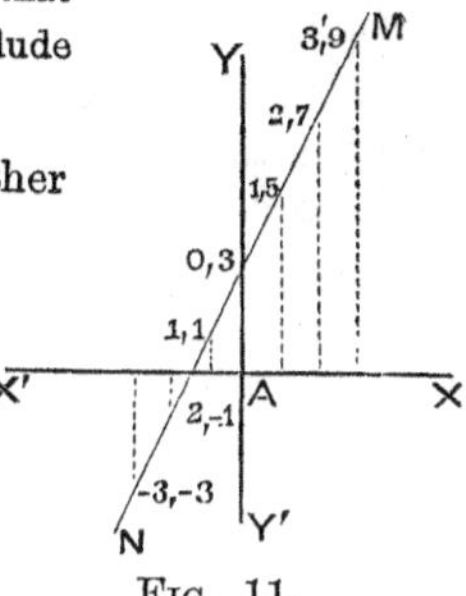

FIG. 11.

SCH.—If any other values be attributed to x, either integral or fractional, positive or negative, and the corresponding values of y deduced, the points thus determined will fall in the line **M N**.

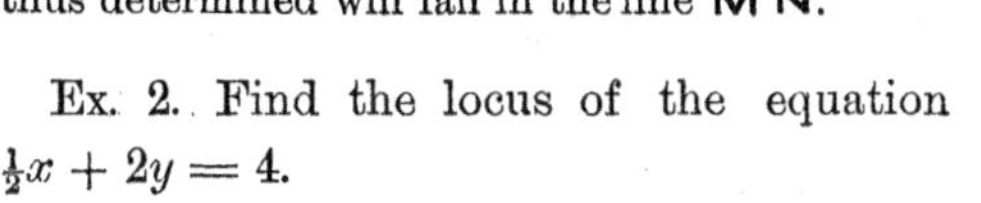

Ex. 2. Find the locus of the equation $\frac{1}{2}x + 2y = 4$.

Result. A straight line cutting the axis of abscissas at 8, 0, and the axis of ordinates at 0, 2.

Ex. 3. Find the locus of the equation $\frac{2}{3}x - 1 = \frac{x - y}{4}$.

Result. A right line passing through 0, 4, and 3, — 1.

24. SCH.—If there is nothing in the nature of the equation to make another course preferable, it is customary to solve it for y, finding the value in terms of x, and constants. If, however, the equation is above the second degree with respect to either of the variables, it is expedient to solve it with reference to the variable which is least involved. Thus, in order to construct $3x - y^2 = 2y^3 - y - 5$, we solve with reference to x, and then substitute arbitrary values for y, finding the corresponding values of x.

25. DEF.—***The Independent Variable*** is the one to which we assign arbitrary values, usually x. The other is called the *Dependent Variable.*

This distinction is made simply for convenience, and is not founded in any difference in the nature of the variables: either variable may be treated as the independent variable.

26. SCH.—There are certain peculiarities of loci, which readily appear from the form of the equation. These should always be noted. Observing them is called *Discussing*, or *Interpreting the Equation.* The following are some of these points:

1st. *The Intersection of the locus with the Axes.* Where the locus cuts the axis of abscissas $y = 0$; hence substituting this value (0) of y, in the equation, and finding the corresponding value or values of x, determines the intersections with the axis of abscissas. In like manner, making $x = 0$, and finding the corresponding values of y, determines the intersections with the axis of ordinates.

2nd. *The Limits between which the locus is comprised, and its continuity or discontinuity between these limits.* These questions are to be determined with respect to each axis. The limits are discovered by determining the greatest and least values of the independent variable which give *real* values to the dependent one. If all values of the independent variable between the limits observed in this way, give *real* values for the dependent variable, the locus is continuous in extent between these limits. If, on the other hand, there are certain values of the former which render the latter *imaginary*, the locus is discontinuous; and the limits of discontinuity are to be observed by finding the limits between which the values of the dependent variable are imaginary.

3rd. *Whether the locus is symmetrical with respect to an axis, or with any line, or not.* The manner of determining this is as follows: If, for each *real* value of one variable, the other has *two* values, numerically equal but with contrary signs, there are points similarly situated on opposite sides of the axis from which the variable having two values is reckoned, and hence the locus is symmetrical with respect to that axis. Again, if there is any line so situated that the values of the intercepts of either of the co-ordinates between it and the locus, on both sides of the line, are equal, the locus is symmetrical with respect to that line.

[NOTE.—There are many other characteristic features of loci which appear more or less immediately from the form of the equation, and some of which will be noticed in a subsequent part of the course. Those now mentioned are sufficient for our present purpose if the pupil becomes perfectly familiar with them. This familiarity can be attained only by a careful study of examples. In fact it is hardly probable that the pupil can understand the full purport of the language of the last scholium until he has solved several examples. After studying a few which follow, he can return and read the scholium again, and be better able to see its meaning.]

Ex. 4. Find the locus of the equation $x^2 + y^2 = 25$.

SOLUTION. $y = \pm\sqrt{25 - x^2}$. For $x = 0$, $y = 5$ and -5. Hence the locus cuts the axis of ordinates at $(0, 5)$ and $(0, -5)$. For $y = 0$, $x = 5$ and -5. Hence the locus cuts the axis of abscissas at $(5, 0)$ and $(-5, 0)$. Again, as every value of x between $+5$ and -5, gives two *real* values for y, numerically equal, but with opposite signs, the locus is symmetrical with respect to the axis of abscissas, and continuous between these limits. In like manner, $x = \pm\sqrt{25 - y^2}$ shows that the locus is symmetrical with respect to the axis of ordinates, and continuous between $y = 5$, and $y = -5$. When x is numerically greater than 5 (either $+$ or $-$), the values of y become *imaginary*. Hence the locus is comprised between the limits $x = 5$, and $x = -5$. From $x = \pm\sqrt{25 - y^2}$, it appears, in like manner, that the limits in the direction of the axis of ordinates are $y = 5$, and -5.

Now assigning to x arbitrary values between $+5$ and -5, we find the following table of values, and points in the locus:

When $x = 1$, $y = \pm\sqrt{24} = \pm 4.9$ nearly; and we have points $(1, 4.9)$ and $(1, -4.9)$;
" $x = 2$, $y = \pm\sqrt{21} = \pm 4.6$ nearly; " " " " $(2, 4.6)$ and $(2, -4.6)$;
" $x = 3$, $y = \pm\sqrt{16} = \pm 4$ " " " " $(3, 4,)$ and $(3, -4)$;
" $x = 4$, $y = \pm\sqrt{9} = \pm 3$ " " " " $(4, 3,)$ and $(4, -3)$;

For negative values of x the following points are found $(-1, 4.9)$ and $(-1, -4.9)$; $(-2, 4.6)$ and $(-2, -4.6)$; $(-3, 4)$ and $(-3, -4)$; $(-4, 3)$ and $(-4, -3)$.

Constructing the points thus determined they are found to be in the circumference of a circle whose radius is 5, and which is symmetrical with the axes, as in *Fig.* 12. It is also to be observed that any values of x, fractional as well as integral, between the limits $x = 5$, and $x = -5$, give values for y which locate points in the same circumference.

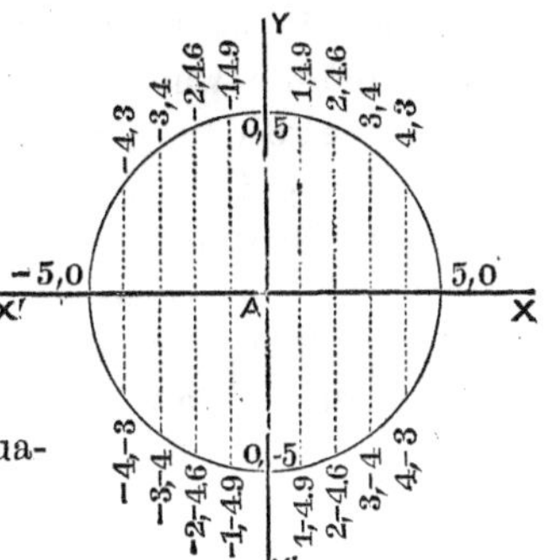

FIG. 12.

Ex. 5. Construct and discuss the equation $9y^2 + 4x^2 = 36$.

SOLUTION.—Solving the equation for y, $y = \pm \frac{2}{3}\sqrt{9 - x^2}$. We now observe that for $x = 0$, $y = \pm 2$; therefore the locus cuts the axis of ordinates at (0, 2) and (0,—2). In like manner, making $y = 0$, $x = \pm 3$; and hence the locus cuts the axis of abscissas at (3, 0) and (— 3, 0). Again, for each value of x which renders $9 - x^2 > 0$, *i. e.*, for each value between $x = 3$, and -3, y is *real* and has two values, numerically equal, but with contrary signs; therefore the locus is symmetrical with reference to the axis of abscissas, and continuous between the limits $x = 3$, and $x = -3$. Beyond these values of x, y becomes imaginary, and the locus is entirely comprised within $x = 3$ and $x = -3$ along the axis of abscissas. In a similar manner from $x = \pm \frac{3}{2}\sqrt{4 - y^2}$, it appears that the locus is comprised between $y = 2$ and $y = -2$, and is symmetrical and continuous with respect to the axis of ordinates.

Finding the values of y corresponding to a sufficient number of arbitrarily taken values of x, so as to enable me to sketch the curve, we have the following table of values:

For $x = 0$,	$y = \pm 2$,	giving the points	a, a'	in *Fig.* 13;
" $x = .5$,	$y = \pm 1.97$,	" " "	b, b'	" "
" $x = 1$,	$y = \pm 1.89$,	" " "	c, c'	" "
" $x = 1.5$,	$y = \pm 1.73$,	" " "	d, d'	" "
" $x = 2$,	$y = \pm 1.49$,	" " "	e, e'	" "
" $x = 2.5$,	$y = \pm 1.1$	" " "	f, f'	" "
" $x = 2.75$,	$y = \pm .8$,	" " "	g, g'	" "
" $x = 2.9$,	$y = \pm .51$,	" " "	h, h'	" "
" $x = 3$,	$y = 0$	" " "	B	" "

Since the equation contains only the square of x, negative values of x give the same values for y as positive values do, and the portion of the curve on the left of the axis of ordinates is symmetrical with that on the right.

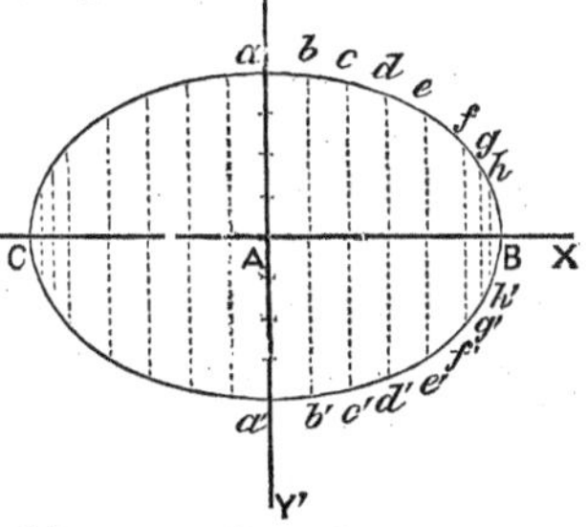

FIG. 13.

Finally, locating the points, as made known in the table, and a similar set of points on the left of the axis of ordinates, we have an ellipse whose axes are 6 and 4, *Fig.* 13.

Ex. 6. What is the locus of $y^2 = 2x - 6$?

Ans.—It cuts the axis of abscissas at 3, 0, and lies wholly to the right of this point, extending indefinitely in two branches, one above the axis of abscissas and one below it; and the two are symmetrical with this axis. *Fig.* 14. The branch **BM** extends indefinitely in the 1st angle, and **BM′** in the fourth. The locus is known as a *Parabola.*

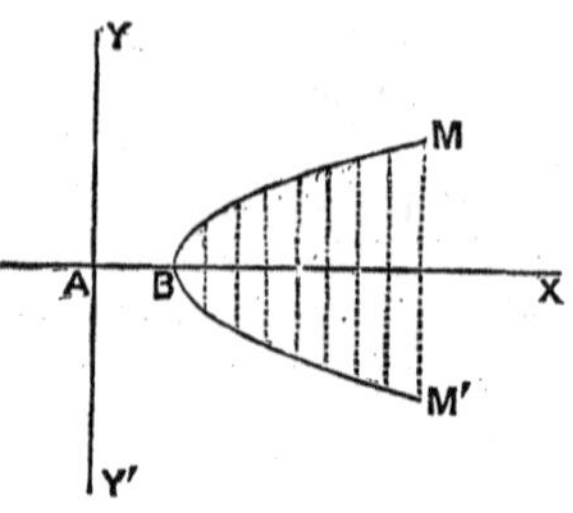

Fig. 14.

Exs. 7 to 10. Construct the following equations: $\frac{1}{2}x + 2y = 4$; $2x + 3y = 0$; $8x^2 + 5y^2 = 12$; $y^2 - 6y + x^2 = 16$.

Ex. 11. Find the locus of $x^2 - y^2 = 10$.

Ans.—The locus is represented in *Fig.* 15. It is discontinuous between $x = \sqrt{10}$ and $x = -\sqrt{10}$; but to the right and left of these points, it extends indefinitely. It is symmetrical with respect to both axes. The curve is known as an *Hyperbola.*

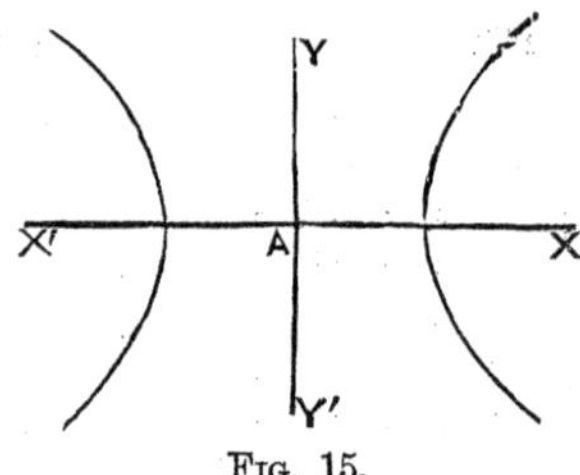

Fig. 15.

Exs. 12 to 23. Construct and discuss the following: $y^2 = 16 - x^2$; $y^2 = 10x - x^2$; $y^2 = 12x$; $x^2 - 6x + 9 + y^2 + 10y = 0$; $25(y + 4)^2 + 16(x - 5)^2 = 400$; $y^2 = 4 + 2(x - 3)^2$; $y^2 = x^2 - 4$; $y^2 = 8x^2 - x^3 + 5$; $xy = 16$; $y^2 = x^2 - x^4$; $y^2 = x^4 - x^2$; $y^2 = x^4 - x^3$.

Ex. 24. Construct and discuss the equation $x = \log y$.

Results.—Assuming x as given in the following table of values (any convenient values of x may be taken), the values of y can be found from a table of logarithms.

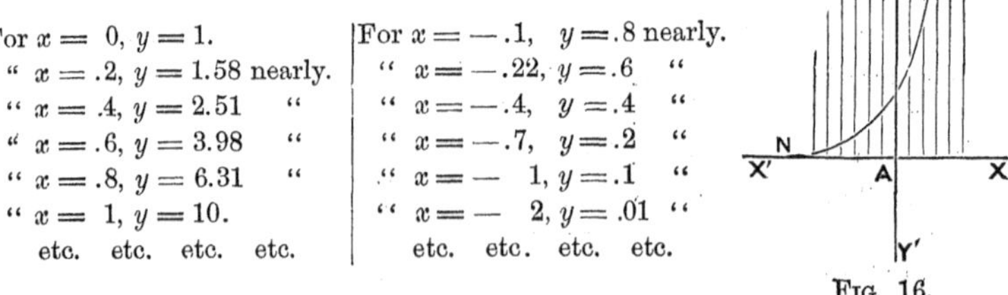

For $x = 0$, $y = 1$.	For $x = -.1$, $y = .8$ nearly.
" $x = .2$, $y = 1.58$ nearly.	" $x = -.22$, $y = .6$ "
" $x = .4$, $y = 2.51$ "	" $x = -.4$, $y = .4$ "
" $x = .6$, $y = 3.98$ "	" $x = -.7$, $y = .2$ "
" $x = .8$, $y = 6.31$ "	" $x = -1$, $y = .1$ "
" $x = 1$, $y = 10$.	" $x = -2$, $y = .01$ "
etc. etc. etc. etc.	etc. etc. etc. etc.

Fig. 16.

Locating these values, we have the curve **MN**, *Fig.* 16, which is called the *Logarithmic Curve.* It lies wholly above the axis of abscissas, as negative numbers have no logarithms. It extends on both sides of the axis of ordinates, and cuts it at (0, 1,) a point through which all logarithmic curves pass, in whatever system the logarithms be taken, since $\log 1 = 0$ in all systems. The curve extends indefinitely to the right and to the left; but the portions are not symmetrical.

Ex. 25. Construct $x = \log y$, assuming 2 as the *base* of the system of logarithms; giving $y = 2^x$.

The values are, $x = 0$, $y = 1$; $x = 1$, $y = 2$; $x = 2$, $y = 4$; $x = 3$, $y = 8$; etc. Also, $x = -1$, $y = .5$; $x = -2$, $y = .25$; $x = -3$, $y = \frac{1}{8}$; $x = -4$, $y = \frac{1}{16}$, etc.

QUERIES.—Locating this curve on the same axes with the preceding, what common point do they possess? Does the right hand branch of this lie to the right, or to the left of the former? Does the left hand branch approach the axis of abscissas more rapidly, or less rapidly in the latter than in the former? What makes these differences? How would it be with a base 100?

Ex. 26. Construct and discuss $y = \sin x$.

SUGS.—The unit arc is a portion of the circumference equal to the radius. This arc is 57.3° nearly; since, radius being unity, the semi-circumference is 3.1416, and $\frac{180°}{3.1416} = 57.3°$ nearly. Hence the following table.

For $x =$		$y =$	For $x =$		$y =$
For $x = 0° =$	0,	$y = 0$	For $x = 180° =$	3.14,	$y = 0$
" $x = 10° =$	.17,	$y = .17$	" $x = 190° =$	3.31,	$y = -.17$
" $x = 20° =$	.35,	$y = .34$	" $x = 200° =$	3.49,	$y = -.34$
" $x = 30° =$	.52,	$y = .50$	" $x = 210° =$	3.66,	$y = -.50$
" $x = 40° =$	.70,	$y = .64$	" $x = 220° =$	3.84,	$y = -.64$
" $x = 50° =$	.87,	$y = .77$	" $x = 230° =$	4.01,	$y = -.77$
etc. etc.	etc.	etc.	etc. etc.	etc.	etc.

This curve is called the *Sinusoid.* Where does it cut the axis of abscissas? Is it limited? What are the limits of y? What is the meaning of $x = -10°$, $x = -20°$, etc.?

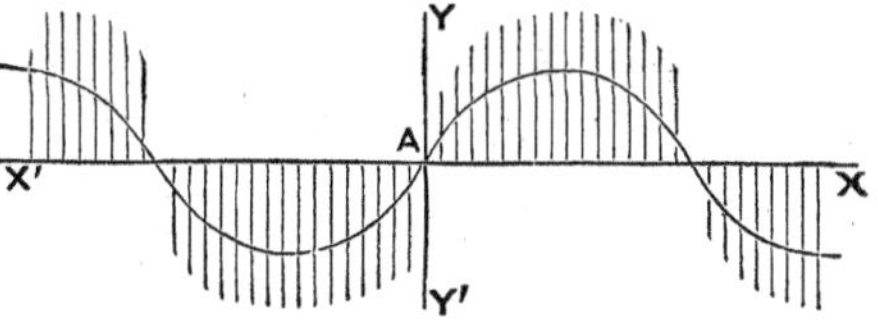

FIG. 17.

Exs. 27 to 33. Construct $y = \tan x$; $y = \cot x$; $y = \cos x$; $y = \text{versin}\, x$; $y = \text{coversin}\, x$; $y = \sec x$; $y = \text{cosec}\, x$.

SCH.—These loci can be constructed with sufficient accuracy without the numerical computations. Thus, taking the Ex. $y = \tan x$, draw a circle **ON**, *Fig.* 18, with any convenient radius. Divide a quadrant, as **MN**, into equal parts, each so small that for practical purposes the chord and arc may be considered equal.

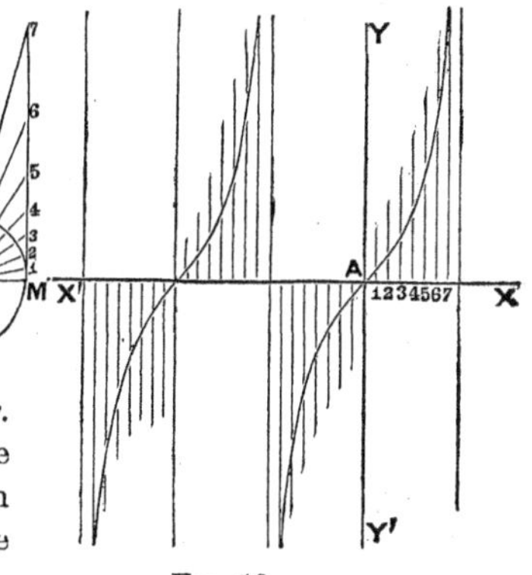

FIG. 18.

Estimating the tangents and arcs from **M**, and having drawn the tangents as in the figure, lay off the arcs on the axis of abscissas. At the extremity of **A1** lay off an ordinate equal to tangent **M1**, etc., etc. There are an infinite number of similar infinite branches to this curve.

On the figure used for getting the tangents, when the arc passes 90° the tangents (and hence the ordinates) become negative. Strictly speaking, negative values of x would be obtained by measuring the arcs on the circle from **M** downward, or from left to right; so that, from $x = 0$ to $x = -90°$, the tangents (and hence the ordinates) are negative. From $x = -90°$ to $x = -180°$, the tangents (and hence the ordinates) are positive. Where do the branches cut the axis of abscissas? At what values of x do the ordinates become infinite?

SECTION III.

The Point in a Plane.

27. Def.—***The Equations of a Point*** are the algebraic expressions which determine its position.

---

28. *Prop.* *The Equations of a Point in a plane are* x = a, *and* y = b, *in which the signs of* a *and* b *are general.*

Dem.—If, as in *Fig.* 19, we make **AB** $= a$, and through **B** draw **DE** parallel to **YY'**, every point in **DE** will have its abscissa equal to a. In like manner make **AC** $= b$, and draw **FG** parallel to **XX'**, and every point in **FG** will have b for its ordinate. Hence the point **P** has a for its abscissa, and b for its ordinate; and since two straight lines can meet in only one point, **P** is the *only* point which has these co-ordinates. Therefore $x = a$, and $y = b$, determine the position of a point. Q. E. D.

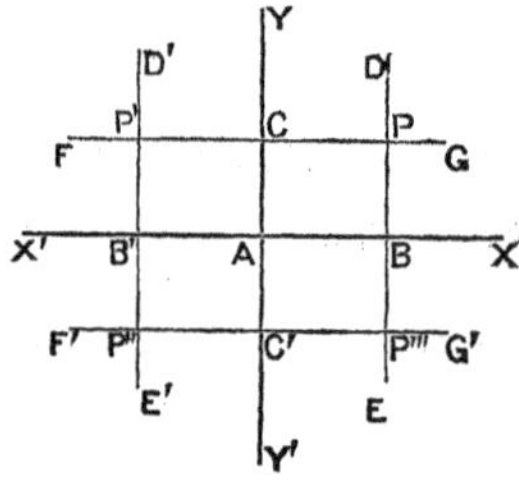

Fig. 19.

Sch. 1.—If we have $x = -a$, and $y = b$, **P'** is the point. If $x = -a$, and $y = -b$, **P''** is the point, etc.

Sch. 2.—If $x = a = 0$, and $y = b$, the point is in the axis of ordinates. If $x = a$, and $y = b = 0$, the point is in the axis of abscissas. $x = 0$, $y = 0$ characterizes the origin.

Sch. 3.—A point is usually designated by simply naming its co-ordinates, the abscissa being mentioned first. Thus the point (m, n) is the same as the point $x = m$, and $y = n$.

Exs. 1 to 6. Locate the points $x = -3$, $y = 4$; (5, —7); (0, —5); (0, 4); (0, 0); (6, 0).

Exs. 7 to 10. How are the points $(5, \frac{0}{0})$; $(\frac{0}{0}, -6)$; $(\frac{0}{0}, m)$; $(-n, \frac{0}{0})$ situated?

Answer to the first.—In a line parallel to the axis of ordinates and at a distance 5 from it. Any point in this line fulfills the conditions, since $y = \frac{0}{0}$, *i. e.*, is indeterminate.

Exs. 11, 12. Construct the triangle whose vertices are $(-3, 4)$; $(5, -1)$; and $(2, -6)$. Also the triangle whose vertices are $(0, 3)$; $(-5, 0)$; and $(0, 0)$.

Exs. 13, 14. What figure is that the vertices of whose angles are $(2, 3)$; $(2, 8)$; $(7, 8)$; and $(7, 3)$? What figure is that the vertices of whose angles are $(2, 9)$; $(-8, 9)$; $(-8, 1)$; and $(2, -1)$?

---

29. Prop. *The Distance between two points in a plane is* $\sqrt{(x' - x'')^2 + (y' - y'')^2}$, *in which* $(x', y',)$ *and* (x'', y'') *are the points.*

DEM.—Let the points (x', y') and $(x'' y'')$ be represented by P′ and P″, as in *Fig.* 20, and the distance between them, P′P″, by D. Draw P″D parallel to AX. Then $P''D = x' - x''$, and $P'D = y' - y''$. From the right angled triangle P′ P″ D, we have, $D = \sqrt{(x' - x'')^2 + (y' - y'')^2}$. Q. E. D.

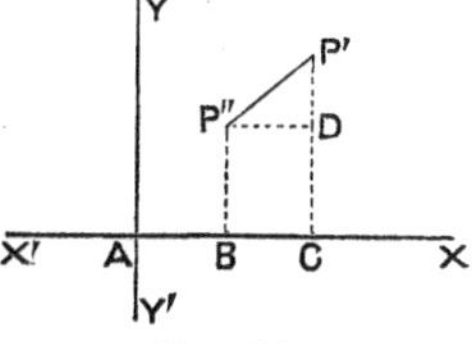

FIG. 20.

COR.—*If either of the points, as* P″, *is at the origin, its co-ordinates are* 0, 0, *and* $D = \sqrt{x'^2 + y'^2}$.

QUERIES.—When P″ is in the axis of abscissas and at the right of the origin, what is the formula? The same with P″ at the *left* of the origin, give $D = \sqrt{(x' + x'')^2 + y'^2}$. If P′ is in the 1st angle and P″ in the 3rd, what is the formula? If P′ is in the axis of abscissas and P″ in the axis of ordinates? If one is in the 2nd and the other in the 4th angle?

SCH.—Observe that the formula $D = \sqrt{(x' - x'')^2 + (y' - y'')^2}$ is strictly general, only noticing carefully the effect of the position of the points, upon the *signs* of their co-ordinates. Thus for a point P″, in the 4th angle, we have x'', and $-y''$; which, substituted in the formula, gives for P′ in the 1st angle and P″ in the 4th, $D = \sqrt{(x' - x'')^2 + (y' + y'')^2}$.

EXAMPLES.—Find the distances between the following points taken two and two: $(3, 5)$; $(2, 6)$; $(-3, -2)$; $(-1, 4)$; $(-2, -1)$; $(-5, -7)$; $(-3, 0)$; $(0, -4)$; $(0, 0)$; $(-5, 0)$.

SECTION IV.

The Right Line in a Plane.

30. Def.—***The Equation of a Locus*** is an equation which expresses the relation between the co-ordinates of every point in the locus.

---

31. *Prop.* *The Equation of a Right Line passing through two given points is* $y - y' = \frac{y' - y''}{x' - x''}(x - x')$, *in which* (x, y) *is any point in the line, and* (x', y') *and* (x'', y'') *are the given points.*

Dem.—Let MN be any right line referred to the rectangular axes XX′, YY′. Let P be any point in the line, and designate its co-ordinates, AD and PD, by x and y. Let P′ and P″ be the *given* points whose co-ordinates are x', y', and x'', y'', respectively. Now drawing P′E and P″F parallel to AX, the triangles PEP′ and P′FP″ are similar, and give PE : P′F :: P′E : P″F. But $PE = y - y'$, $P'F = y' - y''$, $P'E = x - x'$, and $P''F = x' - x''$; hence, substituting these values, we have

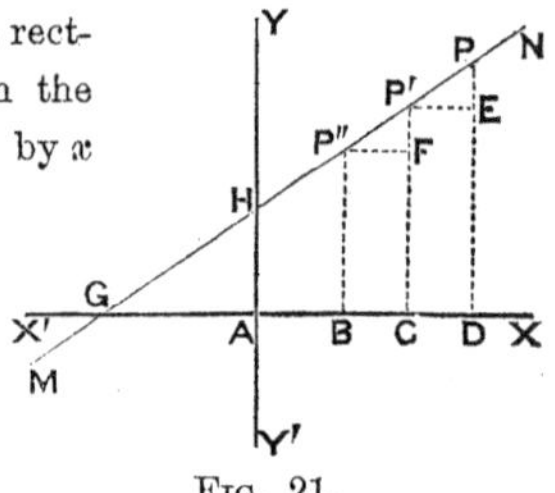

Fig. 21.

$y - y' : y' - y'' :: x - x' : x' - x''$, or $y - y' = \frac{y' - y''}{x' - x''}(x - x')$. Q. E. D.

Cor. 1.—Since $P'P''F = NGX$, and $\frac{y' - y''}{x' - x''} = \frac{P'F}{P''F} =$ $\tan P'P''F$, we have $\frac{y' - y''}{x' - x''} =$ *the tangent of the angle which the line makes with the axis of abscissas.*

If $x' = x''$, $\frac{y' - y''}{x' - x''} = \frac{y' - y''}{0} = \infty$, which being the tangent of 90°, shows that the line is perpendicular to the axis of abscissas. This is as it should be, since if $x' = x''$, the points P′ and P″ are equally distant from YY′, and hence MN is perpendicular to XX′. If $y' = y''$, $\frac{y' - y''}{x' - x''} = \frac{0}{x' - x''} = 0$, which being the tangent of 0°, shows that the line is parallel to (makes *no* angle with) the axis of abscissas. This is as it should be, since by observing the figure, it appears, that when $y' = y''$, MN is parallel to XX′.

32. Cor. 2.—***The Equation of a Right Line passing through one given point.*** If $x' = x''$, and $y' = y''$, we have

$y - y' = \frac{0}{0}(x - x')$, and, by putting the indeterminate expression, $\frac{0}{0}, = a$, $y - y' = a(x - x')$. This is the equation of a straight line passing through a given point, since the conditions, $x' = x''$, $y' = y''$, make P′ and P″ coincide. The a is indeterminate, as it should be, since, through *one* given point, an indefinite number of straight lines can be drawn.

33. Cor. 3.—***The Common Equation of a Right Line.*** If in $y - y' = a(x - x')$, we make $x' = 0$, and designate the corresponding value of y' by b, so that the given point shall be the point in which the line cuts the axis of ordinates, we have, after reduction, $y = ax + b$, which is the *common equation* of the straight line. In this equation a is the tangent of the angle which the line makes with the axis of abscissas, and b the distance from the origin to where the line intersects the axis of ordinates.

Sch. 1.—*Discussion of the Equation* $y = ax + b$. If b be $+$, the line cuts the axis of ordinates above the origin; if $-$, below; if 0, at the origin. In the latter case, we have $y = ax$, as *the equation of a right line passing through the origin.* If a be $+$, the line makes an acute angle with the axis of abscissas, (*i. e.*, it inclines to the right, as the lines in *Fig.* 22), the tangent of an acute angle being $+$. If a be $-$, the angle is obtuse, (*i. e.*, the line inclines to the left, as in *Fig.* 23), since the tangent of an obtuse angle is $-$. If $a = 0$ the line is parallel to the axis of abscissas, and if $a = \infty$, it is perpendicular, as will readily appear.

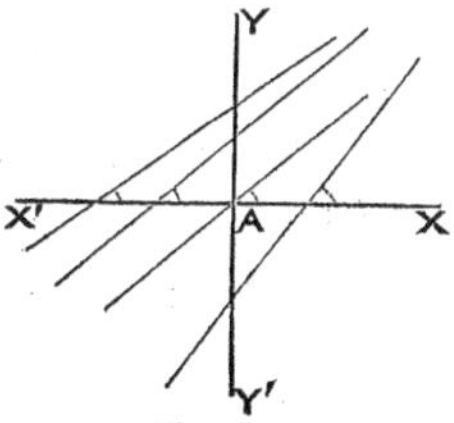

Fig. 22.

Sch 2.—If we solve $y = ax + b$ for x, we have $x = \frac{1}{a}y - \frac{b}{a}$, in which $\frac{1}{a}$ is the tangent of the angle which the line makes with the axis of ordinates, since, in *Fig.* 21, the angle AHG = NHY = 90° — NGX. In this form, $\frac{b}{a}$ is the distance on the axis of abscissas from the origin to where the line cuts it (AG), since the base of a right-angled triangle is equal to the perpendicular divided by the tangent of the angle at the base.

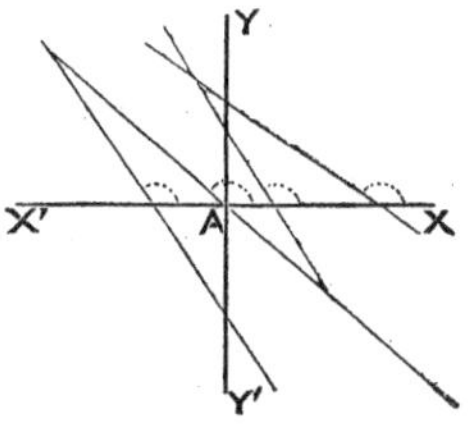

Fig. 23.

34. Cor. 4.—***The Equation of a Right Line referred to oblique axes.*** If the axes are oblique, we still have the same forms, but in this case $\frac{y' - y''}{x' - x''}$, or a, signifies the ratio of the sines of the angles which the line makes with the axes, since the sines of the angles of a plane triangle are to each other as the sides opposite. Thus in *Fig.* 24, **P'F** (or $y'-y''$) : **P''F** (or $x'-x''$) :: **AH** : **AG** :: sin **AGH** : sin **AHG.** Putting β for the angle included by the axes, and α for the angle which the line makes with the axis of abscissas, we get

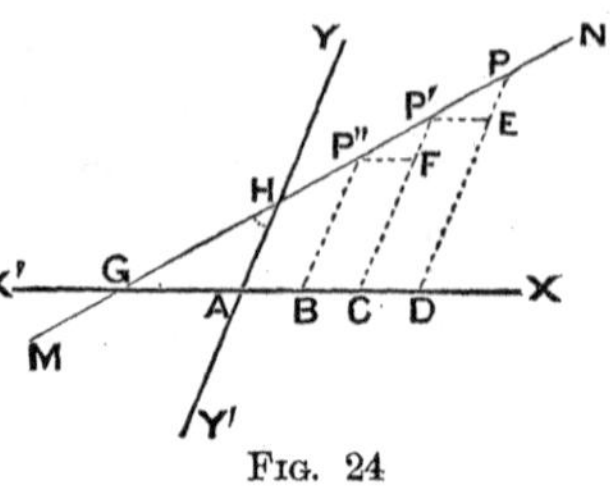

FIG. 24

$\frac{y'-y''}{x'-x''} = \frac{\sin\alpha}{\sin(\beta-\alpha)}$, and, finally, $y = \frac{\sin\alpha}{\sin(\beta-\alpha)}x + b$, as *the equation of a right line referred to oblique axes.*

Ex. 1. Construct the equation $y = 2x + 3$.

SOLUTIONS.—There are three methods of solution. 1st. *By any two points.* As it is known to be an equation of a right line from its form, if any two points be determined, as in the last section, the position of the line will be known. For example, for $x = 3$, $y = 9$, and for $x = -2$, $y = -1$; whence, locating these points and drawing a line through them, we have the construction.

2nd. *By the intersections with the axes.*— This is only a modification of the 1st method, merely making $y = 0$, whence $x = -1\frac{1}{2}$, and making $x = 0$, whence $y = 3$, constructing these intersections, and passing a line through them. (The pupil should execute the figures.)

3rd. *By means of the tangent of the angle which the line makes with the axis of abscissas.* Since $b = 3$, we may lay off **AC** = 3 above the origin, and thus determine **C** as a point in the line. Through **C** drawing **CE** parallel to **AX** and constructing the angle **NCE** so that its tangent shall be 2 (by taking **CD** any convenient length and erecting the perpendicular **FD** = 2**CD**), the line **NM** is the one sought.—Or, having located **C**, take **AG** = $\frac{1}{2}$**AC**, whence tan **AGC** = $\frac{AC}{AG}$ = 2, will give the construction. Or, again, drawing any line making with **XX'** an angle whose tangent is 2, and drawing a line parallel to it through **C**, the latter will be the line sought.

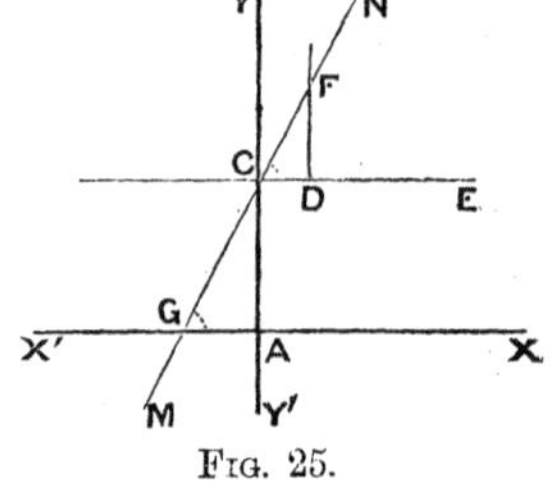

FIG. 25.

SCH.—If the tangent were —, **CD** would be laid off to the left of **C**, or the perpendicular **FD** let fall below **D**.

Ex. 2. Produce the equation of a line passing through (—3, 5), and (2, —1).

SOLUTION.—Here $x' = -3$, $x'' = 2$, $y' = 5$ and $y'' = -1$. Now, substituting these values in $y - y' = \frac{y'-y''}{x'-x''}(x - x')$, and reducing to the form $y = ax + b$, we have $y = -1.2x + 1.4$. The pupil should construct this equation, and then verify the result by locating the points (—3, 5), and (2,—1), observing that, if the work is correct, they will fall in the line. Algebraically, we verify the result by substituting in the equation $y = -1.2x + 1.4$, successively for x and y, (—3, 5), and (2, —1), each of which must satisfy the equation, as it expresses the relation between the co-ordinates of *any* point in the line. Substituting, we get $5 = 3.6 + 1.4$, and $-1 = -2.4 + 1.4$, both of which are correct.

Ex. 3. What angle does the line which passes through the points (3, 5), and (—7, 2) make with the axis of abscissas?

Ans., 16° 42′ nearly.

Ex. 4. Produce the equation of a line passing through the point (2, —3), and making an angle with the axis of abscissas whose tangent is 4. *Ans.*, $y = 4x - 11$.

Ex. 5. Produce the equation of a line passing through (—1, 0), and (—4, —5), construct by the 3rd method, and verify the equation by locating the points.

Ex. 6. Construct the triangle the equations of whose sides are $y = \frac{3}{4}x + 3$, $y = -\frac{4}{5}x + 4$ and $y = \frac{1}{2}x - 1$.

Ex. 7. What is the equation of a line which cuts the axis of ordinates at 3 above, and the axis of abscissas at 5 to the left of the origin? (Notice that this is a case of a line passing through two points.)

Ans., $y = \frac{3}{5}x + 3$.

Ex. 8. What line is $y = 0.x$? What is $x = 0.y$? How is $y = 0.x + 4$ situated? How $y = 0.x - 5$?

Ex. 9. Find the angles which the following lines make with the axis of abscissas: viz., the line passing through (3, 5), and (—1, —4); through (5, —2), and (5, 3); through (—3, 2), and (7, 2). How are these lines severally situated?

---

35. Prop. *Every Equation of the First Degree between two variables is an equation of a right line.*

Dem.—Every such equation may be put in the form $Ay + Bx + C = 0$, in which A and B are the collected coefficients of y and x, and C is the sum of the absolute terms. By transposition and division we have $y = -\frac{B}{A}x - \frac{C}{A}$. Now putting $-\frac{B}{A} = a$ and $-\frac{C}{A} = b$, there results the known form $y = ax + b$. Q. E. D.

Sch.—If B and A have like signs, the line makes an obtuse angle with the axis of abscissas; and if they have unlike signs, it makes an acute angle. If $B = 0$ the line is parallel to the axis of abscissas, and if $A = 0$ it is perpendicular. If A and C have like signs, the line cuts the axis of ordinates below, and, if unlike, above the origin. If $C = 0$ the line passes through the origin. *In general, if an algebraic equation has no absolute term, the locus passes through the origin.* (Why ?)

Ex. 1. Reduce $\frac{x-y}{3} - 4 = 2x - \frac{1}{2}y - \frac{1-x}{2}$ to the form $Ay + Bx + C = 0$, and describe the line according to the suggestions in the preceding scholium.

Ans.—The equation is $y - 13x - 21 = 0$. $A = 1$, $B = -13$, and $C = -21$. As A and B have unlike signs the line makes an acute angle with the axis of abscissas, the tangent of which is 13. It cuts the axis of ordinates *above* the origin at a distance of 21.

Ex. 3. In like manner discuss $3 - \frac{2-y}{5} = \frac{x+y}{3}$; $\frac{x+2y}{3} - 10y = \frac{6x+y}{5}$; $\frac{x-y}{3} + 2 = y + \frac{3x+y}{9}$.

Ex. 4. Construct the figure the equations of whose sides are $2y + 2x = 3x + 3 + y$; $\frac{x+y}{2} - 1\frac{1}{2} = 2x - 6 - y$; $3y + 2x - 6 = \frac{2y-x}{3} + 1$; and $x + y = -3$. What is the figure inclosed?

SECTION V.

Of Plane Angles, and the Intersection of Lines.

36. Prop. *The expression for the value of an angle included between two lines is* $\tan V = \frac{a' - a}{1 + aa'}$, *in which* V *is the angle included by the lines, and* a *and* a' *are the tangents of the angles which the lines make with the axis of abscissas.*

Dem.—Let **MN** and **M'N'**, *Fig.* 26, be two lines whose equations are respectively $y = ax + b$ and $y = a'x + b'$. Now **CBX** being exterior to the triangle **BCD**, we have **DCB** = **CBX** − **CDB**, or by trigonometry

$$\tan \mathbf{DCB} = \frac{\tan \mathbf{CBX} - \tan \mathbf{CDB}}{1 + \tan \mathbf{CBX} \times \tan \mathbf{CDB}}$$

But **DCB** $= V$, tan **CBX** $= a'$, and tan **CDB** $= a$. $\therefore \tan V = \frac{a' - a}{1 + aa'}$. Q. E. D.

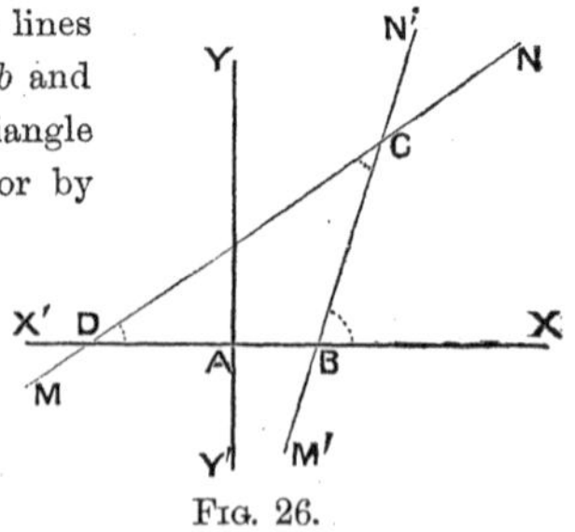

Fig. 26.

SCH.—In applying this formula to any particular example, we may obtain two results, numerically equal, but with opposite signs. Thus, if the two lines are $y = 2x + 4$, and $y = 3x - 5$, and we let $a' = 2$, and $a = 3$, we have $\tan V = \frac{2-3}{1+6} = -\frac{1}{7}$. But, if we let $a = 2$, and $a' = 3$, we have $\tan V = \frac{3-2}{1+6} = \frac{1}{7}$. This ambiguity is as it should be, since the two lines form, in general, two equal acute, and two equal obtuse angles with each other; and as these angles are supplements of each other, they have tangents numerically equal but with *opposite signs*.

---

37. Prob. *To find the equation of a line which makes any required angle with a given line.*

SOLUTION.—Let $y = ax + b$ be the equation of the given line, $y = a'x + b'$ be that of the required line, and m the tangent of the required angle. As the relative directions of the lines depend solely upon a, a', and m, the problem consists in finding the unknown a', in terms of the given tangents a and m. But $m = \frac{a' - a}{1 + aa'}$ by the preceding proposition; whence $a' = \frac{a + m}{1 - am}$; and $y = \frac{a + m}{1 - am}x + b'$ is the equation of the required line.

SCH.—In this form b' is undetermined, as it should be, since there may be an indefinite number of lines which will satisfy the condition, all having the same inclination to the axis of abscissas, but cutting the axis of ordinates at different points.

COR. 1.—*If the required line is to pass through a given point* (x', y'), *we have* $y - y' = \frac{a + m}{1 - am}(x - x')$.

38. COR. 2.—*If the required line is to be parallel to the given line,* $m = 0$, *and we have* $a' = a$. The equations then become $y = ax + b'$, and $y - y' = a(x - x')$, both of which lines are parallel to $y = ax + b$.

39. COR. 3.—*If the lines are to be perpendicular to each other,* $m = \infty$. $\therefore a' = \frac{a + m}{1 - am} = \frac{m}{-am}$* $= -\frac{1}{a}$, or $1 + aa' = 0$, which is called *the equation of the condition of perpendicularity.* The equations of lines perpendicular to $y = ax + b$ will therefore be $y = -\frac{1}{a}x + b'$, and $y - y' = -\frac{1}{a}(x - x')$, the latter passing through (x', y').

---

* The principle upon which this reduction is effected is, that the *finite terms* a and 1 added to the infinites m and $-am$ must be dropped. The axiom is, finites added to infinites do not (appreciably) affect the ratio of the infinites. The word appreciably is thrown in to aid the student's apprehension. It is not required, nor is it strictly correct.

SCH. 2.—Two lines are parallel to each other when the two equations being reduced to the form $y=ax+b$, the coefficients of x are the same in both; and they are perpendicular when these coefficients are reciprocals of each other with opposite signs.

Ex. 1. Find the angle included between $y=-x+2$, and $y=3x-6$.

Result, $\text{Tan } V=\dfrac{3+1}{1-3}=-2$. $\therefore$ The angle is $116°34'$.

Ex. 2. What are the angles of the triangle the equations of whose sides are $2y-5=y-x$; $y+4x=8$, and $y=\frac{1}{5}x$?

Ans. { The tangents of the angles are .6, —21, and 1.5.
The angles are nearly $30°58'$, $92°44'$, and $56°19'$.

Ex. 3. Write the equations of three lines, each parallel to $y=2x-11$, and construct the lines therefrom.

Ex. 4. Write the equation of a line parallel to $y-\frac{1}{2}x=5$ and passing through $(-6, 4)$. Reduce the equation of the parallel to the form $y=ax+b$, and then construct both lines from their equations. Verify the result by constructing the given point; also by observing that the coefficients of x in both equations are equal, and that the co-ordinates of the given point satisfy the equation of the parallel.

Ex. 5. Write the equation of a line passing through $(-\frac{1}{2}, \frac{2}{3})$, and parallel to $\frac{1}{2}x-\frac{1}{3}y=2$. Verify as in Ex. 4.

Result. The equation is $y=\frac{3}{2}x+1\frac{5}{12}$.

Ex. 6. Write the equations of three lines each perpendicular to $\frac{1}{3}y-2x=1$, reduce them to the form $y=ax+b$, and verify the results by construction.

Ex. 7. Write the equation of a line perpendicular to $2y-4=x$, and passing through $(1, -3)$. Verify as before.

Ex. 8. Write the equations of lines perpendicular to $\dfrac{x-y}{2}=\frac{1}{3}x-2$, and severally passing through $(-2, 3)$; $(0, -5)$; $(0, 0)$; and $(-3, 0)$.

Ex. 9. What is the angle included between $y=0.x$, and $y=3x-5$? Between $x=0.y$, and $y=2x+1$?

Ex. 10. What is the equation of a line passing through $(-6, 0)$, and perpendicular to $y=0.x+5$? *Ans.*, $x=0.y-6$.

Ex. 11. What is the equation of a line passing through (—1, 3), and making an angle of 45° with $y = 2x - 5$? *Ans.*, $y = -3x$.

Ex. 12. Produce the equation of a line passing through (—4, —5) and making an angle of 71°34′ with $y = -2x + 7$?

Result, $y = \frac{1}{7}x - 4\frac{3}{7}$, calling tan 71°34′, = 3.

---

40. Prob. *To find the point or points of intersection of two lines.*

Solution.—For a common point the values of x and y are the same in both equations, *and only for such a point.* Therefore, making the equations simultaneous restricts the values of x and y to the required point or points. Consequently, we have only to solve any two given equations for the values of x and y in order to find the point or points in which the loci intersect.

Sch. 1.—The general *formulæ* for the value of the co-ordinates of the point of intersection of two straight lines whose equations are $y = ax + b$, and $y = a'x + b'$, are $x = \frac{b' - b}{a - a'}$, and $y = \frac{ab' - a'b}{a - a'}$. Upon these values we may observe that for $a = a'$, and b and b' unequal, the values of x and y become ∞. This indicates that the lines do not intersect, and hence that they are parallel. Therefore $a = a'$ is *the condition of parallelism of two straight lines.* This may also be seen directly from the meaning of a and a'. As these quantities are the tangents of the angles which the lines make with the axis of abscissas, it follows that when they are equal the lines make equal angles with this axis, and are, therefore, parallel. 2nd. If $b = b'$, and a and a' are unequal, we have $x = 0$, and $y = b = b'$. This is also evident from the meaning of b and b'. Both lines cut the axis of ordinates at the same point. 3rd. If $a = a'$ and $b = b'$, $x = \frac{0}{0}$, and $y = \frac{0}{0}$, and the lines coincide. 4th. If $a = 0$, and $b = 0$, the first equation becomes $y = 0.x + 0$, or the equation of the axis of abscissas, and $x = -\frac{b'}{a'}$, the point of intersection of $y = a'x + b'$ with this axis, as it ought. Sch. 2, Art. 33.

Sch. 2.—Any two equations between two variables being given, if the lines they represent are constructed, and the co-ordinates of the points of intersection measured, we have a graphic solution of the equations.

Ex. 1. Where is the point of intersection of the lines $\frac{1}{2}x - \frac{1}{3}y = 1$, and $y = -2x + 4$? *Ans.*, (2, 0).

Ex. 2. What are the co-ordinates of the vertices of the triangle the equations of whose sides are $y - 2x + 3 = 0$, $\frac{x - y}{3} + 4 = \frac{1}{2}y$, and $x - \frac{1}{4}y = 2$?

Ex. 3. Where does a perpendicular from (—3, 8), to the line $y = \frac{1}{3}x - 5$, intersect the latter? *Ans.*, At $(1\frac{1}{5}, -4\frac{3}{5})$.

Ex. 4. Where does a perpendicular from the origin intersect $2x - 3y = 4$?

Ex. 5. Given $y = \frac{1}{2}x - 3$, $y = -4x - 8$, and $y = -\frac{2}{3}x + 10$, as the equations of the sides of a triangle, required to find where a perpendicular from the angle included between the first two sides, intersects the third side.

Result, At the point 5.5, 6.34, nearly.

Ex. 6. Find the intersections of the loci whose equations are $7(y - x) = 5 - 2x$, and $y^2 + x^2 + 9 = 16 - 6y$, and construct the figure.

Results, At the points (.374 +, .981 +), and (—3.888 +, —2.063 +). The figure is that given in the margin.

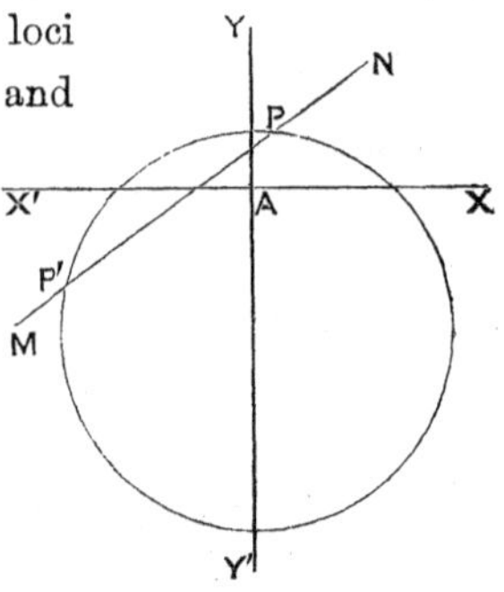

Fig. 27.

Ex. 7. Find the intersections of $y^2 = 10x$, and $x^2 + y^2 = 144$, and construct the loci, thus verifying the solution.

Ex. 8. Find the intersections of $25y^2 + 16x^2 = 1600$, and $16y^2 - 9x^2 + 576 = 0$. *Results,* At (9.12, 3.3), and (—9.12, —3.3), nearly.

Ex. 9. Find the intersections of $x^2 - 8x + y^2 + 6y = 0$, and $y = \frac{1}{2}x + 1$. Also of the first with $3y = 4x$. Also with $y = 3 - x$.

Results.—1st. Imaginary results. No intersections. 2nd. A common point at the origin. 3rd. Two points of intersection.

Sch. 3.—The construction of loci represented by equations affords beautiful illustrations of principles in the theory of equations, concerning the number and character of the roots of an equation.

Ex. 10. Find the intersections of $25y^2 + 4x^2 = 100$, by the following: 1st, $y^2 + x^2 = 9$; 2d, $y^2 + 2y + x^2 = 8$; 3d, $y^2 + 4y + x^2 = 5$; 4th, $y^2 + 10y + x^2 = -16$; and 5th, $y^2 + 12y + x^2 = -27$.

Results.—1st, At (2.44, 1.7); (—2.44, 1.7); (2.44, —1.7); (—2.44, —1.7) which affords an example of 4 *real* roots. 2nd, At (0.2); and ($\pm$ 2.9, $-\frac{34}{21}$), which affords an example of what *seems* to be but *three* roots when there should be *four*. This is explained by the two values of x for $y = 2$, becoming +0 and

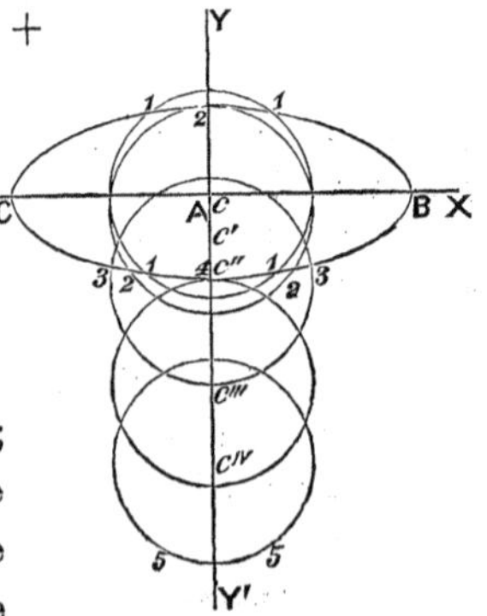

Fig. 28.

—0, or practically, though *not theoretically, one.* 3rd, Gives two real and two imaginary points, illustrating that imaginary roots enter in pairs. 4th, Gives two equal real roots, both 0, and two imaginary, showing a point of contact. 5th, Four imaginary roots, showing no common point, the additional imaginary roots again entering in a pair.

Ex. 11. Find the intersections of $2y^2 - 4xy + 2x^2 - 3y - 2x - 8 = 0$, by $4y^2 + 4x^2 - 11 = 0$. Also by $y^2 + 2y + x^2 - 6x + 6 = 0$. Also by $y^2 + 6y + x^2 - 4x + 9 = 0$.

Results.—By the first in 4 points. By the second in 2 points. By the third not at all. The figure is seen in *Fig.* 29, in which *a a a* is the 1st locus, and 1 1 1, 2 2 2, and 3 3 3, the others, in order.

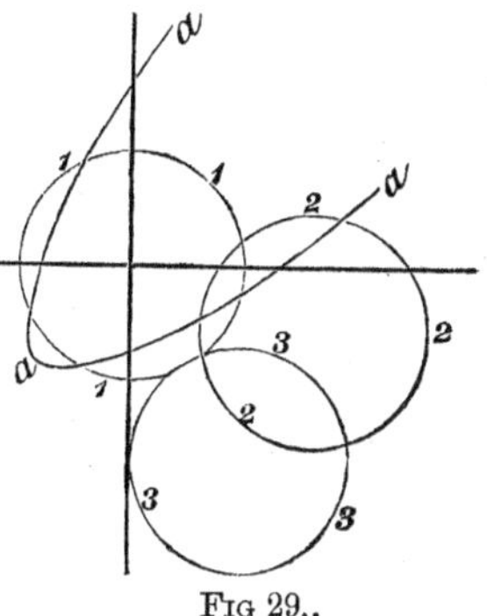

FIG 29..

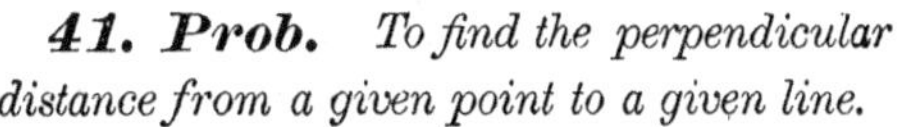
41. Prob. *To find the perpendicular distance from a given point to a given line.*

METHOD OF SOLUTION.—First, find the equation of a line passing through the given point, and perpendicular to the given line (***32, 39***). Second, find the point in which this perpendicular intersects the given line (***40***). The problem then consists in finding the distance between two points (***29.***)

COR.—*To find the distance between two parallels,* write the equation of a line perpendicular to the parallels (***39***), and find its intersections with the parallels. The problem is then the same as (***29***).

Ex. 1. Find the distances of the following points from each of the lines $y = 2x - 3$, and $\frac{1}{2}x - y = -1$, viz., 3, 2; —4, —1; 0, —6; 0, 0.

SOLUTION.—To find the distance from —4, —1, to $y = 2x - 3$, we have for the equation of a line passing through this point and perpendicular to this line $y + 1 = -\frac{1}{2}(x + 4)$, or $y = -\frac{1}{2}x - 3$. The intersection is at 0, —3. The distance between —4, —1 and 0, —3 is $D = \sqrt{16 + 4}$, or $\sqrt{20}$.

Ex. 2. Find the sides, the angles, and the perpendicular distances from the angles to the opposite sides in the triangle the equations of whose sides are $36y - 4x = 45$, $3y + 3 = -x$, and $y = \frac{2}{3}x - 3$.

Results.— The sides are 12.79, 7.44, and 6.79.
The angles are 24°46′, 127°53′, and 27°21′.
The perpendiculars are 5.88, 3.12, and 5.36.

Ex. 3. The vertices of a triangle are at 2, 8 ; —6, 1 ; and 0, —4 ; required the equations of the sides, of the lines drawn from the vertices to the middle of the opposite sides, and of the lines drawn bisecting the angles and terminating in the opposite sides.

SECTION VI.

Of the Conic Sections.

42. *Boscovich's Definition of a Conic Section.*—A Conic Section is a curve, the distance of any point in which from a given point, is to its distance from a given straight line, in a given ratio. If the distance to the point is equal to the distance to the line, the locus is a ***Parabola*** ; if less, an ***Ellipse*** ; if greater, an ***Hyperbola.*** If the distance to the *line* is infinite, the locus is a ***Circle*** ; but if the distance to the *point* is infinite, the locus is a ***Straight Line.***

43. *Prob.* *To construct a Conic Section from Boscovich's definition.*

Solution.—Let F, *Fig.* 30, be the given point, AB the given line, and $m : n$ the given ratio. Through F draw CK perpendicular, and GH parallel to AB. Take FG (= FH) : FC : : $m : n$, and draw CG and CH, producing them indefinitely. Draw a series of parallels to GH, meeting the lines CM and CN. Now with the half of any one of these lines, as LT, for a radius, and the given point, F, as a centre, describe an arc cutting the parallel taken, as at P. Then is P a point in the curve. To prove that P is a point in the curve, join P and F, and draw PR parallel to CK. By similar triangles we then have

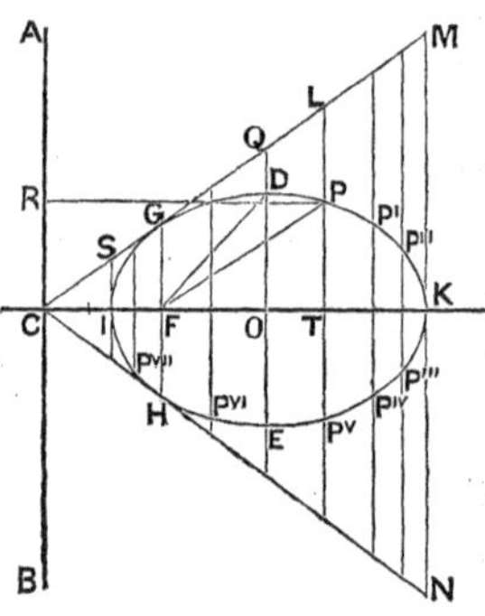

Fig. 30.

LT(= PF) : TC (= PR) : : GF : FC (by construction) : : $m : n$. ∴ PF : PR : : $m : n$. In like manner any required number of points in the curve may be determined, so that by connecting them the curve will be completely drawn. In this figure, as FG $<$ FC the curve is an ellipse. Had FG been taken equal to FC, the curve would have been a *Parabola.* And if FG had been greater than FC, the curve would have been an *Hyperbola.*

44. Defs.—The fixed line, AB, is the ***Directrix.*** The fixed point, F, is the ***Focus.*** CM and CN are the ***Focal Tangents.***

The portion of the perpendicular to the directrix through the focus, CK, intercepted by the curve is the ***Transverse*** or ***Major Axis,*** as IK. The centre of the transverse axis, O, is the centre of the curve.

The perpendicular to the tranverse axis passing through the centre, and limited by the curve (in the ellipse), as DE, is the ***Conjugate,*** or ***Minor Axis.*** The double ordinate passing through the focus, GH, is the ***Latus Rectum, Principal Parameter,*** or the *parameter* to the *transverse axis.* The extremities of the transverse axis, I and K, are the ***Vertices.*** The distances from the focus to the vertices are the ***Focal Distances.*** The ***Eccentricity,*** is the distance from the focus to the centre, divided by the semi-transverse axis, $\frac{FO}{IO}$.

Ex. 1. Construct a parabola whose parameter is 12.

CONSTRUCTION.—Let AB be the directrix. Draw CK perpendicular to it, take F at a distance 6 from the directrix, and through F draw GH parallel to AB. Take GF = HF = 6, and draw CM, and CN through G and H. (The construction is then completed as in the problem above.) To show that any point thus found, as P, is a point in a parabola whose parameter is 12, observe that LT (= PF): CT (= PR) : : GF : CF. But GF = CF = 6, by construction. ∴ PF = PR. Also GH, the parameter, = 2GF = 12.

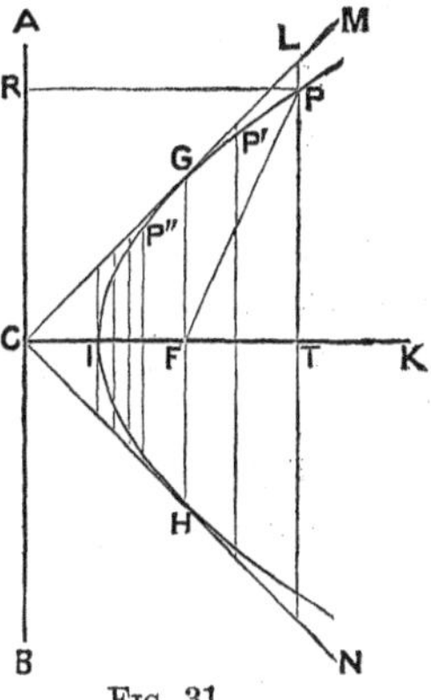

FIG. 31.

QUERIES.—Can the parabola ever return into itself so as to inclose a space? Why? Can portions of this curve lie on both sides of the directrix? Why?

45. COR. 1.—*If* p = *the focal ordinate, i. e., one half the Latus Rectum, the vertex is at* $\frac{1}{2}$p *from the directrix, and the distance,* IF = $\frac{1}{2}$p, *or* $\frac{1}{4}$ *the parameter,* GH.

Ex. 2. Construct an ellipse in which the fixed ratio shall be $\frac{2}{3}$, and the distance from the focus to the directrix 6. Also with the ratio $\frac{1}{2}$ and the distance from the focus to the directrix 5.

QUERIES.—With the same focus and directrix how does varying the ratio affect the form of the curve? With the same ratio and directrix how does varying the position of the focus affect the form of the curve? How does it appear from the first query that when the ratio is 0, the locus is a point?

Ex. 3. Construct an ellipse whose *latus rectum* shall be 6, and the fixed ratio $\frac{3}{7}$.

SUG.—In *Fig.* 30, if **GF** $= 3$ and the characteristic ratio of the curve is $\frac{4}{5}$, what is **CF**?

46. COR. 2—*From* Fig. 30, *by principles of construction it appears that the tangents at the vertices, viz.,* **IS**, *and* **KM**, *are equal, respectively, to the focal distances* **IF**, *and* **FK**. *It also appears that the distance from the focus to the extremity of the conjugate axis in an ellipse,* **FD**, *equals the semi-transverse axis; for* **FD** $=$ **QO** $= \frac{1}{2}($**IS** $+$ **MK**$) = \frac{1}{2}$**IK**.

Ex. 4. Construct an ellipse whose transverse axis shall be 12, and conjugate 10; *i. e., having given the axes, to construct the ellipse.*

Ex. 5. Construct an ellipse whose transverse axis shall be 10, and distance between the foci 8.

Ex. 6. Having given the curve and the transverse axis, to find the foci and directrix of an ellipse.

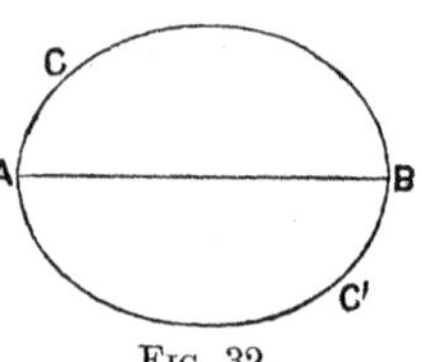

FIG. 32.

SOLUTION.—Let **ACBC'**, *Fig.* 32, be the curve, and **AB** its transverse axis. Bisect the transverse axis with a perpendicular, and the portion of this perpendicular intercepted by the curve will be the conjugate axis. From either vertex of the conjugate axis as a centre, with a radius equal to the semi-transverse axis, describe arcs cutting the transverse axis; these points will be the foci (**46**). As there are two intersections, there are two foci. At each extremity of the transverse axis erect perpendiculars and make them severally equal to the adjacent focal distances, thus obtaining two points in the focal tangent (**46**). Draw the focal tangent, and where it intersects the transverse axis produced, erect a perpendicular to this axis, and this perpendicular will be the directrix.

QUERIES.—How does it appear from the definition of the ellipse, that the curve can not lie on both sides of the directrix? How does it appear that the curve cuts the axis beyond the focus?

Ex. 7. Letting A represent the semi-transverse axis, B the semi-conjugate, $2c$ the distance between the foci, and e the eccentricity, show that

$$e = \frac{c}{A} = \frac{\sqrt{A^2 - B^2}}{A},$$

and hence that $1 - e^2 = \frac{B^2}{A^2}$. How does it appear from this that in the case of the ellipse $e < 1$?

Ex. 8. Construct an ellipse whose transverse axis is 12, and eccentricity $\frac{2}{3}$.

SUG. First find the value of B, which is $4\frac{1}{2}$, nearly.

Ex. 9. Construct an hyperbola.

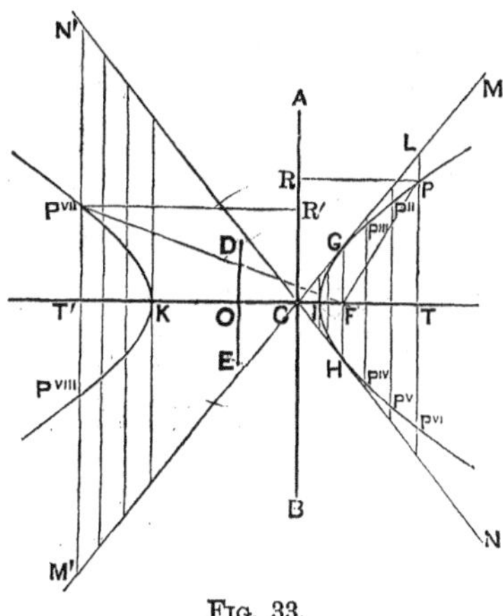

Fig. 33.

Solution.—Let **AB** be the directrix, **F** the focus, and $m : n$ the ratio, in which $m > n$. Through **F** draw **FK** perpendicular to the directrix and **GH** parallel, producing both indefinitely. Take **FG** (= **FH**) : **CF** :: $m : n$. Through **C** and **G** draw **MM**′, and through **H** and **C**, **NN**′, the focal tangents. (The process is exactly analogous throughout, to that pursued in constructing the ellipse, and hence need not be detailed. The student can supply it. It should be noticed, however, that the distance from the focus to any point in the curve, being greater than the distance from the same point to the directrix, there may be (are) points in the curve on the opposite side of the directrix from the focus. These points are determined in the same manner as the others. Thus the point P^{VII} is found by taking **T′N′** as a radius, and from **F** as a centre drawing an arc cutting **T′N′** in P^{VII}. In like manner other points in this branch are located.)

The demonstration is as follows: To prove that any point, as **P**, is in the curve, we have to prove that **PF** : **PR** :: $m : n$; *i. e.*, the distance from any point in the curve to the focus, is to the distance of the same point from the directrix, in a constant ratio ($m : n$), which ratio is greater than 1, in the hyperbola. To prove that the construction gives this proportion, join **P** and **F**, and draw **PR** parallel to **TC**. Now since **PF** = **LT**, and **PR** = **TC**, and by reason of similar triangles, we have **PF** : **PR** :: **LT** : **TC** :: **GF** : **FC** :: $m : n$. In a similar manner any point on the other side of the directrix, found by the method described, as P^{VII}, is shown to be in the curve. Thus $P^{VII}F$ = **N′T′** by construction, $P^{VII}R'$ = **T′C**, and the triangles **CFG** and **CT′N′** are similar; hence $P^{VII}F : P^{VII}R'$ :: **T′N′** : **T′C** :: **GF** : **FC** : $m : n$. Q. E. D.

47. Def's.—***The Axis of the Hyperbola*** is an infinite line drawn through the focus and perpendicular to the directrix, as **TT′**, *Fig.* 33.

The Transverse Axis of the Hyperbola is that portion of the axis of the curve included between the vertices, as **KI**, *Fig.* 33.

The Focal Distances are the distances from the focus to the vertices, as **FI**, and **FK**, *Fig.* 33.

The Conjugate Axis of the Hyperbola is a perpendicular to the transverse axis at its centre, and is limited by an arc drawn from the vertex as a center, with a radius equal to the distance from the focus to the centre. Thus, in *Fig.* 33, **DE** represents the *conjugate axis*, the extremities **D** and **E** being determined by making the distances **DI** and **EI** each equal to **OF**. This definition is a convention adopted for the purpose of rendering more close the analogy between this curve and the ellipse.

A Conjugate Hyperbola is an hyperbola having the conjugate axis of a given hyperbola for its transverse axis, and the transverse axis of the given curve for its conjugate; see Ex. 10, *Fig.* 34. Either of two hyperbolas thus related is conjugate to the other. They are sometimes distinguished as the X hyperbola and the Y hyperbola, each taking the name of the co-ordinate axis upon which its transverse axis lies.

An Equilateral Hyperbola is one which has its conjugate axis equal to its transverse.

Ex. 10. To construct a pair of conjugate hyperbolas whose axes are 8 and 6.

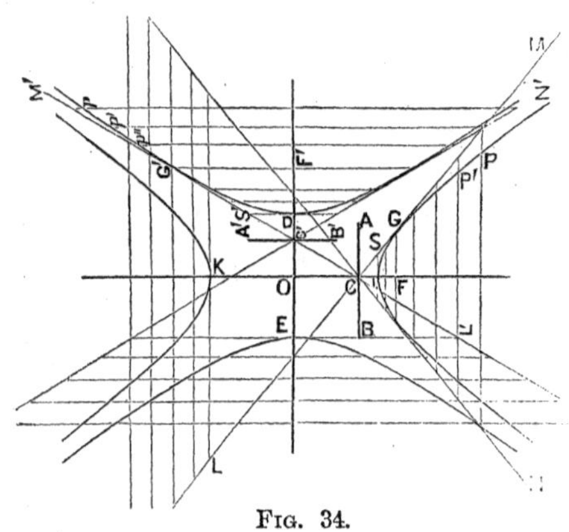

FIG. 34.

SUGS.—Draw two indefinite straight lines at right angles to each other, and take **OI** = **OK** = 4, and **OD** = **OE** = 3. Having constructed the branches on the axis **KI**, *Fig.* 34, as in Ex. 9, take **OF′** = **OF** (which = **ID**), and **F′** is the focus of the conjugate or Y hyperbola. Taking **DS′** = **DF′** and **EL′** = **EF′**, and through **S′** and **L′** drawing a right line, it is one of the focal tangents. Having found the focal tangents the construction proceeds as before.

Ex 11. Construct an hyperbola whose transverse axis is 6, and less focal distance 2. Find also the conjugate axis, focus, and directrix of the conjugate hyperbola.

Ex. 12. Letting e represent the eccentricity of an hyperbola, c the distance from the centre to the focus, A the semi-transverse axis, and B the semi-conjugate, show that

$$e = \frac{c}{A} = \frac{(A^2 + B^2)^{\frac{1}{2}}}{A}, \text{ and hence that } 1 - e^2 = -\frac{B^2}{A^2}.$$ Why is e greater than 1 in the hyperbola?

Ex. 13. What is the eccentricity of an hyperbola whose axes are

10 and 6? What is the eccentricity of an hyperbola whose transverse axis is 12, and less focal distance 3?

Ex. 14. The eccentricity being $1\frac{1}{2}$ and the conjugate axis 4, what is the transverse axis? What the focal distances? What the characteristic ratio (**42**)? *Transverse Axis*, 3.577 +.

---

48. Prop. *Boscovich's ratio and the eccentricity are equal.*

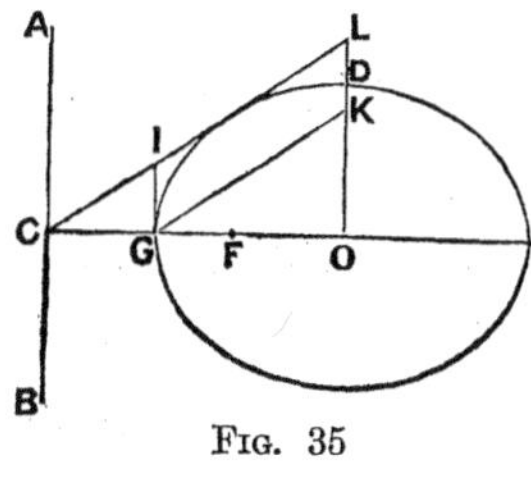

Fig. 35

Dem.—1st. Let AB, *Fig.* 35, be the directrix of an *ellipse*, F the focus, CL the focal tangent, and O the centre. Draw GK parallel to CL. Then since $LO = GO$, and $GF = IG = LK$ (**46**), $LO - LK = KO = GO - GF = FO$. Therefore, $\frac{KO}{GO} =$ the eccentricity (**44**). By definition $\frac{GF}{CG} = \frac{IG}{CG} =$ Boscovich's ratio. Now, by similar triangles, $\frac{IG}{CG} = \frac{KO}{GO}$. Q. E. D.

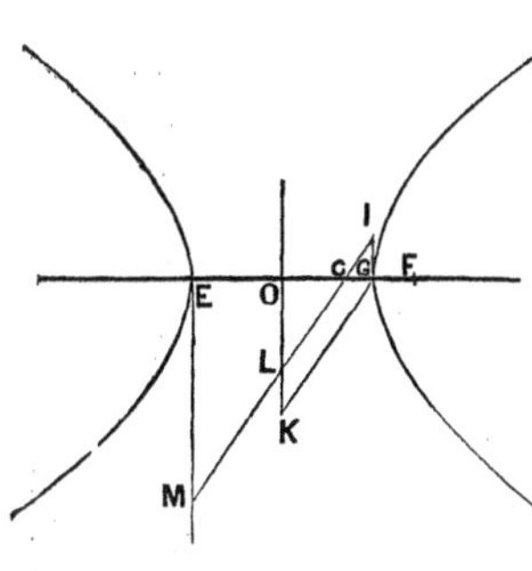

Fig. 36.

2nd. In the *Hyperbola* the demonstration is essentially the same. Thus, in *Fig.* 36, $LO = \frac{EM - IG}{2} = \frac{FE - FG}{2} = GO$. Hence $LO + LK = KO = GO + FG = FO$, and $\frac{KO}{GO} =$ the eccentricity. By definition $\frac{GF}{CG} = \frac{IG}{CG} =$ the characteristic ratio (Boscovich's ratio). Now, by similar triangles, $\frac{IG}{CG} = \frac{KO}{GO}$. Q. E. D.

3rd. In the *Parabola* we may call the eccentricity 1, from analogy; or, better, we may conceive the parabola to be an ellipse with the centre, O, removed to an infinite distance from the vertex, G, *Fig.* 35, whence the fraction $\frac{FO}{GO} = 1$.* Q. E. D.

49. Cor.—*The student should fix in memory the following relations, as they are fundamental, and of frequent use in the reduction of formulæ.*

Letting $A =$ the semi-transverse axis, $B =$ the semi-conjugate axis, $e =$ the eccentricity, and $p =$ the *semi-latus rectum*, we have the following

---

* If the student has difficulty in understanding this statement, let him consider that, O being removed to infinity, the *finite* distance, GF, by which GO *appears* to be greater than FO, is of no *appreciable* value as compared with the terms of the ratio, which are both *infinite*.

FUNDAMENTAL RELATIONS.

From the definition (***42***) and (***48***), we see that, *The distance from any point in the curve to the focus* $\div$ e $=$ *the distance from the same point to the directrix.* Also, *The distance of any point in the curve from the directrix* $\times\, e =$ *the distance of the same point from the focus.*

	Distances.		*Ellipse.*	*Hyperbola.*	*Parabola*
1.	Focus to extremity of Conj.-axis..	=	A	....	∞
	Vertex " "	=	....	Ae	∞
2.	Focus to Centre..................	=	Ae	Ae	∞
3.	Focal Distances..............	=	$A(1 \mp e)$	$A(e \mp 1)$	$\frac{1}{2}p$
4.	Vertices to Directrix............	=	$\frac{A(1 \mp e)}{e}$	$\frac{A(e \mp 1)}{e}$	$\frac{1}{2}p$
5.	Focus to Directrix..............	=	$\frac{A(1-e^2)}{e}$	$\frac{A(e^2-1)}{e}$	p
6.	Centre to Directrix.............	=	$\frac{A}{e}$	$\frac{A}{e}$	∞
7.	$1-e^2$	=	$\frac{B^2}{A^2}$	$-\frac{B^2}{A^2}$	0
8.	Semi-Latus Rectum, p	=	$\frac{B^2}{A}$	$\frac{B^2}{A}$	p

Dem.—For the ellipse see *Fig.* 30, for the hyperbola, *Fig.* 33, and for the parabola, *Fig.* 37.

(1.) *For Ellipse* see (***46***).—*For Hyperbola,* by definition of eccentricity $\frac{FO}{A} = e$. $\therefore$ $FO = Ae =$ the distance from the vertex to the extremity of the conjugate axis, by the definition of the latter (***47***). *For Parabola,* consider the curve as an ellipse with its centre removed to infinity.

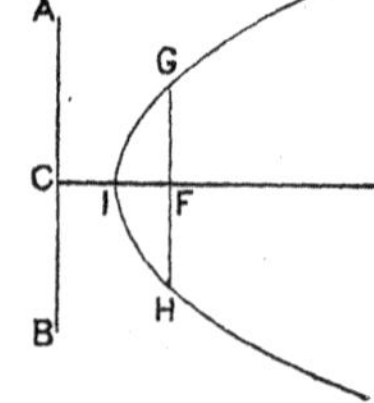

Fig. 37.

(2.) *For Ellipse,* $\frac{FO}{A} = e$ by definition. $\therefore$ $FO = Ae$. *For Hyperbola* and *Parabola,* see above.

(3.) *For Ellipse,* $IF = IO - FO = A - Ae = A(1-e)$. $FK = OK + FO = A + Ae = A(1+e)$. *For Hyperbola,* $IF = FO - IO = Ae - A = A(e-1)$. $FK = FO + OK = Ae + A = A(e+1)$. *For Parabola* see (***45***).

(4.) *For Ellipse,* since I is a point in the curve $IC = \frac{IF}{e} = \frac{A(1-e)}{e}$. Also, $KC = \frac{KF}{e} = \frac{A(1+e)}{e}$. *For Hyperbola,* for same reason $IC = \frac{IF}{e} = \frac{A(e-1)}{e}$; and $KC = \frac{KF}{e} = \frac{A(e+1)}{e}$. *For Parabola,* see (***45***).

(5.) *For Ellipse,* $FC = IF + IC = A - Ae + \frac{A(1-e)}{e} = \frac{A(1-e^2)}{e}$.

For Hyperbola, $FC = IF + IC = Ae - A + \frac{Ae - A}{e} = \frac{A(e^2 - 1)}{e}$. *For Parabola,* $FC = GF = p$, by definition.

(6.) *For Ellipse,* **D** being a point in the curve whose distance from the directrix is **OC**, we have $OC = \frac{FD}{e} = \frac{A}{e}$. *For hyperbola,* $OC = OI - CI = A - \frac{Ae - A}{e} = \frac{A}{e}$. (The distance in the ellipse may be obtained in the same way.) *For Parabola,* same conception as in (1) above.

(7.) *For Ellipse* and *Hyperbola* see *Ex's*. 7 and 12. *For Parabola,* $1 - e^2 = 1 - 1 = 0$.

(8.) *For Ellipse,* $GF = p = FC \times e = A(1 - e^2) = \frac{B^2}{A}$. *For Hyperbola,* $GF = p = FC \times e = A(e^2 - 1) = \frac{B^2}{A}$. *For Parabola,* $GF = p$ by definition.

50. Prob. *To pass a conic section through three given points, so that it shall have a given focus; and to determine its elements; i. e.,* the axes, foci, directrix, eccentricity, etc., if an ellipse or hyperbola, or the *latus rectum* if a parabola.

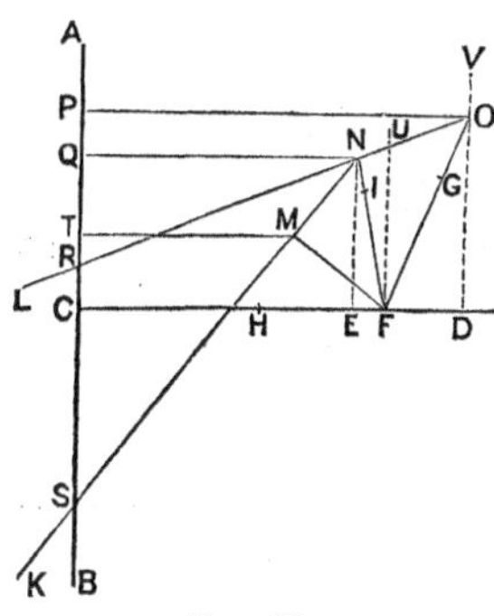

Fig. 38.

Solution.—Let **M**, **N**, and **O**, *Fig.* 38, be the given points, and **F** the given focus. Connect the points with the focus, and draw **ON**, and **NM**, and produce them towards the probable position of the directrix, as to **L** and **K**. Now, take a point **R**, on **OL**, such that **OF** : **NF** :: **OR** : **NR**,* and **R** is a point in the directrix. In like manner, take **NF** : **MF** :: **NS** : **MS**, and **S** is another point in the directrix. Hence the directrix can be drawn.

To prove that a line drawn through **R**, and **S**, as **AB**, is the directrix, we have to show that $\frac{OF}{OP} = \frac{NF}{NQ} = \frac{MF}{MT}$, **OP**, **NQ**, and **MT**, being perpendicular to **AB**. Now **OP** : **NQ** :: **OR** : **NR**. But by construction **OR** : **NR** :: **OF** : **NF**. ∴ **OP** : **OF** :: **NQ** : **NF**, or $\frac{OF}{PO} = \frac{NF}{NQ}$. In like manner **NQ** : **MT** :: **NS** : **MS** :: **NF** : **MF**. ∴ $\frac{NF}{NQ} = \frac{MF}{MT}$. Q. E. D.

To make the numerical computations requires much more labor than to effect the geometrical solution. We may proceed as follows: Having the distances **ON**, **NM**, **OF**, **NF**, and **MF** given in numbers, compute the numerical values of **NR** and **MS** from the proportions used in the construction. The sides of the triangles **OFN** and **NFM** being known, their angles can be found by trigo-

* This construction is effected thus: taking the proportion by division, (**OF** — **NF**), or **OG** : **OF** : : (**OR**—**NR**), or **ON** : **OR**. From this proportion **OR** can be constructed, as the other terms are known.

nometry; whence we get the angle **RNS**, as it equals 180° — (**ONF** + **FNM**). Then, in the triangle **RNS**, we shall have two sides and the included angle; whence the angle **NRS** can be found, and from it **NRQ** becomes known. Now, in the right angled triangle **NRQ**, we know the hypothenuse and one acute angle, and can find **NQ**. Again, letting fall the perpendicular **NE**, forming the triangle **FNE**, we can compute **EF**, since **FN** is known and the angle **FNE** = **FNR** + **RNQ** — 90°. **FC** is therefore known, being equal to **EF** + **NQ**. As $\frac{NF}{NQ}$ is now determined, the ratio e, or, what is the same thing, the eccentricity, is known. Taking a point, as **H**, upon **FC**, such that $\frac{FH}{HC} = \frac{NF}{NQ}$, the vertex is determined. In a similar manner the other vertex of an ellipse or hyperbola can be found. Letting p be half the *latus rectum*, **FU**, it can be found from $\frac{FN}{NQ} = \frac{p}{FC}$.

Ex. 1. Construct a conic section passing through the points **O**, **M**, and **N**, and having **F** for a focus, knowing that **OF** $= 6\frac{1}{4}$, **NF** $= 3\frac{1}{4}$, **MF** $= 2\frac{1}{4}$, **ON** $= \sqrt{18}$, **NM** $= \sqrt{10}$. Let the geometrical construction be given, and also the numerical solution.

The locus is a parabola whose latus rectum is 9.

Ex. 2. Construct and compute as above, when **OF** = 2.08, **NF** 1.08, **MF** = .46, **ON** = 1.12, and **NM** = .87.

Ex. 3. Construct and compute as above, when **OF** = 10, **NF** = 6, **MF** = 3, **ON** = 6, and **NM** = 4.

---

51. Prob. *To produce the general equation of a Conic Section referred to rectangular axes.*

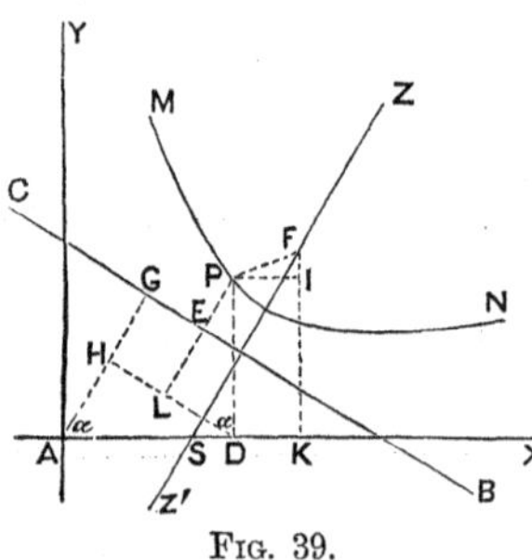

FIG. 39.

SOLUTION.—Let **MN**, *Fig.* 39, be an arc of any conic section, **F** the focus, **CB** the directrix, **ZZ′** the axis of the curve, and **AX** and **AY** the axes of reference. Let **P** be any point in the curve, and its ordinate **PD**: also draw the ordinate of the focus, **FK**. Draw from the origin **AG** perpendicular to **CB**. Draw **DH** parallel, and **PL** perpendicular to **CB**. Join **P** and **F**, and draw **PI** parallel to **AX**.

Let **AG**, the distance from the origin to the directrix, be represented by d; the co-ordinates of the focus, **AK** and **FK**, by m and n respectively; the ratio mentioned in the definition (*42*), $\frac{PF}{PE}$, by e; the angle which the axis of the curve makes with the axis of abscissas, **ZSX** = **GAX** = **LDP**, by α; and the general co-ordinates, **AD** and **PD**, by x and y.

Now, $\overline{PF}^2 = e^2 \cdot \overline{PE}^2$. But $\overline{PF}^2 = \overline{PI}^2 + \overline{FI}^2 = (m - x)^2 + (n - y)^2$. Again, $\overline{PE}^2 = (PL + AH - AG)^2 = (PD \sin \alpha + AD \cos \alpha - d)^2 = (y \sin \alpha + x \cos \alpha - d)^2$. Whence, substituting we have

(*Eq. A*), $(m - x)^2 + (n - y)^2 = e^2(y \sin \alpha + x \cos \alpha - d)^2$.

52. Cor. 1.—*The equation of the ellipse and hyperbola referred to their axes is*

$$y^2 + (1 - e^2)\, x^2 = A^2(1 - e^2),$$

in which A *is the semi-transverse axis, and* e *the eccentricity, or the characteristic ratio* (**42**).

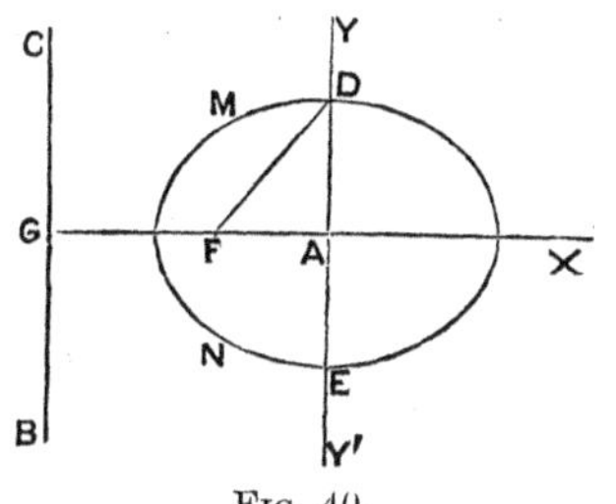

Fig. 40.

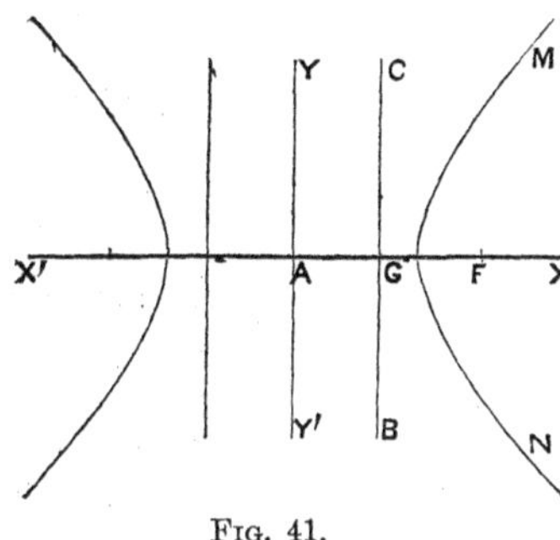

Fig. 41.

Dem.—Let the curve in *Fig.* 39 be conceived to change position so as to assume that in *Fig.* 40 or 41, the axis of the curve coinciding with the axis of abscissas, the origin at the centre, A, F the focus and CB the directrix. As the axis of the curve now coincides with the axis of abscissas, $\alpha = 0$. As the focus falls upon the axis of abscissas, $n = 0$. By consulting (**49**) it will be seen that $m = \mp AF = \mp Ae$, and $d = \mp AG = \mp \frac{A}{e}$*. Whence, substituting these values in *Eq. A*, we have $(\mp Ae - x)^2 + y^2 = e^2\left(x \pm \frac{A}{e}\right)^2 = (ex \pm A)^2$. Expanding and reducing, we have

(*Eq. B*), $y^2 + (1 - e^2)x^2 = A^2(1 - e^2)$.

Q. E. D.

Query.—How is it that the *same* equation is made to represent *two* curves so different from each other?

53. Cor. 2.—*The equation of the ellipse referred to its axes, and in terms of its semi-axes is*

$$A^2y^2 + B^2x^2 = A^2B^2,$$

in which A *and* B *are the semi-axes.*

---

* The upper sign applies to the ellipse, and the lower to the hyperbola.

The same final result may be obtained in the case of the hyperbola by conceiving the axis ZZ' of *Fig.* 39 to have revolved to the left and the branch to fall on the left of the origin. In this case $\alpha = 180°$, $\cos \alpha = -1$, and x is negative.

DEM.—This equation is obtained at once from *Eq. B* by substituting for $1 - e^2$ its value, $\frac{B^2}{A^2}$ (**49**), and clearing of fractions.

54. COR. 3.—*The equation of the hyperbola referred to its axes, and in the terms of its semi-axes, is*

$$A^2y^2 - B^2x^2 = - A^2B^2,$$

in which A *and* B *are the semi-axes.*

DEM.—Substituting in *Eq. B*, $-\frac{B^2}{A^2}$ for $1 - e^2$ (**49**), we have, after clearing of fractions, $A^2y^2 - B^2x^2 = - A^2B^2$.

SCH.—These two equations, viz.,

$A^2y^2 + B^2x^2 = A^2B^2$, for the ellipse, and
$A^2y^2 - B^2x^2 = - A^2B^2$, for the Hyperbola,

are the most important forms of the equations of these loci, and are always meant when their equations are spoken of without specification. They may be called the *Common Forms.*

55. COR. 4.—*The equation of the ellipse referred to its transverse axis and a tangent at the left hand vertex, is*

$$y^2 = \frac{B^2}{A^2}(2Ax - x^2);$$

and of the hyperbola,

$$y^2 = -\frac{B^2}{A^2}(2Ax - x^2).$$

DEM.—In either of the annexed figures, let **AX** and **AY** be the axes, **F** the focus, and **CB** the directrix. Now $a = 0$; $n = 0$; $m = A \mp Ae$; $d = -$ **AG** $= - \frac{A - Ae}{e} = \frac{Ae - A}{e}$ in the ellipse, and $+$ **AG** $= \frac{Ae + A}{e}$ in the hyperbola, hence in general $d = \frac{Ae \mp A}{e}$ (**49**). Substituting these values in *Eq. A* (**51**), and reducing, we have

$$y^2 + (1 - e^2)x^2 - 2A(1 - e^2)x = 0,$$

which is applicable to either locus. For $1 - e^2$ substituting $\frac{B^2}{A^2}$ for the ellipse, and $-\frac{B^2}{A^2}$ for the hyperbola, we have

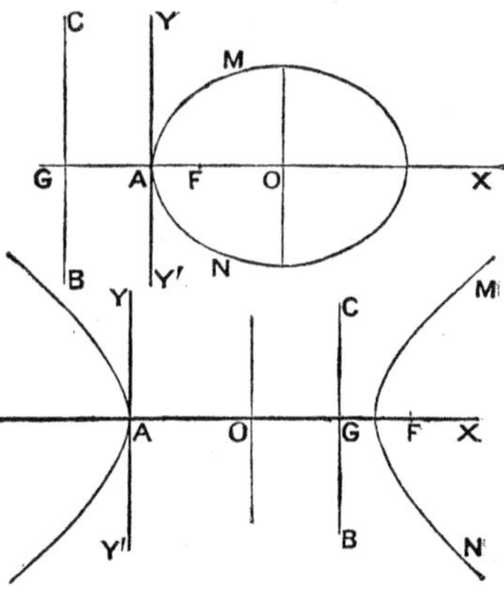

FIG. 42.

$$y^2 = \frac{B^2}{A^2}(2Ax - x^2), \text{ and}$$

$$y^2 = - \frac{B^2}{A^2}(2ax - x^2). \qquad \text{Q. E. D.}$$

56. Cor. 5.—*The equation of the circle referred to a pair of diameters at right angles to each other is*

$$y^2 + x^2 = R^2, \text{ THE COMMON FORM.}$$

When the reference is to a diameter and a tangent at its left hand extremity, the equation of the circle is

$$y^2 = 2Rx - x^2.$$

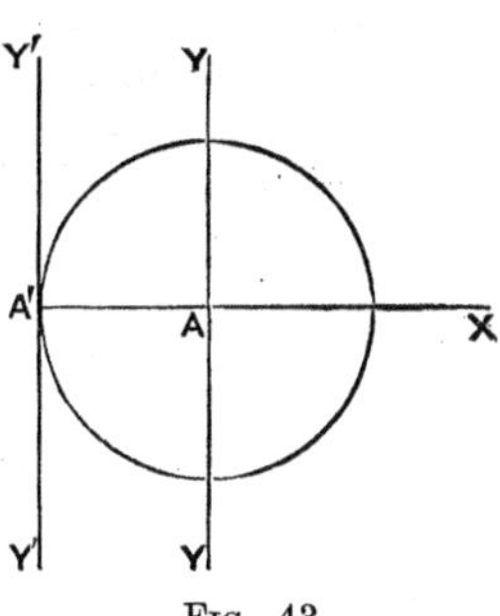

Fig. 43.

Dem.—These two forms are readily produced from $A^2y^2 + B^2x^2 = A^2B^2$, and $y^2 = \frac{B^2}{A^2}(2Ax - x^2)$, by making $A = B = R$, the radius of the circle, and dividing out the common factor R^2. For the first form the origin is at **A**, *Fig.* 43, and for the second form at **A′**.

57. Cor. 6.—*Making* A = B *in the equations of the hyperbola,* Cors. 3rd *and* 4th, *there result* $y^2 - x^2 = -A^2$, *and* $y^2 = -(2Ax - x^2)$, *as equations of the Equilateral Hyperbola* (**47**).

58. Cor. 7.—*To obtain the equation of the hyperbola conjugate to* $A^2y^2 - B^2x^2 = - A^2B^2$, *and referred to the same axes, we have but to change the sign of the absolute term, and write* $A^2y^2 - B^2x^2 = A^2B^2$.

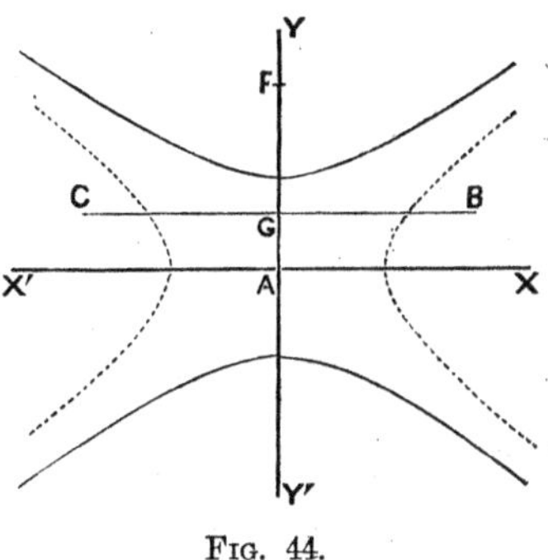

Fig. 44.

Dem.—Conceive the curve *Fig.* 39, to take the position **MN**, *Fig.* 44, the axis of the curve and the focus falling on the axis of ordinates, and the origin being at the centre, **A**. We then have $\alpha = 90^\circ$, $m = 0$, $n =$ **AF** $= Be$, and $d =$ **AG** $= \frac{B}{e}$. Whence, substituting in *Eq. A*, it becomes $x^2 + (Be - y)^2 = e^2\left(y - \frac{B}{e}\right)^2 = (ey - B)^2$; or, expanding and reducing, $x^2 + (1 - e^2)y^2 = B^2(1 - e^2)$. But $1 - e^2 = -\frac{A^2}{B^2}$, which substituted, gives after reduction, $A^2y^2 - B^2x^2 = A^2B^2$. Q. E. D.

59. Cor. 8.—*The equation of the Parabola referred to its axis and a tangent at the vertex, is* $y^2 = 2px$, *in which* 2p *is the latus rectum.*

DEM.—Resuming *Eq. A*, and conceiving the curve situated as in *Fig.* 45, $\alpha = 0$, $n = 0$, $m = \text{AF} = \frac{1}{2}p$, $d = -\text{AG} = -\frac{1}{2}p$, and $e = 1$. Substituting these values, we have $(\frac{1}{2}p - x)^2 + y^2 = (x + \frac{1}{2}p)^2$; or, reducing, $y^2 = 2px$. Q. E. D. This is the *Common Equation* of the parabola.

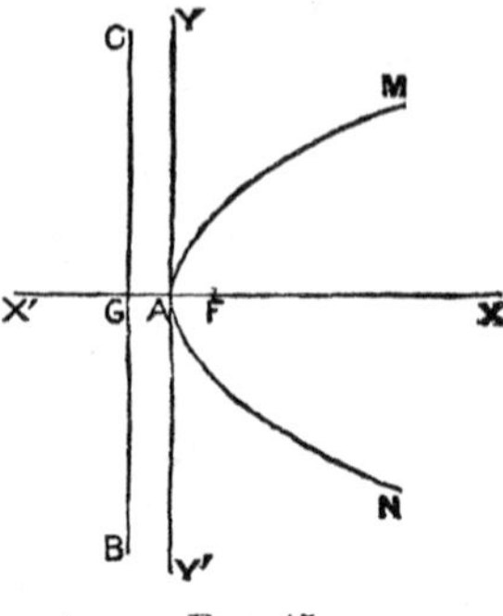

FIG. 45.

60. SCH.—It will be observed that the difference in form between the equation of the ellipse in terms of the semi-axes, and the corresponding equation of the hyperbola may be considered as embraced in the sign of B^2; so that hereafter, in any case when a property of the ellipse is deduced from the equation of that locus—as many will be,—the corresponding property of the hyperbola can be discovered by simply substituting $-B^2$ for B^2 in the result; or if B only is involved, by replacing it by $B\sqrt{-1}$. Our subsequent work may often be much abridged by this means. It is also to be remarked that, if the formula expressing any property of either locus does not contain B, *i. e.*, does not depend upon the conjugate axis, such property is the same in both loci.

---

61. Prop. *Every equation of the second degree, between two variables, is an equation of a conic section.*

DEM.—Resuming *Eq. A*, expanding and collecting terms, we have

$$(1 - e^2\sin^2\alpha)y^2 - 2e^2\sin\alpha\cos\alpha xy + (1 - e^2\cos^2\alpha)x^2 + (2e^2d\sin\alpha - 2n)y$$
$$+ (2e^2d\cos\alpha - 2m)x + (m^2 + n^2 - e^2d^2) = 0. \quad (Eq.\ B.)$$

Representing these coefficients in order by A, B, C, etc., we have—

$$Ay^2 + Bxy + Cx^2 + Dy + Ex + F = 0. \quad (Eq.\ A'.)$$

This is the *Complete Equation* of the second degree between two variables; *i. e.* it contains every variety of terms, with respect to the variables, which such an equation can have.

It now remains to be shown that these coefficients may have such values (by the locus being of a proper species and properly situated) as to cause the equation to take *any* and *every* given form. Dividing *Eq. A'* by F, (any one of the coefficients A, B, C, etc., would do as well) and distinguishing the resulting coefficients by accents, we have $A'y^2 + B'xy + C'x^2 + D'y + E'x + 1 = 0$. Now the *five* coefficients, A', B', C', D', E', depend upon the *five* arbitrary constants, α, m, n, d, and e, in such a way that such values may be assigned to these last quantities, *i. e.*, the locus may be of such species and so situated, as to give the quantities A', B', C', D', E', severally *any* and *all* required values. Hence *every* equation of the second degree between two variables is an equation of a conic section. Q. E. D.

[NOTE.—If the above demonstration seems abstract, let the student consider a special case. For example, let us inquire if $2y^2 - 3xy + 2y - 5x + 4 = 0$, is an equation of a conic section; and, if it is, of what species is the locus, and how situated. Now, dividing this equation through by 4, we have $\frac{1}{2}y^2 - \frac{3}{4}xy + \frac{1}{2}y - \frac{5}{4}x + 1 = 0$. Comparing this with the equation $A'y^2 + B'xy + C'x^2 + D'y + E'x + 1 = 0$, and remembering that $A' = \frac{1 - e^2 \sin^2 \alpha}{m^2 + n^2 - e^2d^2}$, $B' = \frac{-2e^2 \sin \alpha \cos \alpha}{m^2 + n^2 - e^2d^2}$, $C' = \frac{1 - e^2 \cos^2 \alpha}{m^2 + n^2 - e^2d^2}$, $D' = \frac{2e^2d \sin \alpha - 2n}{m^2 + n^2 - e^2d^2}$, and $E' = \frac{2e^2d \cos \alpha - 2m}{m^2 + n^2 - e^2d^2}$, we can write the *five* following equations:

(1). $\frac{1 - e^2 \sin^2 \alpha}{m^2 + n^2 - e^2d^2} = \frac{1}{2}$;

(2). $\frac{-2e^2 \sin \alpha \cos \alpha}{m^2 + n^2 - e^2d^2} = -\frac{3}{4}$;

(3). $\frac{1 - e^2 \cos^2 \alpha}{m^2 + n^2 - e^2d^2} = 0$, the coefficient of x^2 in this special example being 0;

(4). $\frac{2e^2d \sin \alpha - 2n}{m^2 + n^2 - e^2d^2} = \frac{1}{2}$;

(5). $\frac{2e^2d \cos \alpha - 2m}{m^2 + n^2 - e^2d^2} = -\frac{5}{4}$;

Now, from these five *equations*, the values of the *five* quantities α, m, n, e, and d, can be found. But these being known, the species (determined by e) and the situation of the locus are known. As a similar course could be pursued with any particular equation of the second degree between two variables, it is evident that *every* such equation represents some conic section.]

---

62. Prob. *To determine the features of an equation of the second degree between two variables, which characterize the several species of conic sections.*

Solution.—Comparing *Eq. A'* with *Eq. B*, we see that $A = 1 - e^2 \sin^2 \alpha$, $B = -2e^2 \sin \alpha \cos \alpha$, and $C = 1 - e^2 \cos^2\alpha$. Squaring the value of B, and subtracting from this square 4 times the product of A and C, we have

$$B^2 - 4AC = 4e^4 \sin^2 \alpha \cos^2 \alpha - 4(1 - e^2 \sin^2 \alpha)(1 - e^2 \cos^2 \alpha)$$
$$= 4e^4 \sin^2 \alpha \cos^2 \alpha - 4(1 - e^2 \sin^2 \alpha - e^2 \cos^2 \alpha + e^4 \sin^2\alpha \cos^2 \alpha)$$
$$= 4e^4 \sin^2 \alpha \cos^2 \alpha - 4 + 4e^2(\sin^2 \alpha + \cos^2 \alpha) - 4e^4 \sin^2 \alpha \cos^2 \alpha$$
$$= -4 + 4e^2, \text{ since } \sin^2 \alpha + \cos^2 \alpha = 1.$$

$\therefore B^2 - 4AC = 4(e^2 - 1)$.

Now, in the parabola, $e = 1$; in the ellipse, $e < 1$; and in the hyperbola, $e > 1$. Therefore,

$$\left.\begin{array}{l} B^2 - 4AC = 0 \textit{ characterizes the Parabola;} \\ B^2 - 4AC < 0 \textit{ characterizes the Ellipse;} \\ B^2 - 4AC > 0 \textit{ characterizes the Hyperbola.} \end{array}\right\} (D).$$

Ex. 1. Determine the species of the locus of the equation $2y^2 - 3xy + 5x^2 - 2y - 12 = 0$.

Sug.—As the equation is of the second degree the locus is a conic section. Again, in this case, $A = 2$, $B = -3$, and $C = 5$. $\therefore B^2 - 4AC = 9 - 40 = -31 < 0$; and the locus is an ellipse.

Ex. 2. Determine the species of the locus of the equation $y^2 - 5xy - 3x^2 + 2x - 8 = 0$. *The locus is an Hyperbola.*

Ex's 3 to 10. Determine the species of the loci of the following equations: (1) $3y^2 - 2x^2 - 4y + 20 = 0$. (2) $4y^2 - 2y + x = 0$. (3) $4xy = 16$. (4) $2y - 3x = 4(x-5)^2 + (x+y)^2$. (5) $3(x-y)^2 = 2(x+3)$. (6) $y^2 = 3(x-2)$. (7) $y^2 - 5x = 2(x-y)^2 + y$,

The 1st is an Hyperbola; the 2nd, a Parabola; the 3rd, an Hyperbola; the 4th, an Ellipse; the 5th, a Parabola; the 6th, a Parabola; the 7th, an Hyperbola.

SUG'S.- Equations like the 4th must be put in the form $Ay^2 + Bxy + Cx^2 + Dy + Ex + F = 0$, before applying the test. Thus equation (4) becomes $y^2 + 2xy + 5x^2 - 2y - 37x + 100 = 0$. From this $A = 1$, $B = 2$, and $C = 5$. $\therefore B^2 - 4AC = -16 < 0$. In (3), $A = 0$, $B = 4$, and $C = 0$.

63. COR. 1.—*The species of the locus depends solely upon the coefficients* A, B, *and* C.

64. COR. 2.—*The form* $Ay^2 + Cx^2 + Dy + Ex + F = 0$ *embraces all species and varieties of the conic section.*

DEM.—Whatever the locus may be, if the axis of abscissas is assumed parallel to the axis of the locus, $\alpha = 0$. Hence $\sin \alpha = 0$; and B, which equals $-2e^2 \sin \alpha \cos \alpha$, is 0. In like manner, by assuming the axis of abscissas perpendicular to the axis of the locus, $\cos \alpha = 0$; and, consequently, $B = 0$. Q. E. D.

---

65. *Prob.* *To determine the varieties of the ellipse.*

SOLUTION.—As the equation $Ay^2 + Cx^2 + Dy + Ex + F = 0$ includes all species and varieties of the conic sections (**64**), we have only to consider what loci it represents when the condition $B^2 - 4AC < 0$ is fulfilled. This condition can only be fulfilled when A and C have like signs and are both numerically greater than 0; for, as $B = 0$, if A and C have different signs, $B^2 - 4AC$, becomes $+ 4AC$, which is greater than 0. If A or $C = 0$, $B^2 - 4AC = 0$. Now, as the signs of A and C must be alike, we may consider them as always $+$; because, if they were $-$ in a given case, the signs of all the terms of the equation could be changed. Again, since D and E depend upon m and n ($D = 2e^2 d \sin \alpha - 2n$, and $E = 2e^2 d \cos \alpha - 2m$), such values may be given to m and n, *i. e.*, the origin may be so located with reference to the focus, that D and E shall each be 0; nor does this affect the species of the locus, as A and C do *not* depend upon m or n. Hence we learn that the form $Ay^2 + Cx^2 + F = 0$, embraces all the varieties of the ellipse.

On this form we observe that if F is negative, it gives by transposition $Ay^2 + Cx^2 = F$, or the equation of the ellipse referred to its axes, in which $\sqrt{\frac{F}{C}}$ and $\sqrt{\frac{F}{A}}$ are the semi-axes of the curve. If A and C are numerically unequal the axes are unequal, and we have the *Common Ellipse*. If $A = C$, the axes become equal and the locus is a *Circle*. Again, if $F = 0$, $Ay^2 + Cx^2 = 0$, gives $y = \pm \sqrt{-\frac{C}{A}x^2}$, which gives no *real* values except $x = 0$, $y = 0$; and hence represents a *Point* (the origin). Finally, if F is $+$, in the form $Ay^2 + Cx^2 + F = 0$, we have

$y = \pm\sqrt{\frac{-F - Cx^2}{A}}$, in which *all* real values of one variable give imaginary values to the other ; hence the equation has no locus in the plane under consideration.

There are, therefore, 4 varieties of loci embraced in the equation of the second degree between two variables, which fulfill the condition $B^2 - 4AC < 0$, and are hence called varieties of the ellipse ; viz., the *Ellipse proper*, the *Circle*, the *Point*, and the *Imaginary* locus.

---

66. Prob. *To determine the varieties of the hyperbola.*

SOLUTION.—Resuming the equation $Ay^2 + Cx^2 + Dy + Ex + F = 0$, which includes all species and varieties of conic sections (**64**), we observe that the characteristic condition of the hyperbola, $B^2 - 4AC > 0$, can only be fulfilled when A and C have *opposite* signs, and are both numerically greater than 0, since $B = 0$. Also that, as D and E depend upon m and n, and A and C do *not*, such values may be given to m and n, *i. e.*, the origin may be so situated with respect to the focus, that D and E shall each be 0. Hence $Ay^2 - Cx^2 + F = 0$, embraces all the varieties of the hyperbola.

On this form we observe that if F is positive, it gives by transposition $Ay^2 - Cx^2 = -F$, or the equation of the hyperbola referred to its axes, in which $\sqrt{\frac{F}{C}}$ is the semi-transverse, and $\sqrt{\frac{-F}{A}}$ is the semi-conjugate axis. If A and C are numerically unequal these axes are unequal, and we have the *common* form of the *hyperbola*. If $A = C$, the axes become equal and the locus is an *Equilateral Hyperbola*. Again, if F is negative the equation becomes $Ay^2 - Cx^2 = F$, which is the equation of the y hyperbola, since the real axis is on the axis of y , and the imaginary one on the axis of x.* Finally, if $F = 0$, we have $Ay^2 - Cx^2 = 0$, or $y = \pm\sqrt{\frac{C}{A}}x$, which is the equation of two straight lines passing through the origin and making angles with the axis of x, whose tangents are respectively $\sqrt{\frac{C}{A}}$ and $-\sqrt{\frac{C}{A}}$.

There are, therefore, 3 varieties of loci embraced in the equation of the second degree between two variables, which fulfill the condition $B^2 - 4AC > 0$, and are hence called varieties of the hyperbola ; viz., the *Hyperbola with unequal axes*, both on the axis of x, and on the axis of y, the *Equilateral Hyperbola*, and *Two Right Lines* intersecting each other.

---

67. Prob. *To determine the varieties of the parabola.*

SOLUTION.—As the equation $Ay^2 + Cx^2 + Dy + Ex + F = 0$, embraces all species and varieties of the conic sections, we have only to determine what loci it represents when $B^2 - 4AC = 0$. But as $B = 0$, this condition can only be fulfilled by

---

* This form of expression is frequently used instead of "axis of ordinates," and "axis of abscissas."

$A = 0$, or $C = 0$. Now as the equation is symmetrical with respect to x and y, it will be sufficient to examine the case in which $C = 0$, or the form $Ay^2 + Dy + Ex + F = 0$. Remembering that α has been made 0, and that $e = 1$, we find D (which equals $2e^2d \sin \alpha - 2n) = -2n$. This can now be made 0 by taking the axis of the curve for the axis of x, and the equation takes the form $Ay^2 + Ex + F = 0$. But in this case we cannot make $E = (2e^2d \cos \alpha - 2m = 2d - 2m) = 0$, since that would require that $d = m$, which is absurd, since d is the distance from the origin to the directrix, and m is the distance from the origin to the focus. (*Numerically* d may equal m; but $E = 0$ requires that they also have the same sign).

We therefore have to discuss the equation $Ay^2 + Ex + F = 0$, which includes all varieties of the parabola. As F depends upon $m^2 + n^2 - e^2d^2$, and as $n = 0$, and $e = 1$, F may be made 0, by putting $m = -d$, which only requires that the origin be at the vertex. The equation is thus reduced to $y^2 = \pm \frac{E}{A}x$, the $\pm$ sign being given to E, as no restriction has been imposed upon it. This is the common equation of the parabola, in which $\frac{E}{A} = 2p$. The $+$ sign locates the curve at the right of the origin, and the $-$ sign at the left, but both give the same variety. Again, if in $Ay^2 + Bxy + Cx^2 + Dy + Ex + F = 0$, we make $A = 0$, $B = 0$, and $C = 0$, the condition $B^2 - 4AC = 0$ is fulfilled, and the locus is therefore sometimes called a variety of the parabola. This locus is evidently a right line, its equation, $Dy + Ex + F = 0$, being an equation of the first degree between two variables.* Finally, if an equation of the second degree between the two variables can be reduced to either of the forms $y^2 + 2xy + x^2 \pm P(x + y) + S = 0$, $y^2 - 2xy + x^2 \pm P'(x - y) + S' = 0$, or $y^2 \pm 2xy + x^2 + S'' = 0$, the condition $B^2 - 4AC = 0$ is still fulfilled, although the equation may be reduced to the form $y \pm x = m \pm \sqrt{p - q}$, which is the equation of two real or imaginary parallel right lines.

We have, therefore, 4 varieties of the Parabola; viz., the *Common Parabola*, the *Right Line*, *Two Parallel Right Lines*, and *Two Parallel, Imaginary Right Lines.*

68. Cor. 1.—*The eccentricity of the circle is 0, and the directrix is at infinity.*

Dem.—In obtaining the equation of the circle (**65**), we made $\alpha = 0$, and $A = C$. Hence $1 = 1 - e^2$, or $e = 0$. Again, when the ellipse passes into the circle, the foci unite in the centre. Now, calling the distance from any point in the curve to the directrix s, and the radius of the circle R, we have $\frac{R}{s} = 0$ (the distance from any point in the curve to the focus divided by its distance from the directrix equals the eccentricity); whence $s = \infty$.†

---

* In reality this condition is not compatible with our fundamental hypothesis, which requires the equation to be of the second degree. Moreover, the conditions $A = 0$, $B = 0$, and $C = 0$, are inconsistent with the character of the coefficients A, B, and C, inasmuch as they require that $1 - e^2\sin^2\alpha = 0$, $1 - e^2\cos^2\alpha = 0$, and $2e^2\sin\alpha\cos\alpha = 0$, or *three* arbitrary conditions while there are but *two* arbitrary constants, e and α.

† If $R = 0$, the condition $\frac{R}{s} = 0$, is satisfied by *any* value of s. In this case the locus is a

Ex. 1. Determine the species and situation of the locus $y^2+6y-12x+33=0$.

Solution.—To determine the *species* of the locus, observe that the equation is of the second degree between two variables, and fulfills the condition $B^2-4AC=0$. Therefore, the locus is a parabola.

To determine the *situation* of the locus, compare its equation with the general equation of the conic section: viz.,
$(1-e^2\sin^2\alpha)y^2-2e^2\sin\alpha\cos\alpha\,xy+(1-e^2\cos^2\alpha)x^2+(2e^2d\sin\alpha-2n)y+(2e^2d\cos\alpha-2m)x+m^2+n^2-e^2d^2=0.$

As in this example the coefficient of y^2 is 1, divide the general equation through by $1-e^2\sin^2\alpha$ before comparing the coefficients. The five equations from which α, e, m, n, and d are to be found are—

(1) $\dfrac{-2e^2\sin\alpha\cos\alpha}{1-e^2\sin^2\alpha}=0$, or $e^2\sin\alpha\cos\alpha=0$.

(2) $\dfrac{1-e^2\cos^2\alpha}{1-e^2\sin^2\alpha}=0$, or $1-e^2\cos^2\alpha=0$.

(3) $\dfrac{2e^2d\sin\alpha-2n}{1-e^2\sin^2\alpha}=6$, or $e^2d\sin\alpha-n=3-3e^2\sin^2\alpha$.

(4) $\dfrac{2e^2d\cos\alpha-2m}{1-e^2\sin^2\alpha}=-12$, or $e^2d\cos\alpha-m=-6+6e^2\sin^2\alpha$.

(5) $\dfrac{m^2+n^2-e^2d^2}{1-e^2\sin^2\alpha}=33$.

These equations are now to be solved for α, e, m, n, and d. But as the locus is a parabola, $e=1$. Also from $\sin\alpha\cos\alpha=0$, α must be either 0, or 90°. From (2) $\cos\alpha=1$. $\therefore\ \alpha=0$. Making these substitutions in (3), (4), and (5), they become—

(3_1) $n=-3$; (4_1) $d-m=-6$; (5_1) $m^2+n^2-d^2=33$. These equations readily give $m=5$, and $d=-1$.

To *construct* the locus, let **X′X** and **YY′** be drawn at right angles to each other. Locate the focus, **F**, at (5, —3). As $\alpha=0$, draw the axis of the locus parallel to

---

point. There are various views which may be taken of the eccentricity of a right line. Thus, considering it as the limit of the hyperbola, we have the equation $y=\sqrt{\frac{C}{A}}x$, in obtaining which we put $F=m^2-e^2d^2=0$ (**66**). From this $e=\frac{m}{d}$. But $m=Ae$, and $d=\frac{A}{e}$. Hence $e=\frac{Ae^2}{A}=e^2$, and $e=1$. Therefore $y=\sqrt{\frac{C}{A}}x$ is a common limit of the hyperbola and the parabola. Again, if we may be allowed to consider $Dy+Ex+F=0$ (**67**) the equation of a parabola, we reach the same result; viz., $e=1$. If, however, we consider **CB** the directrix, and a focus removed to an infinite distance, and **MN** the line, **PF**, the distance to the focus is ∞, and $e=\frac{\mathbf{PF}}{\mathbf{PD}}=\infty$. But these speculations are rather curious than useful.

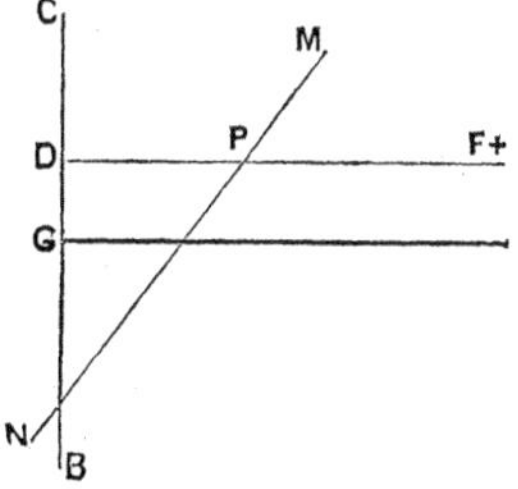

Fig. 46.

$X'X$, and through F. As $d = -1$, take $AG = -1$, and through G, drawing CB perpendicular to $Z'Z$, it is the directrix. Through F draw QR parallel to CB and take $QF = FR = FU$. The construction can now be completed as in (*43*).

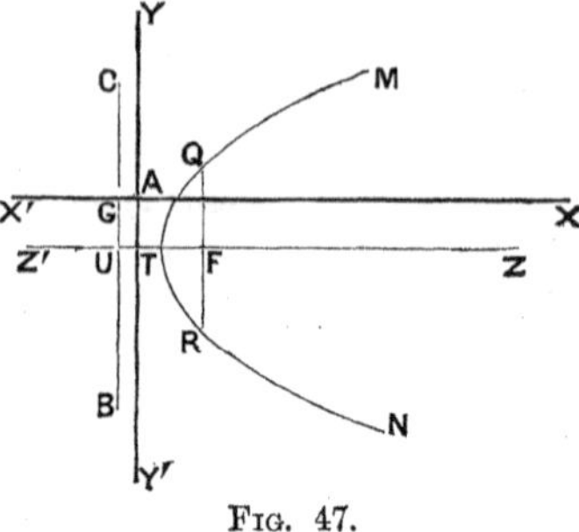

Fig. 47.

2. Determine the species and situation of the locus $y^2 + 2xy + 3x^2 - 4x = 0$.

Solution.—Since $B^2 - 4AC = 4 - 12 = -8 < 0$, the locus is an ellipse. The five equations from which to determine α, e, d, m, and n, are—

$$(1)\quad \frac{-2e^2 \sin\alpha \cos\alpha}{1 - e^2 \sin^2\alpha} = 2\,; \qquad (2)\quad \frac{1 - e^2 \cos^2\alpha}{1 - e^2 \sin^2\alpha} = 3\,;$$

$$(3)\quad \frac{2e^2 d \sin\alpha - 2n}{1 - e^2 \sin^2\alpha} = 0, \text{ or } e^2 d \sin\alpha - n = 0\,; \qquad (4)\quad \frac{2e^2 d \cos\alpha - 2m}{1 - e^2 \sin^2\alpha} = -4\,;$$

$$(5)\quad \frac{m^2 + n^2 - e^2 d^2}{1 - e^2 \sin^2\alpha} = 0, \text{ or } m^2 + n^2 - e^2 d^2 = 0.$$

We observe that (1) and (2) contain only the unknown quantities e and α, and hence are sufficient in themselves to determine these quantities. From (2) we get after clearing of fractions, substituting for $\cos^2\alpha$, $1 - \sin^2\alpha$, and reducing, $\sin\alpha = \frac{1}{2e}\sqrt{e^2 + 2}$; whence, $\cos\alpha = \pm\frac{1}{2e}\sqrt{3e^2 - 2}$. The $+$ sign alone is given to the value of $\sin\alpha$, since, by reckoning the angle from 0 to 180°, we get all possible inclinations of the axis of the locus to the axis of reference. Substituting these values in (1) and reducing, we find $e = .91+$; and consequently, $\sin\alpha = .9239$, and $\cos\alpha = \pm .3827$. $\therefore\ \alpha = 67°\ 30'$, or $112°\ 30'$. To determine whether $\cos\alpha$ is $+$ or $-$, and hence whether α is $67°\ 30'$, or $112°\ 30'$, we consider (1). The denominator of the first member being necessarily $+$, since $1 > e^2 \sin^2\alpha$, the numerator must be rendered $+$, inasmuch as the second member is $+$. But $\cos\alpha$ is the only factor in the numerator which can become $-$ and thus render the product $+$. $\therefore\ \cos\alpha = -.3827$; and $\alpha = 112°\ 30'$. Substituting the values of e, $\sin\alpha$ and $\cos\alpha$ in (3), (4), and (5), they become $(3_1)\ n = .765d$; (4_1), $m = .586 - .317d$; and $(5_1)\ m^2 + n^2 = .82d^2$. Substituting these values of m and n in (5_1), we find $d = .72$, and -3.34. Finally $n = .765d = .55$, and -2.55. $m = .586 - .317d = .36$, and 1.64.

To construct the figure, draw **GG′** through the origin, making an angle of 112° 30′ with the axis of **X**. Locate the foci, **F**, **F′**, at (.36, .55), and (1.64, —2.55). Through one focus, as **F**, draw a line **ZZ′** parallel to **GG′**, and it will pass through the other focus, **F′**, if the work is right. Take **AG** = .72 and **AG′** = —3.34, and draw **CB** and **C′B′** perpendicular to **ZZ′**; these will be the directrices. Now the ratio $e = .91$, the focus **F**, and the directrix **CB**, being known, the curve can be constructed (***43***).

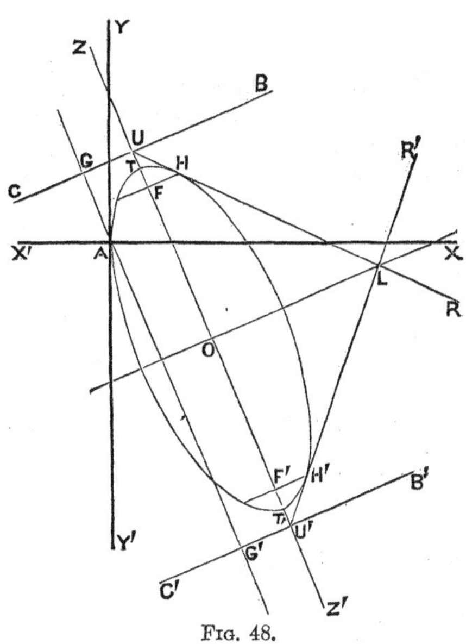

FIG. 48.

3. Determine the character and situation of the locus, $2xy - x + 1 = 0$.

SOLUTION. $B^2 - 4AC = 4$. ∴ The locus in an hyperbola. The five equations from which e, α, m, n, and d are determined are—

(1) $\dfrac{1 - e^2 \sin^2 \alpha}{m^2 + n^2 - e^2d^2} = 0$, or $1 - e^2 \sin^2 \alpha = 0$;

(2) $\dfrac{-2e^2 \sin \alpha \cos \alpha}{m^2 + n^2 - e^2d^2} = 2$, or $-e^2 \sin \alpha \cos \alpha = m^2 + n^2 - e^2d^2$;

(3) $\dfrac{1 - e^2 \cos^2 \alpha}{m^2 + n^2 - e^2d^2} = 0$, or $1 - e^2 \cos^2 \alpha = 0$;

(4) $\dfrac{2e^2d \sin \alpha - 2n}{m^2 + n^2 - e^2d^2} = 0$, or $e^2d \sin \alpha - n = 0$;

(5) $\dfrac{2e^2d \cos \alpha - 2m}{m^2 + n^2 - e^2d^2} = -1$, or $2e^2d \cos \alpha - 2m = e^2d^2 - m^2 - n^2$.

The general equation is divided through by $m^2 + n^2 - e^2d^2$, since in this example the absolute term is 1.

From (1) we have, directly, $\sin \alpha = \dfrac{1}{e}$, discarding the negative root for the reason given in the preceding solution. ∴ $\cos \alpha = \pm \sqrt{1 - \dfrac{1}{e^2}} = \pm \dfrac{1}{e}\sqrt{e^2 - 1}$. Substituting this value of $\cos \alpha$ in (3), we have $e = \sqrt{2} = 1.4142+$. Hence, $\cos \alpha = \pm \sqrt{1 - \dfrac{1}{e^2}} = \pm \sqrt{1 - \dfrac{1}{2}} = \pm \sqrt{\dfrac{1}{2}} = \pm \dfrac{1}{2}\sqrt{2} = \pm .7071+$, and $\alpha = 45°$ or 135°. Substituting the values of e and $\sin \alpha$ in (4), we have $n = \sqrt{2}d$; and by substitution in (2), we find $m = \pm 1$. In this case, it will be seen that if we substitute the + value of $\cos \alpha$ in (2), m becomes imaginary, but the − value gives m real. Therefore, $\cos \alpha = -\frac{1}{2}\sqrt{2}$, and $\alpha = 135°$. Finally, substituting in (5), we find $d = -\frac{1}{4}\sqrt{2}$, and $\frac{3}{4}\sqrt{2}$, or $-.35$, and 1.06; and, consequently, $n = -\frac{1}{2}$, and $\frac{3}{2}$. (The locus is situated as in *Fig.* 49; but the construction is so closely analogous to the preceding, that the student will have no difficulty in effecting it.)

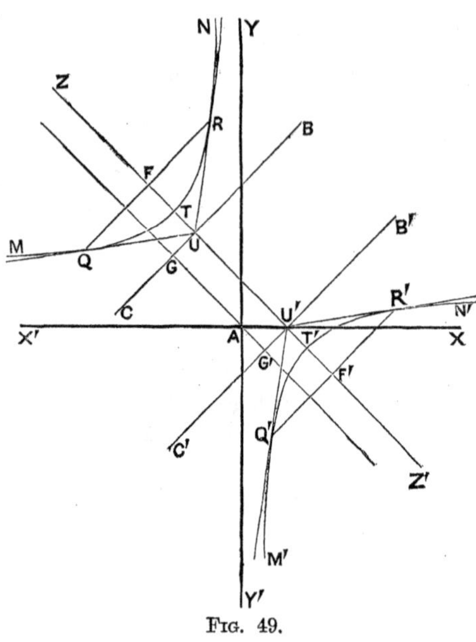

FIG. 49.

4. Determine the character and situation of the locus $y^2 - 2xy + 2x^2 - 2x = 0$.

Results. $e = .924$, $\alpha = 58°17'$, $d = 3.1$ and $-.37$, $m = 1.78$ and $.21$, and $n = 2.27$ and $-.27$.

SCH.—In problems like the preceding it is not admissible to divide the general equation through by a coefficient corresponding to one which is 0 in the particular case, inasmuch as this process would reduce each coefficient to infinity or indetermination.

Thus for *Ex.* 3, should we put the general equation in the form

$$y^2 - \frac{2e^2 \sin \alpha \cos \alpha}{1 - e^2 \sin^2 \alpha} xy + \frac{1 - e^2 \cos^2 \alpha}{1 - e^2 \sin^2 \alpha} x + \frac{2e^2 d \sin \alpha - 2n}{1 - e^2 \sin^2 \alpha} y + \text{etc.};$$

each of the coefficients, when the application was made to the equation $2xy - x + 1 = 0$, would be infinite or indeterminate, since in this example $1 - e^2 \sin^2 \alpha = 0$.

EXERCISES.

[NOTE.—This list of exercises is designed to give the student an opportunity for making an effort to produce the equations himself. Nothing new is developed in them, and the student need not necessarily tarry till he has mastered them all, though by doing so clearness and breadth of view will be promoted. Let every one understand, however, that ability to investigate—to reason for himself—is the proper object toward the attainment of which he should strive.]

1. To produce the equation of the ellipse referred to its own axes and in terms of its semi-axes, directly from the definition, without first obtaining the general equation of the conic section.

SUG'S. $AD = x$, $PD = y$, $AF = Ae$, and $\frac{PF}{PE} = e$. $\overline{PF}^2 = e^2 \times \overline{PE}^2$. $\overline{PF}^2 = y^2 + (Ae + x)^2$, and $\overline{PE}^2 = \left(\frac{A}{e} + x\right)^2$. $\therefore\ y^2 + (1 - e^2)x^2 = A^2(1 - e^2)$. For $1 - e^2$ substituting $\frac{B^2}{A^2}$, we have $A^2y^2 + B^2x^2 = A^2B^2$.

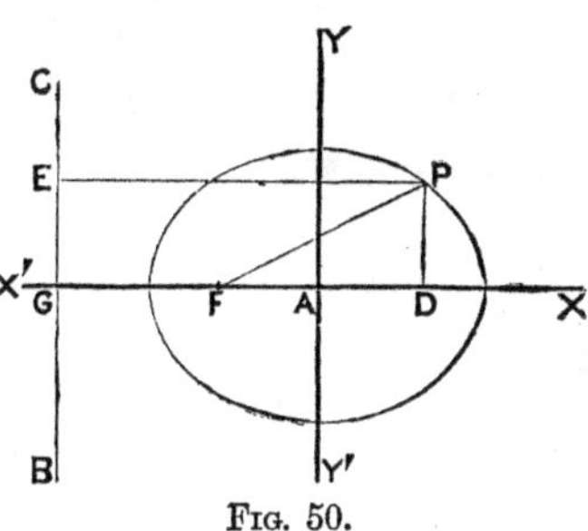

FIG. 50.

2. In like manner produce the equation of the ellipse referred to its transverse axis and a tangent at its left hand vertex, *i. e.*, $y^2 = \frac{B^2}{A^2}(2Ax - x^2)$.

3. Show that the equation of an ellipse referred to its conjugate axis and a tangent at the upper vertex thereof is $x^2 = -\frac{A^2}{B^2}(2By + y^2)$.

SUG'S. $AD = x$, $PD = -y$, $\frac{PF}{PE} = e$, and $\overline{PF}^2 = e^2 \times \overline{PE}^2$. Also $\overline{PF}^2 = (x + Ae)^2 + (B + y)^2$, and $\overline{PE}^2 = \left(\frac{A}{e} + x\right)^2$

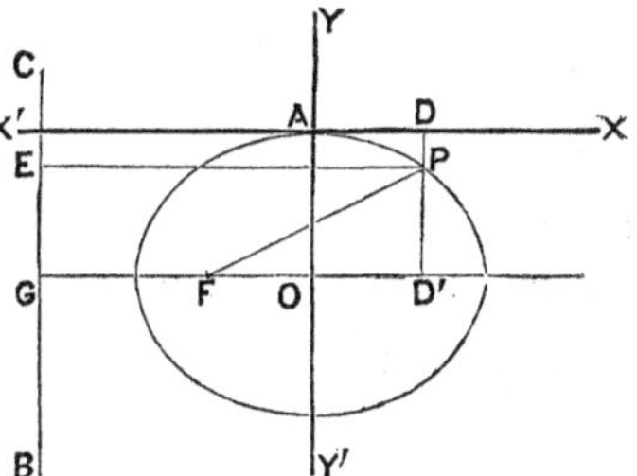

FIG. 51.

4. Produce the common equation of the hyperbola, $A^2y^2 - B^2x^2 = -A^2B^2$, directly as above.

5. In like manner as above produce the common equation of the parabola, $y^2 = 2px$.

6. Show that the equation of an ellipse is $A^2(y - y_1)^2 + B^2(x - x_1)^2 = A^2B^2$, when x_1 and y_1 are the co-ordinates of the centre, and the axes of reference are parallel to the axes of the curve.

SUG'S. $AL = x_1$, $OL = y_1$, $AD = x$, $PD = y$, and $\overline{PF}^2 = e^2 \times \overline{PE}^2$. Also $\overline{PF}^2 = \overline{PR}^2 + \overline{FR}^2$, $PR = y - y_1$, $FR = x - x_1 + Ae$, and $PE = PH - HE = x - x_1 + \frac{A}{e}$. Substituting, $(y - y_1)^2 + x^2 - 2x_1x + 2Aex + x_1^2 - 2Aex_1 + A^2e^2 = e^2x^2 - 2e^2x_1x + 2Aex + e^2x_1^2 - 2Aex_1 + A^2$.

Transposing and collecting terms,

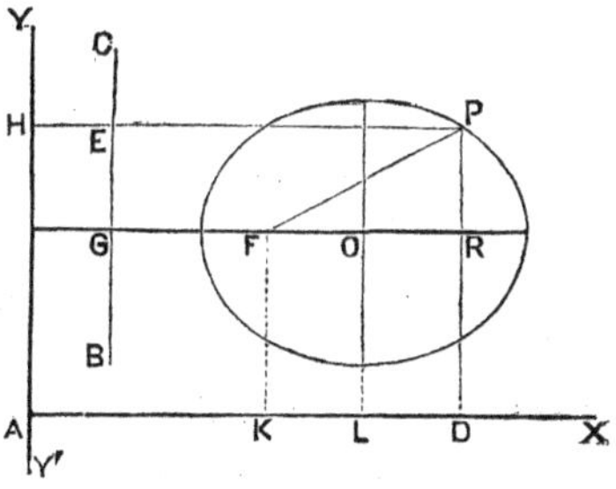

FIG. 52.

$(y-y_1)^2+(1-e^2)x^2-(1-e^2)2x_1x+(1-e^2)x_1{}^2=A^2(1-e^2)$, or

$(y-y_1)^2+(1-e^2)(x-x_1)^2=A^2(1-e^2)$. Putting $\frac{B^2}{A^2}$ for $1-e^2$, we have

$A^2(y-y_1)^2+B^2(x-x_1)^2=A^2B^2$.

7. Deduce from the general equation of the conic section (*51*) the equation of a parabola whose parameter is $2p$, referred to rectangular axes, the axis of the curve falling on the axis of abscissas, and the vertex of the curve being at x_1 to the right of the origin. Also the equation when the vertex is at x_1 to the left of the origin. Also the equation when the vertex is at (x_1, y_1).

The equations are $y^2=2p(x-x_1)$, $y^2=2p(x+x_1)$, and $(y-y_1)^2=2p(x-x_1)$.

69. General Scholium.—It will appear hereafter that the conic sections are formed by the mutual intersection of a plane and a right cone with a circular base. It is from this fact that the name of these curves is derived; and from this as a definition they were formerly studied. The different species arise from different positions of the cutting plane. The plane which gives the parabola lies parallel to one of the elements of the cone, as ONP, and hence cuts but one nappe, and gives but one branch. To produce the ellipse the cutting plane lies between this position and perpendicular to the axis, as RTmn. To produce the hyperbola it lies between parallel to an element and parallel to the axis, as HIK and EGF, and hence cuts both nappes, giving two branches.

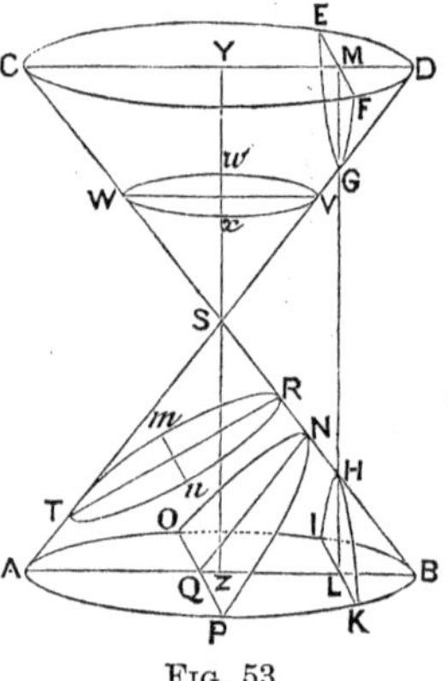

Fig. 53.

Several of the varieties of the conic sections as heretofore considered, may be illustrated by means of this geometrical conception, and their mutual relations more clearly seen. Thus as the plane of the ellipse approaches perpendicularity to the axis, the ellipse approaches the form of a circle into which it passes when the plane becomes perpendicular to the axis. The circle is therefore a variety (or more properly a *limit*) of the ellipse. So also as the plane approaches the vertex, the ellipse diminishes, passing into its limit—*a point*—when the plane passes through the vertex. The hyperbola becomes two intersecting straight lines when the cutting plane passes through the vertex and is not parallel to an element. When it becomes parallel to an element and also passes through the vertex, it gives the limit both of the parabola and the hyperbola, which common limit is a right line. When the cone passes into a cylinder the parabola becomes two parallel right lines, as also the hyperbola may, if it is conceived as produced by a cutting plane perpendicular to the base. If the cutting plane producing the hyperbola is conceived as oblique to the axis, the hyperbola passes into an ellipse when the cone passes into a cylinder.

[NOTE.—Of course the above views are not given as in any sense needed to *confirm* the conclusions of the preceding discussions, but simply to give the student a little further insight into the wonderful harmony which exists between algebraic *formulæ* and geometrical loci.]

---

70. Prop. *Through five points in a plane one conic section may always be made to pass, and but one.*

DEM.—Dividing the general equation $Ay^2 + Bxy + Cx^2 + Dy + Ex + F = 0$ through by F, and distinguishing the new coefficients by accents, we have $A'y^2 + B'xy + C'x^2 + D'y + E'x + 1 = 0$. Now let (x_1, y_1), (x_2, y_2), (x_3, y_3), (x_4, y_4), and (x_5, y_5) be the five given points. Substituting, successively, in the last equation, the co-ordinates of these five points, for the general co-ordinates x and y, there result the five equations

$$A'y_1^2 + B'x_1y_1 + C'x_1^2 + D'y_1 + E'x_1 + 1 = 0;$$
$$A'y_2^2 + B'x_2y_2 + C'x_2^2 + D'y_2 + E'x_2 + 1 = 0;$$
$$A'y_3^2 + B'x_3y_3 + C'x_3^2 + D'y_3 + E'x_3 + 1 = 0;$$
$$A'y_4^2 + B'x_4y_4 + C'x_4^2 + D'y_4 + E'x_4 + 1 = 0;$$
$$A'y_5^2 + B'x_5y_5 + C'x_5^2 + D'y_5 + E'x_5 + 1 = 0;$$

or five arbitrary conditions. This number of conditions is possible, since there are *five arbitrary constants* involved; viz., A', B', C', D', and E'. From these equations, as x_1, y_1, x_2, y_2, x_3, y_3, x_4, y_4, x_5, and y_5, are *known quantities*, the values of A', B', C', D', and E' can be determined. Having found the values of these coefficients, by substituting their values in the general equation $A'y^2 + B'xy + C'x^2 + D'y + E'x + 1 = 0$, there results an equation of the second degree between two variables, or an equation of a conic section. As this equation is satisfied by the co-ordinates of each of the five given points, the locus represented by them passes through these points. Finally, as the five equations are all of the first degree with respect to A', B', C', D', and E', but *one* set of values can be determined for these coefficients. Therefore, *but* one conic section can be made to pass through the five given points. Q. E. D.

SCH. 1.—In this proposition the term Conic Section must be taken in its broadest sense, *i. e.*, as embracing all varieties of these loci, except the so-called imaginary loci.

SCH. 2.—If the five points are so situated that the equation of the locus passing through them lacks some of the terms of a complete equation, it will not do to divide the general equation by the coefficient of such a term. If such an error has been made in the hypothesis in any solution, it will soon appear as the solution proceeds. This case is analogous to the one noticed in the suggestion under *Ex.* 4, page 49.

EX. 1. Produce the equation of a conic section passing through the five points (2, 3), (0, 4), (—1, 5), (—2, —1), and (1, —2), and determine its species.

SOLUTION.—The five equations which determine the coefficients A', B', C', D', and E', are

(1) $9A' + 6B' + 4C' + 3D' + 2E' + 1 = 0$;
(2) $16A' + 4D' + 1 = 0$;
(3) $25A' - 5B' + C' + 5D' - E' + 1 = 0$;
(4) $A' + 2B' + 4C' - D' - 2E' + 1 = 0$;
(5) $4A' - 2B' + C' - 2D' + E' + 1 = 0$.

Solving these equations, we find $A' = -\frac{169}{924}$, $B' = -\frac{220}{924}$, $C' = \frac{89}{924}$, $D' = \frac{445}{924}$, $E' = \frac{113}{924}$. Substituting these values in the general equation $A'y^2 + B'xy + C'x^2 + D'y^2 + E'x + 1 = 0$, clearing of fractions and changing signs, we have $169y^2 + 220xy - 89x^2 - 445y - 113x - 924 = 0$, which is the equation of a conic section passing through the five given points. This locus is an hyperbola, since $B^2 - 4AC > 0$.

Ex. 2. Produce the equation of a conic section passing through the five points (1, 3), (4, —6), (0, 0), (9, —9), and (16, 12), and find its species.

SUGGESTIONS.—As one of the given points is (0, 0), the locus passes through the origin; and hence $F = 0$. The form of the general equation used would, therefore, be $Ay^2 + Bxy + Cx^2 + Dy + Ex = 0$, which divided through by one of the coefficients, as A, gives the form $y^2 + B'xy + C'x^2 + D'y + E'x = 0$. This equation satisfied for the four points (1, 3), (4, —6), (9, —9), and (16, 12), in succession, gives rise to four equations from which the coefficients can be determined.

The locus is a parabola whose equation is $y^2 = 9x$.

Ex. 3. Produce the equation of a conic section passing through (—4, —2), (2, 1), (—6, 3), (0, 0), and (2, —1), and determine its species. *The equation is* $y = \mp \frac{1}{2}x$.

Ex. 4. Produce the equation of a conic section passing through $(3, \sqrt{5})$, $(-2, 0)$, $(-4, -\sqrt{12})$, $(3, -\sqrt{5})$, and $(2, 0)$, and determine its species. *The locus is an equilateral hyperbola.*

Ex. 5. Produce the equation of a conic section passing through $(-\frac{2}{3}, -\frac{1}{3})$, $(2, 1)$, $(\frac{5}{3}, 2)$, $(-\frac{2}{3}, -3)$, and $(\frac{5}{3}, -\frac{2}{3})$ and determine its species.

The locus is an ellipse whose equation is $y^2 - 2xy + 3x^2 + 2y - 4x - 3 = 0$.

Ex. 6. What is the equation of a circle whose radius is 5, referred to rectangular axes, and the origin at the centre? When the origin is on the circumference and the axis of abscissas is a diameter? When the axes are tangent to the circumference?

Equations, $y^2 + x^2 = 25$, $y^2 = \pm 10x - x^2$, and $y^2 + x^2 - 10y - 10x + 25 = 0$.

Ex. 7. What is the equation of an ellipse whose axes are 16 and 10, when referred to its own axes? When referred to its transverse axis and a tangent at the left hand vertex? The corresponding problems in the case of the hyperbola.

Equations, $64y^2 + 25x^2 = 1600$. $64y^2 - 400x + 25x^2 = 0$.
$64y^2 - 25x^2 = -1600$. $64y^2 + 400x - 25x^2 = 0$.

SUGS.—The results of the two preceding examples are readily written from the equations of the respective loci as given in (***53—57***), and are designed to familiarize those most important forms.

Ex. 8. Produce the equation of a parabola referred to rectangular axes, the vertex of the parabola being at (—3, —2), the parameter, 6, and the axis of abscissas parallel to the axis of the curve.

Equation, $y^2 + 4y - 6x - 14 = 0$.

Ex. 9. Produce the equation of an ellipse whose eccentricity is $\frac{2}{3}$, its major axis 18, the centre being at (—2, 3), and the axes of reference being rectangular and parallel to the axes of the curve.

Equation, $9y^2 + 5x^2 - 54y + 20x - 304 = 0$.

Ex. 10. What are the following loci, and what their axes: viz., $9y^2 + 4x^2 = 36$? $7x^2 + 11y^2 = 15$? $100y^2 - 25x^2 = -2{,}500$? $17x^2 - 25y^2 = -116$?

EXERCISES IN PRODUCING THE EQUATIONS OF THE CONIC SECTIONS FROM OTHER DEFINITIONS.

[NOTE.—These exercises may be omitted without destroying the integrity of the course. They are designed simply to lead the student to a more full comprehension of the process of producing an equation of a locus from its definition, a subject of vital importance if one proposes to so master this method of geometrical investigation as to be independent in the use of it.]

1. To produce the common equation of the ellipse from the definition :—*The ellipse is a curve such that the sum of the distances from any point in the curve to two fixed points called the foci, is constant and equal to the major diameter.*

SUGS. $\mathrm{AD} = x$, $\mathrm{PD} = y$, $\mathrm{AB} = A$, $\mathrm{AE} = B$, $\mathrm{AF} = \mathrm{AF}' = c$. Then from the definition $\sqrt{y^2 + (c+x)^2} + \sqrt{y^2 + (c-x)^2} = 2A$. Whence $A^2y^2 + (A^2 - c^2)x^2 = A^2(A^2 - c^2)$. But by definition, $\mathrm{EF} = A$, whence $A^2 - c^2 = B^2$; and we have $A^2y^2 + B^2x^2 = A^2B^2$.

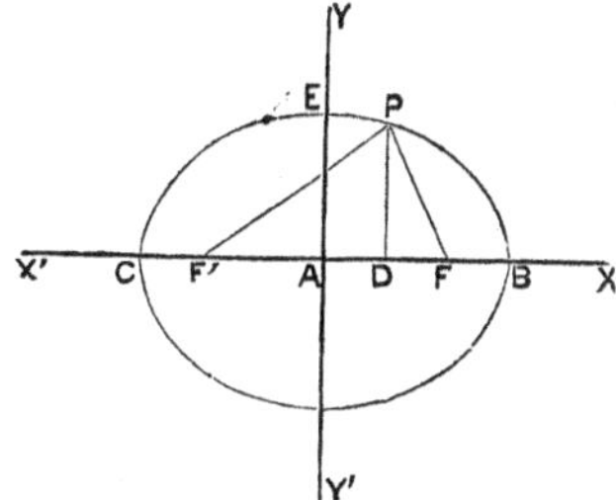

FIG. 54.

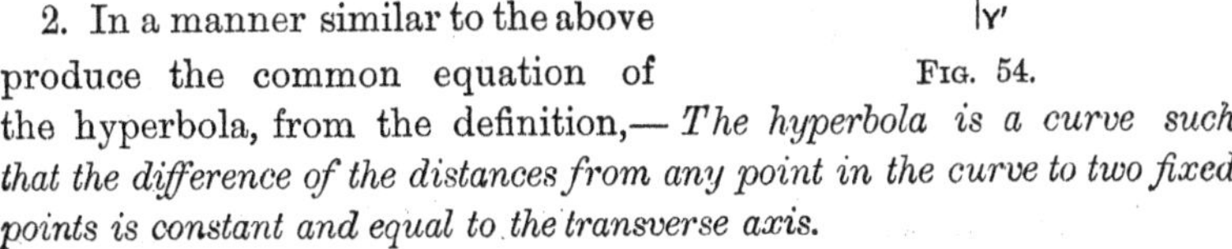
2. In a manner similar to the above produce the common equation of the hyperbola, from the definition,— *The hyperbola is a curve such that the difference of the distances from any point in the curve to two fixed points is constant and equal to the transverse axis.*

SUG'S.—In this case it must be borne in mind that $A^2 + B^2 = c^2$ (**47**), and hence that $A^2 - c^2 = -B^2$. The equation is $A^2y^2 - B^2x^2 = -A^2B^2$, as before produced.

3. To produce the equation of the locus of a point moving so that the square of its distance from a fixed point is in a constant ratio to its distance from a fixed line.

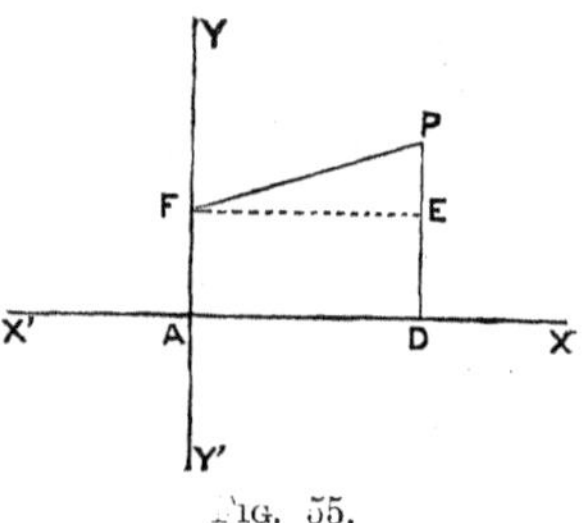

FIG. 55.

SUG'S.—Let the fixed line be taken as the axis of abscissas, and let a perpendicular to it through the fixed point, F, be taken as the axis of ordinates. Let P be any point in the locus. Then $AD = x$, and $PD = y$. As AF is constant, call it a, and let m represent the ratio referred to in the definition in the example. The equation sought is $(y - a)^2 + x^2 = my$, or $y^2 + x^2 - (2a + m)y + a^2 = 0$. This being an equation of the second degree, the locus is a conic section. Again, as $B^2 - 4AC < 0$, it is an ellipse. Finally, as the coefficients of y^2 and x^2 are equal, it is an ellipse with equal axes, or a *circle*.

To determine more fully the situation of this circle, notice that for $y = 0$, $x = \pm\sqrt{-a^2}$, whence we see that, in general, the circle does not cut the axis of x. Making $x = 0$, $y = \frac{2a + m}{2} \pm \sqrt{\frac{4am + m^2}{4}}$. Now, as every value of y in the equation of this locus gives two values of x, numerically equal but with opposite signs, we see that the locus is symmetrical with the axis of y, and that the centre of the circle lies in this axis. But the circle cuts this axis at $\frac{2a + m}{2} + \sqrt{\frac{4am + m^2}{4}}$ and at $\frac{2a + m}{2} - \sqrt{\frac{4am + m^2}{4}}$. Whence the diameter is the difference between these values; and letting r be the radius, we have $r = \frac{1}{2}\sqrt{4am + m^2}$, *i. e.*, the radical part of the root.

In the particular case in which $a = 0$; *i. e.*, when F is at A, the equation becomes $y^2 + x^2 = my$, which is the equation of a circle passing through the origin, and having its centre on the axis of y.

4. In the given right lines **AP**, **AQ**, intersecting at right angles, are taken variable points p, q, such that $\mathbf{A}p : p\mathbf{P} :: \mathbf{Q}q : q\mathbf{A}$; prove that the locus of the intersection of $\mathbf{P}q$, $\mathbf{Q}p$, is an ellipse which touches the right lines in **P** and **Q**.

SUG'S.—Let **AP** and **AQ** be taken as the axes of reference. Call **AP** $= a$, and **AQ** $= b$. Then **AD** $= x$, and **RD** $= y$. From the similar triangles **RPD**, q**PA**, and **R**p**D**, **Q**p**A** obtain the relation between x, y, a, and b. This will be the equation sought. It is, when reduced, $a^2(y-b)^2 + b^2(x-a)^2 = a^2b^2 - abxy$.

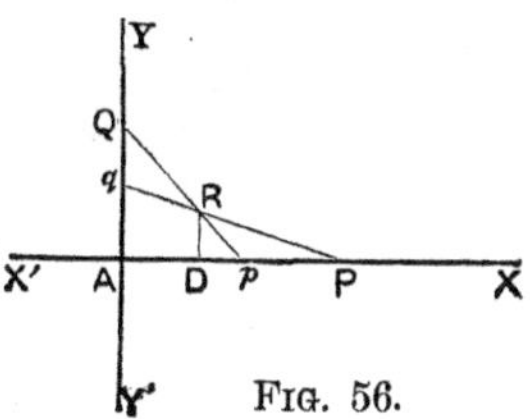

FIG. 56.

SCH.—As this equation is of the second degree, and $B^2 - 4AC = -3a^2b^2$, the locus is an ellipse. As there is a term in xy, the axis of the curve is inclined to the axis of abscissas. For $x = a$, $y = b$ and 0; hence the locus passes through (a, b), and **P**. For $x = 0$, $y = b$; therefore the locus passes through **Q**.

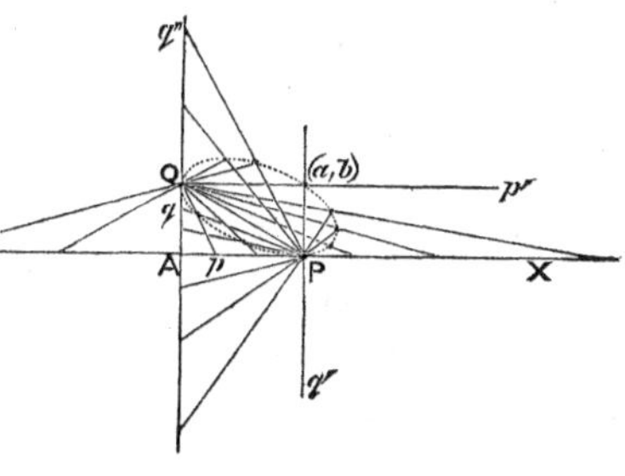

To effect the construction mechanically, take **A**p : p**P** : : **Q**q : q**A**, by composition and alternation, giving **AP** : **AQ** : : **P**p : **A**q. Now assuming p at any point in **AX**, we can find the corresponding value of **A**q. After p passes **P**, **A**q becomes —, and is laid off below **A**. So when **Q**p' passes parallelism with **AX**, **A**p becomes negative.

5. Required the locus of the middle point of a line moving with its extremities in two fixed lines at right angles with each other, while it passes through a fixed point.

SUG'S.—Take the fixed lines as axes of reference. Let **O** be the fixed point, and **CB** the line, the locus of whose centre, **P**, is to be determined. Calling **AD** a, and **OD** b, the equation of the locus is $2xy - ay - bx = 0$, which is the equation of an hyperbola, passing through the origin, since for $x = 0$, $y = 0$; and also passing through **O**, since $x = a$, gives $y = b$. Let the pupil trace the curve.

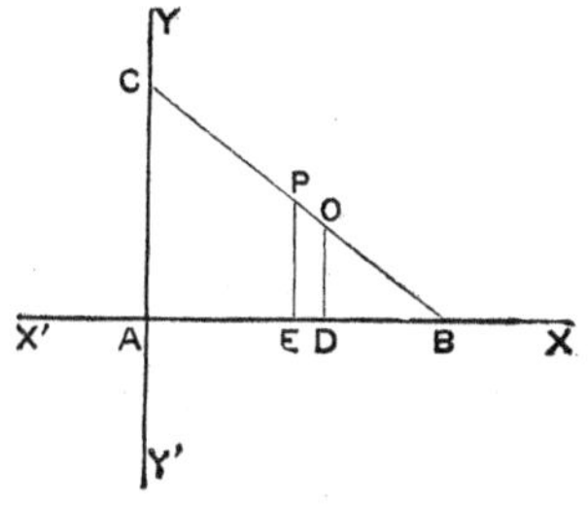

FIG. 57.

6. Required the locus of the point **P**, moving so that $\overline{\mathbf{PD}}^2$, bears a constant ratio to **AD** × **DB**; **A** and **B** being fixed points. What is the locus when this ratio is 1?

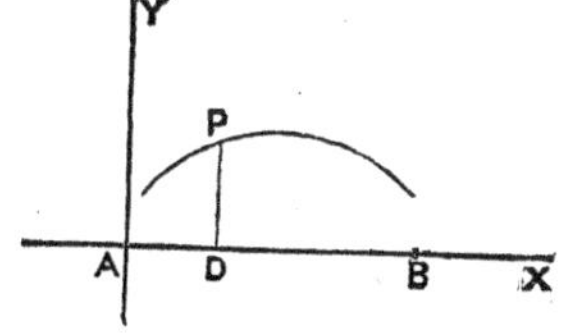

FIG. 58.

The distance **AB** *being called* 2a, *and the ratio* m, *the equation is* $y^2 = mx(2a - x)$, *whence the locus is seen to be an ellipse. If* m = 1, *it is a circle.*

7. If **P** moves in *Fig.* 58, so that $\overline{\mathbf{PD}}^2$ bears a constant ratio to **AD**, what is the locus?

SECTION VII.

Equations of Higher Plane Curves.

71. One variable is called a ***Function*** of another variable when it depends upon that other variable for its value. Thus the ordinate of a curve is a function of the abscissa.

72. Functions are classified as ***Algebraic*** and ***Transcendental;*** and the latter are subdivided into ***Trigonometric,*** and ***Circular, Logarithmic,*** and ***Exponential.***

73. An Algebraic Function is one which involves only the elementary methods of combination, viz., addition, subtraction, multiplication, division, involution and evolution. Thus in $y = ax^2 - 3x^3$, and in all the equations hitherto discussed in this chapter, y is an algebraic function of x, except 24–33, Sec. II.

74. A Trigonometrical Function is one which involves sines, cosines, tangents, cotangents, etc., as $y = \sin x$, $y = \sin x \tan x$, etc.

75. A Circular Function is one in which the concept is an arc (in the trigonometrical the concept is a right line). These are written thus: $y = \sin^{-1}x$, read "y equals *the arc* whose sine is x"; $y = \tan^{-1}x$, read "y equals *the arc* whose tangent is x."

Notice that in the expression $y = \tan^{-1}x$, it is the *arc* which we are to think of, while in the expression $x = \tan y$ it is the *tangent,* which is a right line. Trigonometrical functions are right lines; circular functions are arcs. These functions are mutually convertible into each other; thus, $y = \sin^{-1}x$, is equivalent to $x = \sin y$, the only difference being that in the former we think of the *arc,* the sine being given to tell *what* arc, and in the latter, we think of its *sine,* the arc being given to tell *what* sine.

The circular functions $y = \sin^{-1}x$, $y = \cos^{-1}x$, $y = \sec^{-1}x$, etc., are often called the *Inverse Trigonometrical Functions.*

76. A Logarithmic Function is one which involves logarithms; as $y = \log x$, $\log^2 y = 3 \log ax$, etc.

77. An Exponential Function is one in which the variable occurs as an exponent; as $y = a^x$, $z = x^y$, etc.

78. *Higher Plane Curves* are loci whose equations are above the second degree, or which involve transcendental functions. As it has already been shown that loci of the equations of the 1st degree are right lines, and that loci of the 2nd degree are conic sections, it follows that all other plane loci are higher plane curves. The former are called *Lower Plane Loci.*

Of course the variety of higher plane loci is infinite. We can consider but a few, and these simply as specimens.

79. *An Algebraic Curve* is one whose equation contains only algebraic functions. *A Transcendental Curve* is one whose equation contains transcendental functions; when converted into algebraic forms their degree is infinite.

THE CISSOID OF DIOCLES.

80. DEF. If pairs of equal ordinates be drawn to the diameter of a circle, and through one extremity of this diameter and the point in the circumference through which one of the ordinates is let fall, a line be drawn, the locus of the intersection of this line and the equal ordinate, or that ordinate produced, is the ***Cissoid of Diocles.***

81. *Prob.* *To construct the Cissoid.*

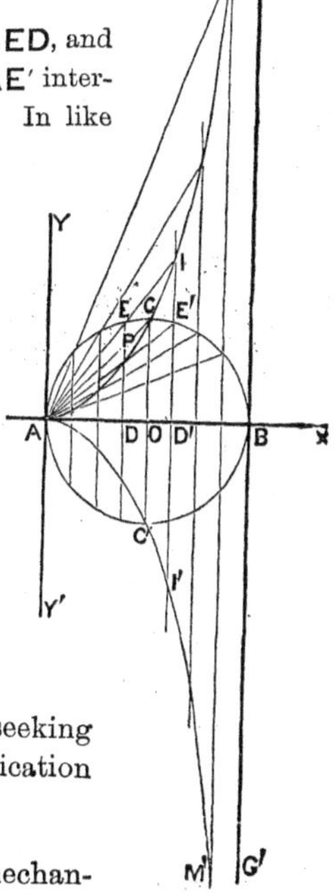

FIG. 59.

SOLUTION.—Let AB be the diameter of a circle; and ED, and E'D' be equal ordinates. Through A and E' draw AE' intersecting ED in P. Then is P a point in the locus. In like manner draw AE and produce it till it meets E'D' produced in I. Then is I a point in the locus. In the same way other points are found both above and below AB. There are, therefore, two branches of the locus ACM and AC'M', symmetrical with respect to the diameter AB. These branches evidently meet at A, pass through the extremities of the diameter CC', and have GG' as a common asymptote.

SCH. 1.—The name Cissoid is from the Greek and signifies ivy-form. It was applied to the curve, probably, from its resemblance to the graceful outline presented by a growth of ivy upon a wall. The locus was invented by the Greek geometer whose name it bears, while he was seeking the solution of the celebrated problem of the Duplication of the Cube.

SCH. 2.—Sir Isaac Newton gave the following mechanical method of describing this locus: Let AB be the diameter of the circle from which the curve would be described by the definition; at the centre **O** erect the perpendicular **OL**, and take **AD** = **AO** = **OB**. Now take a rectangular ruler **FEC**, whose leg **CE** = **AB**, and while the extremity **C** moves in the line **OL**, let the leg **FE** slide through the fixed point **D**, then will the middle point of **CE**, **P**, describe the cissoid. [The demonstration will afford a good exercise for the student.]

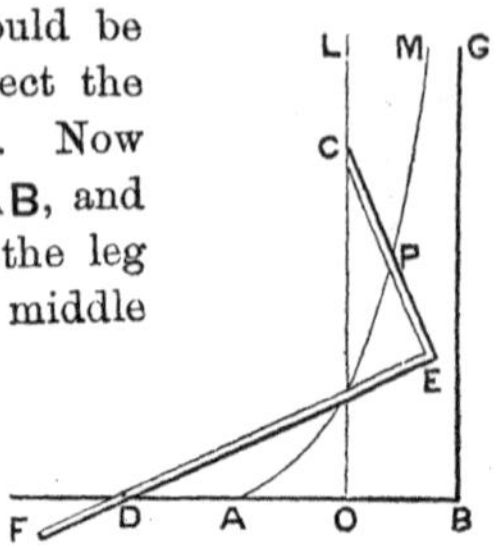

Fig. 60.

Sch. 3.—This curve is also the locus of the vertex of a common parabola rolling upon an equal parabola.

---

82. Prob.—*To produce the equation of the Cissoid of Diocles.*

Solution.—In Fig. 59 let **AX** and **AY** be the axes of reference, **AB** $= 2a$, the diameter of the circle referred to in the definition, and **P** any point in the curve. Then **AD** $= x$, and **PD** $= y$. Draw through **P** the ordinate **ED**, and also draw the equal ordinate **E′D′**. **APE′** is a straight line by definition. We now have **AD** : **PD** :: **AD′** : **E′D′**, or $x : y :: 2a - x : \sqrt{(2a-x)x}$. Squaring and reducing, $x^2 : y^2 :: 2a - x : x$. $\therefore y^2 = \frac{x^3}{2a - x}$ is the equation sought.

Sch. 1.—*Since* $y = \pm\sqrt{\frac{x^3}{2a - x}}$, *every real + value of* $x < 2a$ *gives two real and numerically equal values of* y, *with contrary signs. Hence the locus is symmetrical with respect to the axis of* x. *For* $x = 2a$, $y = \pm\infty$, *whence the branches are infinite, and* **GG′** *is an asymptote to both branches. For all values of* $x > 2a$, *and for* x *negative,* y *is imaginary. Therefore the locus is comprised between the limits* $x = 0$, $x = 2a$.

Sch. 2.—*By the Duplication of the Cube is meant finding the edge of a cube which shall have twice the volume of a cube whose edge is given.* To effect this by means of this curve, let **AM** be any cissoid, **AB** the diameter of the circle which pertains to it, and **O** the centre of that circle. Take **CO** = 2**OB**, and draw **CB**. Let fall from the point **P**, where **CB** cuts the curve, the perpendicular **PK**. Then **PK** = 2**BK**. Now a cube described on **PK** is twice one described on **AK**; for since **PK** $= y$, **AK** $= x$, and **KB** $= 2a - x$, we have $\overline{PK}^2 = \frac{\overline{AK}^3}{KB} = \frac{\overline{AK}^3}{\frac{1}{2}PK}$, or $\frac{1}{2}\overline{PK}^3 = \overline{AK}^3$. $\therefore$ $\overline{PK}^3 = 2\overline{AK}^3$. Finally, let a be the edge of *any* given cube; find a_1 so that $a : a_1 :: AK : PK$, whence $a^3 : a_1^3 : \overline{AK}^3 : \overline{PK}^3$. But $\overline{PK}^3 = 2\overline{AK}^3$. $\therefore$ $a_1^3 = 2a^3$.

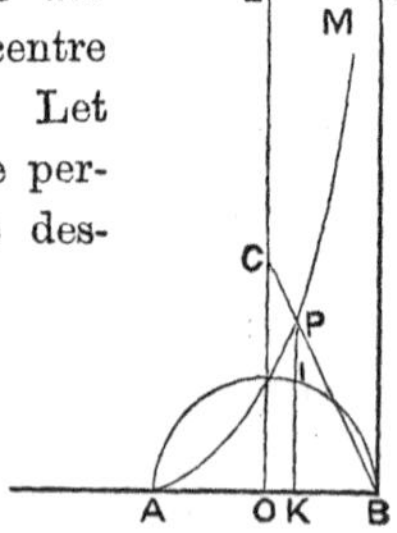

Fig. 61.

By taking **CO** $= 3$**OB** and proceeding in a similar manner, we can triplicate the cube; or in the same way obtain the edge of a cube of any given number of times the volume of a given cube. (The pupil may show that $\overline{IK}^3 = 2\overline{KB}^3$; also that $\overline{AK}^3 = 2\overline{IK}^3$.)

THE CONCHOID OF NICOMEDES.

83. DEF.—***The Conchoid of Nicomedes*** is the locus of a point in a line which revolves on and slides in a fixed pivot, so as to allow a constant portion of the line to project beyond a fixed right line.

84. Prob. *To construct the Conchoid of Nicomedes.*

SOLUTION.—Let **O** be the fixed point, or pivot, **X'X** the fixed line, and **AB** the constant portion of the revolving line. Draw a convenient number of radiating lines through **O**, and on each lay off above **X'X** the distances **C**1, **FP**, **E**6, etc., equal to **AB**. Then will 1, 2, 3, 4, etc , be points in the locus ; and **MBN** will be the conchoid.

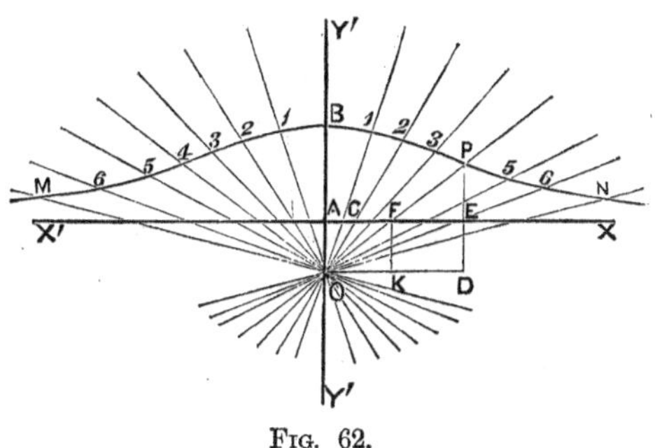

FIG. 62.

SCH.—This locus is readily drawn by mechanical means. Let **X'X** and **YY'** be two bars fixed at right angles to each other. Let any one of the radiant lines, as **OP**, represent a ruler, grooved on the under side so as to slide on the head of a pin fixed in the bar **YY'**, at **O**. Let there be a fixed pin on the under side of the ruler, as at **F**, which can slide in a groove on the upper side of the bar **X'X**. Now, placing the groove in the ruler on the head of the pin at **O**, and the pin in the ruler, in the groove in **X'X**, any point in the ruler, as **P**, will describe the conchoid.

85. Prob. *To produce the equation of the Conchoid of Nicomedes.*

SOLUTION.—Let **P**, *Fig.* 62, be any point in the locus referred to the axes **XX'**, **YY'**; and let its co-ordinates **AE** and **PE**, be x and y. Let **AB** $= a$, and **AO** $= b$. Produce **PE** till it meets **OD** drawn parallel to **AX**. Now, by similar triangles, **PE** : **PD** :: **EF** : **OD** ; or $y : y + b :: \sqrt{a^2 - y^2} : x$. Squaring, $y^2 : (y + b)^2 :: a^2 - y^2 : x^2$. $\therefore\ x^2y^2 = (y + b)^2(a^2 - y^2)$.

SCH. 1.—Since $x = \pm \sqrt{a^2 - y^2}\left(\frac{y + b}{y}\right)$, for every positive value of y, numerically less than a, x has two numerically equal values with opposite signs ; which values increase as y diminishes, and for $y = 0$, $x = \pm\ \infty$.

∴ This portion of the locus is symmetrical with respect to the axis of y, and has the axis of x for a common asymptote of its two branches. Again, as all negative values of y, not numerically greater than a, give numerically equal values of x with opposite signs, there is a portion of the locus below the axis of x, which is also symmetrical with respect to the axis of y. To discover the form of this portion, 1st consider $a > b$. Then for $y = -a$, or $-b$, $x = 0$, but for values of y between these two limits, x has two numerically equal values with opposite signs; hence the locus between these two limits is an oval symmetrical with respect to the axis of y. For y numerically less than b, and negative, the values of x increase numerically till, at $y = 0$, they become $\pm \infty$; hence between **O** and the axis of abscissas there are two infinite branches, symmetrical with respect to the axis of y, and having the axis of x as a common asymptote. 2nd. When $a = b$ the oval disappears. These forms are described mechanically by taking the point on the moving ruler *below* the fixed line.

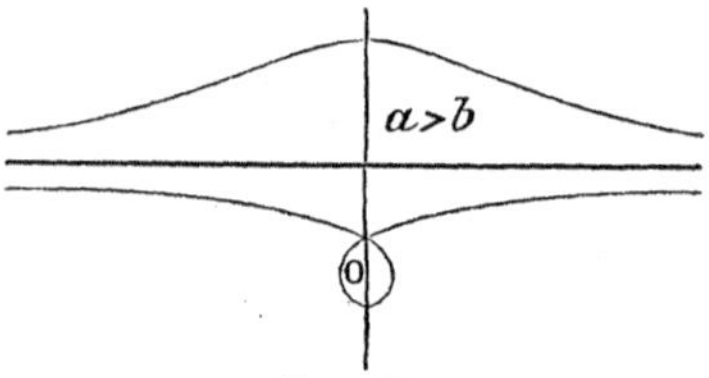

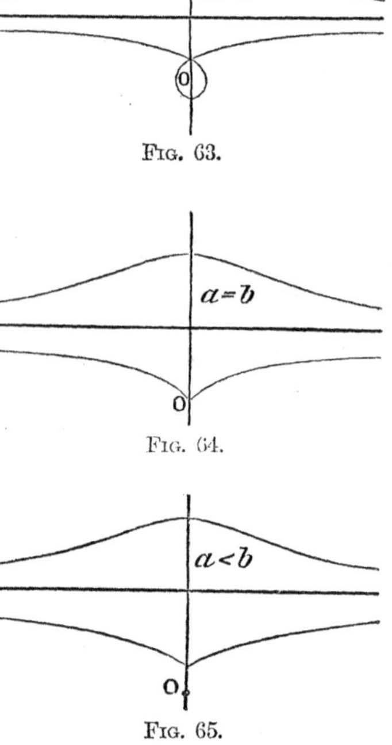

FIG. 63.

FIG. 64.

FIG. 65.

SCH. 2.—When $b = 0$, the equation becomes $x^2y^2 = y^2(a^2 - y^2)$; or $x^2 + y^2 = a^2$. This is the equation of the circle, as it evidently should be.

SCH. 3.—This curve was invented by the geometer whose name it bears, for a purpose similar to that subserved by the cissoid. The problem of the *Duplication of the cube* and the *Trisection of an angle* had been shown to be identical, as both depend upon the insertion of two means in a continued proportion between two extremes. Thus, letting a and b be the extremes, it is required to find x and y, so that $a : x : y : b$; *i. e.*, $a : x :: x : y$, and $x : y :: y : b$. This problem, viz., the insertion of two means between two extremes, is effected by the cissoid. In the cissoid, *Fig.* 59, **ED** and **AD′** are the two means between **AD** and **ID′**: that is, **AD : ED : AD′ : ID′**. The *Trisection of an angle* by means of the conchoid is effected thus: Let **COM**, *Fig.* 66, be the angle to be trisected. From any point, **D**, in one leg let fall a perpendicular, **DB**, on the other. Take **CB** = 2**DO**, and with **O** as the fixed point, **X′X** as the fixed line, and **CO** as the ruler with the constant portion **CB** projecting beyond **X′X**, construct the arc **CR** of the conchoid. Erect **DP** perpendicular to **X′X**, and draw **PO**. Then is **POC** one-third of **COM**. To prove this, bisect **PH** as at **E**, and draw

DE. Draw also FE parallel to DH. Since PE = EH, PF = FD, and ED = PE = EH = DO. By reason of the isosceles triangles PED, and EDO, we have angle DEO = 2P = 2POC. But DEO = EOD. ∴ 2EOC = EOD, or EOC = ⅓COM.

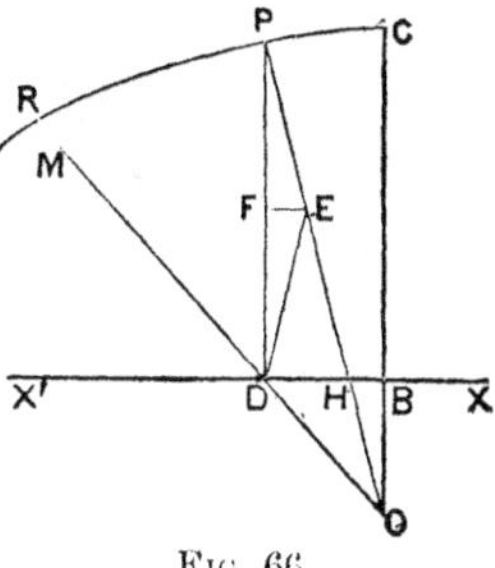

FIG. 66.

[NOTE.—This scholium is by no means necessary to the integrity of the course. It is inserted merely as a matter of interest to the student, giving him a few hints upon a subject which has figured so prominently in the history of geometry. It will afford a good exercise for the student who has time and ability, to demonstrate fully the facts hinted at, and which are not demonstrated above. Thus, let him show why, in the cissoid, AD : ED : AD′ : ID′; also how the insertion of two means enables us to obtain *any* multiple of the cube; also how the conchoid effects the same purpose; and that the two problems are in reality but one.]

THE WITCH OF AGNESI.

86. DEF.—***The Witch of Agnesi*** is the locus of the extremity of an ordinate to a circle, produced until the produced ordinate is to the ordinate itself, as the diameter of a circle is to one of the segments into which the ordinate divides the diameter,—these segments being all taken on the same side.

87. Prob. *To construct the Witch of Agnesi.*

SOLUTION.—Let AB be the circle. Draw a series of parallel ordinates 1 1, 2 2, P′P, etc. To find a point in the locus, take PE : FE : : AB : AE, and P is such a point. In like manner locate other points, as 1, 2, etc.

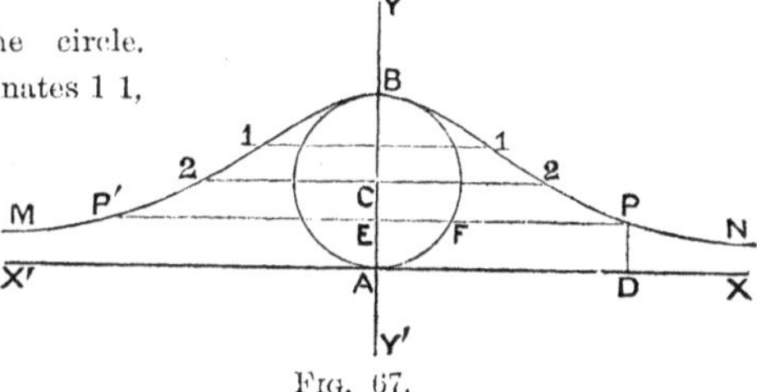

FIG. 67.

88. Prob. *To produce the equation of the Witch.*

SOLUTION.—Letting the axes be as represented in *Fig.* 67, so that P being any point, AD $=x$, PD $=y$, and calling the diameter of the circle, AB $=2a$, the equation is $x^2y = 4a^2(2a-y)$. [Let the student supply the demonstration.]

SCH.—The Witch has but one portion, as represented in the figure; it is symmetrical with respect to the axis YY′, is comprised between $y=0$, and $y=2a$, and has X′X for an asymptote. [Let the student give the proof.]

THE LEMNISCATE OF BERNOUILLI.

89. Def.—***The Lemniscate of Bernouilli*** is a curve such that the product of two lines drawn from any point in it to two fixed points, called the foci, is equal to the square of half the distance between these foci.

---

90. Prob. *To construct the Lemniscate of Bernouilli.*

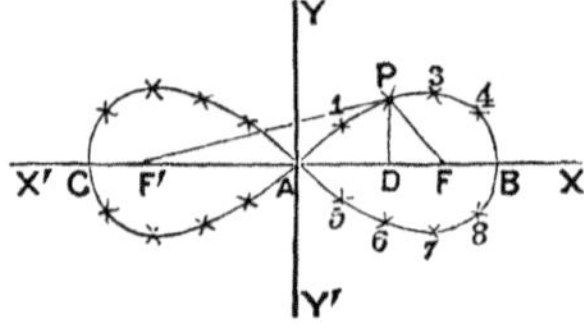

Fig. 68.

Solution.—Let F and F' be the foci. From F' as a centre, with any convenient radius, as F'P, draw an arc, as P6. Find a third proportional to F'P and F'A. Let this be PF. From F as a centre with this third proportional, draw an arc intersecting the former in P and 6. Then will P and 6 be points in the locus; for, by construction $F'P \times PF = AF^2$. In like manner find other points. *Fig.* 69, shows a convenient method of finding these proportionals. GH = F'F, and TG = AF. Since TL : TG : : TG : TI, TL : AF : : AF : TI; and TL and TI are the corresponding radii to be used in locating a point, as P. With one pair of distances four points can be found.

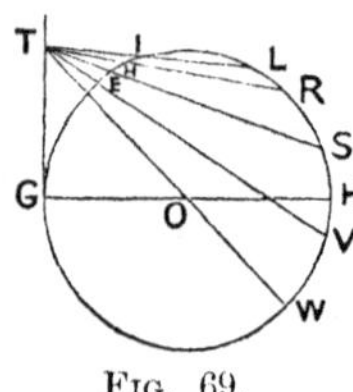

Fig. 69.

---

91. Prob. *To produce the equation of the Lemniscate.*

Solution.—Assuming X'X' and YY' as axes of reference, letting P be any point whose co-ordinates AD and PD, are x and y, and putting AF = AF' $= c$, we have the distance between the two points F' and P, or F'P $= \sqrt{(x+c)^2 + y^2}$. In like manner FP $= \sqrt{(c - x)^2 + y^2}$. Whence by definition $\sqrt{(x + c)^2 + y^2} \times \sqrt{(c - x)^2 + y^2} = c^2$, or $(y^2 + x^2)^2 = 2c^2(x^2 - y^2)$.

Sch. 1.—Let the student observe the symmetry and limits of the curve from its equation. Observe that in the construction F'B = FC = TW. As FB $\times$ F'B $= c^2$, and FB = AB $- c$, and F'B = AB $+ c$, we find that AB $= c\sqrt{2}$. Putting AB $= a = c\sqrt{2}$, $2c^2 = a^2$, whence the equation of the curve in terms of its semi-axis is $(y^2 + x^2)^2 = a^2(x^2 - y^2)$.

Sch. 2.—(To be read on review.) The equation of the Equilateral Hyperbola whose semi-axis is a, and co-ordinates x', y', is $x'^2 - y'^2 = a^2$. The equation of its tangent is $xx' - yy' = a^2$. The equation of a perpendicular from the centre upon the tangent is $x = -\frac{x'}{y'}y$. From these three equations, eliminating x' and y', that is finding the locus of the intersection of a perpendicular from the centre upon the tangent, we find $(x^2+y^2)^2=a^2(x^2-y^2)$. Therefore this lemniscate is the locus of the intersection of a perpendicular from the centre of an equilateral hyperbola upon its tangent, the axes of both loci being coincident.

THE CYCLOID.

92. Def.—***The Cycloid*** is the locus of a point in the circumference of a circle which rolls along a fixed right line.

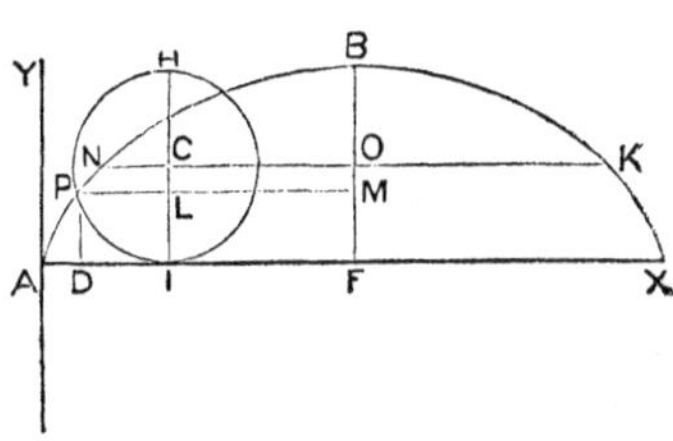

Fig. 70.

Sch.—The cycloid can be constructed mechanically by rolling a wheel, as **HPI**, *Fig.* 70, along the edge of a fixed ruler, as **AX**. A point **P** in the circumference of the wheel describes the cycloid.

93. Def's.—The circle **HPI** is called the ***Generating Circle***, or, simply the ***Generatrix;*** **AX** is the ***Base***, and is equal to the circumference of the generatrix; and **BF**, erected perpendicular to the base at its centre, is the ***Axis***, and is equal to the diameter of the generatrix.

---

94. Prob.—*Having given the cycloid, to put the generating circle in position.*

Solution.—There is given simply the curve **ABX**, *Fig* 70. Draw the base **AX**, and bisect it by the perpendicular **BF**. **BF** is the axis. Bisect the axis by **NK** drawn parallel to the base. Now, to put the generating circle in the position it occupied when the generating point was at **P**, draw from **P** as a centre, with a radius equal to the radius of the generatrix (**BO** or **OF**), an arc cutting **NK**, as at **C**. **C** is the centre of the generatrix.

---

95. Prob.—*To produce the equation of the cycloid referred to its base and a perpendicular at the left hand vertex.*

Solution.—Let **P**, *Fig.* 70, be any point in the cycloid **ABX**, referred to **AY**, and **AX** as axes. Then **AD** $=x$, and **PD** $=y$. Call the radius of the generatrix r. Now **AD** = **AI** − **DI**. But by construction, **AI** = arc **PI** = versin^{-1} **IL**, or versin^{-1}y, *to a radius* r. **DI** = **PL** = $\sqrt{\mathbf{IL}\times\mathbf{LH}}=\sqrt{y(2r-y)}$ $=\sqrt{2ry-y^2}$. $\therefore\ x=\text{versin}^{-1}y-\sqrt{2ry-y^2}$.

Sch.—If y be negative, $\sqrt{2ry-y^2}$ becomes imaginary; hence the curve lies on but one side of the base. For $y=0$, we have $x=\text{versin}^{-1}0=0$,

$2\pi r$, $4\pi r$, etc., etc. Hence we see that there are an infinite number of arcs like AEX, belonging to the curve. This is also apparent from the definition, as each revolution of the generatrix produces one arc, and there is no limit to the number of revolutions. For $y = 2r$, $x = \text{versin}^{-1}(2r) = \pi r$, $3\pi r$, $5\pi r$, etc., etc., as it should from the construction.

---

96. Prob.—*To produce the equation of the cycloid referred to its axis as the axis of abscissas, and a tangent at the vertex of the axis* (**B**, *Fig.* 71), *as the axis of ordinates.*

SOLUTION.—Let PM $= y$, and BM $= x$. Now PM $=$ PL $+$ LM. But PL $= \sqrt{2rx - x^2}$, and LM $=$ IF $=$ AF $-$ AI $=$ the semi-circumference of the generatrix $-$ arc PI. Again, arc PI $= \text{versin}^{-1}(2r - x)$. $\therefore y = \sqrt{2rx - x^2} + \pi r - \text{versin}^{-1}(2r - x)$.

97. COR. PR $=$ *arc* BR, P *being any point in the curve.*

For PR $=$ LM $=$ AF $-$ AI $=$ arc HPI $-$ arc PI $=$ arc HP $=$ arc BR.

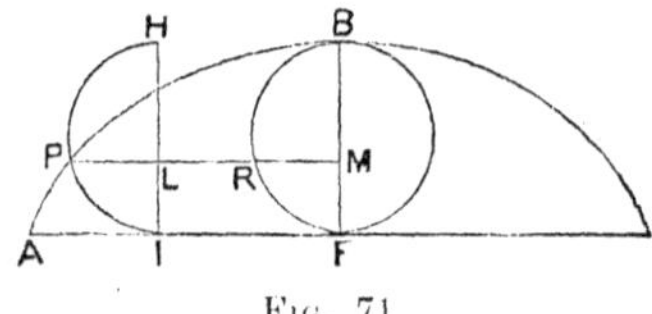

FIG. 71.

SCH. 1.—Considering the equation $x = \text{versin}^{-1}y - \sqrt{2ry - y^2}$, we observe that there are an infinite number of values of x for every value of y. First of all, the term versin^{-1} is ambiguous as to its sign, since a negative arc has the same versed-sine as the numerically equal positive arc. Moreover, whatever a versed-sine may be, there are not only the $+$ and $-$ arcs less than 180°, and also the $+$ and $-$ arcs of 360° $-$ the former, which corresponds to it, but these increased numerically by every multiple of 2π. We are therefore to write the term $\text{versin}^{-1}y$ with the sign $\pm$, and understand that for every value of y it has an infinite number of numerical values, each succeeding value in the series, being numerically 2π greater than the preceding. In the second place the term $-\sqrt{2ry - y^2}$, being a square root is to be written $-(\pm\sqrt{2ry - y^2})$, or $\mp\sqrt{2ry - y^2}$. Writing the equation in this way we have $x = \pm\,\text{versin}^{-1}y \mp \sqrt{2ry - y^2}$. The geometrical significance of these facts is as follows. Let y have any value as PD, *Fig.* 70; then x has 1st, the *positive* value AD and gives the point P, and a numerically equal *negative* value, giving a point P′ similarly situated on the left of AY, if we take $\text{versin}^{-1}y < 180°$; but if we take 360° $-$ this arc, both $+$ and $-$ as the value of $\text{versin}^{-1}y$, we get two other values of x, one $+$ and the other $-$. The former is where PM, *Fig.* 70, produced to the right meets the curve, and the other the corresponding point on the left of AY. Taking for values of $\text{versin}^{-1}y$, the values now considered $+ 2\pi$ simply repeats the curves at a distance 2π both at the right and left. The equation in (**96**) has a similar interpretation.

SCH. 2.—The Cycloid is a transcendental curve, and is next to the Conic Sections in importance among plane loci.

98. It frequently occurs that the equation of a locus can be written immediately from the definition. The Sinusoid is of this character. The definition is, "The Sinusoid is the locus of a point whose abscissa is the arc, while its ordinate is the sine of the arc." Hence the equation is $y = \sin x$. The other simple trigonometrical curves (page 16, Ex's 27—33) are of the same character, as is also the logarithmic curve, $x = \log y$.

99. As has been remarked before, the number of plane curves is infinite. The foregoing have been given as specimens, from which it is hoped that the student will be able to learn how the equations of loci referred to rectangular axes are produced from the definitions of the loci. A great many kinds of curves are suggested by the study of the *Properties* of *Curves*. Some of these will be noticed hereafter. Again, Mechanics and other branches of Physics, give rise to the study of curves, the production of whose equations requires a knowledge of the principles of Natural Philosophy. Thus, the Tractrix is the path described by a weight, **W**, *Fig.* 72, to which a cord, **AW**, is attached, and the extremity, **A**, of the cord, made to pass over the path, **AB**, friction being supposed uniform and perfect. Again, the *Catenary* is the line which a perfectly flexible chain assumes when its ends are fastened at two points, as **A** and **B**, *Fig.* 73, nearer together than the length of the chain. *Caustics* are an interesting class of curves consequent upon the laws of reflected light. The next chapter gives several varieties of curves called *Spirals*.

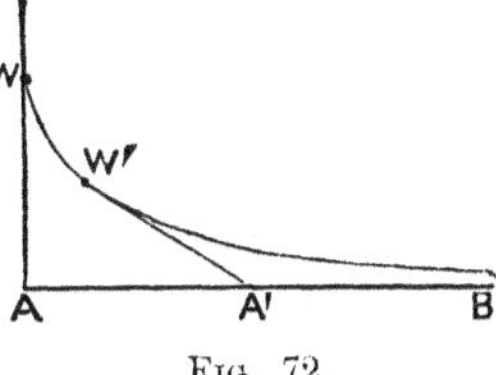

FIG. 72.

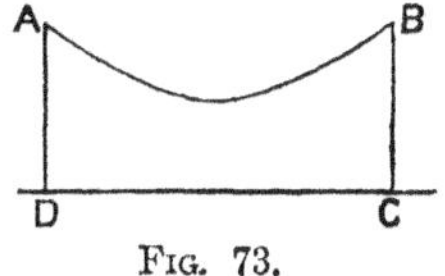

FIG. 73.

CHAPTER II.

THE METHOD OF POLAR CO-ORDINATES.

SECTION I.

Of the Point in a Plane.

100. Prop.—*The Position of any Point in a plane can be designated by giving its Distance and Direction from a fixed point in the plane.* In order to indicate direction, a fixed line has to be assumed.

ILL.—Let **A**, *Fig.* 74, be the fixed point, and **AX** the fixed line. Let r represent the distance from the fixed point to the point to be designated, as **AP**, **AP′**, etc., and θ the angle included between the fixed line and the line from the fixed point to the point to be designated, as **PAX**, **P′AX**, etc., etc. It is evident that by giving all possible values to θ, and r, all points in the plane of the paper may be located. Thus, for **P**, we have $\theta = 35°$, $r = 5$; for **P′**, $\theta = 120°$, $r = 10$; **P″**, $\theta = 195°$, $r = 8$; and for **P‴**, $\theta = 345°$, $r = 11$.

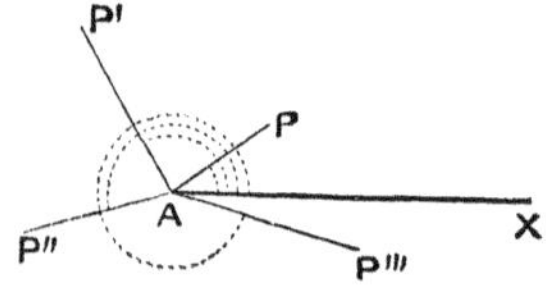

FIG. 74.

101. DEF'S.—***The Pole*** is the assumed fixed point, as **A**. ***The Prime Radius*** (called also the *Initial Line*, and the *Polar Axis*) is the assumed fixed line, as **AX**. ***The Radius vector*** is the distance from the pole to the point to be designated, as **AP**, **AP′**, etc. ***The Variable Angle*** (θ) is the angle indicating the direction of the point from the pole. When θ is reckoned around from right to left, it is called +; when reckoned from left to right, —. The radius is + when estimated in the direction of the extremity of the arc measuring the variable angle; and it is — when estimated in the opposite direction. r and θ are the *Polar Co-ordinates.*

---

102. Prop.—*The Polar equations of a Point are,* r = a, *and* $\theta = b$; *since, by giving suitable values to* r *and* θ, *all points in the plane can be located.*

Ex. 1. Locate $r = 5$, $\theta = \frac{1}{3}\pi$.

SOLUTION.—The radius being 1, π is the semi-circumference. Hence $\frac{1}{3}\pi = 60°$. Now, lay off **PAX** $= \theta = 60°$, *Fig.* 74; and, taking **AP** $= 5$, **P** is the required point.

Exs. 2 to 6. Locate $r=3$, $\theta=\frac{5}{3}\pi$: $r=4$, $\theta=\frac{2}{3}\pi$: $r=6$, $\theta=\frac{4}{3}\pi$: $r=-6$, $\theta=135°$: $r=4$, $\theta=-60°$.

Ex. 7. Show that $r=10$, $\theta=-45°$, is the same point as $r=-10$, $\theta=135°$.

---

103. Prob.—*To find the Distance between two points given by their polar co-ordinates.*

Solution.—Let the co-ordinates of **P**, *Fig.* 75, be (r, θ); and of **P'**, (r', θ'). We are to find **PP'** in terms of r r', θ, and θ'. Now, in the triangle **PAP'**, **AP** $= r$, **AP'** $= r'$ and the included angle **PAP'** $= \theta - \theta'$. Hence, representing **PP'** by D, we have from principles of trigonometry

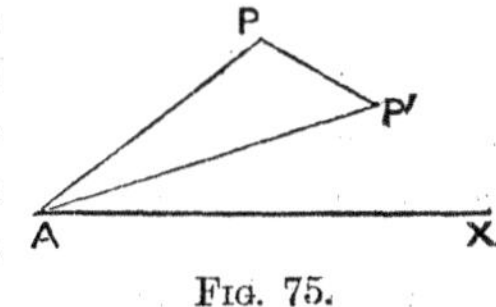

Fig. 75.

$$D = \sqrt{r^2 + r'^2 - 2rr' \cos(\theta - \theta')}, \quad \text{Q. E. D.}$$

Ex. 1. Find the distance between $r=3$, $\theta=\frac{1}{3}\pi$, and $r=4$, $\theta=\frac{5}{6}\pi$.

Ex. 2. Find the distance between $(8, \frac{3}{4}\pi)$, and $(3, \frac{13}{12}\pi)$.

Ex. 3. Find the distance between $(\sqrt{2}, 45°)$, and $(1, 0°)$.

Results in the last three examples, not in order, 7, 1, 5,

SECTION II.

Of the Right Line.

104. Prob.—*To produce the Polar Equation of the Right Line.*

Solution.—The *form* of this equation (like all others) depends upon the *constants* assumed. We will consider two forms.

1st. *When the constants are the length of the perpendicular from the pole upon the line, and the angle which this perpendicular makes with the prime radius.* Thus in *Fig.* 76, let **MN** be any line; **A**, the pole; **AX**, the prime radius; the perpendicular from the pole upon the line, **AD** $= p$; and the angle which the perpendicular makes with the prime radius, **DAX** $= \alpha$. Let **P** be any point in the line **MN**, and its co-ordinates be (r, θ). Now, in the right angled triangle **PAD**, we have **AD** = **AP** cos **PAD**, or $p = r\cos(\theta - \alpha)$;

$$\therefore\ r = \frac{p}{\cos(\theta - \alpha)}. \quad \text{Q. E. D.}$$

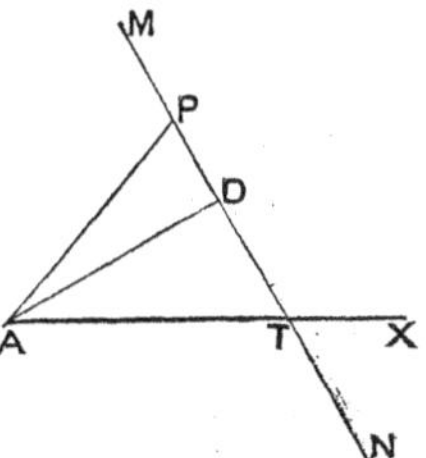

Fig. 76.

2nd. *When the constants are the intercept on the prime*

radius, and the angle which the line makes with the prime radius. In *Fig.* 77, let **MN** be any line referred to the pole, **A**, and the prime radius, **AX**. Represent the intercept, **AT**, by c, and the angle **NTX**, by α. Let **P** be any point in the line, and its co-ordinates **AP** $= r$, and **PAX** $= \theta$. The angle **TPA** $= \theta - \alpha$; and, from the triangle **PTA**, we have **AP** : **AT** :: sin**PTA** : sin **TPA**, or $r : c :: \sin\alpha : \sin(\theta - \alpha)$;

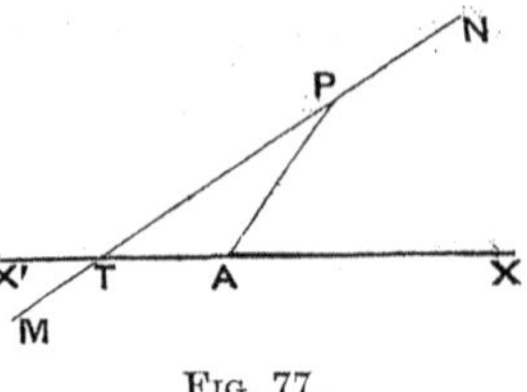

FIG. 77.

$$\therefore\ r = \frac{\sin\alpha}{\sin(\theta - \alpha)}c. \quad \text{Q. E. D.}$$

SCH. 1.—*Discussion of the equation* $r = \frac{p}{\cos(\theta - \alpha)}$. When $\theta = 0$, we have $r = \frac{p}{\cos(-\alpha)}$. This is as it should be, for, when $\theta = 0$, $r =$ **AT**, *Fig.* 76, which, from the triangle **ADT**, is seen to be $\frac{\mathbf{AD}}{\cos \mathbf{DAT}}$, or $\frac{p}{\cos(-\alpha)}$. The — sign of α indicates that the radius vector falls upon the opposite side of the perpendicular from that assumed in producing the equation. . . . When $\theta = \alpha$, $r = \frac{p}{\cos 0} = p$, as it evidently should. . . . When $\theta - \alpha = 90°$, $r = \frac{p}{0} = \infty$. In this case the radius vector becomes parallel to the line, and hence ∞. . . . From $\theta - \alpha = 90°$ to $\theta - \alpha = 270°$, r is negative, as it should be; since, in order to reach the line **MN**, it must be produced *backward, i. e.*, from the pole in a direction opposite to the extremity of the arc θ measured from the prime radius around to the right. . . . From $\theta - \alpha = 270°$ to $\theta - \alpha = 360°$, r is $+$; and at $\theta - \alpha = 360°$, $r = p$, as it should. Also, at $\theta = 360°$, $r = \frac{p}{\cos(-\alpha)} =$ **AT**. . . . When the line **MN** passes through the pole, $r = \frac{0}{\cos(\theta - \alpha)}$, which is 0 for all values of θ except $\theta = (90° + \alpha)$, for which $r = \frac{0}{0}$, *i. e.*, indeterminate. The 0 values of r indicate that we have not to pass any distance from the pole to reach the line; and the $\frac{0}{0}$ value indicates that for *all* values of r, its extremity is in the line **MN**. . . . Finally, when $\alpha = 0$, the line **MN** becomes perpendicular to the prime radius, and its equation is $r = \frac{p}{\cos\theta}$.

SCH. 2.—*Discussion of the form* $r = \frac{\sin\alpha}{\sin(\theta - \alpha)}c$. For $\theta = 0$; $r = -c$; which is evidently correct, as it is reckoned *backward*, and equals c, in length. . . . For all values of $\theta < \alpha$, r is negative, and hence is reckoned backward. . . . For $\theta = \alpha$, $r = \frac{\sin \alpha}{0} c = \infty$, as it should, since it is then parallel to the line. . . . For values between $\theta > \alpha$, and $\theta = 180° + \alpha$, r is positive. . . . At $\theta = 180° + \alpha$, r becomes infinite; and, when θ passes $180° + \alpha$, r is again negative. . . . For $\theta = 180°$, $r = \frac{\sin \alpha}{\sin (180° - \alpha)} c = c$, *i. e.*, **AT**. . . . If the line **MN** passes through the pole, $c = 0$, whence $r = \frac{\sin \alpha \cdot c}{\sin (\theta - \alpha)} = \frac{0}{\sin (\theta - \alpha)} = 0$, for all values of θ except $\theta = \alpha$, in which case $r = \frac{0}{0}$. These results are evidently correct, for in the former cases we have to go 0 distance from the pole in the specified directions, in order to reach the line; and, in the latter (when $\theta = \alpha$) the radius vector falling on the line **MN**, its extremity is equally in the line for all values of r.

Ex. 1. What is the polar equation (first form) of a line the nearest point in which is 6 from the pole, and the perpendicular to which makes an angle of 45° with the prime radius? Where does this line cut the prime radius? For what values of θ is r infinite? What is the value of r for $\theta = 75°$? For $\theta = 15°$? Construct these values of r and verify them by drawing the line.

Ex. 2. What is the polar equation (first form) of a right line perpendicular to the prime radius, and which cuts it at 4 to the left of the pole? What is the value of r when $\theta = 60°$? Why is the sign of r negative in the latter case? Between what values of θ is r positive? What is the value of r when θ is 120°?

The equation is $r = -\frac{4}{\cos \theta}$.

Ex. 3. Give the equation of a line parallel to the prime radius, and 10 above it; also at m below. *The latter equation is,* $r = -\frac{m}{\sin \theta}$.

SECTION III.

Of the Circle.

105. Prob.—*To produce the Polar Equation of a Circle, the pole being in the circumference, and the polar axis being a diameter.*

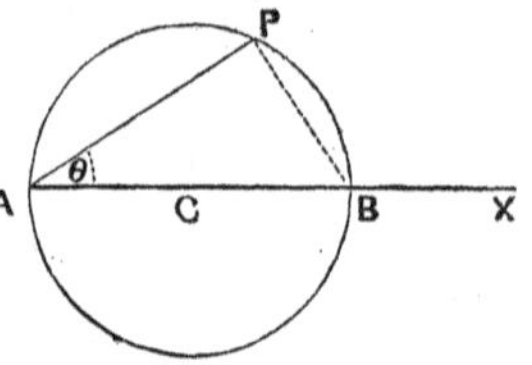

Fig. 78.

Solution.—Let **A**, *Fig.* 78, be the pole; R, the radius of the circle; and **P**, any point in the circumference. Then $\mathbf{AP}=r$, and $\mathbf{PAB}=\theta$. Now, from the right angled triangle **APB**, we have, by trigonometry, $r=2R\cos\theta$. Q. E. D.

Sch. 1.—*Discussion of the Equation.* If $\theta=0$, $r=2R$. If $\theta=\frac{1}{2}\pi$, or $\frac{3}{2}\pi$, $r=0$. For values of θ between $\frac{1}{2}\pi$ and $\frac{3}{2}\pi$, r is negative, indicating that for these values of θ the radius vector must be produced backward to meet the circumference. (The student should observe that the results obtained from the equation, accord with the values as observed from the figure.)

Sch. 2.—If the pole is at the centre, the equation is evidently $r=R$, for all values of θ.

---

106. Prob.—*To produce the General Polar Equation of a Circle.*

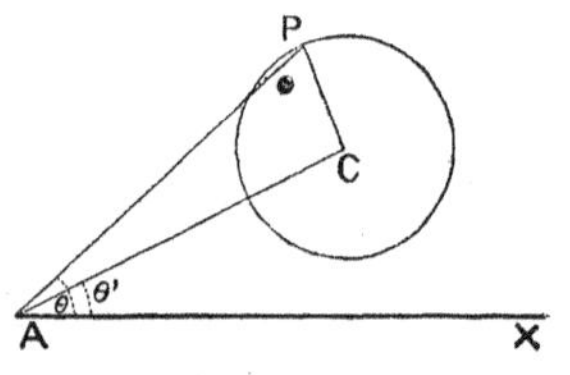

Fig. 79.

Solution.—Let the *constants* be (r', θ'), the co-ordinates of the centre, **C**, *Fig.* 79; and R, the radius of the circle. Let **P** be any point in the circumference, and its co-ordinates, **AP** and the angle **PAX**, be r and θ. Now, since the distance between the points **C** and **P** is constant (R), we have from (**103**) $R=\sqrt{r^2+r'^2-2rr'\cos(\theta-\theta')}$. Whence we have,

$$r^2-2r'\,r\cos(\theta-\theta')=R^2-r'^2. \quad \text{Q. E. D.}$$

Sch.—*Discussion of the Equation.*—Solving the equation for r, we have, $r=r'\cos(\theta-\theta')\pm\sqrt{R^2-r'^2\sin^2(\theta-\theta')}$. This value of r is *real* only for such values of θ as render $r'^2\sin^2(\theta-\theta')<\text{or}=R^2$; *i. e.*, when $\sin(\theta-\theta')$ is numerically $<\text{or}=\frac{R}{r'}$. Now, $\sin(\theta-\theta')=\sin\mathbf{PAC}=\frac{R}{r'}=\frac{\mathbf{PC}}{\mathbf{AC}}$ makes **P** a right angle, and hence, **AP** tangent to the circle. But $\sin^2(\theta-\theta')=\frac{R^2}{r'^2}$ gives $\sin(\theta-\theta')=\pm\frac{R}{r'}$; hence there are *two* positions of r in which it is tangent to the circle. (This is the familiar truth of Geometry, that from any point without a circle *two* tangents can be drawn to the curve.) The condition $\sin(\theta-\theta')=-\frac{R}{r'}$ is satisfied by the lower point of tangency, for which $\theta-\theta'$ becomes —, and hence, $\sin(\theta-\theta')$ is —. Between the limits $\sin(\theta-\theta')=\frac{R}{r'}$ and $-\frac{R}{r'}$, there are two *real* and *unequal* values of r for every value of θ, as evidently should be the case, since between these limits the radius vector meets the curve in two points.... When $\sqrt{R^2 - r'^2 \sin^2(\theta - \theta')} < r' \cos(\theta - \theta')$ the two values of r have the same sign; but when $\sqrt{R^2 - r'^2 \sin^2(\theta - \theta')} > r' \cos(\theta - \theta')$, r has different signs. In the former case the pole is *without* the circle,—in the latter *within*. These facts readily appear by solving the inequality. Thus, in the latter, $R^2 - r'^2 \sin^2(\theta - \theta') > r'^2 \cos^2(\theta - \theta')$, or $R^2 > r'^2 \sin^2(\theta - \theta') + r'^2 \cos^2(\theta - \theta')$. But the latter member reduces to r'^2. $\therefore$ $R > r'$, which puts the pole within the circle.* (For other cases, see examples below.)

Ex. 1. What is the equation of a circle whose radius is 10, and the polar co-ordinates of whose centre are $(15, \frac{1}{2}\pi)$? What values of θ indicate that the radius vector is tangent to the circle? Between what limits of θ is r real? Between what imaginary? What is the position, and what are the values of θ when the radius vector passes through the centre? Construct the figure.

The equation is $r^2 - 30 \sin \theta r = -125$; or $r = 15 \sin \theta \pm \sqrt{100 - 225 \cos^2 \theta}$. *The positions of tangency are* $\cos \theta = \frac{2}{3}$, and $\cos \theta = -\frac{2}{3}$, when $r = 5\sqrt{5}$.

Ex. 2. Give and discuss as above the polar equation of the circle whose centre is at $(8, \frac{1}{4}\pi)$, and whose radius is 10. Does this circle cut the polar axis; and, if so, where? How do you determine this point from the equation?

Equation, $r^2 - 8\sqrt{2}(\sin \theta + \cos \theta)r = 36$. *Condition of tangency,* $\sin \theta + \cos \theta = \frac{3}{4}\sqrt{-2}$. $\therefore$ *The pole is within; as appears also from* $\frac{R}{r'} > 1$. *Cuts the polar axis at* $r = (4 \pm \sqrt{34})\sqrt{2}$.

Ex. 3. Show that the polar equation of a circle is $r^2 - 2r'r \cos \theta = R^2 - r'^2$ when the polar axis passes through the centre and the pole is without or within the circle. (Prove this directly from a figure without reference to the preceding forms.)

Ex. 4. Deduce from the general form in ***106***, the form in ***105***, and also the one in *Ex.* 3.

* This may also be observed from $\sin(\theta - \theta') = \pm \frac{R}{r'}$, which is the condition of tangency. This is possible only when $R < r$, *i. e.*, when the pole is without the circle. When $R = r$, the two tangents become one.

Ex. 5. Let the student show from the annexed figure that in the equation $r = r' \cos(\theta - \theta') \pm \sqrt{R^2 - r'^2 \sin^2(\theta - \theta')}$ the rational part, $r' \cos(\theta - \theta')$, is the chord, **AD**, of the circle described upon **AC** $(= r')$ as a diameter, and that the radical part, $\sqrt{R^2 - r'^2 \sin^2(\theta - \theta')}$, is **PD** = **P'D**, + for **PD** and — for **P'D**. **B** and **B'**, are the points where the radical becomes 0.

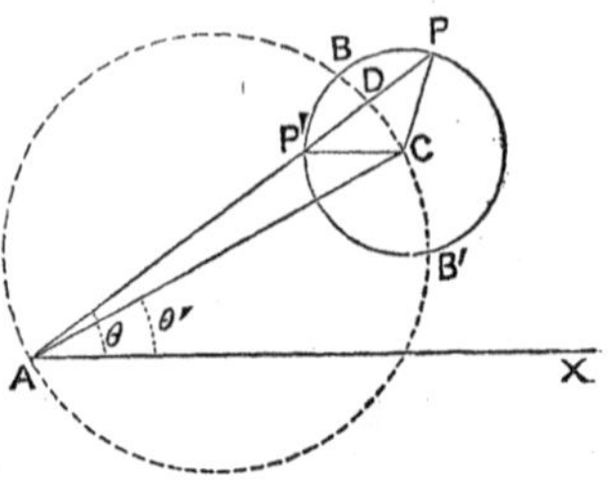

SECTION IV.

Of the Conic Sections.

107. Prob.—*To produce the Polar Equation of a Conic Section.*

Solution.—Let **P** be any point in the curve ; **EC** the directrix ; **A**, the focus and pole ; and **AX**, the axis of the curve and the polar axis. Let $2p$ be the *latus rectum ; e*, Boscovich's ratio, whence $p = \text{CA} \times e$; $\text{AP} = r$, and $\text{PAX} = \theta$. Then $\text{AP} = \text{CD} \times e = (\text{CA} + \text{AD})e$. But $\text{CA} = \frac{p}{e}$, and $\text{AD} = r\cos\theta$; whence, $r = \left(\frac{p}{e} + r\cos\theta\right)e$. $\therefore r = \frac{p}{1 - e\cos\theta}$. Q. E. D.

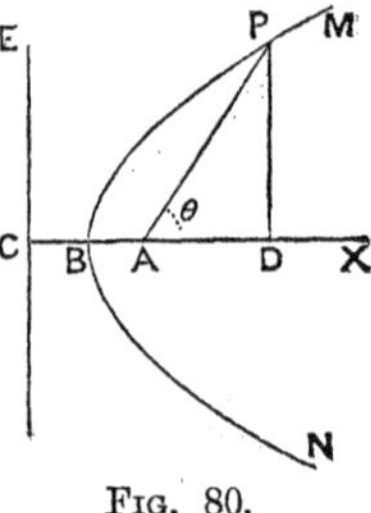

Fig. 80.

108. Cor. 1.—***The Polar Equation of the Parabola.*** *Since in the parabola* e = 1, *its polar equation is*

$$r = \frac{p}{1 - \cos\theta}, \text{ or } r = \frac{p}{\text{versin }\theta}.$$

109. Cor. 2.—***The Polar Equation of the Ellipse and Hyperbola in terms of the semi-transverse axis and eccentricity.*** *Since Boscovich's ratio* (e) *and the eccentricity are the same* (**48**), *and, numerically,* $p = A(1 - e^2)$, *this equation is*

$$r = \frac{A(1 - e^2)}{1 - e\cos\theta}.$$

Sch. 1.—*Discussion of the Polar Equation of the Parabola.*—For $\theta = 0$,

$r = \frac{p}{1 - \cos\theta}$ becomes $r = \frac{p}{1 - 1} = \infty$; *i. e.*, the radius vector falling upon the axis does not meet the curve.....For any value of $\theta > 0$, *however small*, r is finite; which shows that, if a line be drawn from the focus making any angle however small with the axis of the curve, it meets the curve at a finite distance. .... For $\theta = 90°$, $r = p$, as it should. .... For $\theta = 180°$, $r = \frac{p}{1-(-1)} = \frac{p}{1+1} = \frac{1}{2}p$. This is evidently correct, since r becomes **AB**, *Fig.* 80, when $\theta = 180°$. But, in the parabola, **AB** $= \frac{1}{2}p$.....For $\theta = 270°$, $r = p$, as it should. (Let the student discuss in like manner the form $r = \frac{p}{\text{versin } \theta}$.)

SCH. 2.—*Discussion of the equation* $r = \frac{A(1 - e^2)}{1 - e\cos\theta}$ *for the Ellipse.*—For $\theta = 0$, $r = \frac{A(1 - e^2)}{1 - e} = A(1 + e) = A + Ae$; which makes **AC** $= A + Ae$, as it should (**49**).....For the point **P** at the extremity of the conjugate axis, *Fig.* 81, $\cos\theta = \frac{\mathbf{AO}}{\mathbf{AP}} = \frac{Ae}{r}$. Hence $r = \frac{A(1 - e^2)}{1 - \frac{Ae^2}{r}}$, or $r - Ae^2 = A(1 - e^2)$. $\therefore r = A$; *i. e.*, **AP** $= A$ agreeably to (**46**)..... For $\theta = 90°$, $r = A(1 - e^2) = p$..... For $\theta = 180°$, $r = \frac{A(1 - e^2)}{1 + e} = A - Ae$, which is the value of **AB**, as it should be.

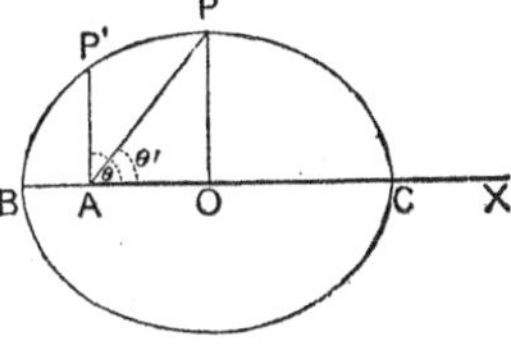

FIG. 81.

SCH. 3.—*Discussion of the Equation* $r = \frac{A(1 - e^2)}{1 - e\cos\theta}$ *for the Hyperbola.*—Remembering that in the hyperbola $e > 1$, we observe that $A(1 - e^2)$ is essentially *negative*, and hence that the *sign* of r depends upon the sign of the denominator, $1 - e\cos\theta$; r being $+$ when $e\cos\theta > 1$, and $-$ when $e\cos\theta < 1$. Now for $\theta = 0$, $r = \frac{A(1 - e^2)}{1 - e} = A + Ae$, which indicates that **A**, *Fig.* 82, is the pole, and **C** is the point located when $\theta = 0$, as **AC** $= A + Ae$ (**49**).....As θ increases from 0, $\cos\theta$ diminishes, which diminishes the *numerical* value of $1 - e\cos\theta$, till $\theta = \cos^{-1}\frac{1}{e}$ (though leaving it negative, and hence r positive), hence r increases and the

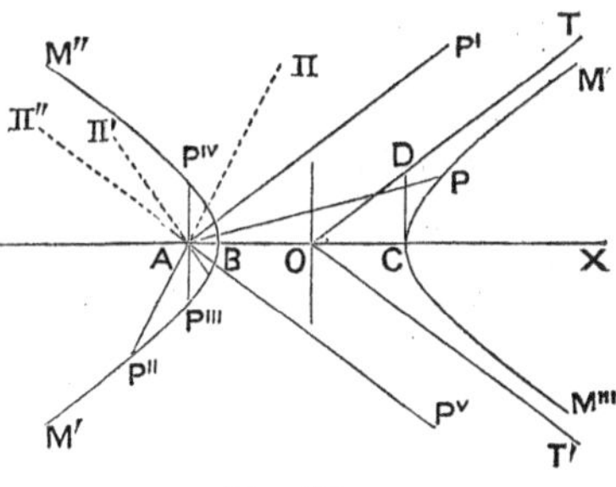

FIG. 82.

arc **CM** is traced. . . . . When $1 - e\cos\theta = 0$, *i. e.*, when $\cos\theta = \frac{1}{e}$, $\left(\text{in which case } \theta = \cos^{-1}\frac{1}{e}\right)$, $r = \infty$. In this position r(**AP**′) becomes parallel to the asymptote **OT**, a line so drawn that **DC** $= B$, whence $\cos$ **DOC** $= \frac{\mathbf{OC}}{\mathbf{OD}} = \frac{A}{Ae} = \frac{1}{e}$. . . . When θ passes the value $\cos^{-1}\frac{1}{e}$, $1 - e\cos\theta$ becomes positive and renders r negative, and the left hand branch begins to be traced, the point moving in the direction **M′BM″**. Thus, when $\theta = \Pi$**AX**, r being negative is reckoned backwards and the point **P″** is located. When $\theta = 90°$, $r = \frac{A(1 - e^2)}{1} = -p$, and **P‴** is located, **AP‴** being equal to $-p$. As θ passes from 90° to 180°, the arc **P‴B** is traced. At $\theta = 180°$, $r = \frac{A(1 - e^2)}{1 - e(-1)} = \frac{A(1 - e^2)}{1 + e} = A - Ae = -$**AB**, as it should. (In like manner let the student observe that as θ passes from 180° to 270°, **BP**^{IV} is traced; at $\theta = 270°$, $r = -p =$ **AP**^{IV}; from $\theta = 270°$ to $\theta = \cos^{-1}\frac{1}{e}$ *in the 4th quadrant*, r remains negative and **P**^{IV}**M″** is traced; at $\theta = \cos^{-1}\frac{1}{e}$ in the 4th quadrant $r = \infty =$ **AP**^{V}, parallel to the asymptote **OT**′; and that, finally, from this value of θ to $\theta = 360°$, r becomes positive again, and the arc **M‴C** is traced.)

Ex. 1. What is the polar equation of a parabola whose principal parameter is 8, the pole being at the focus? What is the length of the radius vector for $\theta = 60°$?

The equation is $r = \frac{4}{1 - \cos\theta}$. *For* $\theta = 60°$, $r = 8$.

Ex. 2. What is the polar equation of an ellipse whose axes are 12 and 8, the pole being at the focus? What are the focal distances; *i. e.*, what are the respective values of r for $\theta = 0$, and for $\theta = 180°$? What is the *semi-latus rectum*; *i. e.*, what is the value of r when $\theta = 90°$?

The equation is $r = \frac{8}{3 - \sqrt{5}\cos\theta}$. *The focal distances are* $6 + 2\sqrt{5}$, and $6 - 2\sqrt{5}$. *The semi-latus rectum is* $2\frac{2}{3}$.

Ex. 3. What is the polar equation of an hyperbola whose transverse axis is 6, and the distance between the foci 10? What is the value of r when $\theta = 0$? When $\theta = 90°$? When $\theta = 180°$? What is the value of θ when $r = \infty$?

The equation is $r = \frac{16}{5\cos\theta - 3}$. *For* $r = \infty$, $\theta = 53°8'$, nearly.

Ex. 4. Show that the polar equation of the parabola becomes $r = \frac{p}{1 + \cos\theta'}$, when the prime radius lies in the position **AC**, *Fig.* 80, and θ' is reckoned from it around to the left (in the usual direction). Show that in this form, for $\theta' = 0$, $r = \frac{1}{2}p$; and for $\theta = 180°$, $r = \infty$.

Sug.—This form is deduced from $r = \frac{p}{1 - \cos\theta}$, by substituting for θ, $\theta' - 180°$, as the initial line is revolved forward 180°.

Ex. 5. A comet is moving in a parabolic orbit around the sun at its focus, and when at 100,000,000 miles from the sun, the radius vector makes an angle with the axis of the orbit of 60°. What is the polar equation of the comet's orbit? How near does it approach the sun? How does it appear in a parabola that $r = 2p$, for $\theta = 60°$?

The equation is $r = \frac{50,000,000}{1 - \cos\theta}$.

SECTION V.

Of Higher Plane Curves.

[Note.—All of this section, except the portion upon Spirals, may be omitted in a shorter course, if thought desirable.]

---

110. Prob. *To produce the Polar Equation of the Cissoid of Diocles.*

Solution.—Let **AM**, *Fig.* 59, be the cissoid, **A**, the pole, **AX**, the polar axis, and **P** any point in the curve; whence $AP = r$, and $PAX = \theta$. Let $AB = 2a$. Now $AP = r = AD \sec\theta = D'B \times \sec\theta$. But $D'B = \frac{\overline{E'B}^2}{AB} = \frac{(2a \sin\theta)^2}{2a} = 2a \sin^2\theta$. Therefore, $r = 2a \sin^2\theta \sec\theta$, or $r = 2a \sin\theta \tan\theta$. Q. E. D.

Sch.—*Discussion of the Equation.* For $\theta = 0$, $r = 0$.... For $\theta = 45°$, $r = a\sqrt{2}$; hence the curve passes through **C**, *Fig.* 59.... For $\theta = 90°$, $r = \infty$.... For $\theta > 90°$ and $< 270°$, r is negative, and the branch, **AM'**, is traced while θ is passing from 90° to 180°, and the branch, **AM**, is traced a second time by the negative radius vector while θ passes from 180° to 270°.... While θ passes from 270° to 360°, r is positive and **AM'** is traced a second time. Therefore the curve is traced twice by one revolution of the radius vector.

111. Prob. *To produce the Polar Equation of the Conchoid of Nicomedes.*

SOLUTION.—In *Fig.* 62, draw FK perpendicular to OD. Let O be the pole, and OD parallel to AX, the polar axis. Let AO $= b$, and AB $= a$. Then P being any point in the curve, we have OP $= r =$ OF + FP $=$ OF $+ a$. But OF $=$ FK $\times$ cosec FOK $= b$ cosec θ. Therefore, $r = b$ cosec $\theta + a$. Q. E. D.

SCH.—*Discussion of the Equation.* For $\theta = 0$, $r = \infty$ .... For $\theta = 90°$, $r = b + a$.... For $\theta = 180°$, $r = \infty$.... For $\theta > 180°$ and $< 360°$, cosec θ is negative, and the lower branch is traced. The student should be careful to observe the several forms, as when $a > b$, $a = b$, $a < b$. (See *85.*)

---

112. Prob. *To produce the Polar Equation of the Lemniscate of Bernouilli.*

SOLUTION.—Using the same notation as in *Fig.* 68, let A be the pole, and AX the polar axis. Let P be *any* point in the curve, and draw AP. Then AP $= r$, and PAX $= \theta$. The following is an outline of the solution :

$$\overline{F'P}^2 = c^2 + r^2 + 2cr \cos \theta. \quad (1).$$

$$\overline{FP}^2 = c^2 + r^2 - 2cr \cos \theta. \quad (2).$$

Multiplying (1) and (2) together, and remembering that F′P $\times$ FP $= c^2$, we have, after a little reduction :

$$r^2 = 4c^2 \cos^2 \theta - 2c^2 = 2c^2(2 \cos^2 \theta - 1).$$

But $2\cos^2 \theta - 1 = \cos 2\theta$. $\therefore r^2 = 2c^2 \cos 2\theta$. Q. E. D.

SCH.—The pupil should discuss this equation as the preceding have been.

[NOTE.—It is frequently more convenient to obtain the polar equation of a curve by transforming its rectilinear equation, according to a process to be explained in a subsequent chapter. We will close this chapter with some account of *Spirals*, a class of curves of much historic interest in consequence of the labor bestowed upon some of them by the old geometricians, and to which the method of polar co-ordinates is specially adapted.]

OF PLANE SPIRALS.

113. DEF'S.—***A Plane Spiral*** is the locus of a point revolving about a fixed point, and continually receding from it in such a manner that the radius vector is a function of the variable angle. Such a curve may cut a right line in an infinite number of points, which would render its rectilinear equation of an infinite degree. Hence, these loci are *transcendental.*

The Measuring Circle is the circle whose radius is the radius vector at the end of one revolution of the generating point in the positive direction.

A Spire is the portion generated by any one revolution of the generating point.

114. *The Spiral of Archimedes* is the locus of a point revolving around and receding from a fixed point so that the ratio of the radius vector to the angle through which it has moved from the polar axis, is constant.

---

115. *Prob.* *To construct the Spiral of Archimedes.*

SOLUTION.—Let A, *Fig.* 83, be the pole, and AX the prime radius. Through A draw any convenient number of indefinite radial lines (say 8) making equal angles with each other. Since θ and r are to vary alike, when $\theta = 0$, $r = 0$, and the spiral begins at the pole. Take any distance, as A1, on Aa, twice this distance, as A2, on Ab, three times the same distance, as A3 on Ac, etc., etc. Then will 1, 2, 3 - - - - - 17 be points in the spiral. Q. E. D.

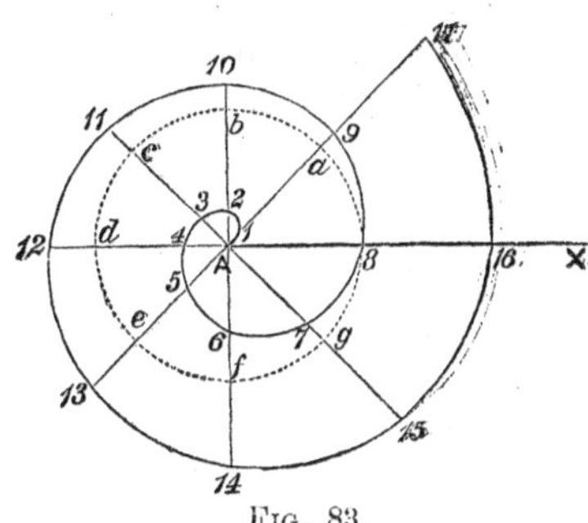

FIG. 83.

ILL.—The dotted line $a\,b\,c\,d\,e\,f\,g$ is the measuring circle, and A1 2 3 - - - - 7 8 is the first spire. 8 9 10 11 - - - - 15 16 is the second spire.

SCH.—The several spires of this spiral are such that, the distance between any two consecutive ones measured on the radius vector being the same, and equal at all points to the radius of the measuring circle.

---

116. *Prob.* *To produce the equation of the Spiral of Archimedes.*

SOLUTION.—Letting a be the ratio of the radius vector to the variable angle, we have $r = a\theta$. Or, otherwise, calling the radius of the measuring circle, A8, *Fig.* 83, 1, for this value of r, $\theta = 2\pi$. Let 3 be any point in the curve, whence A3 represents r, and 3AX, or $8abc = \theta$. Now from the definition, $r : 1 :: \theta : 2\pi$.

$\therefore\ r = \dfrac{\theta}{2\pi}$. Q. E. D.

117. COR.—***The Reciprocal or Hyperbolic Spiral.*** *This Spiral is naturally suggested by the Spiral of Archimedes, as in it the radius vector varies inversely as the variable angle. Hence the equation is*

$$r = \frac{2\pi}{\theta}.$$

CONSTRUCTION.—To construct the Reciprocal Spiral, let A be the pole, and AX the polar axis. Draw any convenient number of radial lines through the pole, making equal angles with each other. Take A1 any convenient length, A2 $= \frac{1}{2}$A1, A3 $= \frac{1}{3}$A1, A4 $= \frac{1}{4}$A1, etc., etc. The points 1 2 3 4 - - - 8 - - - B are points

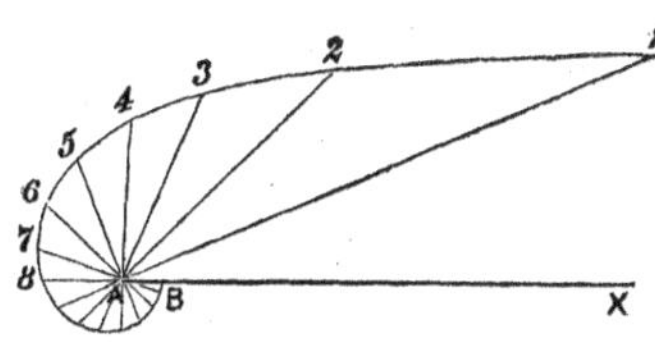

FIG. 84.

in the curve. Since r can become 0 only when $\theta = \infty$, this curve continues to approach the pole as the radius vector revolves, but reaches it only upon an infinite number of revolutions.

118. *The Lituus*.—The equation of this spiral is $r = \frac{2\pi}{\theta^{\frac{1}{2}}}$. Let the student construct it and give the formal definition. The form of the curve is given in *Fig.* 85.

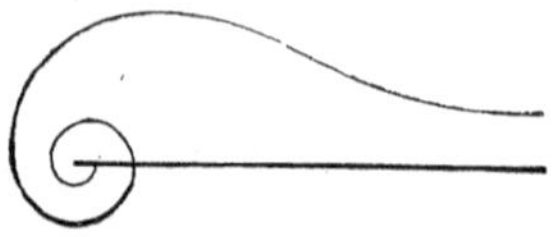

FIG. 85.

119. *The Logarithmic Spiral*.—In this spiral the radius vector increases in a geometrical ratio, while the variable angle increases in an arithmetical ratio. The equation is, therefore, $r = a^\theta$. If a be the base of a system of logarithms, this equation becomes $\theta = \log r$.

CONSTRUCTION.—To construct $r = a^\theta$, let $a = 2$. Then for $\theta = 0$, $r = 1$, which gives the point 0. For $\theta = 1$, *i. e.*, the arc of 57.3° nearly, $r = 2^1 = 2$, which gives the point 1. For $\theta = 2$, *i. e.*, the arc of 114.6° nearly, $r = 2^2 = 4$, which gives the point 2. As θ increases r increases much more rapidly, so that with this small base ($a = 2$), at the end of the first revolution, when $\theta = 6.28+$, $r = 2^{6.28+} =$ more than 64. Hence, at one revolution the radius vector would be 64 times Ao. But, though r increases so very rapidly, it is easy to see that it does not become ∞ till $\theta = \infty$. Again, letting the radius vector revolve in the *negative* direction from AX, so that θ is negative, we have for $\theta = -1 = 0$Aa, $r =$ A$a = 2^{-1} = \frac{1}{2}$. For $\theta = -2$, $r = 2^{-2} = \frac{1}{4}$. Thus, it appears that as the radius vector revolves in this direction it generates a portion of the spiral which at first rapidly approaches the pole, but cannot reach it till $\theta = \infty$ .... Were we to take $a = 10$, the base of the common system of logarithms, the change of r would be so rapid that we could represent but a small arc of the curve.

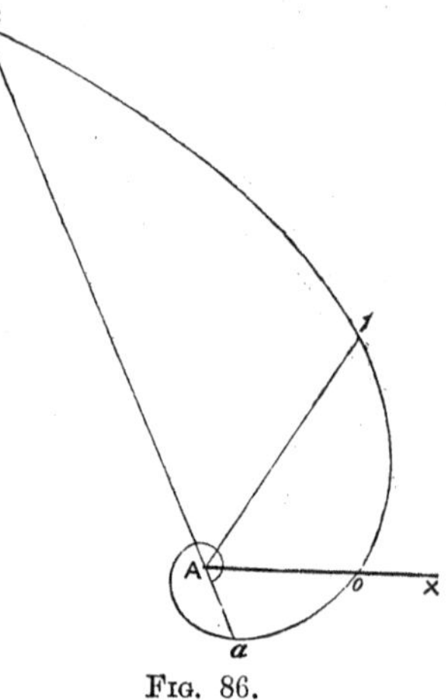

FIG. 86.

CHAPTER III.

TRANSFORMATION OF CO-ORDINATES.

SECTION I.

Methods of Passing from one Set of Rectilinear Axes to Another.

120. Def's.—***Transformation of Co-ordinates*** is the process of changing the reference of a locus from one set of axes to another, or from one system of co-ordinates to another. The problem presents itself under two different aspects which are nearly the converse of each other: 1st, Having given the equation of a locus referred to one set of axes, or system of co-ordinates, to find the equation of the same locus when referred to another set or system. 2nd, Having given the equation of a locus referred to some known axes, to find the position of a new set, to which, when the locus is referred, its equation will take some specified form. The axes, or system, to which reference is made in the given equation, may be called the ***Old,*** or ***Primitive, Axes*** or ***System,*** and those to which the transformation is made, the ***New.***

Ill's.—The equations $x^2 + y^2 = 25$, $y^2 = x(10 - x)$, and $x(x - 4) + y(y - 6) = 12$, may all be considered as equations of the same locus **MON**, *Fig.* 87, but referred, respectively, to the three pairs of axes X_1X_1', Y_1Y_1'; X_1X_1', Y_2Y_2'; X_3X_3', Y_3Y_3'. Now, having given any one of these equations any other of them may be found if we know the position of the new axis with reference to the old. The process is transformation.

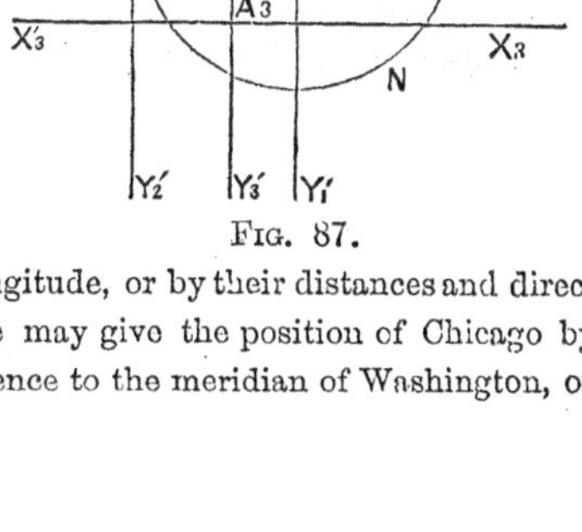

Fig. 87.

Again, we are familiar with various methods of designating particular points on the earth's surface, as by latitude and longitude, or by their distances and directions from a given point. For example, we may give the position of Chicago by stating its latitude and longitude with reference to the meridian of Washington, or by giving its distances from the tropic of Cancer and the meridian of Greenwich, Eng., or, in still another way, by giving its distance and direction from New York city. The process of converting one of these descriptions into any other of them, would furnish an analogy to the process of transformation of co-ordinates. The first two descriptions (considering the earth's surface a plane), would be equations of a point (Chicago) referred to rectangular co-ordinates ; the last would be an example of polar co-ordinates.

We will give one more illustration, as it is of the highest importance that the nature of the problem be understood from the outset. Let the student construct a pair of rectangular axes, $\mathbf{X}_1\mathbf{X}_1'$, $\mathbf{Y}_1\mathbf{Y}_1'$, with the origin $\mathbf{A}_1$, and another pair, $\mathbf{X}_2\mathbf{X}_2'$, $\mathbf{Y}_2\mathbf{Y}_2'$, also rectangular, with the origin at $\mathbf{A}_2$ [the point $(0, -1)$ when referred to the first axes], and the new axis of x, $\mathbf{X}_2\mathbf{X}_2'$, making an angle with the primitive axis of $-45°$, and the new axis of y making an angle of $45°$. Now, upon the first axes, construct $x^2 - 6xy + y^2 - 6x + 2y + 5 = 0$, and upon the second axes construct $y^2 - 2x^2 = 2$, when the two equations will be found to give the same locus. The problem of transformation which affords this illustration may be stated thus ; To transform $x^2 - 6xy + y^2 - 6x + 2y + 5 = 0$, to a new system of rectangular axes having the new origin at $(0, -1)$, and the new axis of x making an angle with the primitive whose tangent is -1.

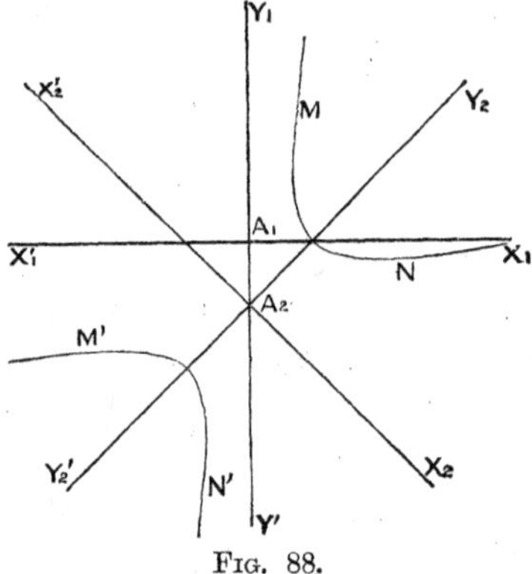

FIG. 88.

To illustrate the second form under which the problem of transformation presents itself, the example of the last paragraph may be stated,—Having given the equation $x^2 - 6xy + y^2 - 6x + 2y + 5 = 0$, as the equation of a locus referred to rectangular axes, required to find the position of a new pair of axes to which, when the locus is referred, its equation will involve no terms in the first power, or in the rectangle of the variables. The result of the solution of this problem would be the determination of the position of new axes as given in the paragraph above.

SCH.—As the *form* (not the *degree*) of the equation of a locus, depends in a large measure upon the situation of the axes, or upon the system used, it will be readily seen that a set of axes in some particular position, or some particular system of co-ordinates, may be best suited to one class of problems, or of loci, and another set or system to another class. It is therefore desirable to be able to pass at will from any one set or system to any other. This transformation is effected by finding the values of the co-ordinates in the given equation in terms of new co-ordinates (and certain constants) and substituting the latter for the former. The methods of doing this we will now explain.

122. Prob.*—To produce the general formulæ for passing from one set of rectilinear co-ordinates to another.*

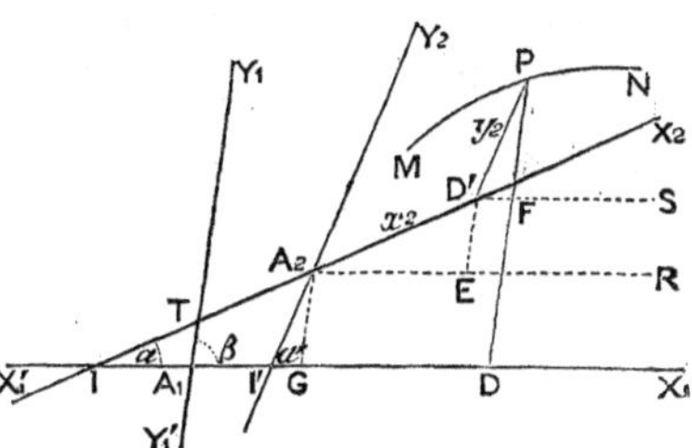

FIG. 89.

SOLUTION.—Let **P**, *Fig.* 89, be any point in a locus **MN** referred to the *Primitive Axes* A_1X_1, A_1Y_1, the co-ordinates being $A_1D = x$, and $PD = y$. Let A_2X_2, A_2Y_2 be the *New Axes*, the co-ordinates of the point **P**, when referred to them being $A_2D' = x_2$, and $PD' = y_2$. Let the angle included between the primitive axes, $Y_1A_1X_1$, be β; the angle which the new axis of x makes with the primitive, X_2IX_1, be α; the angle which the new axis of y makes with the primitive axis of x, $Y_2I'X_1$ be α'; and the co-ordinates of the new origin, A_2, be $A_1G = m$, $A_2G = n$. The problem now is, to find the values of the primitive co-ordinates x, y, in terms of the new co-ordinates x_2, y_2, and the constants m, n, α, α', and β, so that the latter may be substituted for the former in the equation of a locus referred to the primitive axes, and the equation be thus transformed and made to represent the same locus in terms of the new co-ordinates; *i. e.*, referred to the new axes.

We have $x = A_1D = A_1G + A_2E + D'F$. But $A_1G = m$; and from the triangle $A_2D'E$, $A_2E : x_2 :: \sin A_2D'E : \sin A_2ED'$, which becomes $A_2E : x_2 :: \sin(\beta - \alpha) : \sin\beta$, since $A_2D'E = D'ER - D'A_2E = \beta - \alpha$, and $\sin D'EA_2 = \sin D'ER = \sin\beta$ (the sine of an angle equals the sine of its supplement). From this proportion, $A_2E = \dfrac{x_2 \sin(\beta - \alpha)}{\sin\beta}$. In the same manner, from the triangle $PD'F$, we have $D'F = \dfrac{y_2 \sin(\beta - \alpha')}{\sin\beta}$. Substituting these equivalents in the value of x as given above

$$x = m + \frac{x_2 \sin(\beta - \alpha) + y_2 \sin(\beta - \alpha')}{\sin\beta}, \quad (1).$$

Again, $y = PD = A_2G + D'E + PF$. But $A_2G = n$, and from the triangles $A_2D'E$, and $PD'F$, we have as before $D'E = \dfrac{x_2 \sin\alpha}{\sin\beta}$, and $PF = \dfrac{y_2 \sin\alpha'}{\sin\beta}$. Hence

$$y = n + \frac{x_2 \sin\alpha + y_2 \sin\alpha'}{\sin\beta}, \quad (2).$$

123. COR. 1.—*When the New Axes are parallel to the Primitive, the formulæ become*

$$x = m + x_2, \ (1); \ \textit{and} \ y = n + y_2, \ (2);$$

since in such a case $\alpha = 0$, *and* $\alpha' = \beta$; *whence* $\sin\alpha = 0$, $\sin\alpha' = \sin\beta$, $\sin(\beta - \alpha) = \sin\beta$, *and* $\sin(\beta - \alpha') = \sin 0 = 0$.

124. Cor. 2.—*To pass from rectangular axes to oblique, we have*

$$x = m + x_2 \cos \alpha + y_2 \cos \alpha', \quad (1);$$

$$\text{and } y = n + x_2 \sin \alpha + y_2 \sin \alpha', \quad (2).$$

These follow readily from the general formulæ *by observing that in this case* $\beta = 90°$; *whence* $\sin(\beta - \alpha) = \cos \alpha$, $\sin(\beta - \alpha') = \cos \alpha'$, *and* $\sin \beta = 1$.

125. Cor. 3.—*To pass from one set of rectangular axes to another set also rectangular, but not parallel, we have*

$$x = m + x_2 \cos \alpha - y_2 \sin \alpha, \quad (1);$$

$$\text{and } y = n + x_2 \sin \alpha + y_2 \cos \alpha, \quad (2).$$

To deduce these from the general formulæ (**122**), *observe that* $\beta = 90°$, *and* $\alpha' - \alpha = 90°$, *or* $\alpha' = 90° + \alpha$; *whence* $\sin \beta = 1$, $\sin(\beta - \alpha) = \cos \alpha$, $\sin(\beta - \alpha') = \sin(90° - 90° - \alpha) = \sin(-\alpha) = -\sin \alpha$, *and* $\sin \alpha' = \sin(90° + \alpha) = \cos \alpha$.

126. Cor. 4.—*To pass from oblique axes to rectangular, we have*:

$$x = m + \frac{x_2 \sin(\beta - \alpha) - y_2 \cos(\beta - \alpha)}{\sin \beta}, \quad (1);$$

$$y = n + \frac{x_2 \sin \alpha + y_2 \cos \alpha}{\sin \beta}, \quad (2).$$

In this case the $\alpha' - \alpha$ *of the general formula becomes* $90°$, *or* $\alpha' = 90° + \alpha$; *whence* $\sin(\beta - \alpha') = \sin(\beta - 90° - \alpha) = \sin\{-[90° - (\beta - \alpha)]\} = -\sin[90° - (\beta - \alpha)] = -\cos(\beta - \alpha)$, *and* $\sin \alpha' = \sin(90° + \alpha) = \cos \alpha$.

127. Cor. 5.—*When the origin remains the same, and only the direction of the axis is changed,* $m = 0$, *and* $n = 0$, *and we have,*

To pass from one oblique set to another,

$$x = \frac{x_2 \sin(\beta - \alpha) + y_2 \sin(\beta - \alpha')}{\sin \beta}, \quad (1_1)$$

$$y = \frac{x_2 \sin \alpha + y_2 \sin \alpha'}{\sin \beta}; \quad (2_1)$$

To pass from rectangular to oblique axes,

$$x = x_2 \cos \alpha + y_2 \cos \alpha', \quad (1_2)$$

$$y = x_2 \sin \alpha + y_2 \sin \alpha'; \quad (2_2)$$

To pass from one rectangular set to another,

$$x = x_2 \cos \alpha - y_2 \sin \alpha, \quad (1_3)$$

$$y = x_2 \sin \alpha + y_2 \cos \alpha; \quad (2_3)$$

To pass from oblique to rectangular axes,

$$x = \frac{x_2 \sin(\beta - \alpha) - y_2 \cos(\beta - \alpha)}{\sin \beta}, \quad (1_4)$$

$$y = \frac{x_2 \sin \alpha + y_2 \cos \alpha}{\sin \beta} \quad (2_4)$$

SCH.—In (1_4), (2_4) the angles involved are those which the *new* or rectangular axes make with the *primitive* or oblique axis of x. It is sometimes convenient to have *formulæ* for passing from oblique to rectangular axes, in which the angles involved shall be those which the *primitive* or oblique axes make with the *new* or rectangular axis of x. Such *formulæ* may be deduced from the general *formulæ*, but are more readily obtained from (1_2), (2_2), as follows:

Multiplying (1_2) by $\sin\alpha'$, $x\sin\alpha' = x_2\sin\alpha'\cos\alpha + y_2\cos\alpha'\sin\alpha'$,
Multiplying (2_2) by $\cos\alpha'$, $y\cos\alpha' = x_2\cos\alpha'\sin\alpha + y_2\cos\alpha'\sin\alpha'$;
Subtracting, $x\sin\alpha' - y\cos\alpha' = x_2(\sin\alpha'\cos\alpha - \cos\alpha'\sin\alpha)$;
Whence since $\sin\alpha'\cos\alpha - \cos\alpha'\sin\alpha = \sin(\alpha' - \alpha)$ we have

$$x_2 = \frac{x\sin\alpha' - y\cos\alpha'}{\sin(\alpha' - \alpha)}. \qquad (1_5)$$

In like manner eliminating x_2, we have

$$y_2 = \frac{y\cos\alpha - x\sin\alpha}{\sin(\alpha' - \alpha)}. \qquad (2_5)$$

[NOTE.—It will afford the student an excellent geometrical exercise to produce each of the above sets of *formulæ* directly from a figure. The forms in corollaries 1 and 2 are the most important, and should be fixed in memory.]

Ex. 1. Assuming $3x + 5y = 15$ to be the equation of a right line referred to rectangular axes, find the equation of the same line referred to parallel axes whose origin is at (1, 2).

The equation is $3x_2 + 5y_2 = 2$.

Ex. 2. Assuming $2x + 3y = 6$ to be the equation of a right line referred to rectangular axes, find the equation of the same line referred to parallel axes whose origin is at (1, —2). Also to new parallel axes whose origin is at (—3, —4). Also to new parallel axes whose origin is (—6, 6). Construct the given equation and verify the new equations by constructing, on the same figure, the new axes, and observing the position of the line as referred to them. Notice where the line cuts the several axes.

The equations are, 1st, $2x_2 + 3y_2 = 10$; 2nd, $2x_2 + 3y_2 = 24$; and 3rd, $2x_2 + 3y_2 = 0$.

QUERY.—Why do not the coefficients of x and y change in the above transformations?

Ex. 3. Construct the locus $x^2 - 8x + y^2 + 6y = 0$, upon rectangular axes. Then transform the equation by passing to new parallel axes whose origin is (4, —3), and, drawing these axes on the same figure, construct the new equation ($x_2^2 + y_2^2 = 25$) with reference to these axes, observing that the two equations give the same locus.

Ex. 4. Given $4y^2 + 9x^2 = 36$ as the equation of a locus referred to rectangular axes, to transform to new axes with the same origin, the tangents of the angles which the new axes of x and y make with the primitive axis of x, being 3 and —3, respectively. Verify by a construction.

SUG'S.—The *formulæ* are $x = x_2 \cos\alpha + y_2 \cos\alpha'$, and $y = x_2 \sin\alpha + y_2 \sin\alpha'$. In this case $\tan\alpha = 3$, whence $\sin\alpha = \sqrt{\frac{9}{10}}$, and $\cos\alpha = \sqrt{\frac{1}{10}}$. Also $\tan\alpha' = -3$, whence $\sin\alpha' = \sqrt{\frac{9}{10}}$, and $\cos\alpha' = -\sqrt{\frac{1}{10}}$. Introducing these values, the *formulæ* become $x = \sqrt{\frac{1}{10}}(x_2 - y_2)$, and $y = \sqrt{\frac{9}{10}}(x_2 + y_2)$.

The transformed equation is $5x_2^2 + 6x_2y_2 + 5y_2^2 = 40$.

Ex. 5. Given $xy = 16$ as the equation of a locus referred to rectangular axes, to pass to new rectangular axes with the same origin, the new axis of x making an angle of 45° with the primitive axis of x. *Equation*, $x_2^2 - y_2^2 = 32$.

Ex. 6. By the same transformation as in *Ex.* 5, show that $y^4 + x^4 + 6x^2y^2 = 2$, becomes $x_2^4 + y_2^4 = 1$. Construct the locus, and both sets of axes, and observe the position of the locus with respect to the two sets of axes.

Ex. 7. Given the equation $y = ax + b$ (the *common* equation of the straight line), to pass to oblique axes with the same origin.

SUG'S.—The *formulæ* for transforming are $x = x_2 \cos\alpha + y_2 \cos\alpha'$, and $y = x_2 \sin\alpha + y_2 \sin\alpha'$. Substituting and reducing, we have

$$y_2 = \frac{a\cos\alpha - \sin\alpha}{\sin\alpha - a\cos\alpha'} x_2 + \frac{b}{\sin\alpha' - a\cos\alpha'},$$

which is the equation of a right line referred to oblique axes making *any* angles (α, α') with the primitive axis of x. Now, if we desire simply the *form* of the equation of a right line referred to oblique axes, we may consider the new axis of x as coinciding with the primitive, *Fig.* 90, and let the new axis of y make *any* angle, as β, with this. Then $\alpha = 0$, and $\alpha' = \beta$; whence the equation becomes

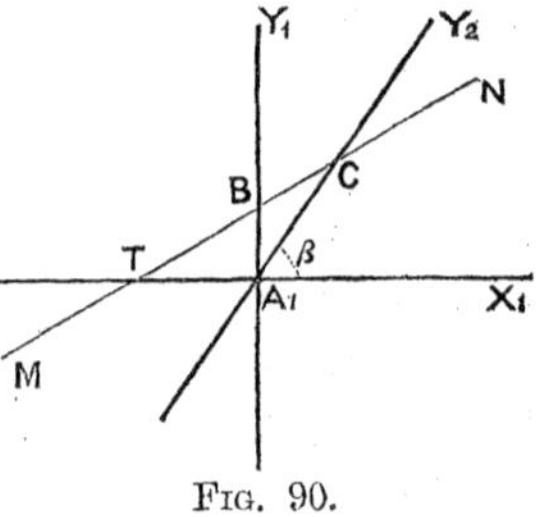

FIG. 90.

$$y_2 = \frac{a}{\sin\beta - a\cos\beta} x_2 + \frac{b}{\sin\beta - a\cos\beta}.$$

Again, letting the angle $NTA_1 = \alpha_1$, $a = \frac{\sin\alpha_1}{\cos\alpha_1}$; whence $\frac{a}{\sin\beta - a\cos\beta} =$

$$\frac{\frac{\sin\alpha_1}{\cos\alpha_1}}{\sin\beta - \frac{\sin\alpha_1}{\cos\alpha_1}\cos\beta} = \frac{\sin\alpha_1}{\sin\beta\cos\alpha_1 - \cos\beta\sin\alpha_1} = \frac{\sin\alpha_1}{\sin(\beta - \alpha_1)}.$$

Also, $\frac{b}{\sin\beta - a\cos\beta} = \frac{b\cos\alpha_1}{\sin(\beta - \alpha_1)} = A_1C$, since $A_1C : A_1B :: \sin CBA_1 : \sin BCA_1$, or $A_1C : b :: \sin(90° + \alpha_1) : \sin(\beta - \alpha_1)$; whence $A_1C = \frac{b\sin(90° + \alpha_1)}{\sin(\beta - \alpha_1)} = \frac{b\cos\alpha_1}{\sin(\beta - \alpha_1)}$. Substituting, we have $y_2 = \frac{\sin\alpha_1}{\sin(\beta - \alpha_1)} x_2 + b'$, calling $A_1C = b'$. (See **34.**)

Ex. 8. Given the equation $y = ax + b$, to find the position of a new set of axes parallel to the primitive, to which, when the locus is referred, its equation shall have no absolute term.

SUG'S.—Substituting for y, $y_2 + n$, and for x, $x_2 + m$, we have $y_2 = ax_2 + am + b - n$. Now, if the new axes can be so situated that $am + b - n = 0$, or $n = am + b$, the condition required will be fulfilled. But n and m are the co-ordinates of the new origin, and the condition $n = am + b$ (n and m being co-ordinates) designates a point in the line $y = ax + b$. Hence the new origin is to be in the line $y = ax + b$. (See **33,** SCH. 1.)

Ex. 9. Transform $A^2y^2 + B^2x^2 = A^2B^2$, to parallel axes with the new origin at $(-A, 0)$. Also to parallel axes with the origin at $(A, 0)$. Also to parallel axes with the origin at $(-m, -n)$. (See **55.**)

Ex. 10. Transform $A^2y^2 + B^2x^2 = A^2B^2$, to oblique axes with the same origin, such that $\tan\alpha \tan\alpha' = -\frac{B^2}{A^2}$, and obtain the result in terms of the diameters lying on the new axes.

SUG'S.—After making the substitutions and collecting terms, we have $(A^2\sin^2\alpha' + B^2\cos^2\alpha')y_2^2 + (A^2\sin^2\alpha + B^2\cos^2\alpha)x_2^2 + 2(A^2\sin\alpha\sin\alpha' + B^2\cos\alpha\cos\alpha')x_2y_2 = A^2B^2$, (1), which is the general equation of an ellipse referred to oblique axes, the origin being at the centre. If these axes are so situated that $\tan\alpha \tan\alpha' = \frac{\sin\alpha\sin\alpha'}{\cos\alpha\cos\alpha'} = -\frac{B^2}{A^2}$, $A^2\sin\alpha\sin\alpha' + B^2\cos\alpha\cos\alpha' = 0$, and the term in x_2y_2 disappears, and the equation is $(A^2\sin^2\alpha' + B^2\cos^2\alpha')y_2^2 + (A^2\sin^2\alpha + B^2\cos^2\alpha)x_2^2 = A^2B^2$. This is the equation of the ellipse referred to the axes required, but it is *not in the terms required;* it is the equation of the ellipse, as BPC, referred to the diameters A_1B_2, A_1D_2, but is in terms of the semi-axes A_1B, A_1G, and the angles α, α', which the new axes make with the primitive axis of x. Thus, P being any point in the curve, A_1E represents x_2, and PE y_2. Now in this equation, when $y_2 = 0$, x_2 becomes A_1B_2. Hence calling A_1B_2 A_1, we have $A_1^2 = \frac{A^2B^2}{A^2\sin^2\alpha + B^2\cos^2\alpha}$,

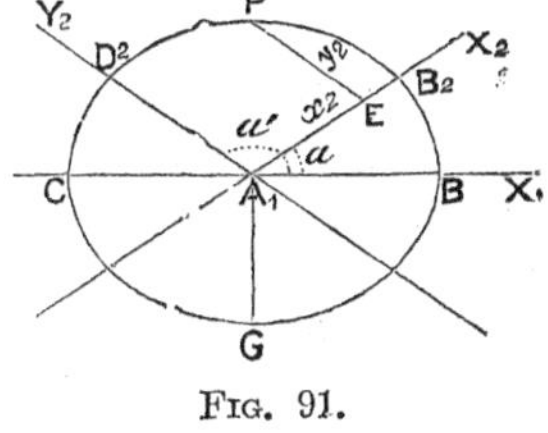

FIG. 91.

or $A^2\sin^2\alpha + B^2\cos^2\alpha = \frac{A^2B^2}{A_1^2}$. In like manner for $x_2 = 0$, y_2 becomes A_1D_2. Hence calling A_1D_2 B_1, we have $B_1^2 = \frac{A^2B^2}{A^2\sin^2\alpha' + B^2\cos^2\alpha'}$, or $A^2\sin^2\alpha' + B^2\cos^2\alpha' = \frac{A^2B^2}{B_1^2}$. Substituting these values of the coefficients of y_2^2, and x_2^2, we have $\frac{A^2B^2}{B_1^2}y_2^2 + \frac{A^2B^2}{A_1^2}x_2^2 = A^2B^2$. Finally, dividing by A^2B^2, and clearing of fractions, $A_1^2y_2^2 + B_1^2x_2^2 = A_1^2B_1^2$. Q. E. D.

SCH.—Diameters so situated as to make $\tan\alpha\tan\alpha' = -\frac{B^2}{A^2}$ are *Conjugate Diameters*, as will appear hereafter. Hence the equation of the ellipse referred to conjugate diameters, and in terms of those diameters, is of the same form as the equation of the curve referred to, and in terms of, its axes.

Ex. 11. Transform $A^2y^2 - B^2x^2 = -A^2B^2$ to an oblique system with the same origin, such that $\tan\alpha\tan\alpha' = \frac{B^2}{A^2}$, and obtain the result in terms of the diameters lying on those axes.

SUG'S.—The student is expected to recognize this as the equation of the hyperbola referred to its own axes. The transformation is in all respects like the above, except that the diameter represented by B_1 is *imaginary, i. e.*, does not meet the real branches of the curve, hence we call $\mathbf{A_1D_2}$, $B_1\sqrt{-1}$, or $(\mathbf{A_1D_2})^2 = -B_1^2$. The equation sought is $A_1^2y_2^2 - B_1^2x_2^2 = -A_1^2B_1^2$.

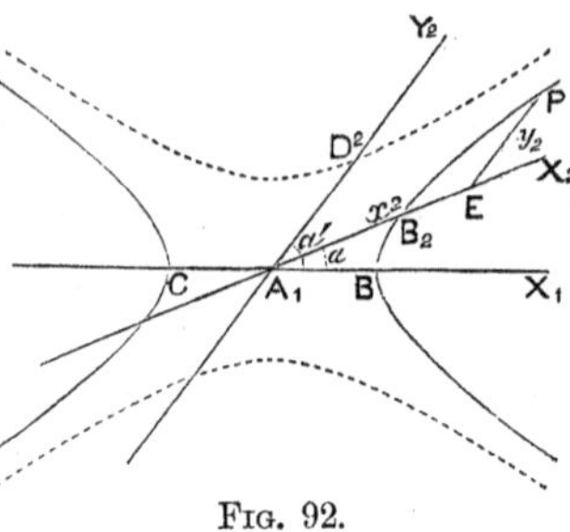

FIG. 92.

SCH.—In each of the two preceding examples, there were given, the equation of the locus, and the position of the new axes, from which to find the form of the new equation. The converse of this problem is important; *i. e.*, Given the equation of the locus, and some specified form of its equation, *to find the position of the new axes.* Thus, for example,—The origin remaining the same, what must be the position of oblique axes, to which, when the ellipse is referred, its equation will take the same form as the common equation. To solve this problem, we first transform $A^2y^2 + B^2x^2 = A^2B^2$, to oblique axes, as in *Ex.* 10, and obtain the form (1) in the suggestions. It then remains to determine what values α and α' must have, *i. e.*, how the new axes must be inclined, to make the equation take the primitive form. Now, the required form has no term in x_2y_2, hence the coefficient of x_2y_2 must be 0, that is, $A^2\sin\alpha\sin\alpha' + B^2\cos\alpha\cos\alpha' = 0$. From this, $\frac{\sin\alpha\sin\alpha'}{\cos\alpha\cos\alpha'} = \tan\alpha\tan\alpha' = -\frac{B^2}{A^2}$; whence we learn that the new axes must be so situated that the rectangle of the tangents of the angles which they make with the primitive axis of x shall be $-\frac{B^2}{A^2}$, in other words they must be *conjugate diameters.* Putting the resulting equation in the form $\frac{A^2\sin^2\alpha' + B^2\cos^2\alpha'}{A^2B^2}y_2^2 + \frac{A^2\sin^2\alpha + B^2\cos^2\alpha}{A^2B^2}x_2^2 = 1$, and making $x_2 = 0$, and $y^2 = 0$, successively, we find that the squares of the new semi-diameters are $\frac{A^2B^2}{A^2\sin^2\alpha' + B^2\cos^2\alpha'}$, and $\frac{A^2B^2}{A^2\sin^2\alpha + B^2\cos^2\alpha}$. This equation and these values refer to any pair of conjugate diameters, as will appear hereafter.

Ex. 12. Find the position of oblique axes to which when $y^2 = 2px$ is referred, the equation will still have the same form.

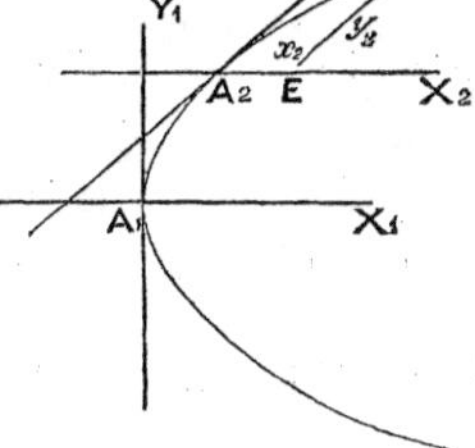

Fig. 93.

Sug's.—Passing to oblique axes in general, the equation becomes

$$(n + x_2 \sin \alpha + y_2 \sin \alpha')^2 = 2p(m + x_2 \cos \alpha + y_2 \cos \alpha');$$

or, expanding and collecting terms

$$\sin^2\alpha' y_2{}^2 + 2\sin \alpha \sin \alpha' x_2 y_2 + \sin^2 \alpha\, x_2{}^2 + \left.\begin{matrix} 2n \sin \alpha' \\ -2p \cos \alpha' \end{matrix}\right| y_2 + \left.\begin{matrix} 2n \sin \alpha \\ -2p \cos \alpha \end{matrix}\right| x_2 + n^2 - 2pm = 0.$$

Now, in order to meet the conditions of the problem,

1st, There must be no absolute term ; hence $n^2 - 2pm = 0$, (1) ;

2nd, There must be no term in y_2 ; hence $2n \sin \alpha' - 2p \cos \alpha' = 0$, (2) ;

3rd, There must be no term in $x_2 y_2$; hence $2 \sin \alpha \sin \alpha' = 0$, (3) ;

4th, There must be no term in $x_2{}^2$; hence $\sin^2 \alpha = 0$, (4).

If these conditions can be fulfilled, and we can discover the position of the new axes which fulfills them, the problem is solved. We observe that as there are but *four* conditions, involving *four* arbitrary constants, α, α', m, and n, these conditions *can* be fulfilled. The first condition, $n^2 - 2pm = 0$, or $n^2 = 2pm$, requires the new origin to be on the curve, since n and m bear the relation of co-ordinates to the curve. The second condition, $2n \sin \alpha' - 2p \cos \alpha' = 0$, or $\dfrac{\sin \alpha'}{\cos \alpha'} = \tan \alpha' = \dfrac{p}{n}$, requires that the new axis of y make an angle with the axis of the curve, whose tangent is p divided by the ordinate (n) of the new origin, which is on the curve. (This makes the new axis of y a tangent to the curve, as will afterwards appear.) The third condition, $2\sin \alpha \sin \alpha' = 0$, gives $\sin \alpha = 0$, since $\sin \alpha'$ is not 0. This requires the new axis of x to be parallel to the primitive axis of x (the axis of the curve), and makes it a diameter. The fourth condition $\sin^2 \alpha = 0$, is fulfilled by the last, and hence requires no further attention. If, therefore, the curve is referred to any diameter, as A_2X_2, and a tangent A_2Y_2 at its vertex, the equation becomes $\sin^2 \alpha' y_2{}^2 = (2p \cos \alpha - 2n \sin \alpha) x_2$; or since $\sin \alpha = 0$, and $\cos \alpha = 1$, $y_2{}^2 = \dfrac{2p}{\sin^2 \alpha'} x_2$, which is the form required. Putting $\dfrac{2p}{\sin^2 \alpha'} = 2p_2$, we have $y_2{}^2 = 2p_2 x_2$.

Sch.—The equation $y_2{}^2 = \dfrac{2p}{\sin^2 \alpha'} x_2$ leaves the problem indeterminate, inasmuch as α' is a function of y_2 ; hence the new origin may be *anywhere* on the curve. In reality the problem furnished four arbitrary constants and required but three conditions (the third and fourth being but one) ; hence we may impose another ; that is, we may put the origin *where we please on the curve.*

Ex. 13. To transform $A^2y^2 - B^2x^2 = -A^2B^2$ so that the hyperbola shall be referred to its asymptotes, *i. e.*, to the produced diagonals of the rectangle drawn on the axes of the curve.

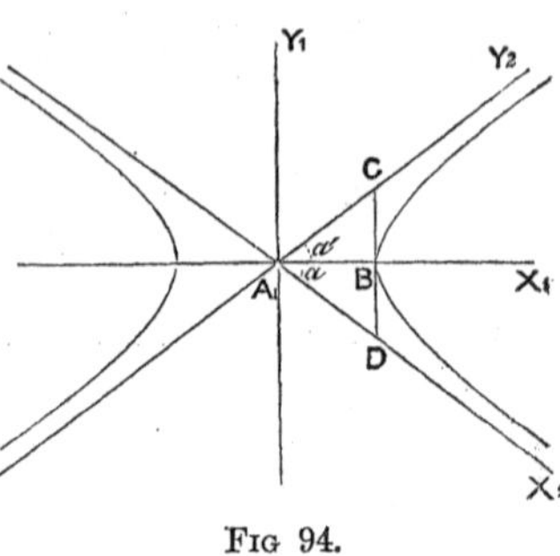

FIG 94.

SUG'S.—Let A_1X_1 and A_1Y_1 be the primitive axes, and the asymptotes A_1X_2, A_1Y_2 be the new axes. As usual, let $X_1A_1X_2 = -\alpha$, and $X_1A_1Y_2 = \alpha'$. Then, since $CB = BD = B$, $A_1B = A$, $A_1C = A_1D = \sqrt{A^2+B^2}$, $\sin\alpha = -\frac{B}{\sqrt{A^2+B^2}}$, $\cos\alpha = \frac{A}{\sqrt{A^2+B^2}}$, $\sin\alpha' = \frac{B}{\sqrt{A^2+B^2}}$, and $\cos\alpha' = \frac{A}{\sqrt{A^2+B^2}}$, the — sign being given to the value of $\sin\alpha$ since α is reckoned around the angular point A_1 from left to right, and $\sin\alpha$ is the sine of a negative arc less than 90°. Putting these values in the *formulæ* for passing from rectangular to oblique axes, we have $x = (x_2 + y_2)\frac{A}{\sqrt{A^2+B^2}}$, and $y = (y_2 - x_2)\frac{B}{\sqrt{A^2+B^2}}$. Now, substituting these values of x and y in the equation to be transformed, there results $\frac{(y_2 - x_2)^2A^2B^2 - (x_2+y_2)^2A^2B^2}{A^2+B^2} = -A^2B^2$; whence expanding and reducing $x_2y_2 = \frac{A^2+B^2}{4}$. Since $\frac{A^2+B^2}{4}$ is constant, we may represent it by c, and write the equation $x_2y_2 = c$. In the case of the equilateral hyperbola $c = \frac{A^2}{2}$, A being the semi-axis.

Ex. 14. To find a system of oblique axes with the origin at the centre, to which when the hyperbola is referred, its equation will take the form $xy = c$.

SUG'S.—The common equation, $A^2y^2 - B^2x^2 = -A^2B^2$, becomes, when we pass to general oblique axes with the same origin

$$\left.\begin{matrix}A^2\sin^2\alpha' \\ -B^2\cos^2\alpha'\end{matrix}\right| y_2^2 + \left.\begin{matrix}2A^2\sin\alpha\sin\alpha' \\ -2B^2\cos\alpha\cos\alpha'\end{matrix}\right| x_2y_2 + \left.\begin{matrix}A^2\sin^2\alpha \\ -B^2\cos^2\alpha\end{matrix}\right| x_2^2 = -A^2B^2.$$

The conditions imposed are, $A^2\sin^2\alpha' - B^2\cos^2\alpha' = 0$, and $A^2\sin^2\alpha - B^2\cos^2\alpha = 0$; whence $\tan\alpha = \frac{\pm B}{A}$, and $\tan\alpha' = \frac{\pm B}{A}$. Now, in order that these values should indicate the positions of different lines, they must be taken with different signs. Thus the new axes are found to make angles with the primitive axis of x whose tangents are $\frac{-B}{A}$, and $\frac{B}{A}$, which relation characterizes asymptotes. The equation then reduces to

$$(2A^2\sin\alpha\sin\alpha' - 2B^2\cos\alpha\cos\alpha')x_2y_2 = -A^2B^2.$$

But the conditions above give $\sin\alpha = \frac{-B}{\sqrt{A^2+B^2}}$, $\cos\alpha = \frac{A}{\sqrt{A^2+B^2}}$, $\sin\alpha' = \frac{B}{\sqrt{A^2+B^2}}$, and $\cos\alpha' = \frac{A}{\sqrt{A^2+B^2}}$. These values substituted in the last equation, give, after reduction, $x_2y_2 = \frac{A^2+B^2}{4}$, which is the same form as found before.

Ex. 15. Letting $x^2 - 6xy + y^2 - 6x + 2y + 5 = 0$ represent a locus referred to rectangular axes, required the equation when the reference is to a new set of rectangular axes with the origin at $(0, -1)$, and the new axis of x makes an angle of $-45°$, or $135°$, with the primitive axis of x.

SUG'S.—The *formulæ* for transformation become, in this case, $x = \sqrt{\tfrac{1}{2}}(x_2 + y_2)$, and $y = \sqrt{\tfrac{1}{2}}(y_2 - x_2) - 1$. The transformed equation is $y_2^2 - 2x_2^2 = 2$ (See *Fig.* 88, and the illustration accompanying it.)

SECTION II.

Methods of Passing from Rectilinear to Polar Co-ordinates, and vice versa.

128. *Prob.*—*To produce the formulæ for passing from a Rectangular to a Polar system of co-ordinates.*

SOLUTION.—Let **P** be any point in a locus **MN** referred to the rectangular axes A_1X_1, A_1Y_1, the co-ordinates of **P** being $A_1D = x$, and $PD = y$, when referred to these axes. Let the pole of the new, or polar system, be A_2, whose co-ordinates are m and n; and let A_2X_2, or A_2X_2', be the polar axis making an angle α with **AX**, or what is the same thing, with A_2K parallel to A_1X_1. Let the polar co-ordinates of **P** be $A_2P = r$, and PA_2X_2, or $PA_2X_2' = \theta$. The angle PA_2K will be $\theta + \alpha$ when the polar axis lies above A_2K, and $\theta - \alpha$, when the polar axis lies below; hence, in general, $PA_2K = \theta \pm \alpha$. Now $x = A_1D = A_1B + A_2H$. But, from the triangle PA_2H, $A_2H = r \cos PA_2K = r \cos(\theta \pm \alpha)$. $\therefore$

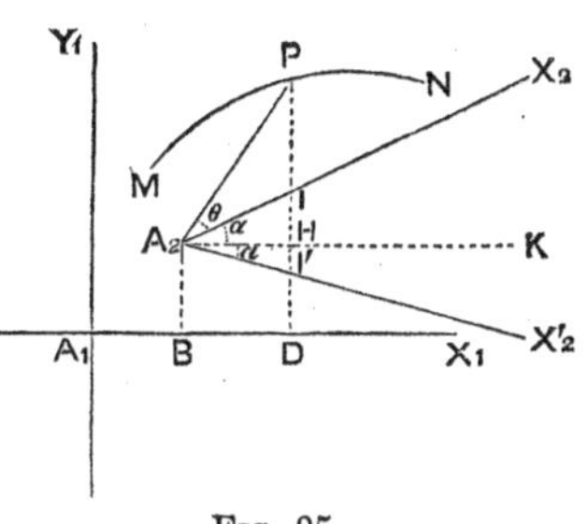

FIG. 95.

$$x = m + r\cos(\theta \pm \alpha), \quad (1).$$

In like manner $y = n + r\sin(\theta \pm \alpha)$, (2), as $y = PD = A_2B + PH$, and $PH = r\sin(\theta \pm \alpha)$.

If the pole is at the primitive origin, $m = 0$, $n = 0$, and

$$x = r\cos(\theta \pm \alpha), \quad (1_1);$$

and

$$y = r\sin(\theta \pm \alpha), \quad (2_1).$$

If the polar axis is parallel to the primitive axis of x, $\alpha = 0$, and the *formulæ* become

$$x = m + r\cos\theta, \quad (1_2);$$

$$y = n + r\sin\theta, \quad (2_2);$$ or, if the pole is at the primitive origin,

$$x = r\cos\theta, \quad (1_3);$$

$$y = r\sin\theta, \quad (2_3).$$

129. Prob.—*To produce the formulæ for passing from a Polar to a Rectangular system of co-ordinates.*

SOLUTION.—From *Fig.* 95 we have $\mathsf{PA_2} = \sqrt{\overline{\mathsf{PH}}^2 + \overline{\mathsf{A_2H}}^2}$, or $r = \sqrt{(y-n)^2 + (x-m)^2}$. From the same triangle we have also $\cos(\theta \pm \alpha) = \frac{x-m}{\sqrt{(y-n)^2+(x-m)^2}}$, and $\sin(\theta \pm \alpha) = \frac{y-n}{\sqrt{(y-n)^2+(x-m)^2}}$, which are the *formulæ* sought.

When the polar axis is parallel to the primitive axis of x, the *formulæ* are $r = \sqrt{(y-n)^2 + (x-m)^2}$, $\cos\theta = \frac{x-m}{\sqrt{(y-n)^2+(x-m)^2}}$, and $\sin\theta = \frac{y-n}{\sqrt{(y-n)^2+(x-m)^2}}$. If at the same time the origin and pole coincide, the *formulæ* are $r = \sqrt{y^2+x^2}$, $\cos\theta = \frac{x}{\sqrt{y^2+x^2}}$, and $\sin\theta = \frac{y}{\sqrt{y^2+x^2}}$.

Ex. 1. Transform $x^2 + y^2 = 5ax$ to polar co-ordinates, the pole being at the origin, and the polar axis coincident with the axis of x.

The equation is $r = 5a\cos\theta$.

SUG.—The *formulæ* are $x = r\cos\theta$, and $y = r\sin\theta$.

Ex. 2. Transform $(x^2+y^2)^2 = a^2(x^2 - y^2)$ to polar co-ordinates, the pole being at the origin, and the polar axis coincident with the axis of x.

The equation is $r^2 = a^2(\cos^2\theta - \sin^2\theta) = a^2\cos 2\theta$.

Ex. 3. Transform $r^2 = a^2\cos 2\theta$ to $(x^2+y^2)^2 = a^2(x^2 - y^2)$. (See last example.)

SUG.—First put the equation in the form $r^2 = a^2(\cos^2\theta - \sin^2\theta)$.

Ex. 4. Under the same conditions as above transform $r^2\cos 2\theta = a^2$ to $x^2 - y^2 = a^2$. Also $xy = a^2$, to $r^2\sin 2\theta = 2a^2$. Also $x^2 + y^2 = (2a - x)^2$, to $r^{\frac{1}{2}}\cos\frac{1}{2}\theta = a^{\frac{1}{2}}$. Also reverse these processes.

Ex. 5. To deduce from $A^2y^2 + B^2x^2 = A^2B^2$, the polar equation of the ellipse, in terms of the transverse axis and eccentricity, the pole being at the left hand focus and the polar axis falling on the axis of the curve.

SUG'S.—The given equation being put in the form $y^2 = (1 - e^2)(A^2 - x^2)$, and the *formulæ* for transformation in the form $x = r\cos\theta - Ae$, and $y = r\sin\theta$, and the substitutions made, we have

$$r^2\sin^2\theta = (1-e^2)(A^2 - r^2\cos^2\theta + 2Aer\cos\theta - A^2e^2).$$

Expanding this and reducing to a known form,

$$r^2 - \frac{2Ae\cos\theta(1-e^2)}{1-e^2\cos^2\theta}r = \frac{A^2(1-e^2)^2}{1-e^2\cos^2\theta};$$ hence

$$r = \frac{Ae\cos\theta(1 - e^2) \pm \sqrt{A^2(1 - e^2)^2(1 - e^2\cos^2\theta) + A^2e^2\cos^2\theta(1 - e^2)^2}}{1 - e^2\cos^2\theta} =$$

$$\frac{Ae\cos\theta(1 - e^2) \pm \sqrt{A^2(1 - e^2)^2}}{1 - e^2\cos^2\theta} = \frac{Ae\cos\theta(1 - e^2) \pm A(1 - e^2)}{1 - e^2\cos^2\theta} =$$

$\dfrac{A(e\cos\theta \pm 1)(1 - e^2)}{1 - e^2\cos^2\theta}$. Now as neither e nor $\cos\theta$ can exceed 1, and as each is generally less than 1, r is positive only for $e\cos\theta + 1$; hence we may reject the $-$ sign in this factor and write $r = \dfrac{A(1 + e\cos\theta)(1 - e^2)}{1 - e^2\cos^2\theta} = \dfrac{A(1 - e^2)}{1 - e\cos\theta}$.

If the pole is taken at the right hand focus, $x = r\cos\theta + Ae$, and $r = \dfrac{A(1 - e^2)}{1 + e\cos\theta}$. (See ***107—109.***)

[NOTE.—There are expedients by which the algebraic reductions in this solution may be simplified; but as our purpose is to exhibit simply the process of transformation, we do not think best to avoid the work by indirect means. Were the object merely to obtain the polar equation, the process of (***107—109***) would be much more simple and elegant.]

Ex. 6. Deduce the polar equation of the parabola, $r = \dfrac{p}{1 - \cos\theta}$, from $y^2 = 2px$. (See ***108.***)

Ex. 7. Transform the equation of the cissoid, $y^2 = \dfrac{x^3}{2a - x}$, to the polar equation. $r = \dfrac{2a(1 - \cos^2\theta)}{\cos\theta}$, or $r = 2a\sin\theta\tan\theta$. (See ***110.***)

GENERAL SCHOLIUM.

The student being now familiar with equations as representatives of loci, is prepared to use them as instruments for the investigation of the properties of their loci. But, in carrying forward this work, the *Calculus* renders great assistance, and for many purposes is indispensable. Therefore before commencing the next chapter, the student must become familiar with the processes of differentiating the various kinds of explicit and implicit functions of a single variable, with successive differentiation, partial differentiation, the development of functions by Maclaurin's and Taylor's theorems, the evaluation of indeterminate forms, and the theory of Maxima and Minima, as treated in the first two chapters of the second part of this volume. Having read those chapters, he can turn back and resume his study of Geometry at this point. Students who do not choose to study the Calculus, may complete their course in this subject by reading Sections XIV. and XV. of the next chapter in this part.

CHAPTER IV.

PROPERTIES OF PLANE LOCI INVESTIGATED BY MEANS OF THE EQUATIONS OF THOSE LOCI.

SECTION I.

Tangents to Plane Loci.

(*a*) BY RECTILINEAR CO-ORDINATES.

130. Def.—***Consecutive Points*** on a line are points nearer to each other than any assignable distance.

Ill.—As we shall have frequent occasion to use this conception, it is of the utmost importance that the definition be clearly comprehended. For example, when we speak of P and P′, *Fig.* 96, as consecutive points, we do not conceive them as absolutely in juxtaposition, *i. e.*, so near each other that there can be no other point between them ; but we mean simply that *we are to reason upon them* as nearer each other than any assignable distance. In short, PP′ is merely to be considered *infinitesimal* in the sense of being less than any assignable distance. So, also, when we speak of PD and P′D′ as consecutive ordinates, we mean that DD′ is *to be treated in the argument* as less than any assignable distance—as infinitesimal.

131. Def.—***A Tangent*** (rectilinear) is a right line passing through two consecutive points of a curve.

Sch.—For many purposes, it is more convenient to speak of the two consecutive points through which a tangent passes, as *one* point, and call it *the point of tangency:* this, of course, is necessarily the case when the expression is in finite terms, as the distinguishing of consecutive points requires infinitesimals. The student has become familiar, in Elementary Geometry, with the conception of a curve as a polygon of an infinite number of infinitesimal sides : the prolongation of one of these sides may be considered a tangent.

132. Cor.—*A Tangent has the same direction as the curve at the point of tangency.*

---

133. *Prop.*—*The first differential coefficient of the ordinate of a curve regarded as a function of the abscissa* $\left(\frac{dy}{dx}\right)$ *is the tangent of the angle which a tangent to the curve makes with the axis of abscissas.*

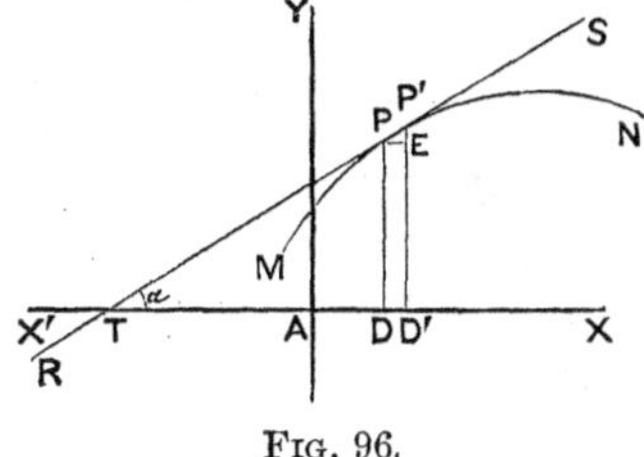

Fig. 96.

Dem.—Let **MN**, *Fig.* 96, be any plane curve whose equation is $y = f(x)$. Let **P** and **P'** be consecutive points, and **PD** and **P'D'** consecutive ordinates. Then is **RS**, drawn through **P** and **P'**, a tangent. Draw **PE** parallel to **X'X**. Since **P** and **P'** are consecutive points, **DD'** and **PE** are contemporaneous infinitesimal increments of x and y, respectively; *i. e.*, **DD'** represents dx, and **P'E** represents dy. Now, letting **STX** be represented by α, we have $\tan\alpha = \tan \mathbf{P'PE} = \frac{\mathbf{P'E}}{\mathbf{PE}} = \frac{dy}{dx}$. Q. E. D.

Ex. 1. What angle does a tangent to the curve $y = x^3 - x^2 + 1$, at the point $x = 2$, make with the axis of x?

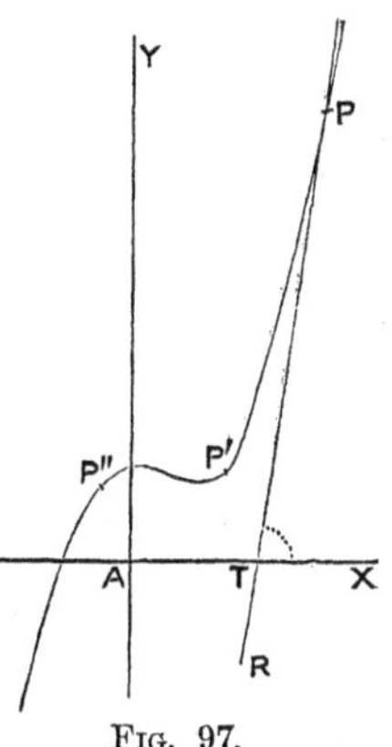

Fig. 97.

Solution. $\frac{dy}{dx} = 3x^2 - 2x$, which, for $x = 2$ becomes $\frac{dy'}{dx'} = 8$, using (x', y') to designate the particular point in distinction from the general point (x, y). ∴ The tangent at the point $x = 2$, makes an angle with the axis of x whose tangent is 8, *i. e.*, an angle of 82° 52′ 30″. The figure is that in the margin, in which **P** is the point of tangency, and **PTX** is the angle whose tangent is 8. ($\text{Tan}^{-1}8 =$ **PTX**.)

Ex. 2. At what point does the curve in the last example run in a direction making an angle of 45° with the axis of x? At what point does it make an angle of 135°? At what point is it perpendicular? At what point parallel?

Sug.—The direction of the tangent being the same as that of the curve, we have simply to find where $\frac{dy}{dx}$, or $3x^2 - 2x$ (which is its general value in this curve) equals 1, 0, —1, or ∞, as these are the tangents of the required angles. Thus for the first we have to find the value of x which satisfies $3x^2 - 2x = 1$. This gives $x = 1$ and $-\frac{1}{3}$. Now, for $x = 1$, $y = 1$ as is found by substituting in the equation of the curve. This point is **P'** in the figure. The curve also runs in the same direction at $(-\frac{1}{3}, \frac{23}{27})$, **P''** in the figure.

Answers, The curve is parallel at (0, 1), and $(\frac{2}{3}, \frac{23}{27})$. It makes an angle of 135° at $x = \frac{1 \pm \sqrt{-2}}{3}$, which being an imaginary point signifies that the curve in *this plane* does *not* make an angle of 135° with the axis of x. It is perpendicular at $x = +\infty$, and $-\infty$; *i. e.*, as the curve extends to the right or left it approaches perpendicularity with the axis of x, but becomes perpendicular only at an infinite distance from the origin.

134. SCH.—To determine at what point on a given curve a tangent must be drawn to make a given angle with the axis of x, —or, what is the same thing, to find a point at which a curve has a given direction,—put the value of $\frac{dy}{dx}$ as derived from the equation of the curve, equal to the tangent of the given angle or direction, and solve this equation in connection with the equation of the locus. To find where the curve is parallel to the axis of x, put the value of $\frac{dy}{dx}$ equal to 0, and solve as before. To find where it is perpendicular, put the value of $\frac{dy}{dx}$ equal to ∞, and solve in the same way.

Ex. 3. At what point on the curve $y^2 = 2x^3$, does a tangent make an angle with the axis of x, whose tangent is 3? At what point is the curve parallel to the axis of x? Where is it perpendicular? What is the direction of the curve at $x = 8$? Construct the figure and observe the agreement of results therewith.

Answers, At (2, 4); at (0, 0); nowhere; $\tan^{-1}(\pm 6)$. The last result indicates two tangents corresponding to $x = 8$, one drawn through (8, 32), and the other through (8, —32).

Ex. 4. At what point in the curve $y^2 = 2x + 3x^2$ must a tangent be drawn to make with the axis of x an angle whose tangent is $\frac{1}{2}$? 1? 2?

Answers. The first two are impossible. How does this appear in the solution, and how from the locus? The tangent of the angle is 2 at ($\frac{1}{3}$, 1), and at (—1, 1).

Ex. 5. At what point on $y = x^3 - 3x^2 - 24x + 85$ is the tangent parallel to the axis of abscissas?

Ans., At (4, 5), and at (—2, 113).

Ex. 6. Find in the curve $y = a + 2(x - b)^{\frac{2}{3}}$, the point at which a tangent is perpendicular to the axis of x. *Result*, At (b, a).

Ex. 7. Under what angle does $y = \frac{x}{1 + x^2}$ cut the axis of abscissas?

SUG'S.—As this curve cuts the axis of x at (0, 0), the question is, What is its direction at that point? Now $\frac{dy}{dx} = \frac{1 - x^2}{(1 + x^2)^2}$, which, for $x = 0$, becomes 1. $\therefore$ The curve cuts the axis of x at an angle of 45°.

Ex. 8. Show that the sinusoid cuts the axis of x alternately at 45° and 135°.

Ex. 9. What angle does the focal tangent of the common parabola make with the axis of x?

135. Cor.—*If the axes are oblique,* $\frac{dy}{dx}$ *signifies the ratio of the sine of the angle which the tangent makes with the axis of* x, *to the sine of the angle which it makes with the axis of* y, *i. e.,* $\frac{\sin \alpha}{\sin (\beta - \alpha)}$. (See ***34.***)

---

136. Prop.—*The general equation of a tangent to any plane curve is*

$$y - y' = \frac{dy'}{dx'}(x - x');$$

in which (x', y') *is the point of tangency, and* x *and* y *are the current co-ordinates of the tangent.*

Dem.—Let **MN**, *Fig.* 98, be any plane curve whose equation is $y = f(x)$, and let **P** be the point of tangency whose co-ordinates are x', y'. Now, the equation of *any* line passing through (x', y') is $y - y' = a(x - x')$ (***32***). But, in order that this line should be tangent to **MN** at **P**, the tangent of the angle **PTD**, which is represented by a in the formula, must be $\frac{dy'}{dx'}$. Hence, substituting, we have $y - y' = \frac{dy'}{dx'}(x - x')$. Q. E. D.

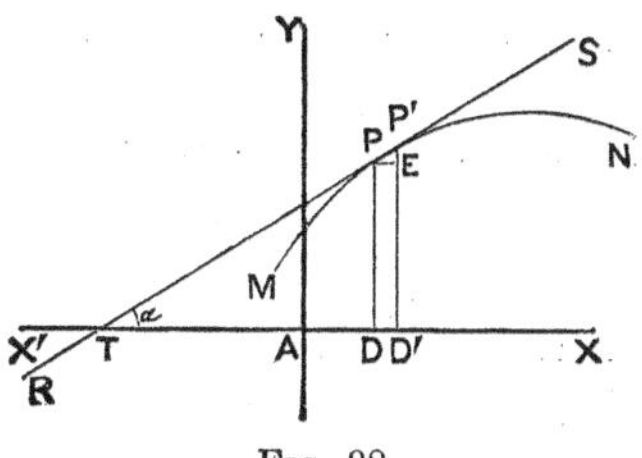

Fig. 98.

Ex. 1. What is the equation of a tangent to an ellipse referred to its axes?

Solution.—The equation of the locus is $A^2y^2 + B^2x^2 = A^2B^2$; whence $\frac{dy}{dx} = -\frac{B^2x}{A^2y}$, which satisfied for the point (x', y') is $-\frac{B^2x'}{A^2y'}$. Substituting this value of $\frac{dy'}{dx'}$ in the general equation of a tangent (***136***), we have $y - y' = -\frac{B^2x'}{A^2y'}(x - x')$. Reducing, this becomes $A^2yy' + B^2xx' = A^2y'^2 + B^2x'^2$. But as (x', y') is a point in the locus $A^2y'^2 + B^2x'^2 = A^2B^2$; hence, finally, $A^2yy' + B^2xx' = A^2B^2$.

Queries.—In $\frac{dy}{dx} = -\frac{B^2x}{A^2y}$, what are x and y co-ordinates of? In $y - y' = \frac{dy'}{dx'}(x - x')$, what are x and y co-ordinates of? What x' y'? Of what degree with respect to the variables is $A^2yy' + B^2xx' = A^2B^2$? Why should it be of this degree (***35***)? What are the variables? Notice that *for the same tangent*, x and y have all values, but x' and y' have fixed values: x', y' are *general, i. e.*, they represent any point in the locus, but they do not represent *all* points at the same time. If in our deductions from $y - y' = \frac{dy'}{dx'}(x - x')$, we had chanced to find $A^2y^2 + B^2x^2$ could we have substituted for it A^2B^2, as we did for $A^2y'^2 + B^2x'^2$? Why not?

Ex. 2. Produce the equation of a tangent to $3y^2+x^2=5$, at $x=1$, and construct, first, the tangent from its equation, and, second, the curve from its equation.

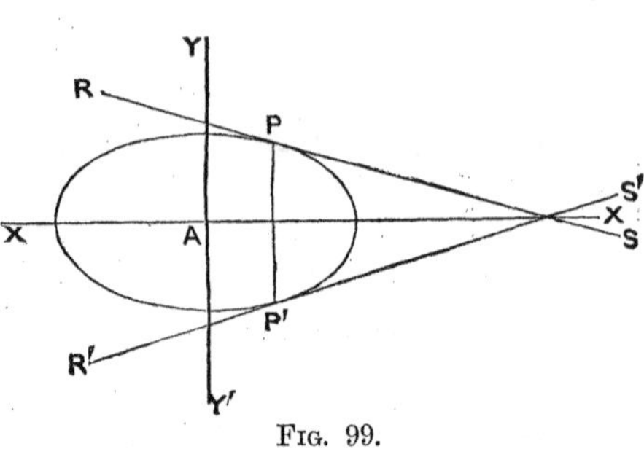

FIG. 99.

SOLUTION.—In this locus $\frac{dy}{dx} = -\frac{x}{3y}$. This is the general value of the tangent of the angle which a tangent to this ellipse makes with the axis of x. For the *particular* point $x=1$ (for which $y=\pm 1.155$, nearly), we have $\frac{dy'}{dx'}=\mp .29$ approximately. Substituting these values in the general equation of a tangent (***136***), we have $y \mp 1.155 = \mp .29(x-1)$, or $y=\mp .29x \pm 1.44$. There are, therefore, two tangents to this locus at $x=1$; one whose equation is $y=-.29x+1.44$, and another whose equation is $y=.29x-1.44$. **RS**, in the figure, represents the former; and **R′S′**, the latter.

Another solution of this example is obtained by observing that the locus $3y^2+x^2=5$ is an ellipse whose semi-axes are $A=\sqrt{5}$, and $B=\sqrt{\frac{5}{3}}$. But the equation of a tangent to an ellipse is $A^2yy' + B^2xx' = A^2B^2$; whence, substituting, we have $5yy' + \frac{5}{3}xx' = \frac{25}{3}$, or $3yy' + xx' = 5$, as the equation of *any* tangent to this ellipse. For the points $(1, \pm 1.155)$ this becomes $y=\mp .29x \pm 1.44$, as before.

Ex. 3. Deduce the equation of a tangent to an hyperbola. Also of a circle. Also of a parabola.

Results. The equation of a tangent to an hyperbola is $A^2yy' - B^2xx' = -A^2B^2$.
" " " " " a circle is $yy' + xx' = R^2$.
" " " " " a parabola is $yy' = p(x+x')$.

Ex. 4. What is the equation of a tangent to the parabola $y^2 = 9x$ at $x=4$? Construct the tangent from its equation, and then construct the parabola as in *Ex.* 2.

For (4, 6) the equation is $y = \frac{3}{4}x + 3$.
Is there another tangent for $x = 4$?

Ex. 5. Produce the equation of a tangent to $3y^2 - 2x^2 = 10$, at $x=4$. Is there more than one tangent? Construct the figure as above. *Equation*, $y = \pm .7127x \pm .8909$.

Ex. 6. What is the equation of a tangent to $y^2 = 4 - x^2$, at $x=3$? Why is the result imaginary?

Ex. 7. What is the equation of a tangent to $y^2 = \frac{x^3}{4 - x}$, at $x = 2$?

Ans., $y = 2x - 2$, and $y = -2x + 2$.

Ex. 8. Show that the equation of the tangent to the Napierian logarithmic curve ($x = \log y$) is $y = y'(x - x' + 1)$. Observe that the ordinate to this curve at any point, is the natural tangent of the angle which the tangent to the curve at that point makes with the axis of abscissas.

Ex. 9. What is the equation of a tangent to an hyperbola referred to its asymptotes ($xy = m$)? *Ans.*, $y = -\frac{y'}{x'}x + 2y'$.

Interpretation of the equation $y = -\frac{y'}{x'}x + 2y'$. If this represents the tangent to an hyperbola referred to rectangular asymptotes, $-\frac{y'}{x'}$, is the tangent of the angle which the tangent to the curve makes with the axis of x; and $2y'$ is the distance from the origin at which the tangent cuts the axis of y, as in all equations of right lines referred to rectangular axes. But in this case the hyperbola is equilateral, since $xy = m$ is the equation of an equilateral hyperbola when the asymptotes are rectangular; or, in other words, no hyperbola but an equilateral one has rectangular asymptotes.....To interpret $y = -\frac{y'}{x'}x + 2y'$ for oblique axes, we observe that $2y'$ is **AO**, *Fig.* 100; and by making $y = 0$ we find that the intercept on the axis of x, **AT**, is $2x'$. Now the coefficient of x, $-\frac{y'}{x'}$, is the ratio of the sine of the angle which the line (tangent) makes with the axis of x to the sine of the angle which it makes with the axis of y, by (***34***). This fact accords with the relations observable from the figure. Thus, in the triangle **AOT**, $\frac{AO}{AT} = \frac{\sin ATO}{\sin AOT}$, or $\frac{y'}{x'} = \frac{\sin ATO}{\sin AOT}$. The minus sign is explained by observing that the line **RS** lies across the 1st angle when **P** is in this angle, and to pass to this position from that in the fundamental figure, *Fig.* 24, the angle **NGX** of that figure becomes **STX** of this, an angle whose sine is +, and equal to sin **STA**. But, by this change of the position of the line, the angle **GHA** of *Fig.* 24, first diminishes to 0 and then re-appears generated from left to right and hence is a negative angle. Therefore sin **AOT** is negative, and we have $\frac{+\sin ATO}{-\sin AOT} = -\frac{y'}{x'}$.

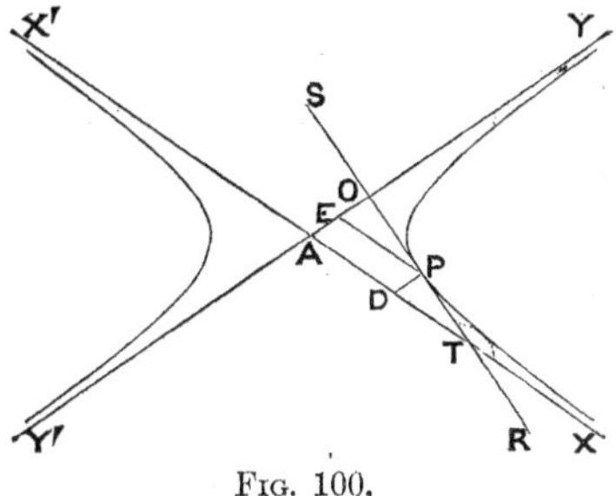

FIG. 100.

Ex. 10. Produce the equation of a tangent to the locus $y^2 = 2x + 3x^2$, at $x = 2$.

Result, There are two tangents, viz.: $y = \pm \frac{7}{4}x \pm \frac{1}{2}$.

Ex. 11. At what angle does the line $y = \frac{1}{2}x + 1$ cut the curve $y^2 = 4x$? *Ans.*, 10° 14′, and 33° 4′.

SUG'S.—Find the point of intersection and the tangent to the curve at this point. Find the angle included between this tangent and $y = \frac{1}{2}x + 1$ by (**36**).

Ex. 12. At what angle does $y^2 = 10x$ intersect $x^2 + y^2 = 144$? *Ans.*, 71° 0′ 58″.

Ex. 13. At what angle does $25y^2 + 16x^2 = 1600$ intersect $16y^2 - 9x^2 = -576$? *Ans.* 61° 58′ 37″.

---

137. Prop.—*The general value of the intercepts* of the axes by a tangent are*

$$X = x' - y'\frac{dx'}{dy'},$$

and

$$Y = y' - x'\frac{dy'}{dx'};$$

in which X is the intercept on the axis of x, *and Y that on the axis of* y, (x′, y′) *being the point of tangency.*

DEM.—The equation of a tangent being $y - y' = \frac{dy'}{dx'}(x - x')$, if we find where this line cuts the axes by making $y = 0$ for the intercept on the axis of x, and finding the value of x; and $x = 0$ and finding the value of y for the intercept on the axis of y, we have the results sought. Thus for $y = 0$, and $x = X$, we have $0 - y' = \frac{dy'}{dx'}(X - x')$, or $X = x' - y'\frac{dx'}{dy'}$. For $x = 0$, and $y = Y$, we have $Y - y' = \frac{dy'}{dx'}(0 - x')$, or $Y = y' - x'\frac{dy'}{dx'}$. Q. E. D.

SCH.—In solving an example we may either apply these *formulæ;* or, first get the equation of the tangent and then make x and y successively $= 0$. This is but an application of (**26**, 1st).

Ex. 1. From $A^2y^2 + B^2x^2 = A^2B^2$, show that a tangent to an ellipse cuts the axes at $X = \frac{A^2}{x'}$, and $Y = \frac{B^2}{y'}$; *i. e., If from any point in an ellipse a tangent and an ordinate be drawn to either axis, half that axis is a mean proportional between the distances of the intersections from the centre.*

---

* This is an abbreviated form of expression for "the distances from the origin to where the curve cuts the axis."

SUG'S.—In the figure, **AT** $= X$, **AD** $= x'$, **PD** $= y'$, **AC** $= Y$, **AB** $= A$, and **AG** $= B$. Hence having obtained $X = \frac{A^2}{x'}$, we have but to put it into the form $X : A :: A : x'$, to observe the truth of the proposition. Also $Y = \frac{B^2}{y'}$ gives $Y : B :: B : y'$.

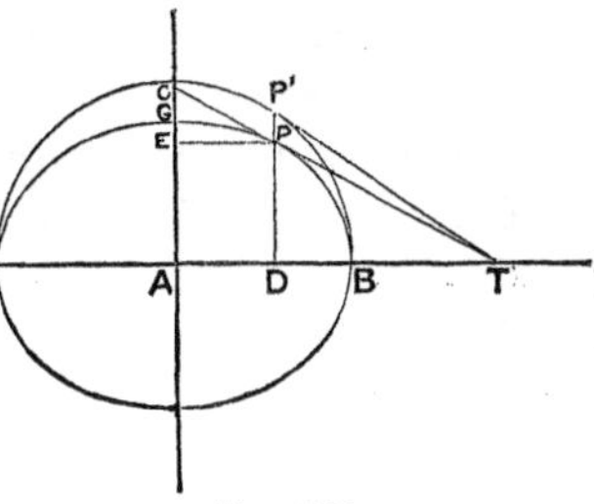

FIG. 101.

138. SCH.—Since $X = \frac{A^2}{x'}$ we see that the intercept of the axis of x does not depend upon the conjugate axis of the ellipse, so that, *if on the same transverse axis, different ellipses be drawn, the intercepts on this axis, by tangents corresponding to the same abscissa are equal.* That is, if x' and A remain the same, **AT** is the same. From this we have a ready method of drawing a tangent to an ellipse geometrically. Thus, let it be required to draw a tangent to the ellipse *Fig.* 101, at the point **P**. Draw a circle (a variety of ellipse) upon the same transverse axis. Draw the ordinate **PD** and produce it to **P'**. Draw a tangent to the circle at **P'**. This fixes the intercept **AT**. Draw a line through **P** and **T** and it is the tangent sought.

[NOTE—The student should make himself perfectly familiar with this, and all methods given for drawing tangents to loci geometrically.]

Ex. 2. Show that in the hyperbola the intercepts on the axes made by a tangent are $X = \frac{A^2}{x'}$, and $Y = -\frac{B^2}{y'}$, and that the proposition in *Ex.* 1, is true also of the hyperbola.

139. SCH.—This principle also affords a method of drawing a tangent to an hyperbola geometrically. From the given point of tangency **P**, let fall the ordinate **PD**; and upon the transverse axis **HB**, and the abscissa **AD**, draw semi-circumferences. From their intersection let fall **LT** a perpendicular upon the axis of x. Draw a line through **P** and **T** and it is tangent to the curve at **P**. *Proof.* Drawing **AL** and **LD**, we have **AD** (or x') : **AL** (or A) :: **AL** (or A) : **AT**. Whence **AT** $= \frac{A^2}{x'}$ and is the intercept made by a tangent at **P**.

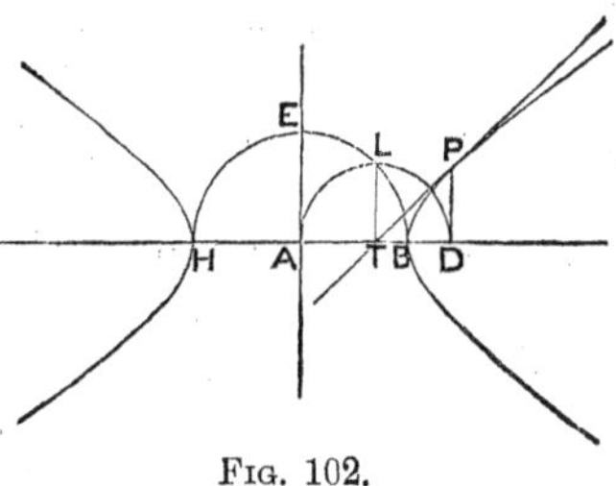

FIG. 102.

Ex. 3. Prove that, if a tangent be drawn to a parabola at any point, the intercept on the axis of x is equal to the abscissa of the point of tangency.

140. SCH.—The principle developed in the solution of this example affords the most simple method of drawing a tangent to the parabola, geometrically. Let it be required to draw a tangent at P, *Fig.* 103. Draw the ordinate PD, take AT = AD, and through P and T draw a line. This will be the tangent required.

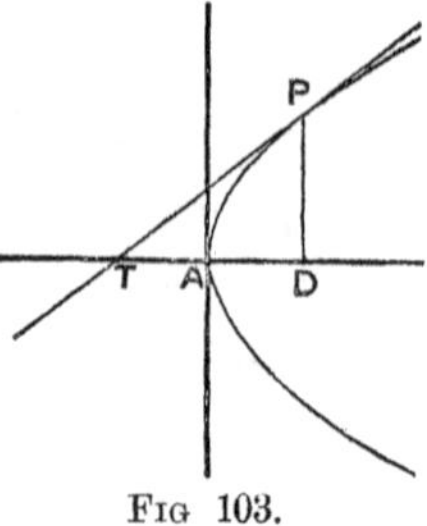

FIG 103.

Ex. 4. To find where the tangent to $y^2x = 4(2 - x)$ (the witch of Agnesi, the radius of the fixed circle being 1), at $x = 2$, cuts the axes.

Results, It cuts the axis of x at $x = 2$, and the axis of y at $y = \infty$, *i. e.*, is parallel to the latter.

[NOTE.—Observe from the last example that a tangent may cut the curve.]

---

141. DEF.—***The Subtangent*** is the portion of the axis of abscissas intercepted between the foot of the ordinate from the point of tangency, and the intersection of the tangent with this axis ; or it may be defined as the projection of the corresponding portion of the tangent upon the axis of abscissas. In each of the three preceding figures **DT** is the subtangent corresponding to **P** as the point of tangency.

142. Prop.—*The general value of a subtangent is*

$$Subt = y'\frac{dx'}{dy'}$$

in which (x′, y′) *is the point of tangency.*

DEM.—In any of the three preceding figures we have from the triangle **PTD**, **DT** = **PD** × cot **PTD**. But **DT** is *subt.*, **PD** $= y'$, and, as tan **PTD** is $\frac{dy'}{dx'}$, cot **PTD** is $\frac{dx'}{dy'}$. $\therefore$ $subt = y'\frac{dx'}{dy'}$. Q. E. D.

Ex. 1. What is the value of the subtangent of $y^2 = 3x^2 - 12$, at $x = 4$?

SUG'S.—For $x = 4$, $y = \pm 6$. $\frac{dy}{dx} = \frac{3x}{y}$. $\therefore$ $y' = \pm 6$, $\frac{dx'}{dy'} = \frac{\pm 6}{12} = \pm \frac{1}{2}$; and *Subt.* $= y'\frac{dx'}{dy'} = 3$. The pupil should construct the figure, if he is not sure that he fully comprehends the example without.

Ex. 2. Find the value of the subtangent of the common parabola. Of the logarithmic curve. *Results,* $2x'$, and m or 1.

Ex. 3. What is the value of the subtangent of $y^2 = 2x^3$ at $x = 2$?
Ans., $\frac{4}{3}$.

Ex. 4. If upon the same transverse axis different ellipses be drawn, prove that the corresponding subtangents are equal.

SUG'S.—The general value is $Subt. = \frac{A^2y'^2}{B^2x'}$ $= \frac{A^2B^2 - B^2x'^2}{B^2x'} = \frac{A^2 - x'^2}{x'}$, a result which does not depend upon B. This truth is illustrated in *Fig.* 104, **DT** being the common subtangent for all the ellipses, corresponding to the same value of x, **AD**. This is essentially the same truth as was brought to light in *Ex.* 1, Art. ***137.***

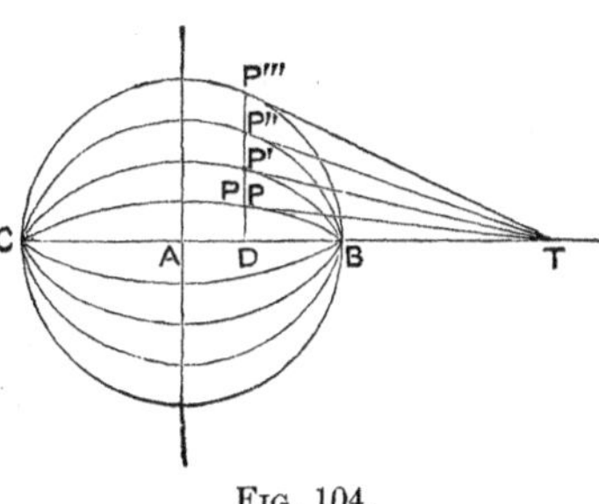

FIG. 104.

Ex. 5. Find the subtangent to the hyperbola referred to oblique asymptotes. *Result*, Subt $= x'$.

SCH.—In *Fig.* 100, **P** being the point of tangency (x', y'), **DT** $=$ *Subt.* $=$ $x' =$ **AD**. Now since **PD** is parallel to **AO**, **PT** $=$ **OP**; *i. e.*, *The intercepts of a tangent to a hyperbola between the point of tangency and the asymptotes are equal.* This affords a method of drawing a tangent when the asymptotes are given. Thus let it be required to draw a tangent at **P**, *Fig.* 100. Draw **PE** parallel to **AX**, and take **AT** $=$ 2**PE**. Through **P** and **T** draw a line and it is the tangent required.

---

143. Prop.—*The general expression for the length of a tangent,* i. e., *for the portion intercepted between the point of tangency and the axis of* x, *is*

$$Tan = y'\sqrt{1 + \frac{dx'^2}{dy'^2}},$$

in which (x', y') *is the point of tangency.*

DEM.—In any one of *Figs.* 101, 102, 103, we have from the right angled triangle **PDT**, $\overline{PT}^2 = \overline{PD}^2 + \overline{DT}^2$, or $PT = \sqrt{\overline{PD}^2 + \overline{DT}^2}$. Now **PT** is *Tan.*, **PD** $= y'$, and **DT** $= y'\frac{dx'}{dy'}$. $\therefore$ $Tan. = \sqrt{y'^2 + y'^2\frac{dx'^2}{dy'^2}} = y'\sqrt{1 + \frac{dx'^2}{dy'^2}}$. Q. E. D.

Ex. 1. What is the length of the tangent to $y^2 = 2x$ at $x = 8$?
Ans., $4\sqrt{17}$.

Ex. 2. Show that

In the ellipse, $Tan = y\sqrt{\frac{Ap^2 - py^2 + Ay^2}{Ap^2 - py^2}}$;

In the hyperbola, $Tan = y\sqrt{\frac{Ap^2 + py^2 + Ay^2}{Ap^2 + py^2}}$;

In the parabola, $Tan = \frac{y}{p}\sqrt{p^2 + y^2}$; p representing the semi-parameter in each case.

---

144. Def.—***An Asymptote*** is a line toward which a curve constantly approaches, but under such a law that they will never meet; or, what is the same thing, that they will meet only at an infinite distance from the origin.

An asymptote is also conceived as a tangent to a curve at an infinite distance from the origin, which yet passes within a finite distance, *i. e.*, cuts one or both axes making finite intercepts.

Ill.—It is quite common for persons encountering this idea for the first time, to repudiate it as an absurdity; but the following illustrations will familiarize it. Let the law of the curve be such that, if ordinates Bb, Cc, Dd, Ee, Ff, Gg, etc., be drawn at *equal* distances from each other, each succeeding ordinate shall be ½ the preceding. It is evident that the curve will continually approach AX, but under such a law that it can never absolutely reach it. (*Practically* such a curve will soon become indistinguishable from the line, that is, will run into it.) AX is an asymptote to this curve.

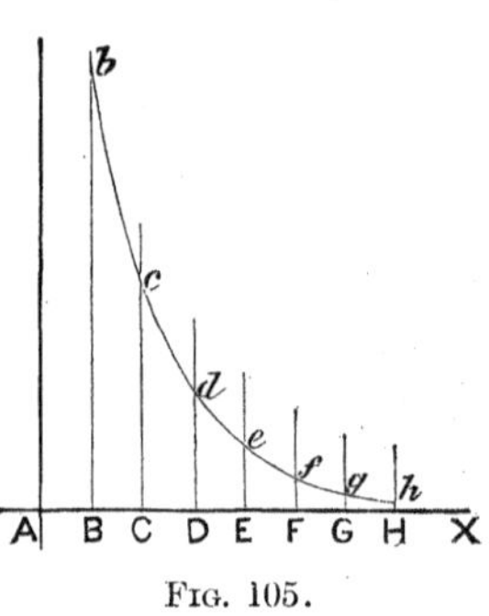

Fig. 105.

In a similar manner two curves may approach each other under such a law that the distance between them shall constantly diminish, and yet the curves never meet. Such curves are asymptotes to each other. Our present purpose embraces only rectilinear asymptotes.

---

145. ***Prob.***—*To determine whether a plane curve has rectilinear asymptotes.*

Solution.—First determine whether the curve has infinite branches. If it has not an infinite branch it cannot have an asymptote, since an asymptote is a tangent at an infinite distance from the origin. Second, if there is an infinite branch, determine the values of the intercepts of the axes by a tangent, $X = x - y\frac{dx}{dy}$, and

$Y = y - x\frac{dy}{dx}$, for x or $y = \infty$. It will be necessary to observe whether both of the variables, or only one of them, vary continuously to infinity, and get the value or the intercepts in terms of that one which does vary continuously. If now one or both of the intercepts thus evaluated is finite, the branch has an asymptote. If both intercepts are infinite, the curve has no asymptote, since the tangent at ∞ does not pass within a finite distance of the origin.

Having ascertained that the branch which is being examined has an asymptote, it remains to determine its position. If the intercepts are both finite and not 0, their values fix the position of the asymptote. If one intercept is finite and the other infinite, the asymptote is parallel to that axis on which the intercept is infinite. Finally, if the intercepts are 0, *i. e.*, if the asymptote passes through the origin, its direction is determined by evaluating $\frac{dy}{dx}$ for a tangent at infinity.

Ex. 1. Examine $y^3 = 6x^2 + x^3$ for asymptotes.

Solution.—Since as x increases from 0 positively, y increases continuously and without limit, is positive and has but one real value, there is an infinite branch extending in the first angle. Now when x is $-$, we have $y^3 = 6x^2 - x^3$, which gives positive values to y till $x^3 = 6x^2$. After $x^3 > 6x^2$, that is after $x > 6$ and negative, y becomes negative and a branch is found extending in the 3rd angle to infinity. Either of these branches may have an asymptote, they may both have the same line for an asymptote, or they may have different asymptotes. To determine what the facts are we find the intercepts made by a tangent. They are, $X = x - y\frac{y^2}{4x + x^2} = \frac{4x^2 + x^3 - y^3}{4x + x^2} = \frac{4x^2 + x^3 - 6x^2 - x^3}{4x + x^2} = \frac{-2x^2}{4x + x^2}$, which for $x = +\infty = -2$; and $Y = y - x\frac{4x + x^2}{y^2} = \frac{2x^2}{(6x^2 + x^3)^{\frac{2}{3}}}$, which for $x = +\infty = 2$. $\therefore$ The branch in the first angle has an asymptote which cuts the axis of y at 2 above the origin, and the axis of x at 2 on the left of the origin. The equation of this line is $y = x + 2$. Finally, as the intercepts have the same values for $x = -\infty$ as for $x = +\infty$, this line is an asymptote to both branches of this locus. The curve is sketched in *Fig.* 106, in which MN is the asymptote.

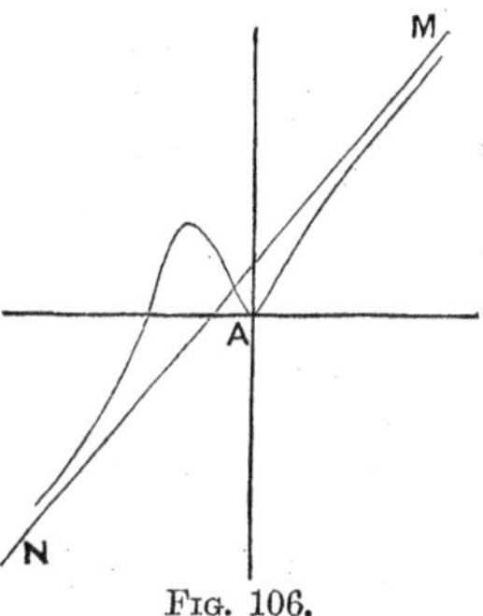

Fig. 106.

Ex. 2. Show that $y^3 = a^3 - x^3$ has an asymptote which is common to its two infinite branches, passes through the origin, and makes an angle with the axis of x of 135°.

Ex. 3. Examine $y^2 = 2x + 3x^2$ for asymptotes.

Ex. 4. Why has $y^2 = x^2 - x^4$ no asymptote?

Ex. 5. Examine $y^2 = ax^3$ for asymptotes.

Ex. 6. Examine the conic sections for asymptotes.

SOLUTION.—An ellipse or a circle can not have an asymptote as neither has an infinite branch. It remains, therefore, to examine the parabola and hyperbola. From $y^2 = 2px$, we have $\frac{dy}{dx} = \frac{p}{y}$; hence $X = x - y\frac{dx}{dy} = \frac{y^2}{2p} - \frac{y^2}{p} = -\frac{y^2}{2p}$, which is $-\infty$ for $y = \infty$. Again $Y = y - x\frac{dy}{dx} = y - \frac{y}{2} = \frac{y}{2}$, which $= \infty$ for $y = \infty$. $\therefore$ The parabola has no asymptote. To examine the hyperbola, we have from $A^2y^2 - B^2x^2 = -A^2B^2$, $\frac{dy}{dx} = \frac{B^2x}{A^2y}$; hence $X = x - y\frac{dx}{dy} = x - \frac{A^2y^2}{B^2x} = \frac{A^2}{x}$, which $= 0$ for $x = \pm\infty$. Also $Y = y - x\frac{dy}{dx} = y - \frac{B^2x^2}{A^2y} = -\frac{B^2}{y}$, which $= 0$ for $y = \pm\infty$. (In this curve both x and y vary continuously to infinity, hence the intercepts may be evaluated in terms of either.) $\therefore$ The hyperbola has two asymptotes, and they both pass through the centre. To determine the direction of the asymptotes we evaluate $\frac{dy}{dx} = \frac{B^2x}{A^2y} = \frac{Bx}{A\sqrt{x^2 - A^2}}$ for $x = \pm\infty$. For $x = +\infty$, $\frac{Bx}{A\sqrt{x^2 - A^2}} = \frac{B}{A}$. For $x = -\infty$, $\frac{Bx}{A\sqrt{x^2 - A^2}} = -\frac{B}{A}$. Whence we learn that the asymptotes are the produced diagonals of the rectangle described upon the axes, as has been stated before.

146. SCH. 1.—If, at successive points along an infinite branch of a curve, we draw tangents, these tangents will either approach some limiting position, or they will not. In the hyperbola, *Fig*, 107, it is evident that the successive tangents **PT**, **P′T′**, **P″T″** are approaching the limiting position **SA**. But in the parabola, *Fig*. 108, it is equally evident, from the way in which the tangents are drawn, that there is no limiting position beyond which a tangent may not pass. Since **AT** = **AD**, **AT′** = **AD′**, **AT″** = **AD″**, the point of intersection with the axis recedes indefinitely as the point of tangency passes to the right. In a similar manner observing the method of drawing a tangent to an hyperbola, *Fig*. 102, it will appear that as **P** recedes, the intersection **L** constantly (but more and more slowly) approaches **E** but can never pass it; and, consequently, that **T** can never pass **A**, however far **P** may recede. From these considerations, an asymptote is seen to be *the limiting position toward which a tangent approaches as the point of tangency recedes to infinity.*

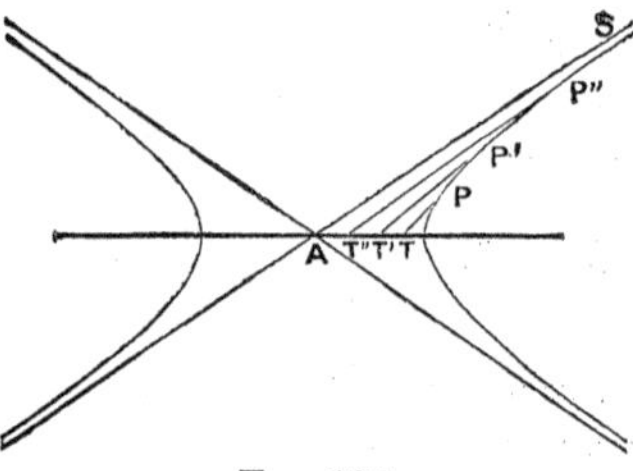

FIG. 107.

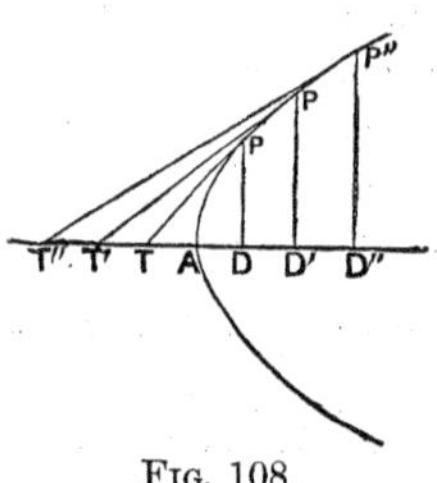

FIG. 108.

147. SCH. 2.—Having found the intercept on the axis of ordinates and the tangent of the angle which the asymptote makes with the axis of abscissas, we can readily write the equation of the asymptote by substituting in $y = ax + b$. Thus the equations of the asymptotes to the hyperbola are $y = \frac{B}{A}x$, and $y = -\frac{B}{A}x$; or $Ay = Bx$, and $Ay = -Bx$.

Ex. 7. Show that $y = -x + \frac{a}{3}$ is the equation of the asymptote to $y^3 = ax^2 - x^3$.

Ex. 8. Show that $x = 2a$, and $y = \pm(x + a)$ are asymptotes to $y^2(x - 2a) = x^3 - a^3$.

Ex. 9. Show that the axis of abscissas ($y = 0x$) is an asymptote to $y(a^2 + x^2) = a^2(a - x)$.

Ex. 10. Show that $x = 2a$ is the asymptote to the cissoid of Diocles.

Ex. 11. Examine $x = \log y$ for asymptotes.

Ex. 12. Examine $y = \tan x$ for asymptotes.

SUG'S.—As this curve is not continuous in the direction $x = \infty$, we evaluate the intercepts for $y = \infty$, for which $x = \frac{1}{2}\pi, \frac{3}{2}\pi, \frac{5}{2}\pi$, etc., and $-\frac{1}{2}\pi, -\frac{3}{2}\pi, -\frac{5}{2}\pi$, etc., $\frac{dy}{dx} = \sec^2 x = 1 + y^2$. $\therefore X = x - \frac{y}{1 + y^2}$, which for $y = \infty, = x = \frac{1}{2}\pi, \frac{3}{2}\pi, \frac{5}{2}\pi$, etc., and $-\frac{1}{2}\pi, -\frac{3}{2}\pi, -\frac{5}{2}\pi$, etc. $Y = y - x(1 + y^2)$, which for $y = \infty, = \infty$. Hence there are an infinite number of asymptotes parallel to the axis of y. (See **23,** *Ex.* 27, SCH., *Fig.* 18.)

Ex. 13. Examine $y = \frac{a^3}{(x - b)^2} + c$ for asymptotes.

SUG'S.—In examining this locus it is necessary to evaluate the intercepts both for $x = \infty$, and $y = \infty$, as there are infinite branches running in both directions. The general form of the locus is given in the figure. The equations of the asymptotes are $x = b$ (the line MN), and $y = c$ (the line M′N′).

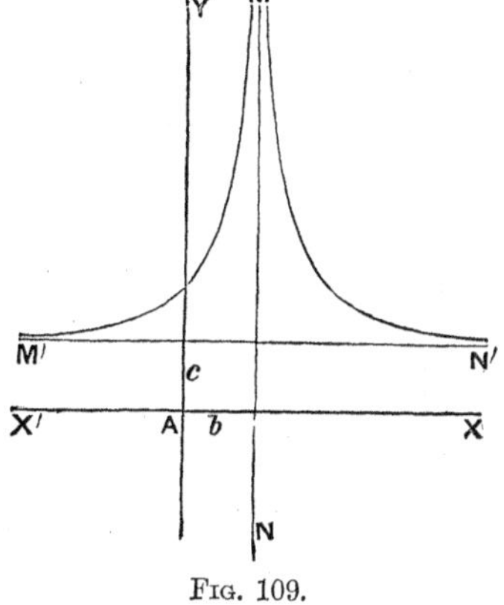

FIG. 109.

148. SCH. 3.—In very many cases there are more expeditious methods than the above for finding asymptotes. It is frequently the case that a simple inspection of the equation of the curve will determine the fact. Thus, in the last example, it is evident that as x increases from 0 to b, y increases, and when $x = b$, y becomes ∞. $\therefore$ This branch of the curve approaches the line MN, parallel to the axis of y, and at a distance b from it, under the law required for an asymptote. So again when x passes $x = b$, it is evident that y grows less, and the curve approaches the axis of x. But, as, when $x = \infty$, $y = c$, this branch extending to the right can never come nearer the axis of x than $y = c$. In like manner when $x = -\infty$, we see that $y = c$. $\therefore$ **M′N′** is an asymptote.

Ex. 14. Determine by inspection the asymptotes to $xy = m$.

Ex. 15. Determine by inspection that $x = a$, and $y = b$, are asymptotes to $xy - ay - bx = 0$.

Sug.—Observe that $y = \frac{bx}{x - a}$, and $x = \frac{ay}{y - b}$.

149. Sch. 4.—An elegant method of examining for asymptotes consists in expanding $y = f(x)$, or $x = f(y)$, into a series, by the binomial theorem, Maclaurin's formula, or some other method, when such development is practicable. This will be best illustrated by an example or two.

Ex. 16. Determine the asymptotes of the locus $x^3 - xy^2 + ay^2 = 0$, by developing $y = f(x)$.

Solution. $y = \pm\sqrt{\frac{x^3}{x - a}} = \pm x\left(1 - \frac{a}{x}\right)^{-\frac{1}{2}} = \pm x\left\{1 + \frac{a}{2x} + \frac{3a^2}{8x^2} +, \text{etc.}\right\}$.

Now, if we take the first two terms of this development we have $y = \pm x \pm \frac{a}{2}$, the equations of two straight lines. Comparing this value of y with the entire value, which is the ordinate of the curve, we see that as x increases the terms following $\frac{a}{2x}$ grow less and less, and consequently that the ordinate of the right line and the corresponding ordinate of the curve become more and more nearly equal; that is, the curve is constantly approaching the lines $y = \pm x \pm \frac{a}{2}$. Now when $x = \infty$ this difference vanishes, as all the terms following $\frac{a}{2x}$, become 0. $\therefore$ The lines $y = \pm x \pm \frac{1}{2}a$ are such that the given curve approaches them constantly but reaches them only at an infinite distance, and are therefore asymptotes. There is also an asymptote at $x = a$, which can be discovered by inspection, as under the last scholium.

Ex. 17. Show by developing $y = f(x)$, that $y = \pm x$ are asymptotes to $y^2 = x^2\frac{x^2 - 1}{x^2 + 1}$.

Ex. 18. Show by developing $y = f(x)$, that $y = \pm(x + a)$ are asymptotes to $y^2 = \frac{x^3 + ax^2}{x - a}$.

Sug.—The value of y developed becomes $y = \pm x(1 + \frac{a}{x} + \frac{a^2}{2x^2} +, \text{etc.})$.

(b) TANGENTS TO POLAR CURVES.

150. It is found most convenient to determine tangents to polar curves by means of the subtangents.

151. Def.—***The Subtangent to a Polar Curve*** is the distance from the pole to the tangent, measured on a perpendicular to the radius vector to the point of tangency. Thus in the figure let **MN** be a curve referred to **P** as its pole, **S** any point in the curve, and **RT** tangent at **S**. Then **PT** drawn through the pole, perpendicular to **PS** and limited by the tangent, is the subtangent.

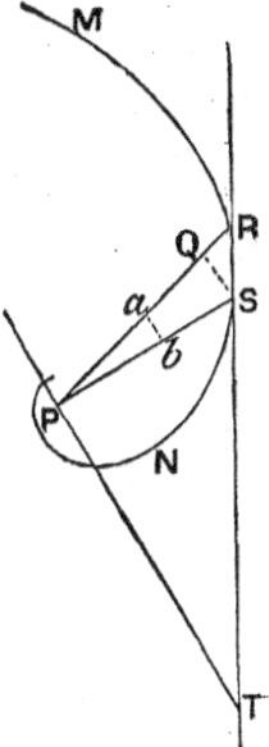

FIG. 110.

---

152. Prop.—*The general value of the subtangent to a polar curve is*

$$\text{Subt.} = \frac{r^2 d\theta}{dr};$$

in which r *is the radius vector and* θ *the variable angle.*

DEM.—In the last figure let **R** be a point on the curve consecutive with **S** (infinitely near it), so that the tangent **RT** is to be considered as coinciding with the curve between **R** and **S**. Draw **PR**, and with radius vector **PS** as a radius draw arc **SQ**, and also with radius **P**b $= 1$, draw *ab* an arc of the measuring circle. Then **RQ** $= dr$, since **RQ** is an infinitesimal increment of the radius vector, contemporaneous with **RS**. So also **RPS**, or $ab = d\theta$. As **SQ** is infinitesimal it may be considered a right line, and it is perpendicular to **PR**. Again, as **R** approaches **S**, the triangle **RQS** approaches similarity with **SPT**; and as it is the relation at the limit that we seek, we are to treat **RQS** as similar to **SPT**. Hence we have **PT** : **PS** :: **QS** : **QR**, or subt. : r :: **QS** : dr. But from the similar sectors **QPS** and a**P**b. we have **QS** $= rd\theta$, and substituting, subt. : $r :: rd\theta : dr$. $\therefore$ subt $= \frac{r^2 d\theta}{dr}$. Q. E. D.

Ex. 1. Find the value of the subtangent to the spiral of Archimedes.

SOLUTION.—The equation is $r = \frac{\theta}{2\pi}$; whence $\frac{d\theta}{dr} = 2\pi$. $\therefore$ Subt $= \frac{r^2 d\theta}{dr} = r^2 \times 2\pi = \frac{\theta^2}{4\pi^2} \times 2\pi = \frac{\theta^2}{2\pi}$.

ILL.—The annexed figure furnishes an illustration. **PT** is subtangent for point **S** and $=$ the square of the numerical value of θ divided by the circumference of the circle whose radius is **PB**. The numerical value of θ in this instance $2\frac{1}{4}\pi$, since in passing from $\theta = 0$ to **S** the radius vector makes $1\frac{1}{8}$ revolutions. But one revolution $= 2\pi$.

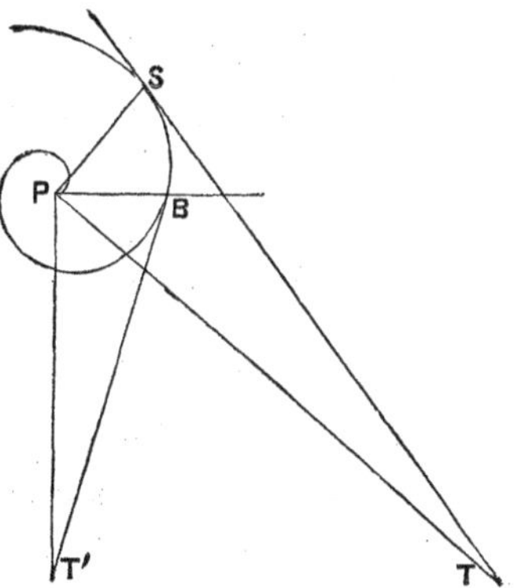

FIG. 111.

Hence $\mathbf{PT} = \frac{(2\frac{1}{4}\pi)^2}{2\pi} = \frac{81}{64} \cdot 2\pi$, or somewhat more than $1\frac{1}{4}$ times the circumference of the measuring circle. If $\mathbf{B}$ is the point of tangency, $\theta = 2\pi$, and subt $\mathbf{PT}' = 2\pi$, or the circumference of the measuring circle. In this spiral the subtangent varies as the square of the measuring arc.

Ex. 2. Prove that in the hyperbolic, or reciprocal spiral the subtangent is constant. What is its value, and what the significance of its sign? Construct the curve and tangents at a few points, and observe the subtangents.

Ex. 3. Find the subtangent to the logarithmic spiral ($r = a^{\theta}$), and show that the angle under which the curve meets the radius vector is constant.

Sug.—The tangent of the angle made by a tangent to the curve and the radius vector is equal to the subtangent divided by the radius vector. In this locus the tangent of the required angle is the modulus of the system of logarithms used in constructing the spiral.

Ex. 4. Prove that $\frac{-r^3}{a^2 \sin 2\theta}$ is the value of the subtangent to the lemniscate of Bernouilli.

---

153. Prob.—*To test polar curves for rectilinear asymptotes.*

Solution.—Any curve which continually revolves around the pole can have no rectilinear asymptote; for with respect to any fixed right line, such a curve will alternately approach and recede. But if for some finite value of θ, r becomes infinite, the curve ceases to revolve around the pole, and will have an asymptote if the tangent at $r = \infty$ passes within a finite distance of the pole; *i. e.*, if the subtangent is finite. Q. E. D.

To construct the asymptote, we observe that the asymptote and radius vector are drawn from a point infinitely distant, to the extremities of a finite subtangent, and hence are to be considered parallel. We therefore determine the value of the subtangent for $r = \infty$, and drawing the radius vector for that value of θ which renders $r = \infty$, erect the subtangent, and through the extremity remote from the pole draw a line parallel to this radius vector. This line will be the asymptote.

Ex. 1. Test the hyperbola for asymptotes by the polar method.

Solution.—The polar equation, when the pole is at F, is $r = \frac{A(1 - e^2)}{e \cos \theta - 1}$. Now for $\cos \theta = \frac{1}{e}$, $r = \infty$; hence if there be an asymptote, it is parallel to $\mathbf{FR}$ so drawn that $\cos \mathbf{RFX} = \frac{1}{e} = \frac{A}{\sqrt{A^2 + B^2}}$. From the equation we have $\frac{d\theta}{dr} = \frac{(e \cos \theta - 1)^2}{Ae(1 - e^2) \sin \theta}$; whence

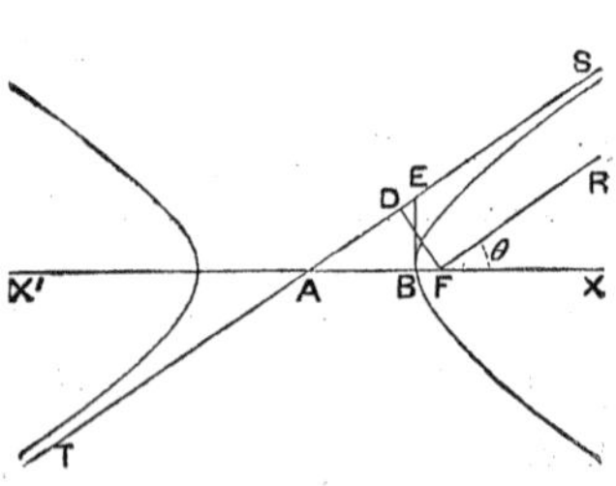

Fig. 112.

$subt. = r^2\frac{d\theta}{dr} = \frac{A^2(1-e^2)^2}{(e\cos\theta-1)^2} \times \frac{(e\cos\theta-1)^2}{Ae(1-e^2)\sin\theta} = \frac{A(1-e^2)}{e\sin\theta}$, which, since for $\cos\theta = \frac{1}{e}$, $\sin\theta = \frac{1}{e}\sqrt{e^2-1}$, $= -A\sqrt{e^2-1} = -B$. There is therefore an asymptote. To construct it draw **FR** making **RFX** $= \cos^{-1}\left(\frac{1}{e}\right)$, **FD** perpendicular to **RF** and $= B$, and through **D** draw **ST** parallel to **RF**. **ST** is the asymptote. Moreover, since $\cos(-\theta) = \cos\theta$, there is another asymptote similarly situated below the polar axis **FX**. Finally, as the angle which a diagonal upon the axes of an hyperbola makes with the axis of x is $\cos^{-1}\frac{A}{\sqrt{A^2+B^2}}$, the asymptotes are parallel to these diagonals; and since **FD** $= B$, **AF** $= \sqrt{A^2+B^2}$, and the asymptotes coincide with the diagonals.

Ex. 2. Show by the polar method that the parabola has no asymptote.

Sug.—In this case for $\theta = 0$, $r = \infty$; but for this value of θ $subt = \infty$. Hence there is no asymptote.

Ex. 3. Show that the hyperbolic spiral has an asymptote parallel to the polar axis and at a distance 2π from it.

Ex. 4. Show that the polar axis is an asymptote to the lituus.

SECTION II.

Normals to Plane Loci.

(a) BY RECTANGULAR CO-ORDINATES.

154. Def.—***A Normal*** to a plane curve is a perpendicular to a tangent at the point of tangency.

155. Prop.—*The general equation of a normal to a plane curve is*

$$y - y' = -\frac{dx'}{dy'}(x - x'),$$

in which (x', y') *is the point in the curve to which the normal is drawn, and* x *and* y *are the general co-ordinates of the normal.*

Dem.—Letting (x', y') represent the point in the curve to which the normal is to be drawn, the equation of a *tangent* through this point is $y - y' = \frac{dy'}{dx'}(x - x')$. Again the equation of *any* line passing through (x', y') is $y - y' = a(x - x')$. Now, in order that the equation for this general line should become the equation of a perpendicular to the tangent, a must $= -\frac{dx'}{dy'}$. $\therefore\ y - y' = -\frac{dx'}{dy'}(x - x')$ is the equation of a normal. Q. E. D.

156. COR.—*The general expression for the tangent of an angle which a normal makes with the axis of abscissas is* $-\frac{dx}{dy}$, (x, y) *being the point in the curve to which the normal is drawn.*

Ex. 1. Produce the equations of the normals to the conic sections.

Results, Ellipse, $y - y' = \frac{A^2y'}{B^2x'}(x - x')$; Circle, $y = \frac{y'}{x'}x$;

Hyperbola, $y - y' = -\frac{A^2y'}{B^2x'}(x - x')$; Parabola, $y - y' = -\frac{y'}{p}(x - x')$.

SCH.—Observe that these equations do not reduce to as simple and symmetrical forms as do those of the tangents to the conic sections. The form of the equation of the normal to the circle shows that the normal of this locus always passes through the centre. It is, of course, the radius.

Ex. 2. What is the equation of the normal to $y^2 = 2x^2 - x^3$ at $x = 1$?

Answer: At $x = 1$, $y = \pm 1$; hence there are two points indicated. The equation of the normal at the former is $y = 2x - 3$, and at the latter $y = -2x + 3$.

Ex. 3. What is the equation of a normal to $y^2 = 6x - 5$ at $y = 5$? What angle does this normal make with the axis of x?

Ex. 4. At what point in the ellipse whose axes are 12 and 8 must a normal be drawn to make an angle of 45° with the axis of x?

Ex. 5. At what point in the witch of Agnesi must a normal be drawn to be perpendicular to the axis of x? To be parallel? To make an angle of 135°?

---

157. DEF.—***The Subnormal*** is the projection of the normal upon the axis of x; or it is the distance from the foot of the ordinate let fall from the point in the curve to which the normal is drawn, to the intersection of the normal with the axis of x.

158. *Prob.*—*To find the general value of the subnormal.*

SOLUTION.—In *Fig.* 113 **PE** is the normal and **DE** the subnormal for the point **P**. Now in the triangle **PDE**, **PD** $= y$, and tan **PED** = (numerically) tan **PEX** $= \frac{dx}{dy}$. $\therefore$ **DE** $= y\frac{dy}{dx}$, by trigonometry.

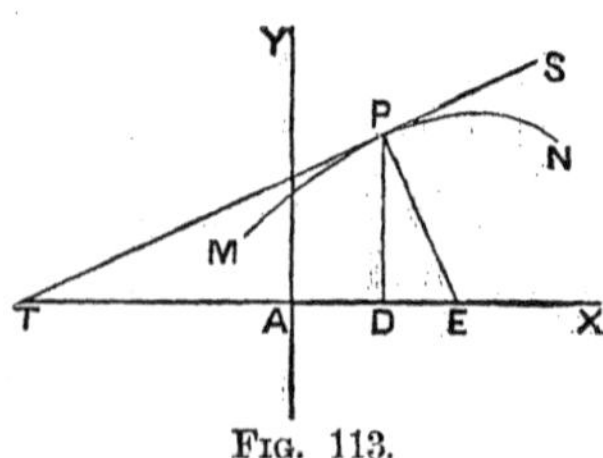

FIG. 113.

Ex. 1. Show that the subnormal to the parabola is constant and equal to half the *latus rectum.* How can a tangent be drawn to the parabola, geometrically, upon this principle?

Ex. 2. Show that the subnormal to the cycloid is $(2ry - y^2)^{\frac{1}{2}}$.

159. Cor.—*Since* DC $=$ PG $= \sqrt{\text{CG} \times \text{GS}} = \sqrt{y(2r - y)} = \sqrt{2ry - y^2}$, *the normal passes through the foot of the vertical diameter of the generating circle for the point to which the normal is drawn. Moreover, since* SPC *is a right angle, the tangent passes through the other extremity of the vertical diameter.*

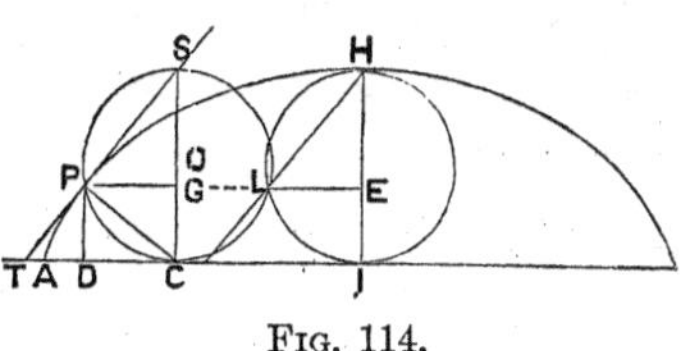

Fig. 114.

160. Sch. 1.—This principle affords a ready method of constructing a tangent to the cycloid geometrically. Let P be the given point through which a tangent is to be drawn. Put the generating circle in position for this point (***94***), and draw the vertical diameter SC. Through S and P draw a right line and it will be the required tangent. Also PC will be the normal to the curve at the point P.

161. Sch. 2.—To draw a tangent which shall make any given angle with the axis of x, draw the generating circle on the axis HI, construct the angle LHI $=$ the complement of the required angle, and through L, the point where this line intersects the circumference of the central generating circle, draw a parallel to the base of the cycloid. Where this parallel cuts the curve P is the required point of tangency. Through this point draw SPT parallel to HL, and it is the tangent required.

---

162. Prob.—*To find the length of the normal, i. e., the portion intercepted between the curve and the axis of* x.

Solution.—In *Fig.* 113, from the right angled triangle, PDE we have PE $= \sqrt{\overline{\text{PD}}^2 + \overline{\text{DE}}^2} = \sqrt{y^2 + y^2\frac{dy^2}{dx^2}} = y\sqrt{1 + \frac{dy^2}{dx^2}}.$ Q. E. D.

Ex. 1. Find the length of the normal in each of the conic sections. What is it in the circle?

Ex. 2. In the cycloid the radius of whose generatrix is 2, what is the length of the normal at $y = 1$? *Ans.*, 2.

163. Cor.—*The normal is but a particular case of a perpendicular to a tangent.*

DEM.—As $y - y' = \frac{dy'}{dx'}(x - x')$ is the equation of a tangent at (x', y'), a perpendicular to this through the point (x'', y'') is $y - y'' = -\frac{dx'}{dy'}(x - x'')$. Making the point through which this perpendicular to the tangent is to pass the point of tangency, the perpendicular becomes a normal, and its equation is $y - y' = -\frac{dx'}{dy'}(x - x')$, since in this case $x'' = x'$, and $y'' = y'$. For a perpendicular from the origin on the tangent we have $y'' = 0$, $x'' = 0$, and $y = -\frac{dx'}{dy'}x$.

Ex. 1. Show that the equation of a perpendicular from the focus of the common parabola upon the tangent is $y = -\frac{y'}{p}(x - \frac{1}{2}p)$.

Ex. 2. Show that the perpendicular distance from the focus of an hyperbola to the asymptote is B.

164. COR.—*The perpendicular from the focus of a parabola upon a tangent meets the tangent in a tangent to the curve at the vertex* (the axis of y).

DEM.—The equation of a tangent is $yy' = p(x + x')$, and of the perpendicular from the focus upon this tangent $y = -\frac{y'}{p}(x - \frac{1}{2}p)$. We have now but to find the intersection of these lines. Equating the values of y, we have $-\frac{y'}{p}(x - \frac{1}{2}p) = \frac{p}{y'}(x + x')$, or $-y'^2(x - \frac{1}{2}p) = p^2x + p^2x'$, or since $y'^2 = 2px'$, $-2pxx' + p^2x' = p^2x + p^2x'$; whence $(p + 2x')x = 0$. Now as x' can not be $-$, $p + 2x'$ can not become $= 0$. Therefore to fulfill the condition $(p + 2x')x = 0$; x must $= 0$. $\therefore$ The point of intersection is at $x = 0$, or in the axis of y. Q. E. D.

(b) NORMALS TO POLAR CURVES.

165. DEF.—***The Subnormal*** to a polar curve is the distance from the pole to the normal, measured on a line perpendicular to the radius vector to the point in the curve to which the normal is drawn. Thus **EP** is the subnormal of the curve **MN** corresponding to **R**, the pole being at **P**.

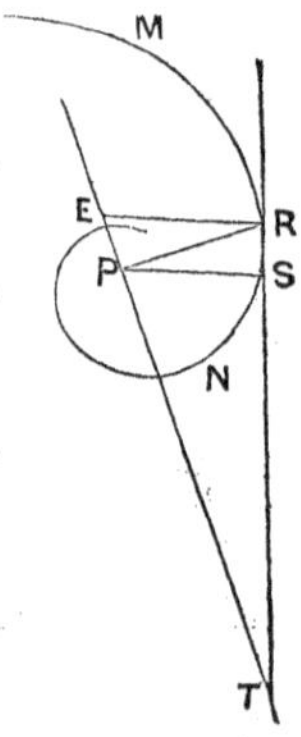

FIG. 115.

166. *Prob.*—*To find the general value of the subnormal to a polar curve.*

SOLUTION. $\mathsf{PT} = r^2\frac{d\theta}{dr}$. $\tan \mathsf{PTR} = \frac{\mathsf{PR}}{\mathsf{PT}} = \frac{r}{r^2\frac{d\theta}{dr}} = \frac{dr}{rd\theta}$.

But $\tan \mathsf{PER} = \frac{1}{\tan \mathsf{PTR}} = \frac{rd\theta}{dr}$. $\therefore$ PE or subnormal $= \frac{\mathsf{PR}}{\tan \mathsf{PER}} = \frac{dr}{d\theta}$. Q. E. D.

Ex. 1. Show that $-\frac{a^2 \sin 2\theta}{r}$ is the value of the polar subnormal of the lemniscate of Bernouilli.

Ex. 2. Show that the subnormal to the logarithmic spiral is $\frac{r}{m}$, m being the modulus of the system; and, consequently, that in the Napierian logarithmic spiral the subnormal always equals the radius vector.

167. Cor. 1.—*The length of a normal to a polar curve is* $\left(\frac{dr^2}{d\theta^2} + r^2\right)^{\frac{1}{2}}$.

168. Cor. 2.—*The length of a perpendicular from the pole upon a tangent is* $p = \frac{r^2}{\left(\frac{dr^2}{d\theta^2} + r^2\right)^{\frac{1}{2}}}$.

Dem.—In *Fig.* 115, let **PS** be the perpendicular from the pole upon the tangent, and consequently parallel to the normal **RE**. From the right angled triangle **PST**, **PS** = **PT** × cos **SPT**. But **PT** (the subtangent) $= \frac{r^2 d\theta}{dr}$; and

$$\cos \mathbf{SPT} = \cos \mathbf{RET} = \frac{1}{\sec \mathbf{RET}} = \frac{1}{(1 + \tan^2 \mathbf{RET})^{\frac{1}{2}}} = \frac{1}{\left(1 + \frac{r^2 d\theta^2}{dr^2}\right)^{\frac{1}{2}}}.$$

$$\text{Whence } p = \frac{\frac{r^2 d\theta}{dr}}{\left(1 + \frac{r^2 d\theta^2}{dr^2}\right)^{\frac{1}{2}}} = \frac{r^2}{\left(\frac{dr^2}{d\theta^2} + r^2\right)^{\frac{1}{2}}}. \quad \text{Q. E. D.}$$

SECTION III.

Direction of Curvature.

(*a*) BY RECTANGULAR CO-ORDINATES.

169. ***Prop.***—*At a point where* $\frac{d^2y}{dx^2}$ *is positive, a curve is concave upward, and where* $\frac{d^2y}{dx^2}$ *is negative the curve is convex upward.*

Dem.—1st. Let ω be the angle which a tangent makes with the axis of x. When the curve is concave upward, as in *Fig.* 116, it is evident that as x increases (as from being the abscissa of **P** to being that of **P'**), ω increases. In other words, if x takes the infinitesimal *increment* dx, the contemporaneous infinitesimal change in ω is

$+d\omega$. Hence when the curve is concave upward, dx and $d\omega$ have the same sign (x and ω are increasing functions of each other).

In a similar manner it is evident that when the curve is convex upward, ω *decreases* as x increases; *i. e.*, if x takes the *increment* dx, the cortemporaneous change in ω is $-d\omega$.

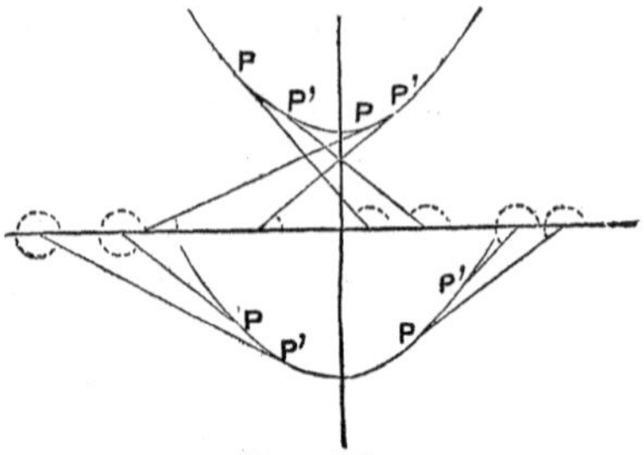

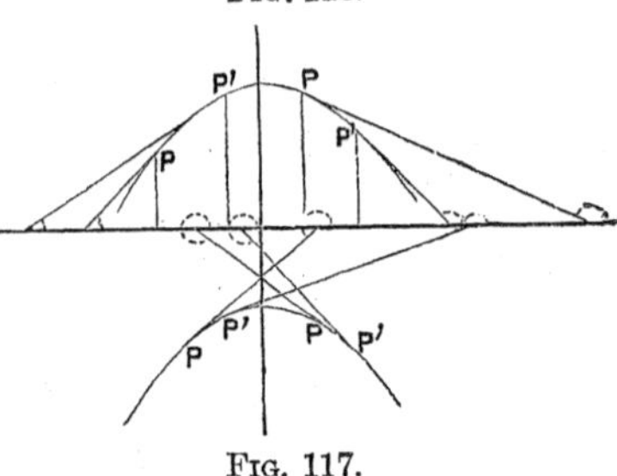

FIG. 116.

2nd. $\dfrac{d^2y}{dx^2} = \dfrac{d \cdot \dfrac{dy}{dx}}{dx} = \dfrac{d \cdot \tan\omega}{dx} = \sec^2\omega \dfrac{d\omega}{dx}$. Now, as $\sec^2\omega$ is always positive, $\dfrac{d^2y}{dx^2}$ is positive when x and ω are increasing functions of each other (when dx and $d\omega$ have like signs), and negative when they are decreasing functions of each other. $\therefore + \dfrac{d^2y}{dx^2}$ indicates that the curve is concave upward, and $-\dfrac{d^2y}{dx^2}$ indicates that it is convex upward. Q. E. D.

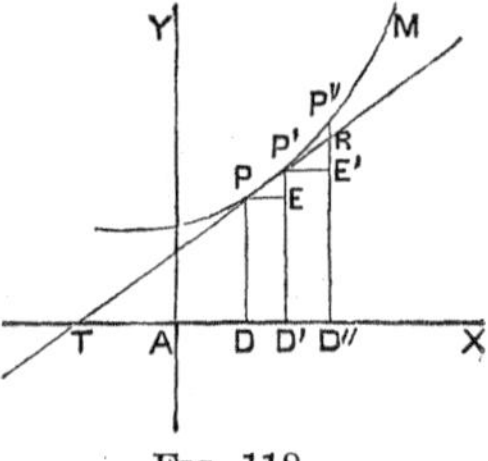

FIG. 117.

Another Demonstration.—Let DD', and $D'D''$, *Fig.* 118, represent consecutive *equal* infinitesimal increments of x, then $P'E$ and $P''E'$ represent contemporaneous infinitesimal increments of y. Represent them respectively by dy_1 and dy_2. The difference between dy_1 and dy_2, is by definition d^2y. But when a curve is concave upward it lies above its tangent. Hence $dy_2 > dy_1$ and $dy_2 - dy_1 = +d^2y$. On the other hand, when the curve is convex upward, as in *Fig.* 119, it lies below its tangent and $dy_2 < dy_1$. Whence $dy_2 - dy_1 = -d^2y$. A similar inspection can easily be made in all cases, both when the curve lies above the axis of x, and when it lies below, and thus the universality of the principle be established. Finally, the sign of $\dfrac{d^2y}{dx^2}$ is the same as that of d^2y, since dx^2 being a square is always positive.

FIG. 118.

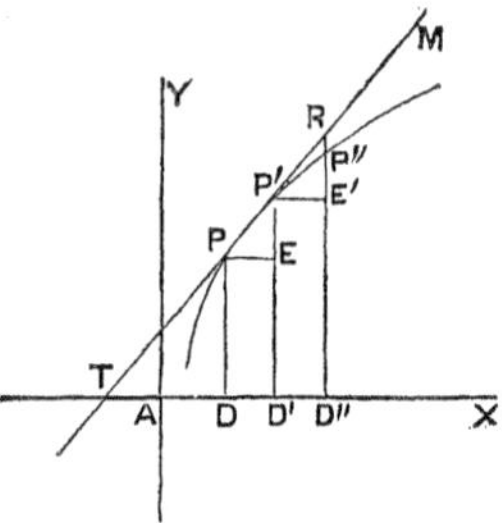

FIG. 119.

170. COR. 1.—*By a course of reasoning entirely similar, taking* y *as the independent variable, it may be shown that* $+\dfrac{d^2x}{dy^2}$ *indicates that a curve is concave to the right, and* $-\dfrac{d^2x}{dy^2}$ *that it is convex to the right.*

171. COR. 2.—*A curve is convex towards the axis of abscissas when* $y\frac{d^2y}{dx^2}$ *is positive, and concave when it is negative.*

DEM.—For points above the axis of x, y is $+$, and if the curve is convex towards the axis (downward) $\frac{d^2y}{dx^2}$ is also $+$; hence $y\frac{d^2y}{dx^2}$ is $+$. For points below the axis y is $-$, and if the curve is convex towards the axis (upward) $\frac{d^2y}{dx^2}$ is $-$; hence $y\frac{d^2y}{dx^2}$ is $+$. Therefore $y\frac{d^2y}{dx^2}$ is always $+$ when the curve is convex towards the axis of x. In a similar manner it may be seen that $y\frac{d^2y}{dx^2}$ is $-$ when the curve is concave towards the axis of x.

Ex. 1. To discover whether $x^2 + y^2 = r^2$ is convex or concave towards the axis of x.

SOLUTION.—From $x^2 + y^2 = r^2$, we have $\frac{d^2y}{dx^2} = -\frac{r^2}{y^3}$. This locus is, therefore, convex upward when y is $+$, and concave when y is $-$. Hence it is always concave to the axis of x.

Ex. 2. Test the following for direction of curvature: $y = b + c(x + a)^2$; and $y = a^2\sqrt{x - a}$.

Results, The first is concave upward; and the second concave towards the axis of x.

Ex. 3. Test the direction of curvature $y = b + (x - a)^3$.

Results, From $x > a$ to $x = \infty$ convex towards the axis of x. From $x < a$ to $x = a - b^{\frac{1}{3}}$, concave. From $x = a - b^{\frac{1}{3}}$ to $x = -\infty$, convex.

Ex. 4. Examine $y = \sin x$; $x = \log y$; $y = \tan x$.

(b) BY POLAR CO-ORDINATES.

172. DEF.—A Polar curve is said to be concave or convex towards its pole at any point, according as the curve at that point does, or does not, lie on the same side of its tangent as the pole.

173. *Prop.*—*A polar curve is concave toward the pole when* $\frac{dr}{dp}$ *is positive, and convex when* $\frac{dr}{dp}$ *is negative; r being the radius vector and p the perpendicular from the pole upon the tangent.*

DEM.—By a simple inspection of (a) *Fig.* 120, it will be seen that r and p are increasing functions of each other when the curve and pole lie on the same side of the tangent; hence $\frac{dr}{dp}$ is $+$. In like manner from (b) it is seen that r and p are decreasing functions of each other when the pole and curve lie on different sides of the tangent; hence $\frac{dr}{dp}$ is $-$.

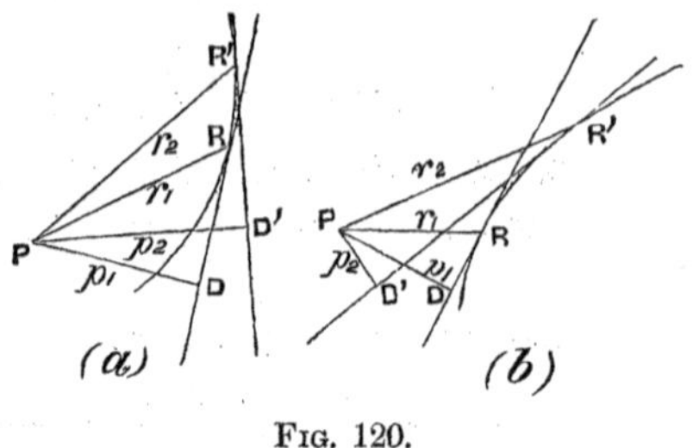

FIG. 120.

174. SCH.—In applying this polar test for direction of curvature, it is necessary that the equation be in terms of p and r. If given in r and θ, θ can be eliminated between the equation of the curve, and $p = \frac{r^2}{\left(\frac{dr^2}{d\theta^2} + r^2\right)^{\frac{1}{2}}}$ (***168***).

Ex. 1. Examine the lituus $\left(r = \frac{a}{\theta^{\frac{1}{2}}}\right)$ with reference to direction of curvature.

SUG'S.—From $r = \frac{a}{\theta^{\frac{1}{2}}}$, $\frac{dr^2}{d\theta^2} = \frac{1}{4}a^2\theta^{-3} = \frac{r^6}{4a^4}$. This substituted in $p = \frac{r^2}{\left(\frac{dr^2}{d\theta^2} + r^2\right)^{\frac{1}{2}}}$, gives $p = \frac{2a^2r}{(r^4 + 4a^4)^{\frac{1}{2}}}$. Whence $\frac{dr}{dp} = \frac{(4a^4 + r^4)^{\frac{3}{2}}}{2a^2(4a^4 - r^4)}$. $\therefore$ This spiral is concave towards the pole for values of r less than $a\sqrt{2}$, and convex for $r > a\sqrt{2}$.

Ex. 2. Show $r = a^{\theta}$ is always concave towards the pole.

SECTION IV.

Singular Points.

175. DEF.—***Singular Points*** of curves are points which possess some property not common to others. Of such points we shall notice: 1st, Points of maxima and minima ordinates; 2nd, Points of inflexion; 3rd, Multiple points; 4th, Cusps; 5th, Isolated or Conjugate points; 6th, Stop points; 7th, Shooting points.

MAXIMA AND MINIMA ORDINATES.

176. Def.—An ordinate is at a maximum when it is greater than the immediately preceding and the immediately succeeding values; and at a minimum when it is less than the immediately preceding and immediately succeeding values.

177. Prob.—*To find the position and values of maxima and minima ordinates.*

Solution.—As $y=f(x)$, this problem is the ordinary one of maxima and minima of functions of a single variable, treated in the Calculus. Hence we find the values of x which render $\frac{dy}{dx}=0$ or ∞, as critical values, *i. e.*, values to be examined, and at which the property exists, if it exist at all. To distinguish between maxima and minima values we have the common test; namely, $+\frac{d^2y}{dx^2}$ characterizes a minimum, and $-\frac{d^2y}{dx^2}$ characterizes a maximum, subject to the conditions discussed in the Calculus. The value or values of y corresponding to the value or values of x found as above, will be the required maxima or minima ordinates.

A Geometrical Solution.—If PD is a maximum, it is evident that at the left of P the tangent makes an acute angle with the axis of x, *i. e.* $\frac{dy}{dx}$ is $+$, and at the right $\frac{dy}{dx}$ is $-$. $\therefore$ $\frac{dy}{dx}=0$ is the point of change from $+$ to $-$, or the point of maximum ordinate. In like manner at the left of a point of minimum ordinate, as P′, $\frac{dy}{dx}$ is $-$, and at the right $+$. $\therefore$ $\frac{dy}{dx}=0$ locates also minimum ordinates.* Finally, since at a point of maximum ordinate the immediately preceding and succeeding ordinates are less, the curve is concave downward, whence we have $-\frac{d^2y}{dx^2}$ characterizing such a point. But, at a point of minimum ordinate, the immediately preceding and succeeding values of y being greater, the curve is convex downward and we have $+\frac{d^2y}{dx^2}$ characterizing this point.

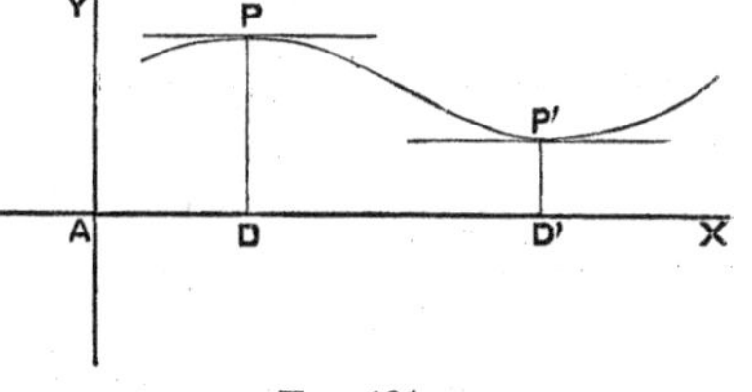

Fig. 121.

178. Sch.—If only the numerical values of the ordinates be considered, when P lies below the axis, $-\frac{d^2y}{dx^2}$ will characterize a minimum, and $+\frac{d^2y}{dx^2}$ a maximum. But a numerical maximum, if $-$, is properly considered a minimum; and a negative numerical minimum, is properly a maximum.

* For $\frac{dy}{dx}=\infty$, see Calculus, p. 91.

Ex. 1. Examine $y = x^3 - 9x^2 + 24x + 16$ for maxima and minima ordinates.

SOLUTION. $\frac{dy}{dx} = 3x^2 - 18x + 24 = 0$. $\therefore x = 4$ and 2, $\frac{d^2y}{dx^2} = 6x - 18$. For $x = 4$, $\frac{d^2y}{dx^2} = 6$; hence $x = 4$ corresponds to a minimum, which value is 32. For $x = 2$, $\frac{d^2y}{dx^2} = -6$; hence $x = 2$ corresponds to a maximum, which value is 36.

[NOTE.—The student should construct the locus, and notice the points. Also substitute values for x a little greater than 4 and a little less, and the same for the point $x = 2$, observing in the results the maxima and minima values of y.]

Ex. 2. Find the location and value of maxima and minima ordinates in the following curves: (1), $y = x^5 - 5x^4 + 5x^3 + 1$; (2), $y = x^3 - 3x^2 - 24x + 85$; (3), $y = 5(x - x^2)$; (4), $y = (2ax - x^2)^{\frac{1}{2}}$; (5), $y = x^4 - 8x^3 + 22x^2 - 24x + 12$; (6), $y = b + (x - a)^3$; (7), $y = x^2(a - x)^2$; (8), in the logarithmic curve; (9), in the curve of tangents; (10), in the cycloid; (11), in the parabola; (12), in the lemniscate of Bernouilli.

POINTS OF INFLEXION.

(a) BY RECTANGULAR CO-ORDINATES.

179. DEF.—***A Point of Inflexion*** is a point where a curve changes direction of curvature for continuously increasing values of x or y. Such a point is also characterized by the fact that the tangent at the point cuts the curve in the point of tangency.

ILL.—In passing from P′ to P″, the curve MN changes direction of curvature, being convex downward at P′, and upward at P″. The point P at which this change occurs is a point of inflexion. The student should not confound a point of inflexion with such a point as P in M′N′. It is true that reckoning *along the curve* from M′ to N′ the curve changes direction of curvature with reference to the axis of x; but not so in reckoning along AX. From D′ to D the curve is both concave and convex towards the axis, and does not *change* at P, but is limited there.

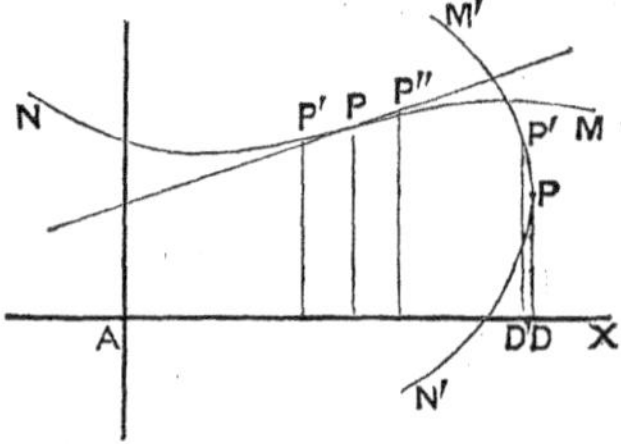

FIG. 122.

180. Prob.—*To determine points of inflexion.*

SOLUTION.—If examined with respect to the axis of x, since, when the curve is convex downward we have $+\frac{d^2y}{dx^2}$, and when concave downward $-\frac{d^2y}{dx^2}$, at the point of inflexion $\frac{d^2y}{dx^2}$ must change sign, and hence must $= 0$, or ∞, $\therefore$ If there be a point of inflexion it is where $\frac{d^2y}{dx^2} = 0$ or ∞. Having determined this point, either construct the curve in the neighborhood of it, or, better, substitute in $\frac{d^2y}{dx^2}$ a value of x a little greater, and one a little less than the critical value, and observe whether $\frac{d^2y}{dx^2}$ really does change sign at the point under consideration.

The precaution in the latter part of this solution is necessary; for, though a varying quantity cannot change sign without passing through 0 or ∞, it does not necessarily change sign upon passing through these values. Thus, let **MN** be a curve whose equation is $y = f(x)$. Now, as x passes from the value **AD** to that of **AD′**, y passes through 0, *but does not change its sign*. In like manner by referring to *Fig.* 109, it will be seen that in the curve there delineated, y passes through ∞ without changing its sign.

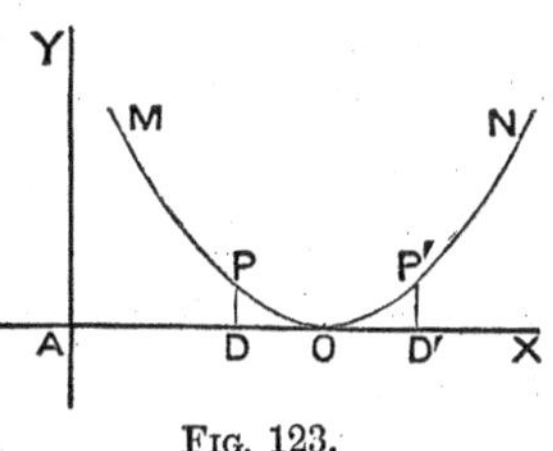

FIG. 123.

Ex. 1. Examine $y = b + (x - a)^3$ for points of inflexion.

SOLUTION. $\frac{d^2y}{dx^2} = 6(x - a) = 0$, gives $x = a$, as a critical point, *i. e.*, one which *may* have the property sought. Now for $x > a$, $\frac{d^2y}{dx^2}$ is $+$; and for $x < a$, $\frac{d^2y}{dx^2}$ is $-$. Therefore there is a point of inflexion at $x = a$. For $x = a$, $y = b$; hence the point of inflexion is (a, b).

Ex. 2. Examine the following for points of inflexion: $a^2y = x^3 - cx^2$; $y = x + 36x^2 - 2x^3 - x^4$; $y = \sin x$; $y = \tan x$; $x = \log y$; the witch of Agnesi.

(*b*) BY POLAR CO-ORDINATES.

181. Prob.—*To test polar curves for points of inflexion.*

SOLUTION.—The equation being put into the form $p = f(r)$, we have seen that for $\frac{dp}{dr}$ $+$, the curve is concave towards the pole, and for $\frac{dp}{dr}$ $-$, it is convex. Therefore $\frac{dp}{dr} = 0$, or ∞, indicates a critical point. If upon examination $\frac{dp}{dr}$ is found to change sign at this point, the point is one of inflexion. Q. E. D.

Ex. 1. Test the lemniscate of Bernouilli for points of inflexion.

SOLUTION.—The equation is $r^2 = a^2 \cos 2\theta$; whence we have $p = \frac{r^3}{\pm a^2}$, and

$\frac{dp}{dr} = \frac{3r^2}{\pm a^2}$. Putting $\frac{3r^2}{\pm a^2} = 0$, $r = 0$. If, therefore, there is a point of inflexion it is at $r = 0$, that is, at the pole. Finally, $\frac{dp}{dr} = \frac{3r^2}{\pm a^2} = \frac{3a^2 \cos 2\theta}{\pm a^2} = \pm 3 \cos 2\theta$, which changes sign for consecutive, real values of r; *i.e.*, when θ passes from 45° to 135° for which change r passes through 0.

Ex 2. Examine the lituus $\left(r = \frac{a}{\theta^{\frac{1}{2}}}\right)$ for points of inflection.

There is a point of inflection at $r = a\sqrt{2}$, $\theta = 28° \ 38' +$.

Ex. 3. Examine $r \frac{a\theta^2}{\theta^2 - 1}$ for points of inflexion.

SOLUTION. $\frac{dr^2}{d\theta^2} = \frac{4r(r-a)^3}{a^2}$. $\therefore p = \frac{ar^2}{(4r^4 - 12ar^3 + 13a^2r^2 - 4a^3r)^{\frac{1}{2}}}$, and $\frac{dp}{dr} = \frac{(6r^2 - 13ar + 6a^2)(-a^2r^2)}{(4r^4 - 12ar^3 + 13a^2r^2 - 4a^3r)^{\frac{3}{2}}} = 0$. Whence $r = 0$, $\frac{3}{2}a$, and $\frac{2}{3}a$. If, therefore, there is a point of inflexion, it must be where r passes through 0, $\frac{3}{2}a$, or $\frac{2}{3}a$. But $\frac{dp}{dr}$ changes sign only with the factor $6r^2 - 13ar + 6a^2$; and this factor does not change sign when r passes through 0, but does at $r = \frac{3}{2}a$ and $\frac{2}{3}a$. (To determine these facts, substitute $r = 0 + h$, and $r = 0 - h$; also $r = \frac{2}{3}a + h$, and $r = \frac{2}{3}a - h$, etc., h being treated as infinitesimal.) $\therefore$ There is a point of inflection at $r = \frac{3}{2}a$. Where $r = \frac{3}{2}a$, $\theta = \sqrt{3}$, or about 99°.26. Where $r = \frac{2}{3}a$, θ is imaginary.

MULTIPLE POINTS.

182. DEF.—There are two species of ***Multiple Points,*** viz., 1st, A point where two or more branches of a curve intersect; 2nd, A point where two or more branches are tangent to each other. The latter are sometimes called *Points of Osculation.* The annexed figures illustrate both species. The first curve has a *triple* point of the first species at **P**, and the second a *double* point of the second species at **P**.

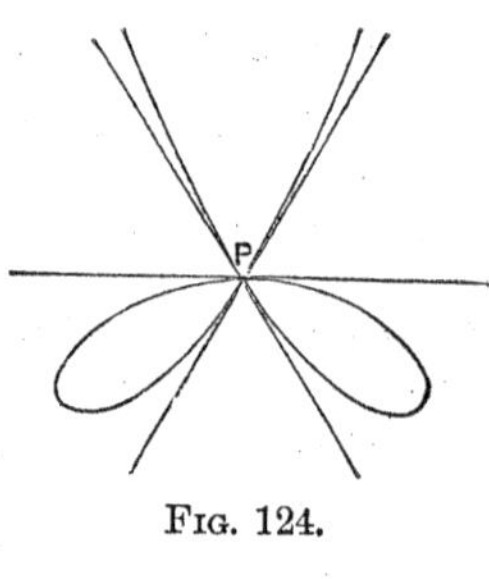

FIG. 124.

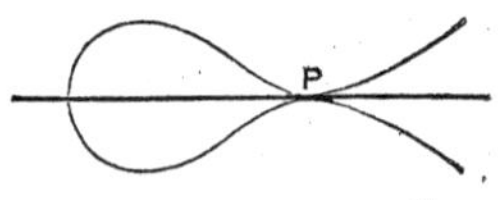

FIG. 125.

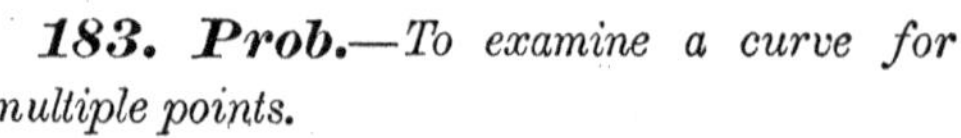

183. Prob.—*To examine a curve for multiple points.*

SOLUTION.—Since two or more branches pass through a multiple point, for $x =$ the abscissa of such a point, y has but one value, while at other points near it, y has two or more values for each value of x. In explicit functions, or in functions of a comparatively simple form, such a point can generally be determined by inspection. Having found a value of x for which y has but one value, and on both sides of which it has two or more, form $\frac{dy}{dx}$, and observe whether it has equal or unequal values at this point. If $\frac{dy}{dx}$ has unequal values the branches of the curve intersect at the point, since their tangents do, and the point is of the first species. If $\frac{dy}{dx}$ has but one value for these values of x and y, the tangents to the branches at the point coincide and the point is of the second species.

When the critical points are not readily determined by inspection, put the equation in the form of an implicit function *without radicals*. Let it be $u = f(x, y) = 0$. Form $\frac{dy}{dx} = -\frac{\frac{du}{dx}}{\frac{du}{dy}}$. Now, as the equation of the locus did not contain radicals, and as differentiation does not introduce them, the only way in which $\frac{dy}{dx}$ can have several values is by taking the form $\frac{0}{0}$. Hence we have $\frac{dy}{dx} = -\frac{\frac{du}{dx}}{\frac{du}{dy}} = \frac{0}{0}$, or $\frac{du}{dx} = 0$, and $\frac{du}{dy} = 0$, from which to determine critical values of x and y, (that is, those values which *may* correspond to multiple points). Solving the equations $\frac{du}{dx} = 0$, and $\frac{du}{dy} = 0$, for x and y, see which of the values found satisfy the equation of the locus. If, at any point thus determined, y has but one real value for the particular value of x, and on both sides of it y has two or more real values, this point is a multiple point. Its species can be determined, as before, by evaluating $\frac{dy}{dx} = \frac{0}{0}$, for the particular values of x and y which locate the point.

Ex. 1. Test for multiple points $y = (x - a)\sqrt{x} + b$.

SOLUTION.—Since $\sqrt{x}$ is both + and −, y has in general two values. But it is evident that for $x = 0$, y has but one value, namely, b; also for $x = a$, y has but one value, b. These are the critical values of x and y. Upon the point $(0, b)$, we observe that the branches do not pass *through* it; since for x negative y is imaginary. Hence $(0, b)$ is *not* a multiple point. But upon the point (a, b) we observe that y has two real values on each side of it. This is therefore a double point. Now $\frac{dy}{dx} = \pm\frac{3x - a}{2\sqrt{x}}$, which for $x = a$ gives $\frac{dy}{dx} = \pm\sqrt{a}$. ∴ The point is of the first species, and the tangents

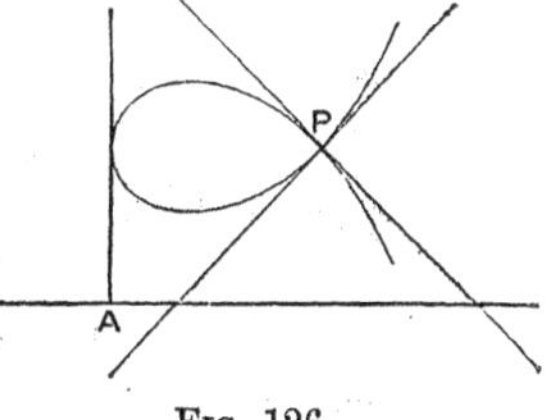

FIG. 126.

to the curve at the point make angles with the axis of x whose tangents are $+\sqrt{a}$, and $-\sqrt{a}$. The form of the curve is given in the figure.

Ex. 2. Examine $y^2 = x^2 - x^4$ for multiple points.

Ex. 3. Examine $x^4 + 2ax^2y - ay^3 = 0$ for multiple points.

SOLUTION.—As it is not easy to discover by inspection all the points to be examined in this case, we will proceed by the second method. We find $\frac{dy}{dx} = -\frac{\frac{du}{dx}}{\frac{du}{dy}} = -\frac{4x^3 + 4axy}{2ax^2 - 3ay^2}$. Whence $4x^3 + 4axy = 0$, and $2ax^2 - 3ay^2 = 0$. These equations give the following critical values: $\begin{cases} x = 0 \\ y = 0 \end{cases}$; $\begin{cases} x = \frac{1}{3}a\sqrt{6} \\ y = -\frac{2}{3}a \end{cases}$; and $\begin{cases} x = -\frac{1}{3}a\sqrt{6} \\ y = -\frac{2}{3}a. \end{cases}$ But of these only the first set satisfy the equation of the curve. The point (0, 0) is, therefore, to be examined. Since none but even powers of x are involved, a change in its sign does not change the form of the function ; hence the form of curve is the same on both sides of the axis of y. As the equation is a cubic, there it at least *one* real root, and hence one branch at least passes through the origin in the plane of the axes. To determine whether the other roots are real or imaginary, and hence whether the other branches lie in the same plane with the axes we might solve the equation. But this is not necessary. We can more readily determine the facts by examining the tangents. Evaluating $\frac{dy}{dx} = -\frac{4x^3 + 4axy}{2ax^2 - 3ay^2}$ for $x = 0$, $y = 0$, we find $\frac{dy}{dx} = 0$, $+\sqrt{2}$ and $-\sqrt{2}$. Therefore there are three tangents, and the point is a triple point of the first species. The curve is that given in *Fig.* 124, (***182***).

Ex. 4. Examine $ay^3 - x^3y - ax^3 = 0$ for multiple points.

SUG'S.—The values arising from $\frac{du}{dx} = 0$, and $\frac{du}{dy} = 0$, are $x = 0$, $y = 0$, and $x = a\sqrt[3]{3}$, $y = -a$. But only the first satisfy the equation of the curve. Evaluating $\frac{dy}{dx} = \frac{3x^2y + 3ax^2}{3ay^2 - x^3} = \frac{0}{0}$ for these values, we find $\left(\text{using } p \text{ for } \frac{dy}{dx}\right)$, $p^3 = 1$, $p^3 - 1 = 0$, or $(p-1)(p^2 + p + 1) = 0$. Whence $p = 1$, or $-\frac{1}{2} \pm \frac{1}{2}\sqrt{-3}$. Hence we see that there is but one tangent in this plane, and therefore but one branch passing through the origin, and no multiple point.

Ex. 5. Show $x^4 + x^2y^2 - 6ax^2y + a^2y^2 = 0$ has a multiple point of the second species at the origin.

CUSPS.

184. Def.—***A Cusp*** is a variety of the second species of double point, in which the osculating branches terminate in the point. Cusps are of two kinds: 1st, When the branches lie on different sides of the tangent; 2nd, When the branches lie on the same side of the tangent.

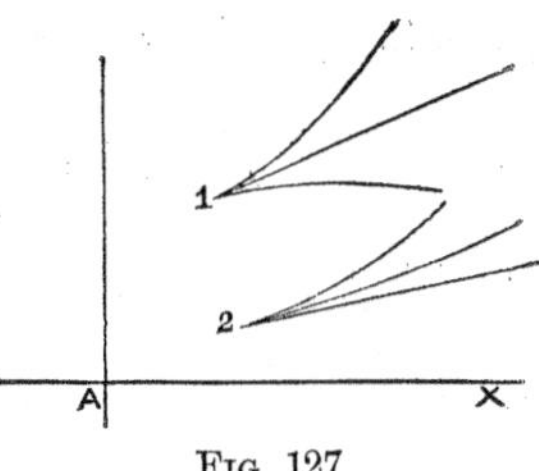

Fig. 127.

185. Prob.—*To examine a curve for cusps.*

Solution.—The process is the same as for multiple points of the second species, the only difference being that the branches stop at the point instead of running through it; and hence that the values of y are real on one side and imaginary on the other.

To ascertain of which kind the cusp is, we may compare the ordinates of the curve in the vicinity of the point, with the corresponding ordinate of the tangent; or, by means of $\frac{d^2y}{dx^2}$, ascertain the direction of curvature; or we may construct the curve about the point. By the first method we discover that the cusp is of the first kind, if the ordinate of the tangent is intermediate in value between the corresponding ordinates of the curve; and that it is of the second kind, if the ordinate of the tangent is less or greater than both the corresponding ordinates of the curve.

If the common tangent is perpendicular to the axis of x, it is best to discuss the cusp with respect to the axis of y, using $\frac{dx}{dy}$, etc.

Ex. 1. Examine $(y-b)^2 = (x-a)^3$ for cusps.

Solution.—We have $y = b \pm (x-a)^{\frac{3}{2}}$, from which we see by inspection that for $x = a$, y has but one value, for $x < a$, y is imaginary, and for $x > a$, y has two real values. Therefore (a, b) is the point to be examined. $\frac{dy}{dx} = \pm \frac{3}{2}(x-a)^{\frac{1}{2}}$ which for $x = a$ becomes ± 0. Hence the two tangents are seen to coincide, their common equation being $y = b$; and there is a cusp. To determine the kind of cusp, we consider the values of the ordinates of the curve for x a little greater than a, as $a+h$, h being an infinitesimal. Substituting, we have $y = b \pm (a+h-a)^{\frac{3}{2}} = b \pm h^{\frac{3}{2}}$. Thus we see that one of the ordinates of the curve, as $SE = b + h^{\frac{3}{2}}, > b$, the corresponding ordinate of the tangent; and the other, as $S'E = b - h^{\frac{3}{2}}, < b$. The cusp is therefore of the first kind.

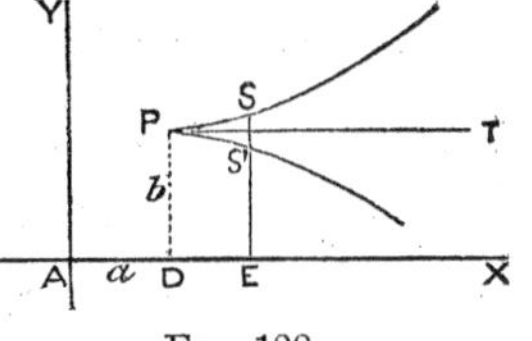

Fig. 128.

Ex. 2. Show that $y = a + x + bx^2 + cx^{\frac{5}{2}}$ has a cusp of the second kind, if the sign of $x^{\frac{5}{2}}$ be considered as ambiguous, and that the equation of the tangent at the cusp is $y = x + a$.

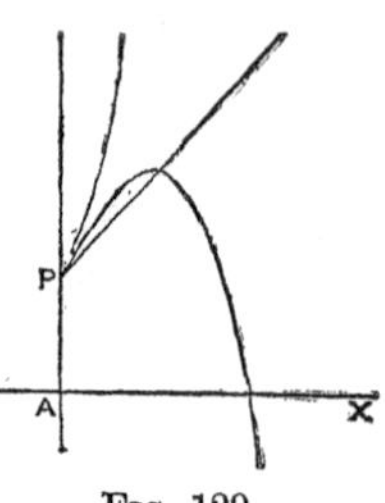

FIG. 129.

SUG.—To determine the kind of cusp, we have $\frac{d^2y}{dx^2} = 2b \pm \frac{15}{4}cx^{\frac{1}{2}}$, both of which values are + for infinitesimal positive values of x. Therefore both branches of the curve are convex downward in the vicinity of the point, and the cusp is of the second kind. The curve has the general form represented in the figure. There is a point of inflexion in the lower branch at $x = \frac{64b^2}{225c^2}$, and it cuts the tangent at $x = \frac{b^2}{c^2}$.

Ex. 3. Show that $cy^2 = x^3$ has a cusp of the first kind at the origin.

Ex. 4. Show that $(y - b - cx^2)^2 = (x - a)^5$ has a cusp of the second kind at $(a, b + ca^2)$.

CONJUGATE POINTS.

186. DEF.—***A Conjugate Point*** is an isolated point the co-ordinates of which satisfy the equation, while in the vicinity of the point, and on each side, real values of one co-ordinate give imaginary values to the other.

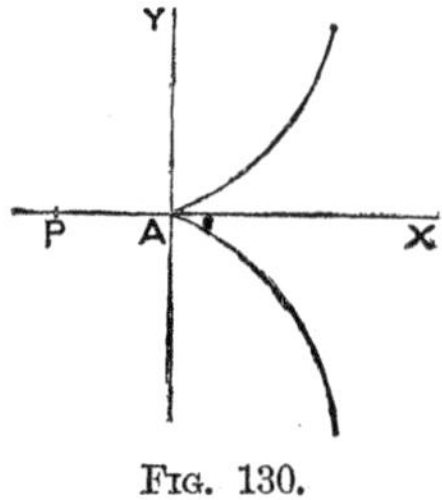

FIG. 130.

ILL.—In the equation $y = (a + x)\sqrt{x}$, if x is negative, y is, *in general*, imaginary; but for the particular value $x = -a$, $y = 0$. Hence **P** is a point in the locus; and as there are no other points in this plane adjacent to it, **P** is an isolated or conjugate point. On the right of the origin any real value of x gives two real, numerically equal values to y, with opposite signs. The curve has therefore two infinite branches on this side, which are symmetrical with respect to the axis of x.

187. Prop.—*At a conjugate point some one or more of the differential coefficients* $\frac{dy}{dx}, \frac{d^2y}{dx^2}, \frac{d^3y}{dx^3}, \frac{d^4y}{dx^4}$, *etc., is imaginary.*

DEM.—Let $y = f(x)$ be the equation of a curve having a conjugate point at (x, y). Then letting h represent an infinitesimal increment or decrement of x, and y' the corresponding value of y, we have $y' = f(x \pm h) =$ an *imaginary* quantity, from the definition (***186***). But

$$y' = f(x \pm h) = y \pm \frac{dy}{dx}\frac{h}{1} + \frac{d^2y}{dx^2}\frac{h^2}{1 \cdot 2} \pm \frac{d^3y}{dx^3}\frac{h^3}{1 \cdot 2 \cdot 3} +, \text{etc.}$$

Now as y and h are both real, to make y' imaginary, some one or more of the coefficients $\frac{dy}{dx}, \frac{d^2y}{dx^2}, \frac{d^3y}{dx^3}$, etc., must be imaginary. Q. E. D.

---

188. Prop.—*Let* $\varphi(x, y) = u = 0$ *be the equation of a curve, freed from radicals; if there is a conjugate point at* (x, y), *the partial differential coefficients* $\frac{du}{dy}$ *and* $\frac{du}{dx}$ *are each equal to* 0, *and* $\frac{dy}{dx} = -\frac{\frac{du}{dx}}{\frac{du}{dy}} = \frac{0}{0}$.

DEM.—Let $\frac{d^ny}{dx^n}$ be the first differential coefficient which is imaginary (***187***). Take the nth derived equation of $u = \varphi(x, y) = 0$, and we have (see Calculus ***112***), $\frac{du}{dy}\frac{d^ny}{dx^n} + \text{- - - - - - - - - -} + \frac{d^nu}{dx^n} = 0$, in which the omitted terms are made up of differential coefficients of u with respect to x and y, and differential coefficients of y with respect to x, of lower orders than the nth. Now, the former are rational, since $u = \varphi(x, y)$ does not contain radicals, and differentiating does not introduce them; and the latter are rational by hypothesis. Hence, in order that the first member of the derived equation may be 0 (which is a rational quantity), $\frac{du}{dy}$ must $= 0$ and thus destroy the imaginary factor $\frac{d^ny}{dx^n}$. Again, $\frac{du}{dx} + \frac{du}{dy} \cdot \frac{dy}{dx} = 0$ (Calculus —); whence as $\frac{du}{dy} = 0$, $\frac{du}{dx} = 0$, and $\frac{dy}{dx} = -\frac{\frac{du}{dx}}{\frac{du}{dy}} = \frac{0}{0}$. Q. E. D.

---

189. Prob.—*To examine a curve for conjugate points.*

SOLUTION.—Since at a conjugate point $\frac{du}{dx} = 0$, and $\frac{du}{dy} = 0$, if we find the values of x and y which satisfy these equations, these values make known the points to be examined; *i. e.*, they are the *critical values*, the same as in the case of multiple points. Having determined the critical values, we may form $\frac{dy}{dx}, \frac{d^2y}{dx^2}, \frac{d^3y}{dx^3}$, etc.; and, if for the co-ordinates of any point under consideration, any one of these coefficients becomes imaginary, that point is a conjugate point.

SCH.—The labor of producing the higher orders of differential coefficients is often so great, that it is better, if $\frac{dy}{dx}$ does not become imaginary, to examine the point by substituting successively $a + h$ and $a - h$ for x in the equation of the curve, a being the value of x to be tested, and h an infinitesimal. If both values of y found in this way are imaginary, the point is a conjugate point.

Ex. 1. Examine $ay^2 - x^3 + 4ax^2 - 5a^2x + 2a^3 = 0$ for conjugate points.

SOLUTION. $\frac{du}{dx} = -3x^2 + 8ax - 5a^2 = 0$, and $\frac{du}{dy} = 2ay = 0$, give $x = a$, $y = 0$, and $x = \frac{5}{3}a$, $y = 0$. Only the first two of these values $(a, 0)$ satisfy the equation of the curve ; hence this point is to be examined. To do this we form $\frac{dy}{dx} = -\frac{\frac{du}{dx}}{\frac{du}{dy}} = \frac{3x^2 - 8ax + 5a^2}{2ay}$. To evaluate this for $x = a$, $y = 0$, we have

$$\frac{dy}{dx} = \frac{6xdx - 8adx}{2ady} = \frac{-2a}{2a}\frac{dx}{dy}, \text{ for } x = a,\ y = 0.$$

Whence $\frac{dy^2}{dx^2} = -1$, or $\frac{dy}{dx} = \sqrt{-1}$. As this is an imaginary quantity, $x = a$, $y = 0$ is a conjugate point.

Ex. 2. Examine $y^2 = x(x + a)^2$ for conjugate points.

There is a conjugate point at $x = -a$, $y = 0$.

Ex. 3. Examine $x^4 - ax^2y - axy^2 + a^2y^2 = 0$ for conjugate points.

There is a conjugate point at $(0, 0)$.

Ex. 4. Examine $(c^2y - x^3)^2 = (x - a)^5(x - b)^6$ for conjugate points, a being greater than b.

SUG'S.—There is a conjugate point at $x = b$, $y = \frac{b^3}{c^2}$. Neither $\frac{dy}{dx}$, nor $\frac{d^2y}{dx^2}$ are imaginary for these values, though $\frac{d^3y}{dx^3}$ is. The better way to solve this, is to find the critical values $x = a$, $y = \frac{a^3}{c^2}$, and $x = b$, $y = \frac{b^3}{c^2}$, as usual. Then substituting in the equation of the curve, we find that both points satisfy the equation, and hence are to be examined. Then substitute in the equation, solved for y, the values $a + h$ and $a - h$. These give real values for y on one side of the point and imaginary values on the other. Hence $x = a$, $y = \frac{a^3}{c^2}$ is not a conjugate point. In the same way substitute in the value of y, $b \pm h$, and y is found to be imaginary on both sides of the point.

SHOOTING POINTS.

190. DEF.—***A Shooting Point*** is a point at which two or more branches of a curve terminate, while each branch has a different tangent at the point.

[NOTE.—This subject is not of sufficient importance to justify an extended discussion. We shall merely give a couple of examples.]

Ex. 1. To show that $y = x \tan^{-1}\frac{1}{x} = x \cot^{-1} x$, has a shooting point at the origin, if we limit the discussion to $x < \pi$.

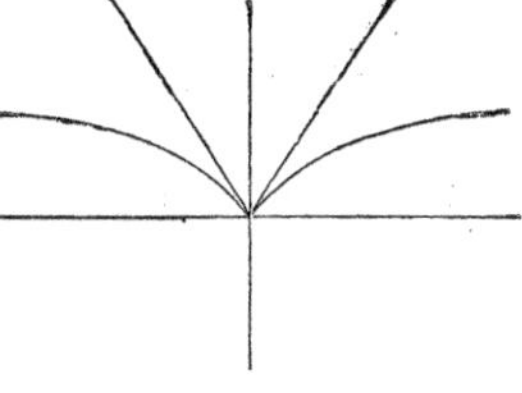

Fig. 131.

Solution.—For $x = 0$, we have $y = 0 \cdot \cot^{-1} 0 = 0 \cdot \frac{1}{2}\pi = 0$; hence the curve has a point in the origin. In the vicinity of the origin, *i. e.*, for very small values of x, x and $\cot^{-1} x$ have the same sign, both being $+$ on the right of the origin, and both $-$ on the left; therefore y is $+$ near the origin and the curve lies above the axis. Moreover, $\frac{d^2y}{dx^2} = -\frac{2}{(1+x^2)^2}$, wherefore the branches on both sides of the origin are concave towards the axis of x, and there is a salient point at the origin, as in the figure. To show that there are two tangents to the curve at this point, and hence that it is not a cusp, to which it bears some resemblance, we form $\frac{dy}{dx} = \cot^{-1} x - \frac{x}{1+x^2}$. This for $x = +0$ is $+\frac{1}{2}\pi$; and for $x = -0$, is $-\frac{1}{2}\pi$.

Ex. 2. To show that $y = \frac{x}{1 + e^{\frac{1}{x}}}$ has a shooting point at (0, 0).

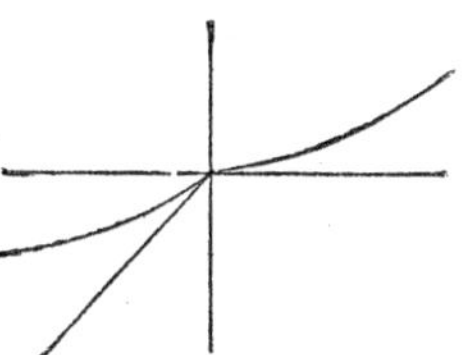

Fig. 132.

Sug's.—For x small and $+$, y is $+$; and for x small and $-$, y is $-$; hence the branches lie as in the figure. Again $\frac{dy}{dx} = \frac{1}{1+e^{\frac{1}{x}}} + \frac{e^{\frac{1}{x}}}{x\left(1+e^{\frac{1}{x}}\right)^2}$, which for $x = +0$, gives $\frac{dy}{dx} = 0$; and for $x = -0$, gives $\frac{dy}{dx} = 1$. Therefore there is a shooting point at (0, 0).

STOP POINTS.

191. Def.—***A Stop Point*** is a point at which a single branch of a curve terminates.

Ex. 1. To show that $y = x \log x$ has a stop point at the origin.

Solution.—In this curve for all $+$ values of x less than 1, y is $-$ and has but one value; for $x = 1$, $y = 0$; and for $x > 1$, y is $+$ and has but one value. There is therefore a single branch of the curve extending to the right from the origin, below the axis of x from $x = 0$ to $x = 1$, and above the axis of x, beyond $x = 1$. But for x negative, y is impossible. $\therefore$ There is a stop point at the origin.

Ex. 2. Show that $y = e^{-\frac{1}{x}}$ has a stop point at the origin.

SECTION V.

Tracing Curves.

192. Def.—***Tracing a Curve*** is discovering from the equation of the curve and its derived functions the general form and leading peculiarities of the curve, and its position with reference to the assumed axes, so that the mind can conceive the locus, or that it may be sketched without going through the details of substituting a series of values, as was done in Section II., Chapter I.

Sch.—While it is practicable to give certain general directions for tracing curves, much has to be to be left to the ingenuity of the student, as the infinite variety of forms of equations renders different methods expedient in different cases. Nor do we know how to trace the loci represented by every form of equation: this would be equivalent to solving equations of all degrees.

---

193. Prob.—*To trace a plane curve given by its equation referred to rectangular axes.*

Method of Solution.—If practicable, put the equation in the form $y = f(x)$. Notice where it cuts the axes. Observe the limits and infinite branches. Examine infinite branches for asymptotes. Find the direction of curvature between established or characteristic points. Determine positions of maxima and minima ordinates. Sometimes it may be serviceable to ascertain the direction of the curve at certain points, as where it cuts the axes, by means of its tangent at those points. Notice the position and character of singular points.

Sch.—In giving the above method of tracing curves, it is not meant that the processes there detailed are necessarily to be gone through with in the order given, nor in fact that they are *all* to be applied in tracing the same curve. These are only means to be used as occasion may require. Again, while these processes are general, and constitute what is usually called "tracing curves," there are other methods better adapted to certain cases. Of these we shall give, in the sequel, three; viz., one when the equation can be put into the form $y = \varphi(x) \pm \psi(x)$, in which $y = \varphi(x)$ is a diametral locus to that represented by the entire equation; another by transformation from one set of rectilinear co-ordinates to another; and a third by passing from one system of co-ordinates to another, as from rectilinear to polar. But we will first attend to a few examples by the general method.

Ex. 1. Trace the curve $y^2 = ax^2 + bx^3$.

Solution.—We have $y = \pm x\sqrt{a + bx}$. On this we observe that for $x = 0$,

$y=0$. $\therefore$ The curve passes through the origin. For $y=0$, $x=0$, or $-\frac{a}{b}$. $\therefore$ The curve cuts the axis of x also at $-\frac{a}{b}$.

For all negative values of x between 0 and $-\frac{a}{b}$, y is real, but beyond $-\frac{a}{b}$ in a negative direction y is imaginary. $\therefore$ $x=-\frac{a}{b}$ is the limit of the curve in this direction. But for all positive values of x, or for all values of $x > -\frac{a}{b}$, y has two numerically equal, real values, affected with opposite signs. $\therefore$ The curve is symmetrical with respect to the axis of x, and has two infinite branches extending to the right.

Again $\frac{dy}{dx} = \pm \frac{a+\frac{3}{2}bx}{\sqrt{a+bx}}$, which for $x=-\frac{a}{b}$, becomes $\frac{dy}{dx} = \pm\infty$, and for $x=0$, $\frac{dy}{dx} = \pm\sqrt{a}$. $\therefore$ At $\left(-\frac{a}{b}, 0\right)$ the curve cuts the axis of x perpendicularly, and at (0, 0) it cuts it in two directions, viz., at $\tan^{-1}(+\sqrt{a})$, and $\tan^{-1}(-\sqrt{a})$. This also shows that (0, 0) is a multiple point, a double point.

Examining for direction of curvature, we have $\frac{d^2y}{dx^2} = \pm \frac{4ab+3b^2x}{4(a+bx)^{\frac{3}{2}}}$, which is $\pm$ between 0 and $-\frac{a}{b}$, and $\pm$ between 0, and $+\infty$. $\therefore$ At the left of the origin, the curve is concave towards the axis of x, and at the right, convex.

We have a maximum and a minimum ordinate at $x=-\frac{2a}{3b}$, $y=\pm\frac{2a}{9b}\sqrt{3a}$, as appears by solving the equation $\pm\frac{a+\frac{3}{2}bx}{\sqrt{a+bx}}=0$.

It only remains to examine the infinite branches for asymptotes.

$$X = x - y\frac{dx}{dy} = \frac{\frac{1}{2}bx^2}{a+\frac{3}{2}bx} = \infty, \text{ for } x=\infty \text{; and}$$

$$Y = y - x\frac{dy}{dx} = \frac{\mp\frac{1}{2}bx^2}{\sqrt{a+bx}} = \mp\infty, \text{ for } x=\infty.$$

Therefore there are no asymptotes.

From this investigation the curve is readily conceived to have the form given in the figure, which is constructed assuming $a=3b$.

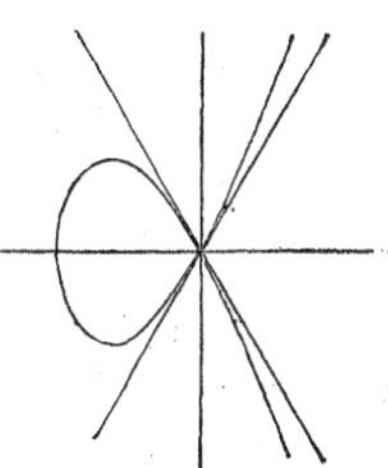

FIG. 133.

Ex. 2. Trace the curve $y^2=a^2x^3$.

Results. The curve is symmetrical with respect to the axis of x; extends only to the right; is convex to the axis of x; has two infinite branches; has a cusp of the first kind at the origin, with the axis of x for the common tangent; and has no asymptote.

Ex. 3. Trace the curve $y=\frac{x}{1+x^2}$.

Results. The curve cuts the axes at the origin under an angle of $\frac{1}{4}\pi$; has one infinite branch extending to the right above the axis of x, and another extending to the left below this axis; has a maximum ordinate at $x = +1$, and a minimum at $x = -1$; has the axis of x as an asymptote to both branches; has points of inflexion at (0, 0), and at $x = \sqrt{3}$, and $x = -\sqrt{3}$; between the latter points is concave towards the axis of x, and beyond them is convex.

Ex. 4. Trace $y^3 = a^3 - x^2$.

Ex. 5. Trace $(y - x^2)^2 = x^5$.

Ex. 6. Trace $ay^2 - x^3 + bx^2 = 0$.

Results. The curve cuts the axis of x at right angles at $(b, 0)$; has a conjugate point at the origin; has points of inflexion at $x = \frac{4}{3}b$; is concave to the axis of x from $x = b$ to $x = \frac{4}{3}b$, and convex beyond; has two infinite symmetrical branches without asymptotes.

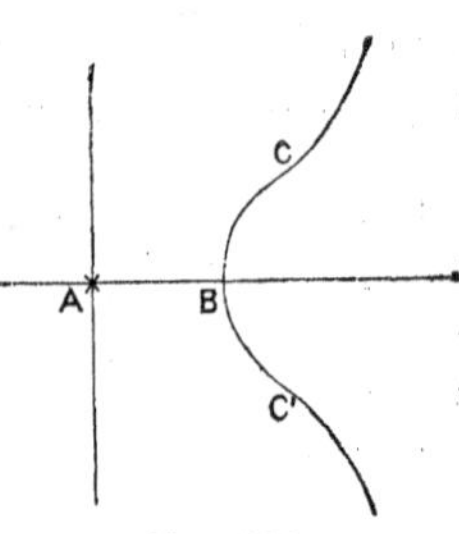

FIG. 134.

Ex. 7. Trace $ay^2 - x^3 + (b - c)x^2 + bcx = 0$.

The form of the curve is given in the figure. Observe that when $c = 0$ this locus becomes identical with the preceding, which is sometimes called the campanulate (bell shaped) parabola.

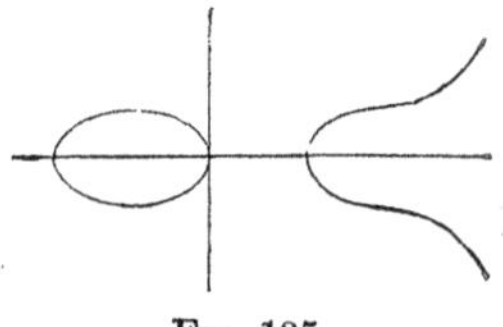
FIG. 135.

Ex. 8. Trace the Folium of Des Cartes, whose equation is $y^3 - 3axy + x^3 = 0$.

Ex. 9. Trace $y^3 = 2ax^2 - x^3$.

Ex. 10. Trace $y^2 = \dfrac{x^3}{x - a}$. Examine the curve for asymptotes, for maxima and minima ordinates, for cusps, for direction of curvature, and points of inflexion.

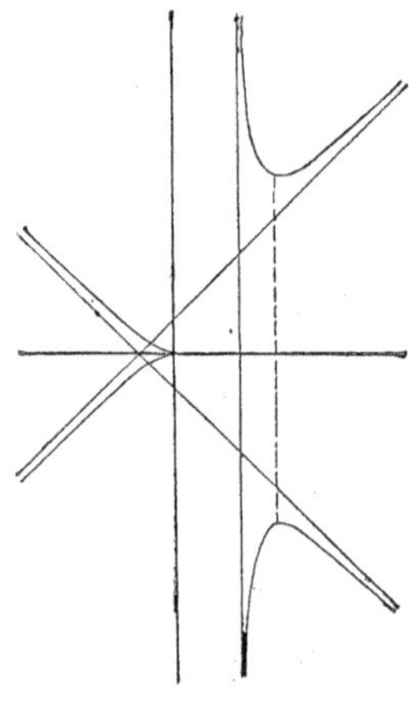
FIG. 136.

Ex. 11. Trace $y^2 = \frac{x^3 + x^2}{x - 1}$. Examine the curve for asymptotes, for limits, and for maxima and minima ordinates.

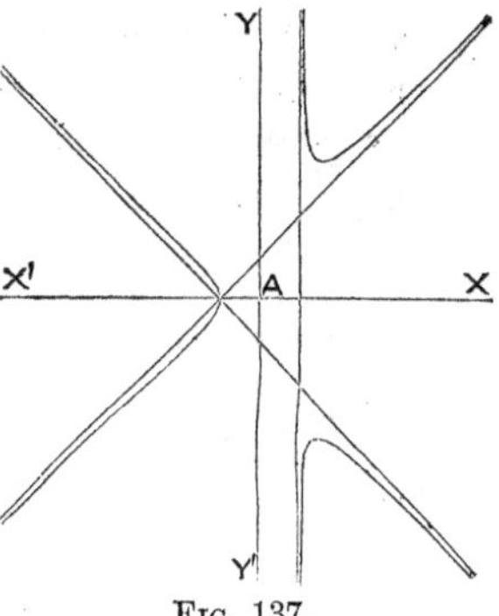

FIG. 137.

---

194. Prob.—*To trace a curve of the second order, that is, the locus of* $Ay^2 + Bxy + Cx^2 + Dy + Ex + F = 0$, *by direct inspection of its equation.*

SOLUTION.—One method of solving this problem has been given on pages 46—49. The present method is given as a good algebraic exercise, and in illustration of the remark in the preceding scholium upon equations which take the form $y = \varphi(x) \pm \psi(x)$.

Solving the equation for y we have

$$y = -\frac{1}{2A}(Bx + D) \pm \frac{1}{2A}\sqrt{(B^2 - 4AC)x^2 + 2(BD - 2AE)x + (D^2 - 4AF)}.$$

1st. If we construct the straight line of which $y = -\frac{1}{2A}(Bx + D)$ is the equation (let it be represented by **MN** in the figure), any value of x (as **AD**) which locates a point (as **P**) in this line, locates, in general, two points (**P'**, **P''**) in the curve, on opposite sides of the line and equally distant from it, this distance being the radical part of the value of y. Therefore $y = -\frac{1}{2A}(Bx + D)$, is a diameter of the locus.

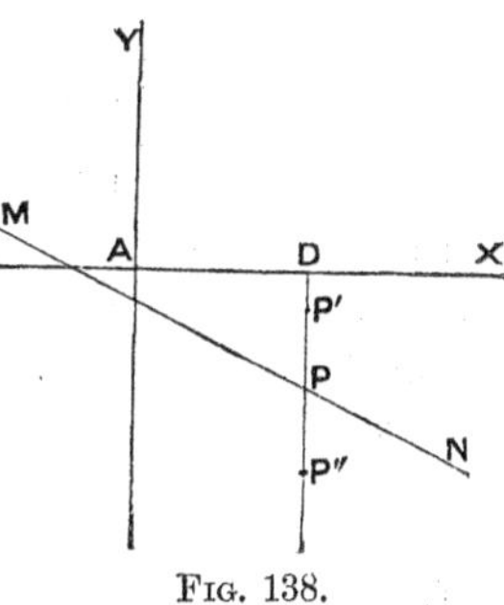

FIG. 138.

2nd. For such value or values of x as render the radical 0, y has but one value, and at this point, or these points, the locus cuts its diameter. Hence $(B^2 - 4AC)x^2 + 2(BD - 2AE)x + (D^2 - 4AF) = 0$ determines where the locus cuts the diameter $y = -\frac{1}{2A}(Bx + D)$. In general, this gives two values of x, indicating that the locus cuts its diameter in two points, as in the ellipse and hyperbola. But if $B^2 - 4AC = 0$, the equation becomes $2(BD - 2AE)x + (D^2 - 4AF) = 0$, which gives only one point of intersection, as in the parabola, a result which agrees with the fact that $B^2 - 4AC = 0$ characterizes a parabola (**62**). Locating the point, or points, at which the curve cuts its diameter, we know, if there are two points, and the curve is an ellipse, that it lies between these limits, or, if an hyperbola, beyond. These facts will readily appear by observing whether intermediate values of x give real or imaginary values to y. Thus the limits of the curve appear.

3rd. If the locus is an ellipse, the values of y midway between the two values of

x which correspond to the extremities of the diameter, make known a diameter parallel to tangents at the extremities of the former, and hence determine the circumscribed parallelogram. Thus the situation of the ellipse becomes known.

4th. If the locus is an hyperbola, we can determine a few values of y corresponding to values of x without the limits, and thus locate the curve. It is often expedient to find the intersections with the axes.

5th. If the locus is a parabola, having determined its diameter and vertex, a few values of x will make known sufficient points to enable us to sketch the curve. The intersections with the axes may also be of service.

Ex. 1. Trace the curve whose equation is $y^2 - 2xy + 2x^2 + 2y + x + 3 = 0$.

SUG'S.—Since $B^2 - 4AC < 0$, the locus is an ellipse. Solving for y, we have

$$y = x - 1 \pm \sqrt{-x^2 - 3x - 2};$$

whence $y = x - 1$ is a diameter, which we construct. $\sqrt{-x^2 - 3x - 2} = 0$, gives $x = -1$, and -2, the limits of the curve. Between these limits y is real, and without them it is imaginary. For $x = -1\frac{1}{2}$, $y = -2$, and -3. Thus we find the circumscribed parallelogram.

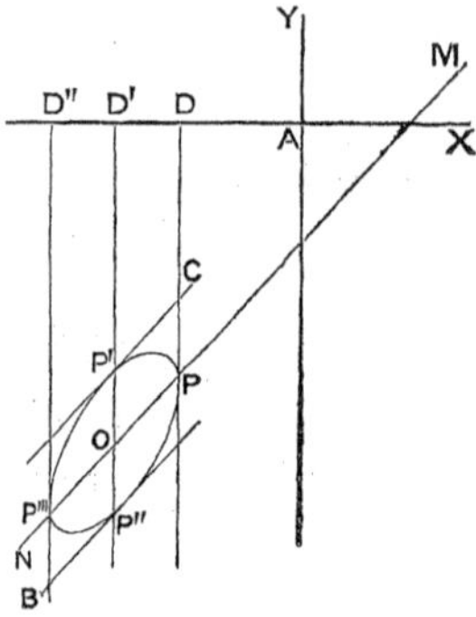

FIG. 139.

Ex. 2. Trace the curve whose equation is $y^2 + 2xy - 2x^2 - 4y - x + 10 = 0$.

SUG'S.—As $B^2 - 4AC > 0$ the locus is an hyperbola. $y = -x + 2 \pm \sqrt{3(x^2 - x - 2)}$. $y = -x + 2$ locates **NM**. From $3(x^2 - x - 2) = 0$, we find **P** and **P'''**, at $x = 2$, and -1. Between these values y is imaginary; hence the locus lies beyond these points to the right and left. Putting $y = 0$, we have $-2x^2 - x + 10 = 0$, whence $x = 2$, and $-2\frac{1}{2}$, and the curve cuts the axis of x at **C** and **B**. For $x = 4$, $y = 3 \cdot 5$ and $-7 \cdot 5$ nearly, and we have 1 and 2. In like manner as many points as we wish may be found; but with the diameter and intersections with the axis, little or nothing more is necessary in order to form a pretty definite idea of the situation of the curve.

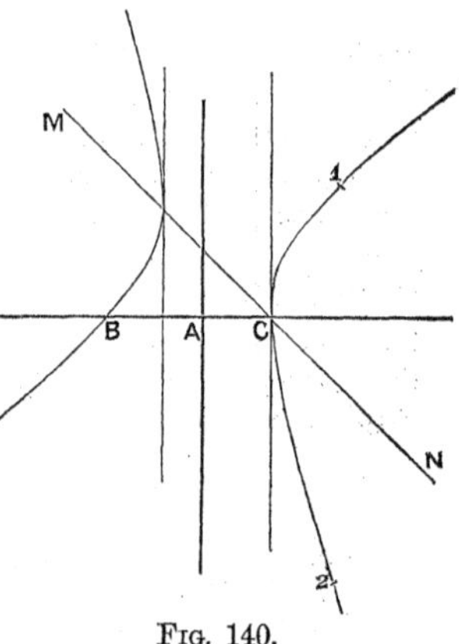

FIG. 140.

Ex. 3. Trace the locus $y^2 - 2xy + x^2 - 4y + x + 4 = 0$.

SUG'S.—Since $B^2 - 4AC = 0$, the locus is a parabola, $y = x + 2$ is the equation of a diameter. For $x = 0$, $y = 2$. For x negative, y is imaginary. For $x = 3$, $y = 8$, and 2.

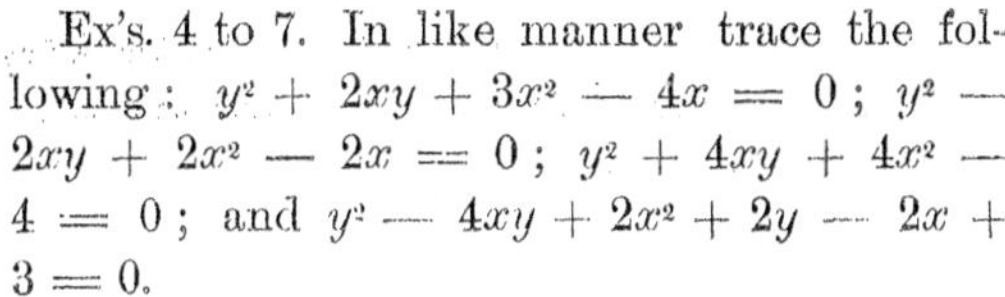

Ex's. 4 to 7. In like manner trace the following: $y^2 + 2xy + 3x^2 - 4x = 0$; $y^2 - 2xy + 2x^2 - 2x = 0$; $y^2 + 4xy + 4x^2 - 4 = 0$; and $y^2 - 4xy + 2x^2 + 2y - 2x + 3 = 0$.

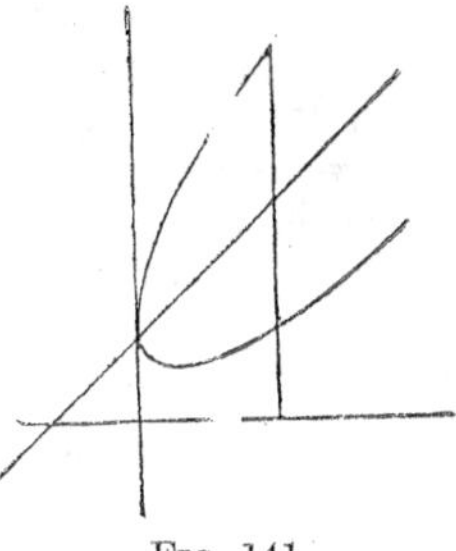

FIG. 141.

---

195. Prob.—*To trace a locus of the second order by means of transformation of co-ordinates.*

SOLUTION.—We will illustrate this method by an example. The method itself is altogether too tedious for practical purposes, but is highly important as giving a clear view of a process which we have occasion to use for other purposes. Let us trace the locus whose equation is $y^2 + 2xy + 3x^2 - 4x = 0$.

This is an ellipse, since $B^2 - 4AC < 0$. We will find its equation when referred to its own axes. This requires transformation from one rectangular system to another. The *formulæ* for this transformation are $x = x_1 \cos\alpha - y_1 \sin\alpha + m$, and $y = x_1 \sin\alpha + y_1 \cos\alpha + n$. Substituting these in the equation, we have

$$\begin{array}{r|r|r|r|r|r|l}
\cos^2\alpha & y_1^2 + 2\sin\alpha\cos\alpha & x_1y_1 + \sin^2\alpha & x_1^2 + 2n\cos\alpha & y_1 + 2n\sin\alpha & x_1 + n^2 & \\
-2\sin\alpha\cos\alpha & -2\sin^2\alpha & +2\sin\alpha\cos\alpha & +2m\cos\alpha & +2m\sin\alpha & +2mn & \\
+3\sin^2\alpha & +2\cos^2\alpha & +3\cos^2\alpha & -2n\sin\alpha & +2n\cos\alpha & +3m^2 & =0. \\
 & -6\sin\alpha\cos\alpha & & -6m\sin\alpha & +6m\cos\alpha & -4m & \\
 & & & +4\sin\alpha & -4\cos\alpha & &
\end{array}$$

(Eq. A.)

As the required form of the equation is $Ay^2 + Bx^2 + F = 0$, we desire to eliminate the terms containing x_1y_1, and y_1 and x_1. To find the direction of the new axes, *i. e.* to determine the value of α, and to find the position of the new origin. *i. e.*, to determine the values of m and n, which will effect this reduction, we place the coefficients of the terms to be eliminated each equal to 0, and solve the resulting equations. These equations are

(1) $2\sin\alpha\cos\alpha - 2\sin^2\alpha + 2\cos^2\alpha - 6\sin\alpha\cos\alpha = 0$;

(2) $2n\cos\alpha + 2m\cos\alpha - 2n\sin\alpha - 6m\sin\alpha + 4\sin\alpha = 0$;

(3) $2n\sin\alpha + 2m\sin\alpha + 2n\cos\alpha + 6m\cos\alpha - 4\cos\alpha = 0$.

From (1) we find $\sin\alpha = .92388$, and $\cos\alpha = -.38268$, whence $\alpha = 112° 30'$.

From (2) we have $\tan\alpha = \dfrac{m+n}{n+3m-2}$, or $-2.414 = \dfrac{m+n}{n+3m-2}$; and from (3), $2.414 = \dfrac{n+3m-2}{m+n}$. Solving these equations we find $m = 1$, and $n = -1$.

Substituting these values of $\sin\alpha$, $\cos\alpha$, m, and n in (Eq. A.), we have, after reduction,

$$3.414y_1^2 + .5858x_1^2 - 2 = 0,$$

as the equation of the ellipse referred to its own axes. This gives the axes as 3.7, and 1.53.

To locate the curve we have but to construct the new origin at $(1, -1)$ as A_1, and drawing A_1X_1 making an angle of $112° 30'$ with the primitive axis x, make A_1Y_1 perpendicular to it, and on these axes construct an ellipse whose axes are 3.7, and 1.53.

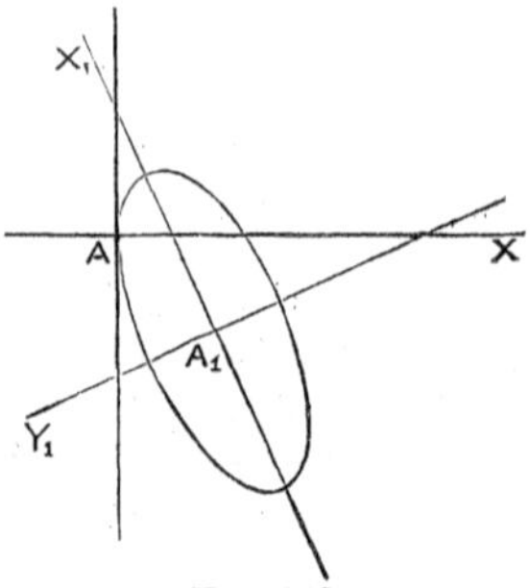

FIG. 142.

Ex. Trace by means of transformation of co-ordinates the locus whose equation is $x^2 - 6xy + y^2 - 6x + 2y + 5 = 0$.

Results. The new origin (the centre of the hyperbola) is at $(0, -1)$. The transverse axis, which is the new axis of x makes an angle of 135° with the primitive; and the transformed equation is $2y^2 - 4x^2 - 4 = 0$.

---

196. Prob.—*To trace a Polar curve.*

METHOD OF SOLUTION.—1st. Assign such values to θ as give easily determined values of r: these will usually be such as $0, \frac{1}{2}\pi, \pi, \frac{3}{2}\pi, 2\pi$, etc.; or, if some multiple of θ is involved in the equation, like parts of these values. Thus if $\sin 2\theta$ is involved, making $\theta = 0°$, $\sin 2\theta = 0$; if $\theta = 15°$, $\sin 2\theta = \frac{1}{2}$; if $\theta = 45°$, $\sin 2\theta = 1$, etc. Construct these points. This will often be sufficient to determine the locus.

2nd. Form $\frac{dr}{d\theta}$ and observe when (for what values of r and θ) r and θ are increasing functions of each other and when decreasing. When $\frac{dr}{d\theta} = 0$ the point is an apsis, *i. e.* one at which the curve is at right angles to the radius vector: at such a point r is a maximum or minimum. Thus in the ellipse when the pole is taken at the focus the vertices of the transverse axis are apsides.

3rd. Examine the curve for asymptotes, direction of curvature, points of inflexion, and any other peculiarities which may be suggested at this stage of the proceeding.

Ex. 1. Trace the lituus $r = \frac{a}{\theta^{\frac{1}{2}}}$.

SOLUTION.—The unit angle being that whose arc equals radius is about $57°.3$. Now letting $a = 1$, and $\theta = 1, 2, 3, 4, 5$, and 6, successively, we get $r = \pm 1, \pm .7, \pm .58, \pm .5, \pm .45, \pm .41, \pm .4$, etc. Locating the positive values, we get the points 1, 2, 3, etc.; and locating the negative values we have $-1, -2, -3$, etc. The two branches are symmetrically equal.

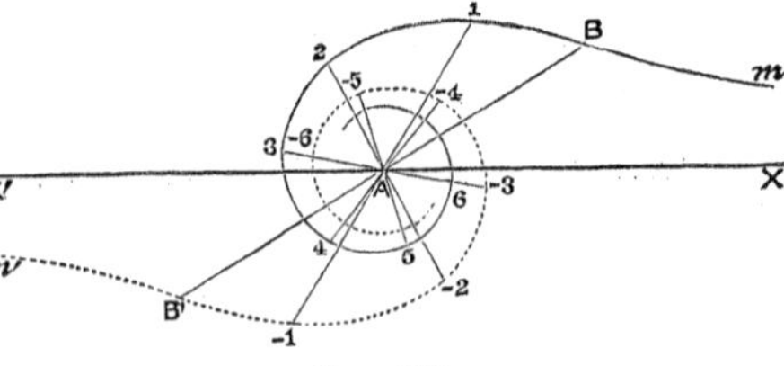

FIG. 143.

Again $\frac{dr}{d\theta} = -\frac{r^3}{2}$, a being $=1$; whence it appears that r and θ are decreasing functions of each other throughout all their values, and the curve makes an infinite number of revolutions around the pole, commencing from ∞ when $\theta = 0$, and reaching the pole when $\theta = \infty$. $\frac{dr}{d\theta} = -\frac{r^3}{2} = 0$, gives $r = 0$. The curve cuts the radius vector obliquely, being parallel to it at ∞, and approaching perpendicularity as r approaches 0, or θ approaches ∞. The pole is an apsis.

Since for $\theta = 0$, $r = \infty$, the subtangent $\frac{r^2 d\theta}{dr} = -\frac{2}{r}$, is 0 for $\theta = 0$, and the polar axis is an asymptote.

To discuss the direction of curvature, we obtain the equation of the spiral in terms of the perpendicular from the pole upon the tangent. This equation is $p = \frac{2r}{(r^4+4)^{\frac{1}{2}}}$; whence $\frac{dp}{dr} = \frac{2(4-r^4)}{(r^4+4)^{\frac{3}{2}}}$. There is a point of inflexion at $r = \pm\sqrt{2}$, $\theta = \frac{1}{2}$, **B** and **B'**. From **B** to the right this branch is convex toward the pole; and from **B** toward the left it is concave, as appears from considering the sign of $\frac{dp}{dr}$, for $r > \sqrt{2}$, and for $r < \sqrt{2}$.

Ex. 2. Trace the locus whose polar equation is $r = a \sin 3\theta$.

SOLUTION.—If $\theta = 0^\circ$, 30°, 60°, 90°, 120°, 150°, 180°, successively, $r = 0$, a, 0, $-a$, 0, a, 0.

$\frac{dr}{d\theta} = 3a \cos 3\theta$, which is positive from $\theta = 0$, to $\theta = 30^\circ$, negative from $\theta = 30^\circ$ to $\theta = 90^\circ$, positive from $\theta = 90^\circ$ to $\theta = 150^\circ$, etc. Whence we see that r begins at 0 when $\theta = 0^\circ$, increases till $\theta = 30^\circ$, diminishes as θ passes from 30° to 60°, becomes 0° for $\theta = 60^\circ$, continues to diminish (becoming negative) as θ passes to 90°, becomes $-a$, at 90°, etc. [The pupil should trace r through an entire revolution, in both positive and negative directions.]

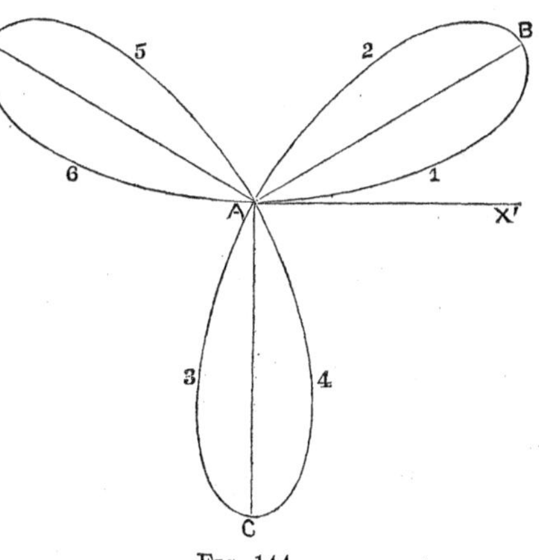

FIG. 144.

$\frac{dr}{d\theta} = 3a \cos 3\theta = 0$, gives apsides at $\theta = 30^\circ$, 90°, and 150°, *i. e.* at **B**, **C**, and **D**, in the figure.

As r never $= \infty$, there is no asymptote.

The equation in terms of the perpendicular upon the pole is $p = \frac{r^2}{(9a^2 - 8r^2)^{\frac{1}{2}}}$, whence $\frac{dp}{dr} = \frac{18a^2 r - 8r^3}{(9a^2 - 8r^2)^{\frac{3}{2}}}$ and the curve is always concave toward the pole.

Ex. 3. Construct the locus whose equation is $x^4 - ax^2y + ay^3 = 0$, by first passing to the polar equation.

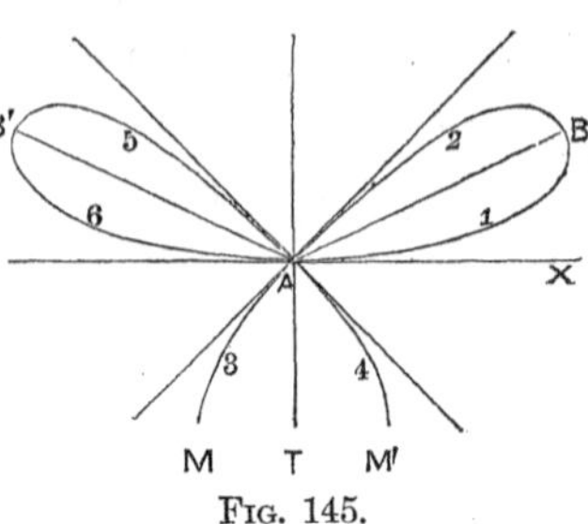

FIG. 145.

SOLUTION.—The polar equation with the pole at the origin is $r = a\dfrac{\sin\theta}{\cos^4\theta}(\cos^2\theta - \sin^2\theta)$. From $\theta = 0^\circ$ to $\theta = 45^\circ$ r is real, finite and passes from 0 to 0. Therefore there is a loop in the first octant. Letting $a = 1$, $\dfrac{dr}{d\theta} = \dfrac{1 - 3\sin^2\theta - 2\sin^4\theta}{\cos^5\theta} = 0$, gives a maximum radius vector for $\theta = 32^\circ$ nearly, $r = .45$.

From $\theta = 45^\circ$ to $\theta = 135^\circ$, r is negative. We will first examine the values of r between $\theta = 45^\circ$ and 90°. This gives a continuous curve in the 6th octant, **A3M**. Putting the equation in the form $r = \sec\theta(\tan\theta - \tan^3\theta)$, we observe that as $\tan\theta > 1$ from $\theta = 45^\circ$ to $\theta = 90^\circ$, r rapidly increases, and becomes $-\infty$ at $\theta = 90^\circ$. The branch in this octant is, therefore, infinite. To ascertain more fully the character of this branch, we form the subtangent Subt $= \dfrac{r^2d\theta}{dr} = \dfrac{\sin^2\theta(\cos^2\theta - \sin^2\theta)^2}{\cos^8\theta} \times \dfrac{\cos^5\theta}{1 - 3\sin^2\theta - 2\sin^4\theta} = \dfrac{\tan^2\theta\sec\theta(1 - 2\sin^2\theta)^2}{1 - 3\sin^2\theta - 2\sin^4\theta}$. Now since between 45° and 90°, $\sin^2\theta$ is between $\frac{1}{2}$ and 1, $\tan\theta$ between 1 and ∞, $\sec^2\theta$ between 2 and ∞, and $\tan\theta$ and $\sec\theta$ increase much more rapidly than $\sin\theta$, it is easy to see that subt. constantly increases and becomes $-\infty$ at $\theta = 90^\circ$. $\therefore$ This is a parabolic branch and approaches to parallelism with **AT**.

Finally, since $r = f(\sin\theta, \cos\theta)$, and only even powers of $\cos\theta$ are involved, the values of r will be repeated in the inverse order as θ passes from 90° to 180°.

Ex. 4. Trace the locus whose equation is $y^2 = \dfrac{x^3 + x^2}{x - 1}$, by passing to the polar equation.

SUG.—The polar equation with the pole at the origin is $r = \dfrac{1}{\cos\theta(1 - 2\cos^2\theta)}$.

(See *Ex.* 11, **193.**)

SECTION VI.

Rate of Curvature.

197. DEF.—***The Curvature*** of a plane curve is its rate of deviation from a tangent, and is measured by the subtenses of indefinitely small but equal arcs.

ILL.—Let **MN** and *mn* be any two circles, **AT** and **AT'** tangents, and **AS** and **AS'** infinitely small but equal arcs. Then will **TS** and **T'S'**, drawn perpendicular to the tangents, be the subtenses which measure the curvature of the arcs of the respective circles ; and we shall have *curvature of* **MN** : *curvature of* *mn* :: **TS** : **T'S'**. That curve is said to have the greatest curvature which deviates most rapidly from its tangent ; thus, in circles, the greater the radius the less the curvature ; *i. e.*, the curvature and radius are inverse functions of each other. It is also evident that the circumference of the same circle has the same curvature at all points ; while in other curves, as the conic sections, the curvature varies at every successive point. In the ellipse the curvature varies from its maximum at the extremities of the transverse axis to its minimum at the extremities of the conjugate axis. In the parabola and hyperbola the curvature is greatest at the vertex and diminishes as the point recedes, becoming 0 at infinity.

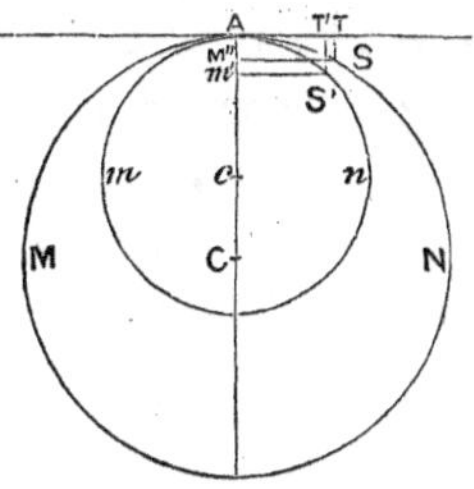

Fig. 146.

It is the object of this section to present a method of measuring curvature, and of comparing the rates of curvature of the same curve at different points, and to ascertain the law of variation. For this purpose a circle is used, called the osculatory circle.

198. Def.—***An Osculatory Circle*** is a circle which has the same curvature as a given curve at a given point ; or, it may be defined as the circle which has the closest contact with a given curve at a given point.

Ill.—Let **BDEC** be an ellipse. If with the centres upon **DC** various circumferences be passed through **D**, it is evident that they will coincide in very different degrees with the ellipse. Some will fall within, and others without. Now the one which coincides most nearly, as in this case **MN**, is the osculatory circle of the ellipse at the point **D**. The arc of the osculatory circle in this case is exterior to the ellipse. The osculatory circle at the vertex, as $m''n''$ is within, and at any other point, as **P**, cuts the ellipse, as will be shown hereafter.

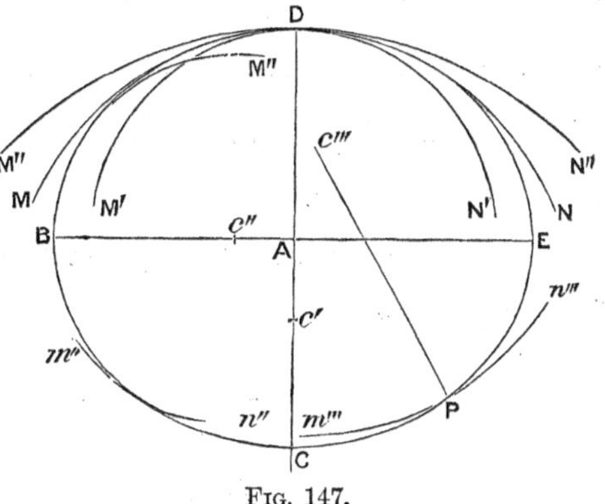

Fig. 147.

199. Def.—***The Radius of Curvature*** is the radius of the osculatory circle ; ***The Centre of Curvature*** is the centre of the osculatory circle ; and the point of closest contact is the point of osculation.

200. Def.—***Contact.*** Let **MN** and **M'N'** be two curves whose equations are respectively $y=f(x)$ and $y'=\varphi(x')$. Suppose the curves to have a common point **P**, so that for $x=x'=$ **AD**, $y=y'=$ **PD**. Now if x and x' take the infinitesimal increment **DD'**, which we will represent by h, designating the corresponding values of y and y', by Y and Y' (**SD'** and **S'D'**), we have

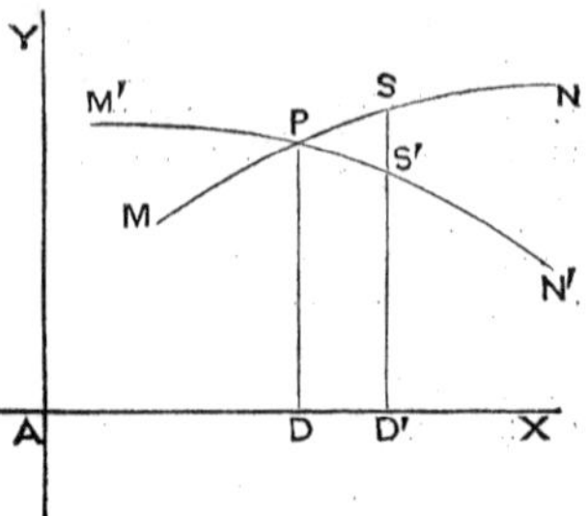

Fig 148.

$$\text{Y}=f(x+h)=y+\frac{dy}{dx}h+\frac{d^2y}{dx^2}\frac{h^2}{2}+\frac{d^3y}{dx^3}\frac{h^3}{2\cdot3}+\frac{d^4y}{dx^4}\frac{h^4}{2\cdot3\cdot4}+\text{, etc.};$$

$$\text{and Y}'=\varphi(x'+h)=y'+\frac{dy'}{dx'}h+\frac{d^2y'}{dx'^2}\frac{h^2}{2}+\frac{d^3y'}{dx'^3}\frac{h^3}{2\cdot3}+\frac{d^4y'}{dx'^4}\frac{h^4}{2\cdot3\cdot4}+\text{, etc.}$$

Subtracting the second of these equations from the first, we have

$$\text{Y}-\text{Y}'=(y-y')+\left(\frac{dy}{dx}-\frac{dy'}{dx'}\right)h+\left(\frac{d^2y}{dx^2}-\frac{d^2y'}{dx'^2}\right)\frac{h^2}{2}+\left(\frac{dy^3}{dx^3}-\frac{d^3y'}{dx'^3}\right)\frac{h^3}{2\cdot3}+\text{, etc.}$$

Now the contact of these curves will evidently be closer as Y — Y' (**SS'**) is less. We may therefore notice the following degrees of conformity:

1st. If in the case of any two loci whose equations $y=f(x)$ and $y'=\varphi(x')$, there is no value of $x=x'$ which renders $y=y'$, there is no common point.

2nd. If for $x=x'$, $y=y'$ and the differential co-efficients are unequal, the contact is the slightest, and is mere ***Intersection.***

3rd. If in addition to $y=y'$, $\frac{dy}{dx}=\frac{dy'}{dx'}$, and the succeeding coefficients are unequal, the contact is closer than before and is called ***Tangency.*** This is called contact of the *First Order.*

4th. If we have at the same time $y=y'$, $\frac{dy}{dx}=\frac{dy'}{dx'}$, and $\frac{d^2y}{dx^2}=\frac{d^2y'}{dx'^2}$, and the succeeding coefficients unequal, the contact is of the *Second Order*, etc.

201. Sch.—A geometrical elucidation of this subject is obtained by considering that "an infinitesimal element of the curve commencing from a given point, being straight, is coincident with the tangent line at that point; and the next element of the curve, being inclined at an angle to the former one, deviates from the tangent. Now let the two consecutive elements be of equal lengths, and from the extremity of the second let a perpendicular be drawn to the tangent: as this perpendicular is longer or shorter, the curve will deviate more or less from the tangent, that is, be more or less bent."* Again, in general a rectilinear tangent is considered as having *two points* in common with a curve, and a circle *three*, since through the three consecutive points one circumference, and only one, can be passed.

202. DEF.—***A Parameter,*** as the term is used in this and similar discussions, is an arbitrary constant entering into an equation of a locus, but which is made variable by hypothesis. Thus in the equation $y = ax + b$, a and b are constants as ordinarily considered, that is have the same values throughout the same discussion. Again, they are *arbitrary* constants, since they may have any values. Finally, we may consider how a straight line changes position when a and b vary continuously. In this case a and b are called *parameters.*

---

203. *Prop.*—*If one curve be given in species, magnitude, and position, that is entirely given, and a second given only in species, in general the highest order of contact possible is equal to the number of parameters in the equation of the second curve less one.*

ILL.—As this proposition usually seems to the learner quite abstract, we will give a familiar illustration of its meaning before proceeding to its demonstration. Let $9y^2 + 4x^2 = 36$ be the first locus. The species is *ellipse;* the magnitude is determined by the value of the axes 6 and 4 : the form of the equation determines the position of the locus. Thus this curve is given in species, magnitude and position, or entirely given. Constructing it we have the ellipse in the figure.

Let the second equation be that of a circle in its general form, viz., $(x - m)^2 + (y - n)^2 = r^2$, in which m, n, and r, are arbitrary constants, which we propose to treat as variables, thus making them parameters. It is evident that the closeness of contact of these two curves will depend on two things, the value of the radius, and the position of the centre; but the position of the centre depends upon the values of m and n. Hence the closeness of contact depends upon the values of the *three* parameters m, n, and r. Thus if **P** be the common point, by locating the centre at **C**, and using **CP** as radius, it is evident that the contact is much closer than when **C**′ is the centre and **C**′**P** the radius. There is therefore *some position of the centre* and *some value of the radius* which will give the circle closer contact than any other. Moreover it is evident that we have given the widest possible opportunity for varying the contact, by taking that form of the equation of the circle which has the three parameters m, n, and r. We will now give the demonstration.

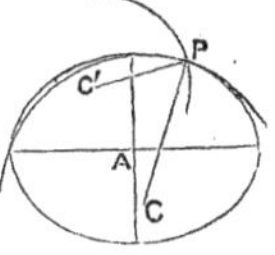

FIG. 149.

DEM.—Let $y = f(x)$, and $y' = \varphi(x')$ be the equations of the loci. In order that we may make $y = y'$ for some value of $x = x'$ we must have liberty to impose *one* arbitrary condition (*i. e.*, to vary the second locus in at least one respect), but this requires one parameter. If, in addition to this parameter, there is a second (*i. e.*, if we can vary the curve in another respect) we can impose another arbitrary condi-

---

* Price's Infinitesimal Calculus.

tion, as $\frac{dy}{dx} = \frac{dy'}{dx'}$, and so on for any number of parameters. Hence we see that *one* parameter makes intersection possible; *two* make tangency or *contact* of the *first* order possible; *three* contact of the *second* order, etc.

204. Cor. 1.—*The right line can have in general no higher order of contact than the first (tangency), since its equation* y = ax + b *has but two parameters* a *and* b.

205. Cor. 2.—*As the equation of the circle in its general form has but three parameters, it can in general have no higher order of contact than the second.*

206. Cor. 3.—*The parabola can have contact of the third order, and the ellipse and hyperbola of the fourth.*

207. Sch.—This discussion assumes that $y = f(x)$, which is given in all respects, is of such a character as to allow of any degree of contact. Of course the possibilities of contact are limited as much by one of the loci as by the other. Thus, if the first locus were a circle and the second an ellipse, the contact could not in general be above the second order, although the ellipse has a possible contact of the fourth order with other curves. Again, in this discussion we have said "in general," since exceptions occur at certain singular points. Some of these will be noticed hereafter. Thus far we have given the broader view of osculation, although for the practical purpose of the measurement of curvature we might limit our view to the circle, as we shall do in the following propositions.

---

208. Prob.—*To produce the general differential formulæ for the value of radius of curvature and the co-ordinates of the centre of curvature of any plane curve, in terms of the co-ordinates of the given curve.*

Solution 1.—Let $y = f(x)$ be the equation of the given locus, and $(x' - m)^2 + (y' - n)^2 = r^2$ the equation of the circle. Now as the equation of the circle contains three arbitrary constants, m, n, and r, we may impose three conditions and find the values of these constants which fulfill them. The conditions requisite for the closest contact which a circle can have, are, for $x = x'$, $y = y'$, $\frac{dy}{dx} = \frac{dy'}{dx'}$, and $\frac{d^2y}{dx^2} = \frac{d^2y'}{dx'^2}$. These therefore are the conditions to be imposed, and from which the values of m, n, and r are to be obtained. In any given case it will be sufficient to find the values of y, $\frac{dy}{dx}$, and $\frac{d^2y}{dx^2}$, in the equation of the locus, and also the values of y', $\frac{dy'}{dx'}$, and $\frac{d^2y'}{dx'^2}$, in the general equation of the circle, and equating the corresponding values find from the three equations thus formed the values of m, n and r.

But for practical purposes, general *formulæ* are more convenient. These are readily produced, as follows :

Differentiating the equation of the circle twice in succession we have

(1) $(x'-m)+(y'-n)\frac{dy'}{dx'}=0$

(2) $1+\frac{dy'^2}{dx'^2}+(y'-n)\frac{d^2y'}{dx'^2}=0.$

In these equations and the general equation of the circle

(3) $(x'-m)^2+(y'-n)^2=r^2,$

we can now substitute the values of y, $\frac{dy}{dx}$, and $\frac{d^2y}{dx^2}$ as obtained from the equation of the given locus considering $x=x'$, and have

(4) $(x-m)+(y-n)\frac{dy}{dx}=0,$

(5) $1+\frac{dy^2}{dx^2}+(y-n)\frac{d^2y}{dx^2}=0,$

(6) $(x-m)^2+(y-n)^2=r^2.$

In order to solve these equations for r, m, and n, we get from (5)

(7) $y-n=-\frac{1+\frac{dy^2}{dx^2}}{\frac{d^2y}{dx^2}}$, which substituted in (4) gives

(8) $x-m=\frac{\left(1+\frac{dy^2}{dx^2}\right)\frac{dy}{dx}}{\frac{d^2y}{dx^2}}.$ Substituting these values in (6) and reducing we have

(9) $r=\pm\frac{\left(1+\frac{dy^2}{dx^2}\right)^{\frac{3}{2}}}{\frac{d^2y}{dx^2}}$, which is the formula for *radius of curvature.*

The co-ordinates of the centre (m and n) are written at once from (8) and (7). They are

(10) $m=x-\frac{\left(1+\frac{dy^2}{dx^2}\right)\frac{dy}{dx}}{\frac{d^2y}{dx^2}}$, and

(11) $n=y+\frac{1+\frac{dy^2}{dx^2}}{\frac{d^2y}{dx^2}}.$ Q. E. D.

SOLUTION 2.—Let **P**, **P′**, and **P″**, *Fig.* 150, be three consecutive points through which the curve **MN**, whose equation is $y=f(x)$, and the osculatory circle mn whose equation is $(x'-m)^2+(y'-n)^2=r^2$, pass, and between which they coincide. **PP′**, and **P′P″** are then to be considered straight lines. Pass a circumference through these three points by erecting perpendiculars at the middle points of the chords **PP′**, and **P′P″**. These perpendiculars, **CD** and **CD′**, are consecutive normals. *Hence*

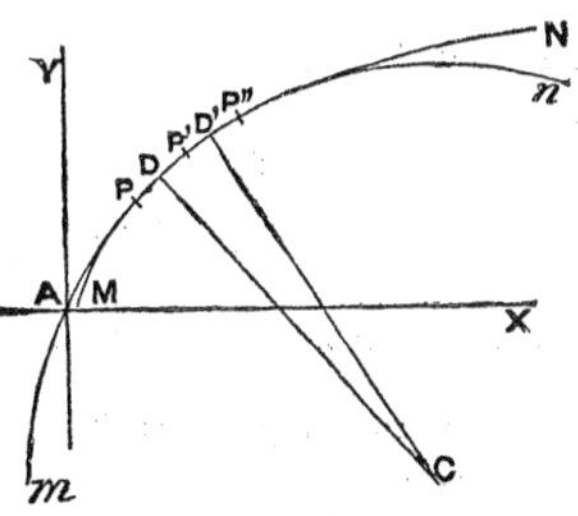

FIG. 150.

the centre of the osculatory circle may be conceived as the intersection of two consecutive normals.

Having premised the above fact, let **MN**, *Fig.* 151 be a curve whose equation is $y=f(x)$. Let **PC** and **P′C** be two consecutive normals. Then is **C** the centre of osculation, and **PC** $= r$, the radius of curvature. Again, let s represent the length of the curve, and ψ, the angle at the centre of the osculatory circle to radius unity. As **P** and **P′** are consecutive points, **PP′** $= ds$, **PL** $= dx$, **P′L** $= dy$, and $ce = d\psi$ are contemporaneous infinitesimal elements of s, x, y, and ψ respectively. Moreover, drawing **EH** parallel to **P′C**, the angle **PHE** $=$ **PCP′** $= d\psi$ is the corresponding infinitesimal element, of the angle which the normal makes with the axis of x; or $d\psi = d.\tan^{-1}\left(-\frac{dx}{dy}\right)$, as $\tan^{-1}\left(-\frac{dx}{dy}\right)$ is the angle a normal makes with the axis of x. Now since $ce = d\psi$ is an arc at a unit's distance from the centre, and **PP′** $= ds$, is the corresponding arc at r from the centre, we have

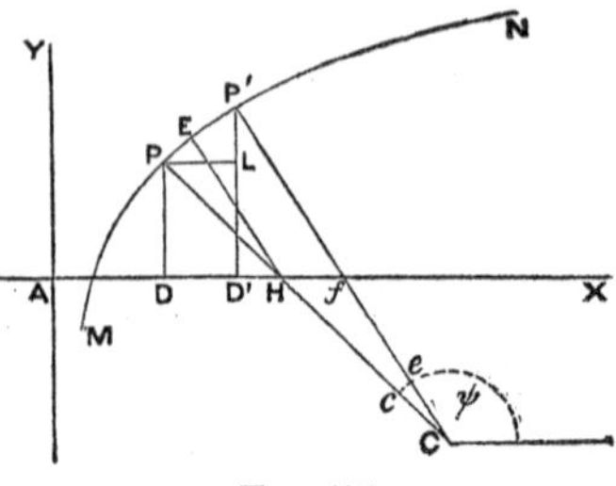

Fig. 151.

(1) $ds = -r d\psi$, or $r = -\frac{ds}{d\psi}$, the $-$ sign signifying that s is a decreasing function of ψ.

But $ds = \pm\sqrt{dy^2 + dx^2}$; and differentiating $\tan^{-1}\left(-\frac{dx}{dy}\right)$ with respect to x,

$$d\psi = d\ \tan^{-1}\left(-\frac{dx}{dy}\right) = \frac{\frac{d^2y\,dx}{dy^2}}{1+\frac{dx^2}{dy^2}} = \frac{d^2y\,dx}{dy^2+dx^2}.$$ Substituting these values of ds and $d\psi$ in (1), we have $r = \pm\frac{(dy^2+dx^2)^{\frac{3}{2}}}{d^2y\,dx} = \pm\frac{\left(1+\frac{dy^2}{dx^2}\right)^{\frac{3}{2}}}{\frac{d^2y}{dx^2}}$, as before.

209. Sch.—Since the numerator of the value of r, equals $\frac{ds^3}{dx^3}$, and x and s are increasing functions of each other, it is always to be regarded as $+$; whence we see that the sign of r depends upon the sign of $\frac{d^2y}{dx^2}$. Therefore r is to be considered $+$ when the curve is convex downward, and $-$ when it is convex upward (***169***).

Ex. 1. Find the radius of curvature, and the co-ordinates of the centre of osculation in the common parabola.

Solution.—We have $\frac{dy}{dx} = \frac{p}{y}$, and $\frac{d^2y}{dx^2} = -\frac{p^2}{y^3}$. $\therefore r = \frac{(y^2+p^2)^{\frac{3}{2}}}{p^2}$, neglecting the sign. Again $m = x - \frac{\left(1+\frac{dy^2}{dx^2}\right)\frac{dy}{dx}}{\frac{d^2y}{dx^2}} = \frac{\frac{3}{2}y^2+p^2}{p}$, and $n = y + \frac{1+\frac{dy^2}{dx^2}}{\frac{d^2y}{dx^2}} = -\frac{y^3}{p^2}$.

210. COR. 1.—*The radius of curvature at the vertex of the common parabola is half the latus-rectum, since at this point* $y = 0$, *and* $r = \frac{(y^2 + p^2)^{\frac{3}{2}}}{p^2} = p$.

211. COR. 2.—*The radius of curvature in the common parabola varies as the cube of the normal, since* normal $= (y^2 + p^2)^{\frac{1}{2}}$, *and* $r = \frac{(\text{normal})^3}{p^2}$.

Ex. 2. What is the radius of curvature of a parabola whose latus-rectum is 9, at $x = 3$? What are the co-ordinates of the centre of curvature? What are they at the vertex? Construct such a parabola with the osculatory circles in position.

Answer. For $x = 3$, $r = \mathbf{CP} = 16.04$; $m = \mathbf{AE} = 13\frac{1}{2}$; $n = \mathbf{EC} = -6.91$. At the vertex $r = \mathbf{AC'} = 4\frac{1}{2}$; $m = \mathbf{AC} = 4\frac{1}{2}$, and $n = 0$.

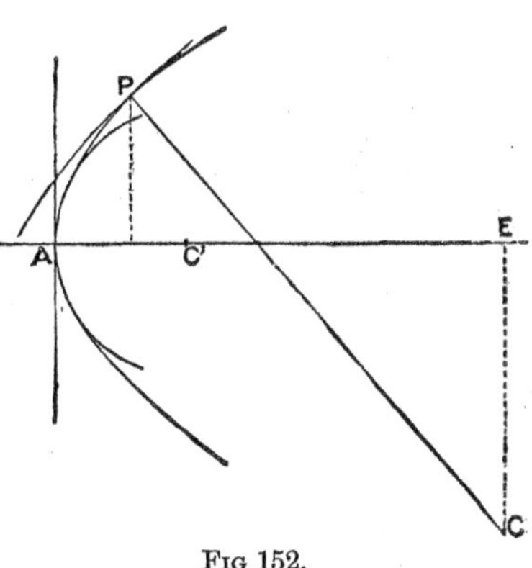

FIG 152.

Ex. 3. Find the radius of curvature of the ellipse, and the co-ordinates of the centre of curvature.

SUGGESTIONS. $\frac{dy}{dx} = -\frac{B^2x}{A^2y}$, and $\frac{d^2y}{dx^2} = -\frac{B^4}{A^2y^3}$; whence $r = \frac{(A^4y^2 + B^4x^2)^{\frac{3}{2}}}{A^4B^4}$; $m = x - \frac{x(A^4y^2 + B^4x^2)}{A^4B^2} = \frac{c^2x^3}{A^4}$, ($c^2$ being $= A^2 - B^2$); and $n = y - \frac{y(A^4y^2 + B^4x^2)}{A^2B^4} = -\frac{c^2y^3}{B^4}$.

212. COR. 1.—*The radius of curvature at the extremities of the transverse axis of an ellipse is half the latus-rectum, or* $\frac{B^2}{A}$; *and at the extremities of the conjugate axis, it is* $\frac{A^2}{B}$.

213. COR. 2.—*The radius of curvature in an ellipse varies as the cube of the normal, since* normal $= \frac{1}{A^2}\sqrt{A^4y^2 + B^4x^2}$, *giving* $r = \frac{(\text{normal})^3A^2}{B^4}$.

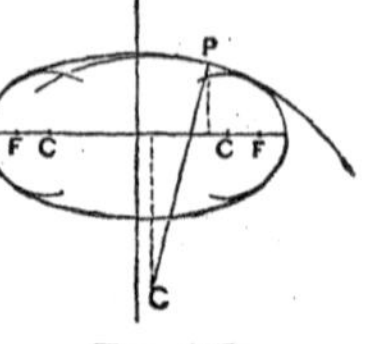

FIG. 153.

Ex. 4. Find the radius of curvature at $x = 2$, and also at the vertices of the axes of the ellipse whose axes are 8 and 4. Find also the centre of curvature, and construct the osculatrices.

Results. At $x = 2$(**P**), $(.375, -3.9)$, *i. e.* **C**, is the centre of curvature and $r = 5.86$(**PC**). At the vertices of the transverse axis **C** is the centre of curvature and $r = 1$.

214. SCH.—The centre of curvature being the intersection of two consecutive normals, it is always in the normal drawn to the point of osculation. Hence having found the value of r in any given case, if we can draw the normal geometrically, it is not necessary to find the co-ordinates of the centre of curvature in order to draw the osculatrix. If, however, we do not know how to draw the normal geometrically, the co-ordinates of the centre of curvature give a point in it, whence it can be drawn.

Ex. 5. Find the radius of curvature of logarithmic curve, $x = \log y$.

$$r = \frac{(m^2 + y^2)^{\frac{3}{2}}}{my}.$$

Ex. 6. Find the radius of curvature in the cubical parabola, $y^3 = a^2x$.

$$r = \frac{(9y^4 + a^4)^{\frac{3}{2}}}{6a^4y}.$$

Ex. 7. Find the radius of curvature of the curve $y = x^3 - x^2 + 1$, where it cuts the axis of y, and also at the point of minimum ordinate. How does it appear from the operation that the curve is concave towards the axis of x at the former point and convex at the latter? (See *Fig* 97.)

At the first point $r = -\frac{1}{2}$; at the second $r = \frac{1}{2}$.

Ex. 8. Find the radius of curvature of the locus $y^2 = 6x^2 + x^3$. How does it appear that this locus is always concave towards the axis of x? (See *Fig.* 106.)

$$r = \frac{\{y^4 + (4x + x^2)^2\}^{\frac{3}{2}}}{-8x^2y}.$$

Ex. 9. Prove that in the cycloid the radius of curvature equals twice the normal. Construct a cycloid and upon this principle draw the osculatory circle at several points. What is the radius of curvature at the points where the cycloid meets its base? What at the vertex?

---

215. Prop.—*At a point of inflexion a rectilinear tangent to a curve has contact of the second order.*

DEM.—Let $y = f(x)$ be the equation of the curve, and $y' = ax' + b$ be the equation of a right line. At a point of tangency in general we have for $x = x'$, $y = y'$, and $\frac{dy}{dx} = \frac{dy'}{dx'}$. But at a point of inflexion $\frac{d^2y}{dx^2} = 0$. Also in the equation of the right line $\frac{d^2y'}{dx'^2} = 0$. $\therefore \frac{d^2y}{dx^2} = \frac{d^2y'}{dx'^2}$, and we have the conditions of contact of the second order. Q. E. D.

---

216. Prop.—*At points of maximum and minimum curvature of any plane curve, the osculatory circle has contact of the third order.*

DEM.—At such points $\frac{dr}{dx} = 0$. Now differentiating $r = \frac{\left(1 + \frac{dy^2}{dx^2}\right)^{\frac{3}{2}}}{\frac{d^2y}{dx^2}}$ we have

$$\frac{dr}{dx} = \frac{\frac{3}{2}\left(1 + \frac{dy^2}{dx^2}\right)^{\frac{1}{2}} \times 2\frac{dy}{dx}\left(\frac{d^2y}{dx^2}\right)^2 - \frac{d^3y}{dx^3}\left(1 + \frac{dy^2}{dx^2}\right)^{\frac{3}{2}}}{\left(\frac{d^2y}{dx^2}\right)^2} = 0;$$

Whence $\frac{d^3y}{dx^3} = \frac{3\frac{dy}{dx}\left(\frac{d^2y}{dx^2}\right)^2}{1 + \frac{dy^2}{dx^2}}$.

But in the circle we have found $y - n = -\frac{1 + \frac{dy^2}{dx^2}}{\frac{d^2y}{dx^2}}$. Differentiating this, and finding the value of $\frac{d^3y}{dx^3}$, we have $\frac{d^3y}{dx^3} = \frac{3\frac{dy}{dx}\left(\frac{d^2y}{dx^2}\right)^2}{1 + \frac{dy^2}{dx^2}}$. Therefore as the third differential coefficient is the same in the circle as at a point of maximum or minimum curvature of any plane curve, the contact is of the third order at such points. Q. E. D.

217. COR.—*The contact of the osculatory circles at the vertices of the conic sections is closer than at other points, a fact which is also apparent in the construction.*

---

218. Prop.—*When contact is of an even order the loci intersect; but when of an odd order they do not.*

DEM.—Let $Y = f(x)$ and $y = \varphi(x)$ be the equations of the two loci. Then the difference of their ordinates corresponding to $x \pm h$ is $Y' - y' =$

$$\left(\frac{dY}{dx} - \frac{dy}{dx}\right)\frac{(\pm h)}{1} + \left(\frac{d^2Y}{dx^2} - \frac{d^2y}{dx^2}\right)\frac{(\pm h)^2}{2} + \left(\frac{d^3Y}{dx^3} - \frac{d^3y}{dx^3}\right)\frac{(\pm h)^3}{2 \cdot 3} + \left(\frac{d^4Y}{dx^4} - \frac{d^4y}{dx^4}\right)\frac{(\pm h)^4}{2 \cdot 3 \cdot 4}$$

$+$, etc. Now, when the order of contact is *even*, the first term of this difference which does not reduce to 0, and which fixes the sign of the sum of the series, contains an *odd* power of $\pm h$; and hence $Y' - y'$ is positive for $+ h$, and negative for $-h$, showing that the loci intersect at the point. If, on the other hand, the contact is of an *odd* order, the first term which does not reduce to 0 contains an even power of $\pm h$; and hence does not change sign with h, and one of the curves lies within the other, as in tangency. Q. E. D.

219. Cor.—*The osculatory circle always cuts a conic section except at points of maximum and minimum curvature.*

---

220. Prob.—*To produce the formula for radius of curvature in terms of Polar Co-ordinates.*

Solution.—We will produce this formula by transformation of co-ordinates, as the process affords both a good exercise in transformation, and also in changing the independent variable. In order to distinguish between radius of curvature and radius vector, let the former be represented by R, and the latter by r.

We have already seen that $R = \frac{(dx^2 + dy^2)^{\frac{3}{2}}}{d^2y\ dx}$ (*208*). But this formula was produced on the assumption that x was equicrescent, and hence $d(dx) = 0$. To give it the more general form, we have only to remember that $d^2y = d\left(\frac{dy}{dx}\right)dx = \frac{d^2y\ dx - d^2x\ dy}{dx}$; and hence that the general formula is

$$(1)\quad R = \frac{(dx^2 + dy^2)^{\frac{3}{2}}}{d^2y\ dx - d^2x\ dy}.$$

The equations for transformation are $y = r\sin\theta$, and $x = r\cos\theta$. Considering θ equicrescent (*i. e.* as the independent variable) and differentiating, we get

$$dy = dr\sin\theta + r\cos\theta\, d\theta,$$
$$dy^2 = dr^2\sin^2\theta + 2r\sin\theta\cos\theta\, dr\, d\theta + r^2\cos^2\theta\, d\theta^2,$$
$$d^2y = d^2r\sin\theta + 2\cos\theta\, dr\, d\theta - r\sin\theta\, d\theta^2,$$
$$dx = dr\cos\theta - r\sin\theta\, d\theta,$$
$$dx^2 = dr^2\cos^2\theta - 2r\sin\theta\cos\theta\, dr\, d\theta + r^2\sin^2\theta\, d\theta^2,$$
and $$d^2x = d^2r\cos\theta - 2\sin\theta\, dr\, d\theta - r\cos\theta\, d\theta^2.$$

$$\therefore\ (dx^2 + dy^2)^{\frac{3}{2}} = (dr^2 + r^2d\theta^2)^{\frac{3}{2}},\text{ and}$$
$$d^2y\, dx - d^2x\, dy = 2(\sin^2\theta + \cos^2\theta)dr^2d\theta - rd\theta(\sin^2\theta + \cos^2\theta)d^2r + (\sin^2\theta + \cos^2\theta)r^2d\theta^3$$
$$= 2dr^2\, d\theta - r\, d\theta\, d^2r + r^2\, d\theta^3.$$

Whence, substituting, we have

$$R = \frac{(dr^2 + r^2d\theta^2)^{\frac{3}{2}}}{2dr^2d\theta - rd\theta d^2r + r^2d\theta^3} = \frac{\left(\frac{dr^2}{d\theta^2} + r^2\right)^{\frac{3}{2}}}{2\frac{dr^2}{d\theta^2} - r\frac{d^2r}{d\theta^2} + r^2}.\quad \text{Q. E. D.}$$

221. Cor.—*Since the length of a normal to a polar curve is* $\left(\frac{dr^2}{d\theta^2} + r^2\right)^{\frac{1}{2}}$, (*167*), *representing the normal by N, we have*

$$R = \frac{N^3}{2\frac{dr^2}{d\theta^2} - r\frac{d^2r}{d\theta^2} + r^2}.$$

Ex. 1. Find the radius of curvature of the logarithmic spiral, $r = a^{\theta}$.

SOLUTION. $\frac{dr}{d\theta} = a^{\theta} \log a$, and $\frac{d^2r}{d\theta^2} = a^{\theta} \log^2 a$.

$$\therefore R = \frac{(a^{2\theta} \log^2 a + r^2)^{\frac{3}{2}}}{2a^{2\theta} \log^2 a - ra^{\theta} \log^2 a + r^2} = \frac{(a^{2\theta} \log^2 a + r^2)^{\frac{3}{2}}}{a^{2\theta} \log^2 a + r^2} = (a^{2\theta} \log^2 a + r^2)^{\frac{1}{2}} = N,$$

the polar normal. [The first reduction is made by remembering that $r = a^{\theta}$.]

Ex. 2. Find the radius of curvature of the lemniscate of Bernouilli, $r^2 = a^2 \cos 2\theta$.

SOLUTION. $\frac{dr}{d\theta} = -\frac{a^2 \sin 2\theta}{r} = -\frac{a \sin 2\theta}{\sqrt{\cos 2\theta}}$,

$$\frac{dr^2}{d\theta^2} = \frac{a^2 \sin^2 2\theta}{\cos 2\theta},$$

and $$\frac{d^2r}{d\theta^2} = -\frac{2a \cos^2 2\theta + a \sin^2 2\theta}{\cos^{\frac{3}{2}} 2\theta}.$$

Substituting these values, we have

$$R = \frac{\left(\frac{a^2 \sin^2 2\theta}{\cos 2\theta} + a^2 \cos 2\theta\right)^{\frac{3}{2}}}{\frac{2a^2 \sin^2 2\theta}{\cos 2\theta} + \frac{2a^2 \cos^2 2\theta + a^2 \sin^2 2\theta}{\cos 2\theta} + a^2 \cos 2\theta}$$

$$= \frac{\left(\frac{a^2}{\cos 2\theta}\right)^{\frac{3}{2}}}{\frac{3a^2 \sin^2 2\theta + 3a^2 \cos^2 2\theta}{\cos 2\theta}} = \frac{a^3}{3a^2\sqrt{\cos 2\theta}} = \frac{a^2}{3r}.$$

SECTION VII.

Evolutes and Involutes.

222. DEF.—***An Evolute*** of a curve is the locus of the centre of curvature. The primary curve is called the ***Involute.***

ILL.—If **MN** be a plane curve, and the centre of curvature, **C**, be determined for any point, **P**; then, as **P** passes along the curve to **P′**, **P″**, **P‴**, etc., the centre of curvature will describe another curve, as **C**, **C′**, **C″**, **C‴**, etc. **M′N** being thus described is the *evolute* of **MN**; and **MN** is the *involute* of **M′N**.

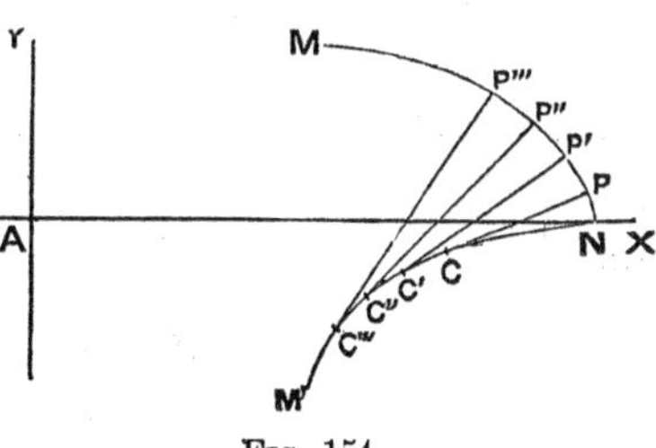

FIG. 154.

223. Prob.—*Given the equation of a plane curve, to find the equation of its evolute.*

SOLUTION.—Let $y = f(x)$ be the equation of the given curve, as MN *Fig.* 154. Now the co-ordinates of the centre of the osculatory circle are the co-ordinates of the evolute. Hence, if we combine the equations

$$m = x - \frac{\left(1 + \frac{dy^2}{dx^2}\right)\frac{dy}{dx}}{\frac{d^2y}{dx^2}}, \quad \text{and} \quad n = y + \frac{1 + \frac{dy^2}{dx^2}}{\frac{d^2y}{dx^2}},$$

with $y = f(x)$, and eliminate x and y there will result an equation between m and n, the co-ordinates of the evolute, which is therefore its equation.

SCH.—The equation of the locus, $y = f(x)$ is needed in connection with the values of m and n, only when these values contain both x and y.

Ex. 1. Find the equation of the evolute of the common parabola.

SOLUTION.—We have $\frac{dy}{dx} = \frac{p}{y}$, and $\frac{d^2y}{dx^2} = -\frac{p^2}{y^3}$. Whence

$$m = x - \frac{\left(1 + \frac{dy^2}{dx^2}\right)\frac{dy}{dx}}{\frac{d^2y}{dx^2}} = \frac{3y^2 + 2p^2}{2p},$$

$$\text{and } n = y + \frac{1 + \frac{dy^2}{dx^2}}{\frac{d^2y}{dx^2}} = -\frac{y^3}{p^2}.$$

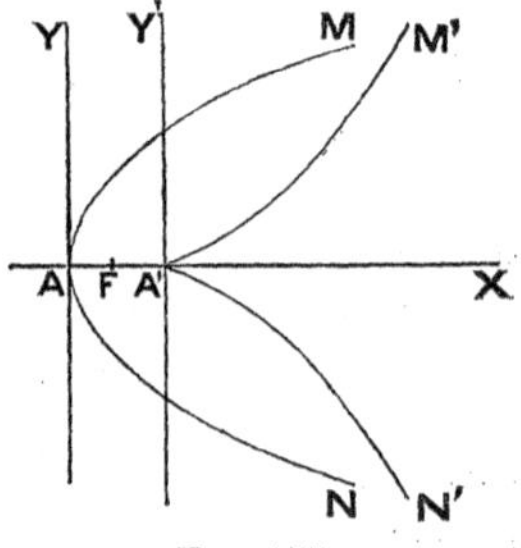

FIG. 155.

Now from $m = \frac{3y^2 + 2p^2}{2p}$, and $n = -\frac{y^3}{p^2}$, if we eliminate y we obtain, after a little reduction, $n^2 = \frac{8}{27p}(m - p)^3$, which is the equation sought. Tracing the curve we find M′A′N′, *Fig.* 155. If we transfer the origin to A′, the equation becomes $n^2 = \frac{8}{27p}m^3$.

224. SCH.—This locus is called the *Semi-cubical Parabola*, any curve having infinite branches and no rectilinear asymptotes being called a parabola.

Ex. 2. Find the evolute of the circle.

SUG'S.—The equations to be solved are

$$m = x - \frac{r^2}{y^2}\frac{y^3}{r^2}\frac{x}{y} = 0,$$

$$n = y - \frac{r^2}{y^2}\frac{y^3}{r^2} = 0,$$

and $x^2 + y^2 = r^2$. Whence $m = 0$ and $n = 0$, for all values of x and y, and the evolute is a point, the centre. This is evidently correct, since all normals (radii) of the circle meet at the centre.

Ex. 3. Find the evolute of the ellipse.

SUG'S.—We have $m = \frac{x^3c^2}{A^4}$, $n = -\frac{y^3c^2}{B^4}$, and $A^2y^2 + B^2x^2 = A^2B^2$, from which to eliminate x and y and find an equation between m and n, the co-ordinates of the evolute. The equation sought is $A^{\frac{2}{3}}m^{\frac{2}{3}} + B^{\frac{2}{3}}n^{\frac{2}{3}} = (A^2 - B^2)^{\frac{2}{3}}$. The evolute is of the form **CC″C′C‴**, *Fig.* 156.

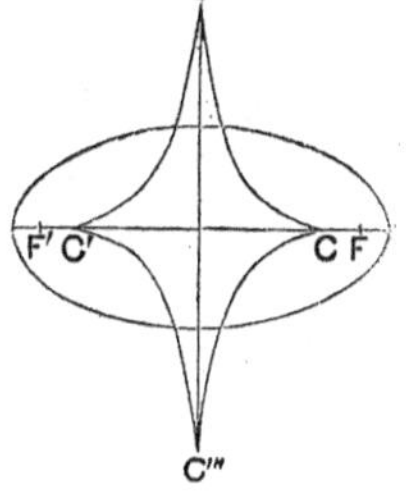

FIG. 156.

Ex. 4. Produce the equation of the evolute to the cycloid.

SUG'S.—We have $\frac{dy}{dx} = \frac{\sqrt{2ry - y^2}}{y}$, and $\frac{d^2y}{dx^2} = -\frac{r}{y^2}$. Whence $m = x + 2\sqrt{2ry - y^2}$, and $n = -y$. $\therefore y = -n$, and $x = m - 2\sqrt{-2rn - n^2}$. Substituting these in the equation of the cycloid, we have $m = \text{vers}^{-1}(-n) + \sqrt{-2rn - n^2}$.

225. COR.—*The evolute of a cycloid is an equal cycloid.*

DEM.—The equation $x = \text{vers}^{-1}(-y) + \sqrt{-2ry - y^2}$ is the equation of a cycloid referred to its highest point **A**, as the origin and having a tangent at that point as the axis of abscissas and the axis of the cycloid for the axis of ordinates. This will readily appear by producing the equation under these conditions. Thus in *Fig.* 157 **AD** = **AE** + **ED** = **CH** + **FP** = **CB** − **HB** + **FP** = arc **HPE** − arc **HP** + **FP** = arc **EP** + **FP**.

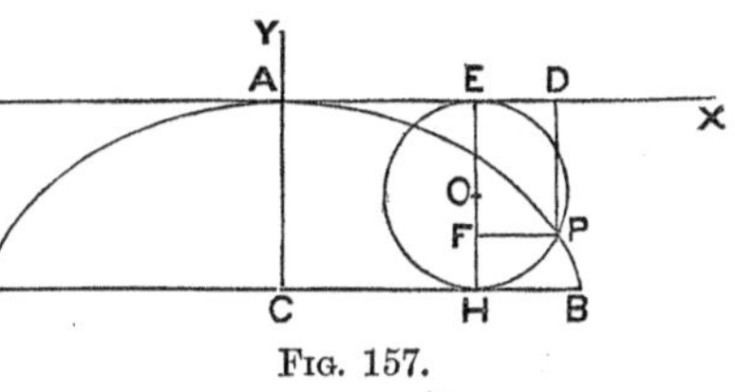

FIG. 157.

But **AD** $= x$, **PD** $= -y$, arc **EP** $=$ vers^{-1} **EF** $=$ vers^{-1} **PD** $= \text{ver}^{-1}(-y)$, and **FP** $= \sqrt{\mathbf{EF} \times \mathbf{FH}} = \sqrt{\mathbf{PD} \times \mathbf{FH}}$ $= \sqrt{(-y)(2r - \mathbf{PD})} = \sqrt{(-y)(2r + y)} =$ $\sqrt{-2ry - y^2}$. Hence $x = \text{vers}^{-1}(-y) +$ $\sqrt{-2ry - y^2}$.

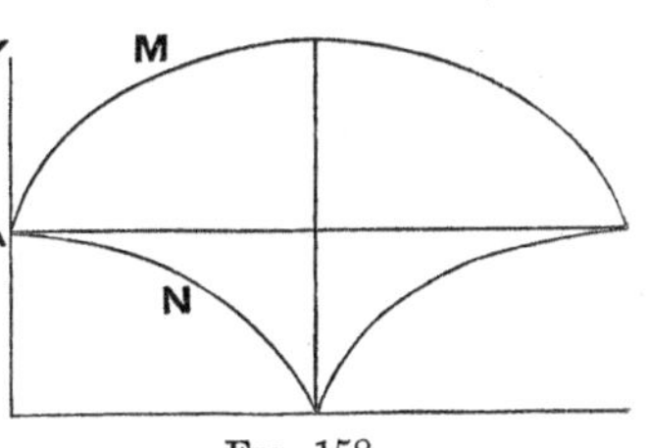

FIG. 158.

Thus we see that **M** *Fig.* 158 being a cycloid whose equation is $x = \text{vers}^{-1}y -$ $\sqrt{2ry - y^2}$, **N** is its evolute whose equation is $m = \text{vers}^{-1}(-n) + \sqrt{-2rn - n^2}$, referred to **A** as its origin. This equation is satisfied for none but negative values of n, and gives $m = 0$, $2\pi r$, $4\pi r$, etc., for $n = 0$; and also for $n = -2r$, $m = \pi r$, $3\pi r$, etc., as it should.

SCH.—The student will readily discern the character of the evolute of the cycloid from the property that the radius of curvature is always twice the normal. Thus if the two circles C, and C′ roll along the bases AX and A′X′ at equal rates so as to keep their centres in the same vertical line P′ will describe the evolute as P does the involute.

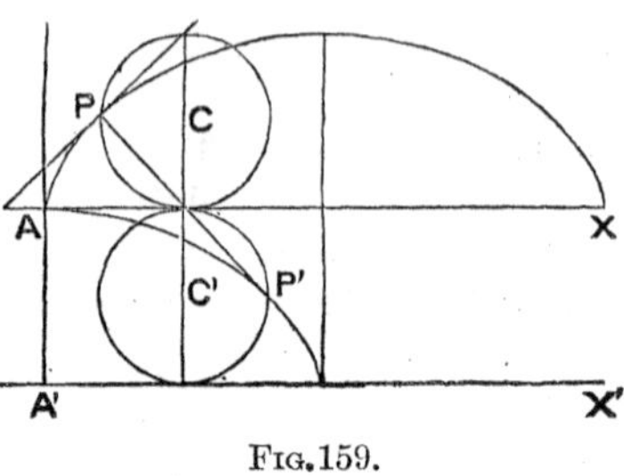

FIG. 159.

---

226. Prop.—*A (produced) normal to an involute is tangent to the evolute, the point of tangency is the centre of curvature, and consequently the normal thus produced is the radius of curvature.*

DEM.—Let (m, n) be any point in the evolute of AM, from it draw a normal to AM, and let (x, y) be the point at which it is normal. The equation of this normal is
$$y - n = -\frac{dx}{dy}(x - m),$$
or
$$x - m + \frac{dy}{dx}(y - n) = 0 \ (1).$$
Now as the point (m, n) changes position (x, y) also changes, and to observe the law of change we differentiate (1) for x, y, m and n as variables. This gives

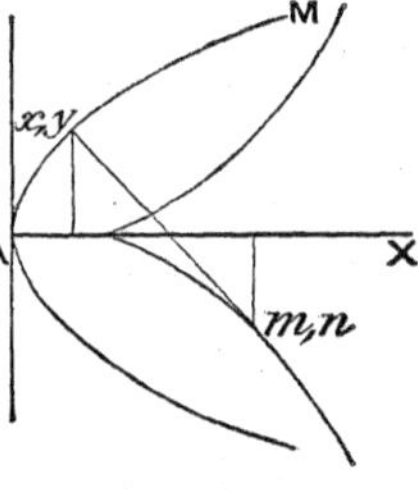

FIG. 160.

$$dx - dm + \frac{dy^2 - dndy}{dx} + (y - n)\frac{d^2y}{dx} = 0,$$
or
$$1 + \frac{dy^2}{dx^2} + (y - n)\frac{d^2y}{dx^2} - \frac{dm}{dx} - \frac{dndy}{dx^2} = 0. \quad (2)$$

But as (m, n) is in the evolute we have (**208**) $y - n = -\dfrac{1 + \frac{dy^2}{dx^2}}{\frac{d^2y}{dx^2}}$; whence $1 + \dfrac{dy^2}{dx^2} + (y - n)\dfrac{d^2y}{dx^2} = 0$. Therefore dropping these terms, (2) becomes
$$-\frac{dm}{dx} - \frac{dn\,dy}{dx^2} = 0, \text{ and } -\frac{dx}{dy} = \frac{dn}{dm}.$$

Hence the equation $y - n = -\dfrac{dx}{dy}(x - m)$, which is the equation of a normal to the involute at (x, y), may be written
$$(y - n) = \frac{dn}{dm}(x - m),$$
which is the equation of a tangent to the evolute at (m, n). Q. E. D.

227. COR.—*The radius of curvature and the arc of the evolute vary by equal increments; that is, the arc of the evolute between two centres of curvature equals the difference between the corresponding radii of curvature.*

DEM.—Since the radius of curvature is a tangent to the evolute it coincides with the arc between two consecutive points. Thus **P** and **P′** being consecutive points on the involute, the radius at **P** is to be considered as having the two consecutive points **C** and **C′** common with the evolute to which it is tangent; and as **P** passes to **P′**, the radius of curvature so changes position as to have the consecutive points **C′** and **C″** common, and to coincide with the curve between them. Thus it appears that the radius of curvature and the arc of the evolute vary by equal increments.

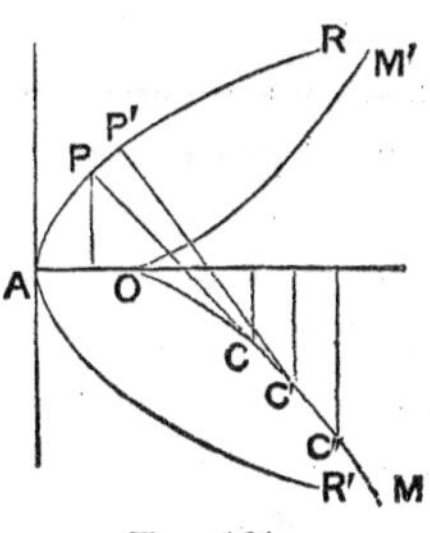

FIG. 161.

228. SCH.—From these relations it is easy to see how an involute may be described mechanically from its evolute. For example, to draw a parabola, make a pattern of the form **AOCM** *Fig.* 161, the edge **OCM** being the arc of an evolute to the required parabola, and **AO** $=p$, Fasten a cord at **M** and, wrapping it around the edge of the pattern, fasten a pencil to the free end at **A**. Keeping the string tight, move the pencil along as from **A** to **P**, **P′**, **R**, and it will describe the parabola which is an involute to **OM**. In like manner any curve can be described by means of a pattern of its involute. The cycloid and ellipse are drawn with special facility by this method. Thus, for the ellipse, take a thin rectangular board **ABED**, and upon it fasten two patterns **ACOD**, and **BC′OE**, the edges **CO** and **C′O** being the evolute. Then fastening at **O**, one end of a string whose length is **ACO**, the free end will describe the semi-ellipse as it is moved from **A** to **B**. Upon this principle attempts have been made to make a pendulum vibrate in the arc of a cycloid.

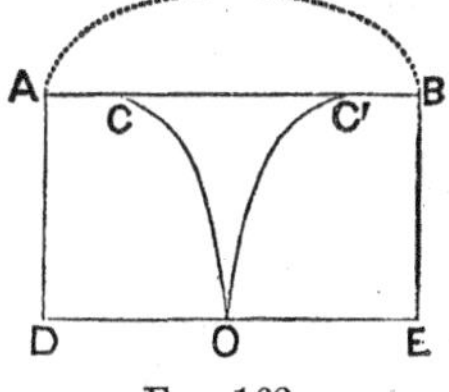

FIG. 162.

229. COR.—*Every curve has one and only one evolute; but every evolute has an infinite number of involutes, since every point in the string describes an involute as the string unwraps from the evolute.*

SECTION VIII.

Envelopes to Plane Curves.

230. DEF.—***An Envelope*** is the locus of the intersection of consecutive lines, or curves, represented by a given equation, when one or more of its parameters are made variable.

ILL.—Let $(x-m)^2+y^2-r^2=0$ be the equation of the locus whose envelope is required. Let m be the (variable) parameter. Let $r = \mathbf{B1}$, so that $\mathbf{B}aa'$ shall be one position of the given locus, which in this case is a circle. Now suppose m to take an infinitesimal increment dm, putting the centre at 2, and giving $\{x - (m + dm)\}^2 + y^2 - r^2 = 0$ as the equation of the consecutive locus.

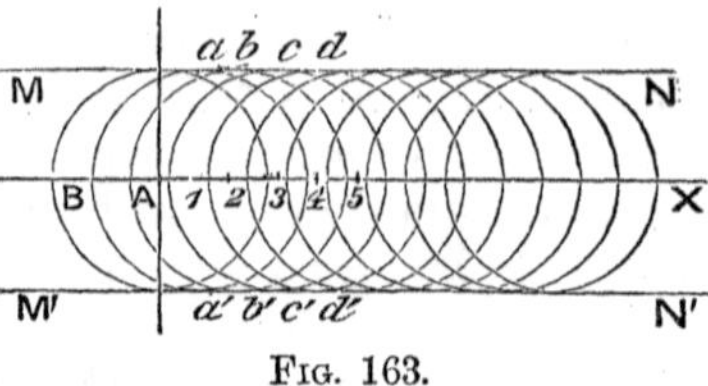

FIG. 163.

The intersections of these loci, as a, a', are points in the envelope. Again, let m take another infinitesimal increment, as 2 3, then b and b' are points in the envelope. In like manner the intersections of 3 and 4, 4 and 5, etc., etc., give points in the envelope. The envelope in this case is evidently the two parallel right lines **MN**, **M'N'**.

Were we to make r vary at the same time as m, the form of the envelope would be changed, and would depend upon the relative rates of change of r and m. Of course, the student will understand that the points of intersection a, b, c, d, etc., are only *in* the envelope when 1 2, 2 3, 3 4, etc., are *infinitesimal;* in other words, the envelope is the limit toward which these consecutive intersections approach as the increments 2 3, 3 4, etc., diminish.

---

231. Prob.—*To find the equation of the envelope of a given locus.*

SOLUTION.—Let $F(x, y, m) = 0$ be the equation of the given locus. The consecutive locus will be $F(x, y, m + dm) = 0$, or $F(x, y, m) + d_mF(x, y, m) = 0$. If we now combine the equation of the locus with this equation of its consecutive, eliminating m, we shall determine the locus of the intersection, *i. e.*, the envelope. But since $F(x, y, m) = 0$, the equation of the consecutive can always be reduced to $d_mF(x, y, m) = 0$. Hence in practice we simply combine the equation of the locus with its first differential equation eliminating the parameter, thus obtaining the envelope. Q. E. D.

SCH.—It is of course possible that the consecutive loci may not intersect; as, for example, $x^2 + y^2 = r^2$, when r is made variable.

Ex. 1. Find the envelope of $y^2 = m(x - m)$.

SOLUTION.—Differentiating with reference to m, we have $0 = dm(x - m) - mdm$, or $0 = xdm - 2mdm$. Whence $m = \frac{1}{2}x$. Combining this with $y^2 = m(x - m)$, so as to eliminate m, *i. e.*, substituting $\frac{1}{2}x$ for m, we have $y = \pm \frac{1}{2}x$, as the equation of the envelope.

ILL.—The geometrical significance of this operation will be readily seen by constructing a few parabolas on the same axis, giving to m *slightly* differing values. The consecutive intersections, a, b, c, a', b', c', etc., will evidently approach the two straight lines **AM**, **AM'** as the difference between the consecutive values of m is made less; therefore these lines are the envelope of the parabola $y^2 = m(x - m)$, or the series of consecutive parabolas of which four are represented in the figure.

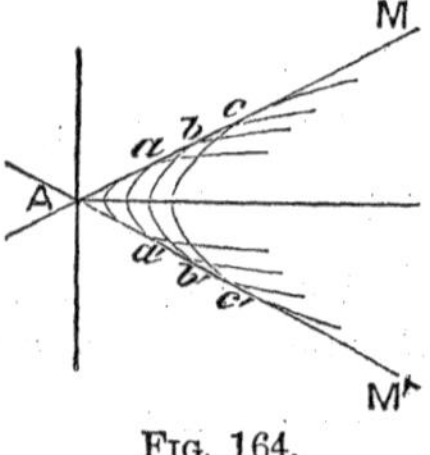

FIG. 164.

Ex. 2. Find the envelope of $y = ax + \frac{m}{a}$, a being the parameter. Construct a figure illustrating the result.

The envelope is $y^2 = 4mx$.

Ex. 3. A line of fixed length slides between two fixed lines at right angles to each other; required the envelope.

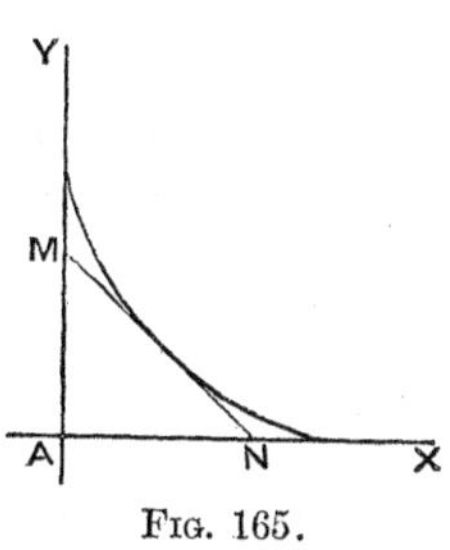

Fig. 165.

Solution.—Let the axes **AX** and **AY** be the fixed lines at right angles to each other, between which the line **MN** of fixed length, as c, slides. Let **AM** $= b$, and **AN** $= a$, whence tan **MNX** $= -\frac{b}{a}$, and the equation of **MN** is $y = -\frac{b}{a}x + b$, or $\frac{y}{b} + \frac{x}{a} = 1$. (1). We have also $a^2 + b^2 = c^2$ (2). The most *direct* method (not the most expeditious) would now be, to find the value of a or b, from (2), and substitute it in (1), which would then have but a single parameter and its envelope could be found as before. But the following method is less tedious: Differentiating (1) and (2) we have $\frac{ydb}{b^2} + \frac{xda}{a^2} = 0$ (3); and $ada + bdb = 0$ (4). Whence by eliminating da and db between (3) and (4), we have $\frac{x}{a} = \frac{a^2y}{b^3}$; which substituted in (1) gives after reduction $b^3 = c^2y$. Similarly $a^3 = c^2x$. Substituting these values in (1), there results $y^{\frac{2}{3}} + x^{\frac{2}{3}} = c^{\frac{2}{3}}$, the equation sought.

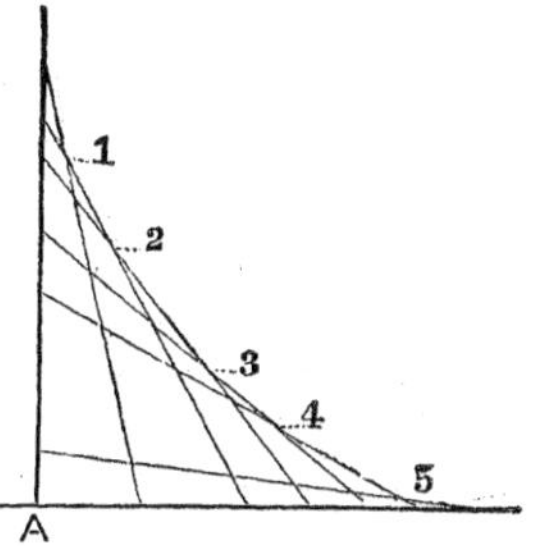

Ill.—This locus is readily sketched by drawing **MN** in slightly changed positions, and noting the intersections of consecutive lines, as in *Fig.* 166.

Fig. 166.

Sch.—This locus is a variety of *Hypocycloid*, a kind of curve generated by a point in the circumference of a circle rolling on the concave arc of (within) a fixed circle. In this variety the radius of the fixed circle is 4 times that of the generatrix.

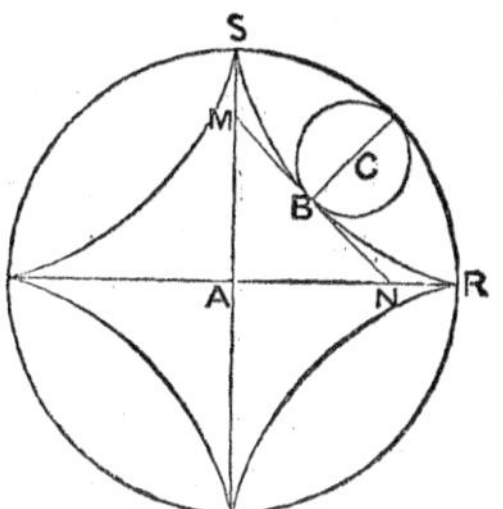

Fig. 167.

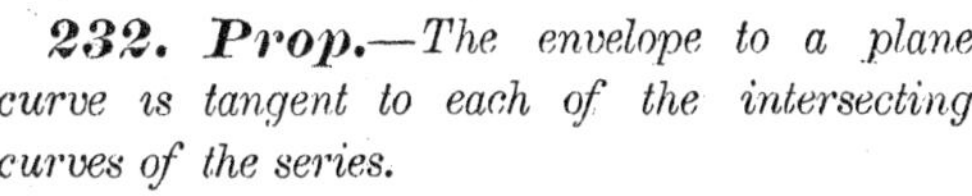

232. *Prop.*—*The envelope to a plane curve is tangent to each of the intersecting curves of the series.*

Dem.—Let $F(x, y, m) = 0$ (1), be the given locus. But from $d_mF(x, y, m) = 0$ (2), we have $m = \varphi(x, y)$; whence the equation of the envelope becomes $F\{x, y, \varphi(x, y)\} = 0$ (2_1). If now $\frac{dy}{dx}$ is the same for both the locus and its envelope, it follows that they have a common tangent, wherever they have a common point. From (1) we get by differentiating

$$\frac{dF(x, y, m)}{dx} + \frac{dF(x, y, m)}{dy} \cdot \frac{dy}{dx} = 0, \text{ from which } \frac{dy}{dx} = -\frac{\dfrac{dF(x, y, m)}{dx}}{\dfrac{dF(x, y, m)}{dy}}.$$

Differentiating (2_1) we have

$$\frac{dF\{x, y, \varphi(x, y)\}}{dx} + \frac{dF\{x, y, \varphi(x, y)\}}{dy} \cdot \frac{dy}{dx} + \frac{dF\{x, y, \varphi(x, y)\}}{d\varphi(x, y)} \cdot \left(\frac{d\varphi(x, y)}{dx}\right) = 0.$$

But as by (2) $d_m F\{x, y, \varphi(x, y)\} = 0$, and $\varphi(x, y) = m$, this becomes $\dfrac{dF(x, y, m)}{dx} +$

$\dfrac{dF(x, y, m)}{dy} \cdot \dfrac{dy}{dx}$, whence $\dfrac{dy}{dx} = -\dfrac{\dfrac{dF(x, y, m)}{dx}}{\dfrac{dF(x, y, m)}{dy}}$, the same as in the equation of the locus.

Ex. 4. Find the locus to which the hypotenuse of a right angled triangle of constant area is always tangent.

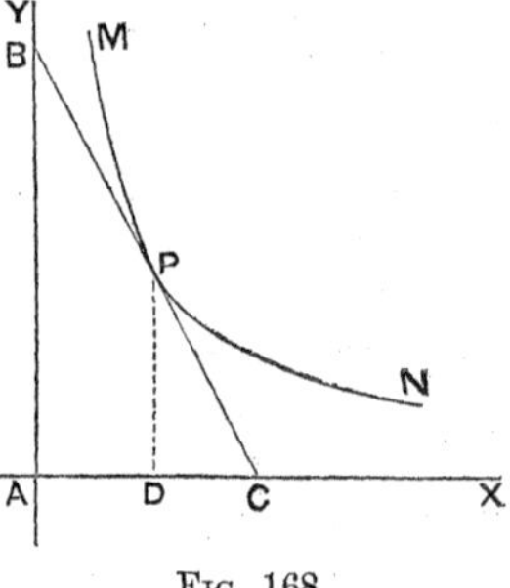

FIG. 168.

SOLUTION.—Let the constant area **ABC** $= a$, the parameter **AC** $= m$. Then **AB** $= \dfrac{2a}{m}$, $\tan$ **BCX** $= -\dfrac{2a}{m^2}$, and the equation of **BC** is $y = -\dfrac{2a}{m^2}x + \dfrac{2a}{m}$. The equation sought is $xy = \dfrac{a}{2}$, the equation of an hyperbola.

Ex. 5. What is the envelope of an ellipse which retains its axes in the same right lines, but varies in eccentricity so that $AB =$ a constant, m?

SUG'S.—Since $AB = m$, the equation of the ellipse is $A^4y^2 + m^2x^2 = A^2m^2$; in which A is the parameter. The equation of the envelope is $xy = \frac{1}{2}m$, an equilateral hyperbola referred to its asymptotes.

As will appear hereafter, the area of an ellipse is πAB. Hence the area of the above locus is constant.

Ex. 6. From every point in the circumference of a circle, pairs of tangents are drawn to another circle. Find the locus to which the chord connecting corresponding points of tangency is constantly tangent.

SOLUTION.—Letting **C** be the centre of the first and **A** of the second circle, it is evident that as **P** moves around the circle **P′P″** will change its position. The envelope of **P′P″** is required. Let **CP** $= r$, and **AP″** $= r'$. Let **P** be designated as (m, n), **P′** as (m', n'), and **P″** as (m'', n'').

The equation of the locus **P′P″** whose envelope is required is $y - n' =$

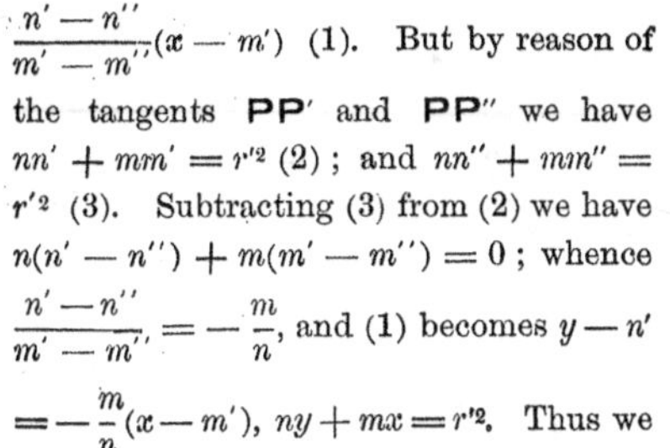

$\frac{n'-n''}{m'-m''}(x-m')$ (1). But by reason of the tangents **PP'** and **PP''** we have $nn'+mm'=r'^2$ (2); and $nn''+mn''=r'^2$ (3). Subtracting (3) from (2) we have $n(n'-n'')+m(m'-m'')=0$; whence $\frac{n'-n''}{m'-m''}=-\frac{m}{n}$, and (1) becomes $y-n'=-\frac{m}{n}(x-m')$, $ny+mx=r'^2$. Thus we find the equation of the given locus **P'P''** to be $ny+mx=r'^2$ (4).

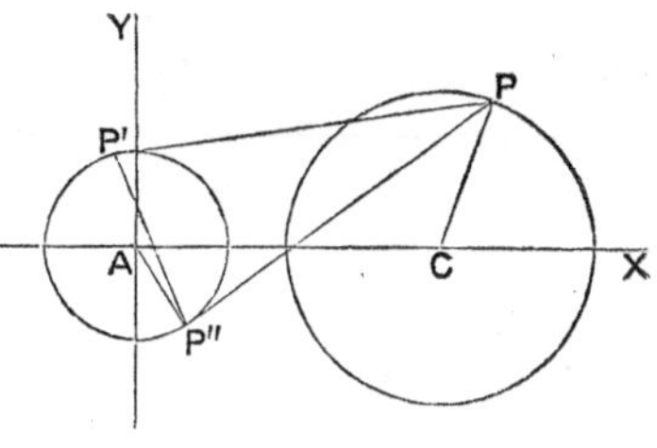

Fig. 169.

Again, if we let the distance between the centres of the circles **AC** be represented by a, we have the relation between n and m in the equation

$$n^2+(m-a)^2=r^2 \quad (5).$$

The problem then is to find the envelope of (4), when the relation between n and m is that given in (5). Differentiating (4) and (5) considering m and n as variables, we have $y\frac{dn}{dm}+x=0$, $\frac{dn}{dm}=-\frac{x}{y}$, and $n\frac{dn}{dm}+m-a=0$. $\therefore$ $-\frac{nx}{y}+m-a=0$, $my-nx=ay$ (6).

Finally eliminating m and n between (4) (5), and (6), and reducing, we have $r^2y^2+(r^2-a^2)x^2+2ar'^2x=r'^4$, as the equation of the envelope.

Hence the envelope is a conic section. When $a=0$ it is a circle; when $a<r$, an ellipse; when $a=r$ a parabola; and when $a>r$, an hyperbola.

---

233. *Prob.*—*An infinite number of parallel right lines meet a given curve on the same side; and where each meets the curve a line is drawn making an angle with the parallel which is bisected by the normal at that point. Required the envelope of the line.*

Solution.—Let **MN** be the given curve, **PO** one of the parallels, **PQ** the normal, and **PG** the locus whose envelope is sought. Our first purpose is to find the equation of **PG**. Let $y'=\varphi(x')$ be the equation of **MN**, and v the tangent of the constant angle **PSX**. Since **P**, whose co-ordinates are x', y', is a point in **PG**, the equation has the form $y-y'=a(x-x')$ in which a is the tangent of **PDX**. Now **PDX** = **PQX** − **DPQ** = **PQX** − **QPS** = **PQX** − (**PSX** − **PQX**) = 2**PQX** − **PSX**. Therefore $\tan \mathbf{PDX}=\frac{\tan 2\mathbf{PQX}-\tan \mathbf{PSX}}{1+\tan 2\mathbf{PQX}\tan \mathbf{PSX}}$. Again, as **PQ** is normal to $y'=\varphi(x')$, $\tan \mathbf{PQX}=-\frac{dx'}{dy'}$; whence $\tan 2\mathbf{PQX}=\frac{2\tan \mathbf{PQX}}{1-\tan^2 \mathbf{PQX}}=\frac{-2\frac{dx'}{dy'}}{1-\frac{dx'^2}{dy'^2}}$. Substituting, and

Fig. 170.

introducing v for tan **PSX**, we have

$$a = \frac{\frac{dx'^2}{dy'^2}v - v - 2\frac{dx'}{dy'}}{1 - \frac{dx'^2}{dy'^2} - 2\frac{dx'}{dy'}v}.$$

Putting $\frac{dx'}{dy'} = p$, for convenience, the equation of **PG** becomes

$$y - y' = \frac{p^2v - v - 2p}{1 - p^2 - 2pv}(x - x').$$

From this equation, its first differential equation, and the equation of the curve $y' = \varphi(x')$, if x', and y' be eliminated, the resulting equation between x and y will be the equation of the envelope sought. But the difficulties of elimination are often insurmountable. We give two cases which are readily solved.

234. Cor. 1.—*If* **OP** *is parallel to* **AX**, $v = 0$, *and the equation* **PG** *becomes*

$$y - y' = \frac{2p}{p^2 - 1}(x - x').$$

235. Cor. 2.—*If* **OP** *is perpendicular to* **AX**, $v = \infty$, *and the equation becomes*

$$y - y' = \frac{1 - p^2}{2p}(x - x').$$

236. Sch.—It is a well known property of light that its rays impinging upon a reflecting surface are thrown off so as to make the angle between the reflected ray and the normal, equal to that between the incident ray and the normal. In consequence of this law, when the rays of the sun, which are practically parallel, are reflected from a curved surface, the intersections of the consecutive reflected rays produce a luminous curve, called a *Caustic*, which is an example of the envelope discussed in the problem. The annexed figure affords an illustration. Let **NAN′** be a section of a circular cylindrical mirror, made perpendicular to its axis. Let 1 to 11 be rays of light parallel to the axis of the mirror **AC**. The envelope of the reflected rays is the caustic curve **NM**. **MN** shows the lower branch of the caustic, the rays not being represented. This curve may be seen inside of a ring lying

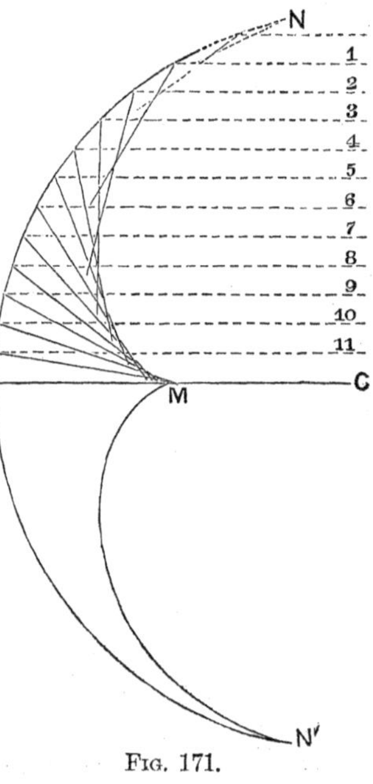

Fig. 171.

on a table in the light. It is familiar to the milkman, as "the cow's foot in the milk," which is the caustic formed upon the smooth surface of the milk in a bright tin pail, by reflection of the light from the inside of the pail.

Ex. 1. To produce the equation of the caustic when the incident rays are parallel to the axis of a parabolic reflector.

SOLUTION.—We have $y'^2 = 4mx'$, and $\frac{dx'}{dy'} = \frac{y'}{2m}$, $4m$ being the parameter of the parabola. Substituting this value of p, and for x', $\frac{y'^2}{4m}$, in the equation (**234**), we have after reduction

$$yy'^2 - 4m^2y + 4m^2y' = 4my'x \quad (1).$$

Differentiating (1) with respect to the parameter y', gives $y' = \frac{2mx - 2m^2}{y}$. Substituting this in (1), and reducing we have

$x = m \pm \sqrt{-y^2}$, as the equation of the caustic. This can only be satisfied for $y = 0$, $x = m$; whence we see that the caustic is a point, the focus.

Ex. 2. To find the caustic to the circle when the incident rays are parallel to the axis of x.

SOLUTION.—Equation of circle $y'^2 + x'^2 = r^2$. $\therefore \frac{dx'}{dy'} = -\frac{y'}{x'}$. Equation of reflected ray (**234**) $y - y' = \frac{2p}{p^2 - 1}(x - x')$, becomes $y - y' = -\frac{2x'y'}{y'^2 - x'^2}(x - x')$,

or $x - x' = \frac{y'^2 - x'^2}{2x'y'}(y' - y)$, or, $x - x' = \frac{1}{2}\left(\frac{y'}{x'} - \frac{x'}{y'}\right)(y' - y)$ (1).

Differentiating (1) with respect to x', we have

$$-1 = \frac{1}{2}\left\{\left(\frac{-\frac{x'^2}{y'} - y'}{x'^2} - \frac{y' + \frac{x'^2}{y'}}{y'^2}\right)(y' - y) - \left(\frac{y'}{x'} - \frac{x'}{y'}\right)\frac{x'}{y'}\right\},$$

Whence $-2 = -\left(\frac{x'^2 + y'^2}{y'x'^2} + \frac{y'^2 + x'^2}{y'^3}\right)(y' - y) - 1 + \frac{x'^2}{y'^2}$,

or $\quad -1 = -\frac{r^2}{y'^3}\left(\frac{y'^2}{x'^2} + 1\right)(y' - y) + \frac{x'^2}{y'^2}$,

or $\quad \frac{x'^2}{y'^2} + 1 = \frac{r^2}{y'^3}\left(\frac{y'^2}{x'^2} + 1\right)(y' - y)$,

or $\quad \frac{x'^2 + y'^2}{y'^2} = \frac{r^2}{y'^3}\left(\frac{y'^2 + x'^2}{x'^2}\right)(y' - y)$; which divided by $\frac{x'^2 + y'^2}{x'^2}$ gives

$\frac{x'^2}{y'^2} = \frac{r^2}{y'^3}(y' - y)$. From which we find $y' = r^{\frac{2}{3}}y^{\frac{1}{3}}$.

To find x', substitute in (1) for $y' - y$ its value $\frac{y'x'^2}{r^2}$, and we have

$x - x' = \frac{y'^2 - x'^2}{2y'x'}\frac{y'x'^2}{r^2} = \frac{y'^2 - x'^2}{2r^2}x'$. But $x'^2 = r^2 - y'^2$ whence $x - x' =$

$\frac{-r^2 + 2y'^2}{2r^2}x'$, or $x' = \frac{2r^2x}{r^2 + 2y'^2}$. Putting $r^{\frac{2}{3}}y^{\frac{1}{3}}$ for y', this becomes $x' = \frac{2r^{\frac{4}{3}}x}{r^{\frac{4}{3}} + 2y^{\frac{2}{3}}}$.

Finally, squaring these values of y', and x' and substituting in $y'^2 + x'^2 = r^2$, we have $r^{\frac{4}{3}}y^{\frac{2}{3}} + \dfrac{4r^{\frac{4}{3}}x^2}{(r^{\frac{2}{3}} + 2y^{\frac{2}{3}})^2} = r^2$, which is the equation of the caustic sought.

[NOTE.—At this stage of the course the student will need to acquaint himself with the elements of the Integral Calculus, as given in Chap. III. of the second part of this volume.]

SECTION IX.

Rectification of Plane Curves.

(*a*) BY MEANS OF RECTANGULAR CO-ORDINATES.

237. DEF.—***To Rectify a Curve*** is to find its length. The term arises from the conception that a right line is to be found which has the same length as the curve.

238. ***Prop.***—*The formula for the rectification of plane curves is*

$$dz = \sqrt{dx^2 + dy^2};$$

in which z *represents the length of the curve, and* x *and* y *the general co-ordinates.*

DEM.—Let **MN** be any plane curve, **AD** $= x$, and **PD** $= y$ be any co-ordinates, and let **DD'** represent dx; then will **P'E** represent dy, and **PP'**, dz; *i. e.*, dx, dy, and dz will represent contemporaneous infinitesimal increments of the co-ordinates and the arc. From the right angled triangle **PEP'** we have at once $dz = \sqrt{dx^2 + dy^2}$. Q. E. D.

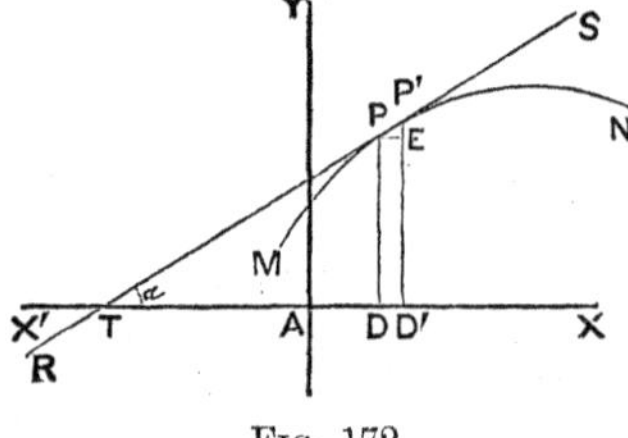

FIG. 172.

239. SCH.—To apply this formula to any particular curve, we have simply to find dx or dy from the equation of the proposed curve, substitute it in the formula, and then integrate between proper limits.

Ex. 1. Rectify the semi-cubical parabola whose equation is $y^2 = ax^3$. (See evolute of common parabola.)

DEM.—Differentiating $y^2 = ax^3$, we have $dy = \dfrac{3ax^2}{2y}dx$, whence $dy^2 = \dfrac{9a^2x^4}{4y^2}dx^2 = \dfrac{9a^2x^4}{4ax^3}dx^2 = \frac{9}{4}axdx^2$. Substituting this value in the formula for rectification, it becomes $dz = (dx^2 + \frac{9}{4}axdx^2)^{\frac{1}{2}} = (1 + \frac{9}{4}ax)^{\frac{1}{2}}dx$. Integrating we have $z =$

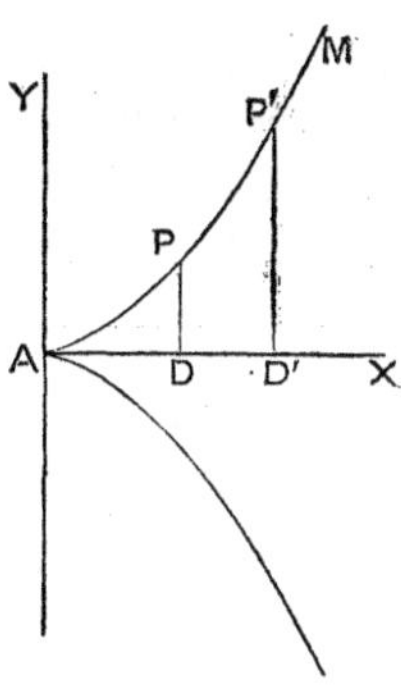

FIG. 173.

$\frac{8}{27a}(1 + \frac{9}{4}ax)^{\frac{3}{2}} + C$. To determine C we may reckon the length of the curve from the origin **A**, whence for $x = 0$, $z = 0$, and we have $0 = \frac{8}{27a} + C$, or $C = -\frac{8}{27a}$. The *corrected* integral is therefore $z = \frac{8}{27a}[(1 + \frac{9}{4}ax)^{\frac{3}{2}} - 1]$. To illustrate this result consider **AM** the curve whose length is to be found, or, which is to be rectified. As this curve is infinite in extent, we can only inquire for the length of some specified portion of it. Let it be required to find the length of the curve between the origin and the point **P** whose abscissa we will call b. Substituting this value of x, we have arc $\mathbf{AP} = z = \frac{8}{27a}[(1 + \frac{9}{4}ab)^{\frac{3}{2}} - 1]$. Were it required to find the length of some other arc, as **PP′**, we should integrate between the limits $x =$ **AD**, and $x =$ **AD′**. Thus, let $\mathbf{AD} = b$ and $\mathbf{AD'} = c$. Resuming the *indefinite* or general integral $z = \frac{8}{27a}(1 + \frac{9}{4}ax)^{\frac{3}{2}} + C$, substituting successively $x = b$, and $x = c$, and subtracting the former result from the latter, we get for the length of the arc **PP′**, the *definite* integral

$$z = \frac{8}{27a}[(1 + \frac{9}{4}ac)^{\frac{3}{2}} - (1 + \frac{9}{4}ab)^{\frac{3}{2}}].$$

Ex. 2. Rectify the common parabola.

SOLUTION.—From $y^2 = 2px$ we have $dx^2 = \frac{y^2dy^2}{p^2}$; whence $dz = \frac{1}{p}(p^2 + y^2)^{\frac{1}{2}}dy$. To integrate this apply formula **C** of reduction, and we have

$$z = \frac{y\sqrt{p^2 + y^2}}{2p} + \frac{p}{2}\int\frac{dy}{\sqrt{p^2 + y^2}}. \quad \text{But}$$

$$\int\frac{dy}{\sqrt{p^2+y^2}} = \log[y + \sqrt{p^2+y^2}] + C. \quad \therefore z = \frac{y\sqrt{p^2+y^2}}{2p} + \frac{p}{2}\log[y + \sqrt{p^2+y^2}] + C.$$

Estimating the arc from the vertex, which is the origin, we have $C = -\frac{p}{2}\log p$; and the corrected integral is $z = \frac{y\sqrt{p^2 + y^2}}{2p} + \frac{p}{2}\log\left[\frac{y + \sqrt{p^2 + y^2}}{p}\right]$.

SCH.—Instead of integrating as above, we may expand $(p^2 + y^2)^{\frac{1}{2}}$ by Maclaurin's or the Binomial theorem, and then integrate each term separately, obtaining $z = (y + \frac{1}{2}\cdot\frac{1}{3}\cdot\frac{y^3}{p^2} - \frac{1}{2}\cdot\frac{1}{2}\cdot\frac{1}{2}\cdot\frac{1}{5}\cdot\frac{y^5}{p^4} + \frac{1}{2}\cdot\frac{1}{2}\cdot\frac{3}{2}\cdot\frac{1}{2}\cdot\frac{1}{3}\cdot\frac{1}{7}\cdot\frac{y^7}{p^6} - \text{etc.}) + C.$

Ex. 3. Rectify the circle.

SOLUTION.—From $x^2 + y^2 = r^2$ we have $dy^2 = \frac{x^2}{y^2}dx^2$, hence $dz = \left(\frac{x^2 + y^2}{y^2}\right)^{\frac{1}{2}}dx$

$dz = -\frac{r dx}{(r^2 - x^2)^{\frac{1}{2}}}$, and $z = r \sin^{-1}\frac{x}{r} + C$. But this is only a restatement of the problem and is of no use in the solution. We shall have to integrate in some other way. We may write

$$dz = r(r^2 - x^2)^{-\frac{1}{2}} dx = \left(1 - \frac{x^2}{r^2}\right)^{-\frac{1}{2}} dx = r(1 - x_1{}^2)^{-\frac{1}{2}} dx_1,$$

putting $\frac{x}{r} = x_1$ for convenience. Expanding by Maclaurin's or the Binomial theorem, and integrating each term separately, we have

$$z = r\left(x_1 + \frac{x_1{}^3}{2 \cdot 3} + \frac{3x_1{}^5}{2 \cdot 4 \cdot 5} + \frac{1 \cdot 3 \cdot 5x_1{}^7}{2 \cdot 4 \cdot 6 \cdot 7} + \text{etc.}\right) + C.$$

Restoring x this becomes

$$z = r\left(\frac{x}{r} + \frac{x^3}{2 \cdot 3r^3} + \frac{3x^5}{2 \cdot 4 \cdot 5r^5} + \frac{1 \cdot 3 \cdot 5x^7}{2 \cdot 4 \cdot 6 \cdot 7r^7} + \text{etc.}\right) + C.$$

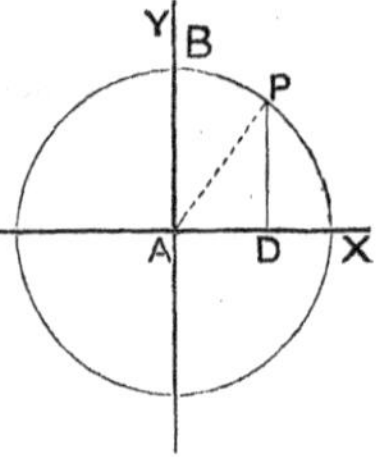

Fig. 174.

To determine C, reckon the arc from the axis of ordinates (**B**), whence for $x = 0$ $z = 0$, and therefore $C = 0$. Then making $x = r$ we have the length of the quadrant **BX**, $z = r\left(1 + \frac{1}{2 \cdot 3} + \frac{1 \cdot 3}{2 \cdot 4 \cdot 5} + \frac{1 \cdot 3 \cdot 5}{2 \cdot 4 \cdot 6 \cdot 7} + \text{etc.}\right)$

Representing the sum of the series in the parenthesis by $\frac{1}{2}\pi$, we have $z =$ a quadrant $= \frac{1}{2}r\pi$, and the whole circumference $= 2\pi r$. Letting D be the diameter, this becomes $D\pi$, whence it appears that π, the sum of the series above, is the ratio of the circumference of a circle to its diameter. By extending the terms in this series sufficiently, reducing each to a decimal fraction and adding, we find $\pi = 3.1415926+$. For practical purposes π is usually taken as 3.1416, and for still ruder approximations as $3\frac{1}{7}$.

240. Cor. 1.—*The circumferences of circles are to each other as their radii, or as their diameters.*

241. Sch.—The quantity π has not only fundamental importance in geometry, but has great historic interest. Upon it depend both the method of obtaining the circumference of a circle, of a given radius, and the area of a circle, as well as many other problems. The ancients sought with much diligence to discover its value. Archimedes (287 B. C.) found it to be between $3\frac{10}{71}$ and $3\frac{1}{7}$. Metius (1640) gave a nearer approximation in the fraction $\frac{355}{113}$. In 1853 Mr. Rutherford presented to the Royal Society of London a computation by Mr. W. Shanks of Houghton-le-Spring, extending the decimal to 530 places. The following is its value to 50 places:

3.141 592 653 589 793 238 462 643 383 279 502 884 197 169 399 375 10.

Ex. 4. Rectify the cycloid.

Solution.—We have $dx^2 = \frac{y^2 dy^2}{2ry - y^2}$, whence

$$\int dz = (2r)^{\frac{1}{2}} \int (2r - y)^{-\frac{1}{2}} dy = -2(2r)^{\frac{1}{2}}(2r - y)^{\frac{1}{2}} + C.$$

Reckoning the arc from the origin $C = 4r$, and the corrected integral is $z =$

$-2(2r)^{\frac{1}{2}}(2r-y)^{\frac{1}{2}}+4r$. Making $y=2r$, $z=\frac{1}{2}$ the cycloidal arc $=4r$, whence the entire arc of the cycloid is seen to be 4 *times the diameter of the generatrix.*

242. Cor. 2.—*Any arc of a cycloid, estimated from the vertex, is equal to twice the corresponding chord of the generatrix.*

Dem.—Resuming the indefinite integral $z=-2(2r)^{\frac{1}{2}}(2r-y)^{\frac{1}{2}}+C$, if we estimate the arc from **B**, where $y=2r$, we have $C=0$; and the corrected integral is $z=-2(2r)^{\frac{1}{2}}(2r-y)^{\frac{1}{2}}$. This is the length of any arc estimated from **B**, as **BP**, **PD** being y. But in the right angled triangle **BEC**, $\mathbf{BE}=\sqrt{\mathbf{BC}\times\mathbf{BG}}=\sqrt{2r(2r-y)}$. $\therefore$ arc **BP** $=2$ times chord **BE**. Q. E. D.

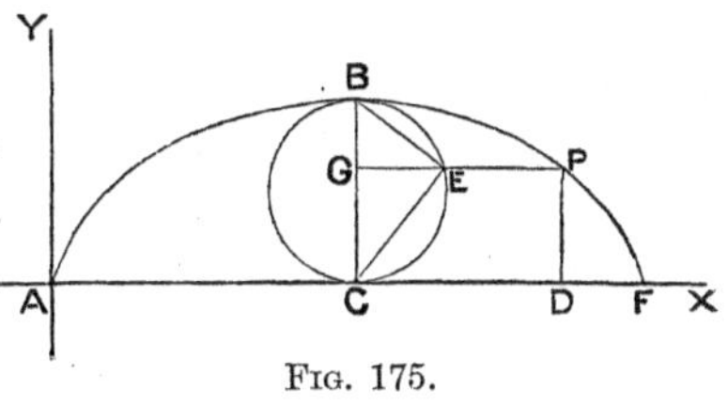

Fig. 175.

Sch.—Both the fact that the length of the cycloid equals 4 times the diameter of the generatrix, and that any arc estimated from the vertex equals 2 times the corresponding chord of the generatrix, are readily observed from the manner in which the curve is described from its evolute. Thus the radius of curvature at the vertex is 2 times the diameter of the generatrix. But this is the arc of the evolute. So also as the string unwinds from the evolute, the radius of curvature is seen to be the arc of the cycloid and equal to twice the corresponding chord of the generatrix. (See *Fig.* 159.)

Ex. 5. Rectify the hypocycloid, whose equation is $x^{\frac{2}{3}}+y^{\frac{2}{3}}=a^{\frac{2}{3}}$. (See Fig. 167.) *The length of the entire curve is* 6*a*.

Ex. 7. Rectify the ellipse.

Solution.—For this purpose the equation

$$y^2+(1-e^2)x^2=A^2(1-e^2)$$

is most convenient. From this we have $dy^2=\frac{x^2}{y^2}(1-e^2)^2dx^2$; whence

$$dz=\sqrt{dx^2+\frac{x^2}{y^2}(1-e^2)^2dx^2}=\frac{dx\sqrt{y^2+x^2(1-e^2)^2}}{y}$$

$$=dx\sqrt{\frac{(1-e^2)(A^2-x^2)+(1-e^2)^2x^2}{(1-e^2)(A^2-x^2)}}=dx\sqrt{\frac{A^2-x^2+x^2-e^2x^2}{A^2-x^2}}$$

$$=dx\sqrt{\frac{A^2-e^2x^2}{A^2-x^2}}=\frac{Adx}{\sqrt{A^2-x^2}}\sqrt{1-\frac{e^2x^2}{A^2}}.$$

But this form can only be integrated approximately. Developing $\left(1-\frac{e^2x^2}{A^2}\right)^{\frac{1}{2}}$, we have $dz=\frac{Adx}{\sqrt{A^2-x^2}}\left\{1-\frac{e^2x^2}{2A^2}-\frac{e^4x^4}{2\cdot4\cdot A^4}-\frac{3e^6x^6}{2\cdot4\cdot6A^6}-\text{etc.}\right\}$, whence

$$\int dz=A\int\frac{dx}{\sqrt{A^2-x^2}}-\frac{e^2}{2A}\int\frac{x^2dx}{\sqrt{A^2-x^2}}-\frac{e^4}{2\cdot4A^3}\int\frac{x^4dx}{\sqrt{A^2-x^2}}-\frac{3e^6}{2\cdot4\cdot6A^5}\int\frac{x^6dx}{\sqrt{A^2-x^2}}$$

$-$ etc. Now integrating each of these terms separately we have $z = A\sin^{-1}\frac{x}{A} -$
$\frac{e^2}{2A}\left(\frac{A^2}{2}\sin^{-1}\frac{x}{A} - \frac{x}{2}\sqrt{A^2 - x^2}\right) - \frac{e^4}{2\cdot 4A^3}\left[\frac{3A^2}{4}\left(\frac{A^2}{2}\sin^{-1}\frac{x}{A} - \frac{x}{2}\sqrt{A^2 - x^2}\right) - \frac{x^3}{4}\sqrt{A^2 - x^2}\right]$
$- \frac{3e^6}{2\cdot 4\cdot 6A^5}\left\{\frac{5A^2}{6}\left[\frac{3A^2}{4}\left(\frac{A^2}{2}\sin^{-1}\frac{x}{A} - \frac{x}{2}\sqrt{A^2 - x^2}\right) - \frac{x^3}{4}\sqrt{A^2 - x^2}\right] - \frac{x^5}{6}\sqrt{A^2 - x^2}\right\}$
$-$ etc., $+ C$.

If we estimate the arc of the ellipse from the extremity of the conjugate axis, we have for $x = 0$, $z = 0$; whence substituting, we find $C = 0$.

Again, making $x = A$, and observing that $\sin^{-1}1 = \frac{\pi}{2}$, we have

$$z_1 = \frac{\pi A}{2} - \frac{e^2}{2A}\left(\frac{A}{2}\cdot\frac{\pi A}{2}\right) - \frac{e^4}{2\cdot 4A^3}\left[\frac{3A^2}{4}\left(\frac{A}{2}\cdot\frac{\pi A}{2}\right)\right] - \frac{3e^6}{2\cdot 4\cdot 6A^5}\left\{\frac{5A^2}{6}\left[\frac{3A^2}{4}\left(\frac{A}{2}\cdot\frac{\pi A}{2}\right)\right]\right\} - \text{etc.}$$

Uniting and multiplying by 4, we have for the entire circumference of the ellipse,

$$4z_1 = 2\pi A\left(1 - \frac{e^2}{2\cdot 2} - \frac{3e^4}{2\cdot 2\cdot 4\cdot 4} - \frac{3\cdot 3\cdot 5e^6}{2\cdot 2\cdot 4\cdot 4\cdot 6\cdot 6} - \text{etc.}\right).$$

This series converges more or less rapidly as the eccentricity is greater or less, but is always converging.

(b) RECTIFICATION BY MEANS OF POLAR CO-ORDINATES.

243. Prop.—*The formula for rectifying polar curves is*

$$dz = (r^2 d\theta^2 + dr^2)^{\frac{1}{2}}$$

DEM.—Let **A** be the pole of the curve **MN**, **AP** any radius vector, and **AP′** the consecutive position of the radius vector, so that **PAP′** $= d\theta$, θ being the variable angle. Let z represent any arc of the curve, and r the radius vector, and with **A** as a centre and radius **AP**, draw **PD**. Then **PP′** $= dz$, and **P′D** $= dr$, are infinitesimals of z and r respectively, and contemporaneous with $d\theta$. Now from the right angled triangle **P′DP**, right angled at **D**, we have **P′P** $= \sqrt{\overline{PD}^2 + \overline{P'D}^2}$. But $d\theta$ being the arc measuring **PAP′** at a unit's distance from **A**, **PD** $= rd\theta$; whence, substituting dz for **PP′**, dr for **P′D**, and $rd\theta$ for **PD**, we have $dz = (r^2d\theta^2 + dr^2)^{\frac{1}{2}}$. Q. E. D.

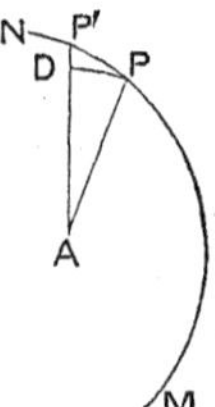

FIG. 176.

Ex. 1. Rectify the logarithmic spiral, $\log r = \theta$.

SOLUTION. $d\theta^2 = \frac{M^2dr^2}{r^2}$, $\therefore dz = (M^2 + 1)^{\frac{1}{2}}dr$, and $z = (M^2 + 1)^{\frac{1}{2}}r + C$. If the arc be reckoned from the pole, so that $z = 0$ when $r = 0$, the constant $C = 0$, and the corrected integral is $z = (M^2 + 1)^{\frac{1}{2}}r$. If we take the Naperian logarithm of r, we have $z = 2^{\frac{1}{2}}r$. Now when $\theta = 0$ $r = 1$, and $z = 2^{\frac{1}{2}}$. But by tracing this curve we see that if **AB** *Fig.* 177, represent the value of $r(=1)$ when $\theta = 0$, there are an infinite number of spires between this

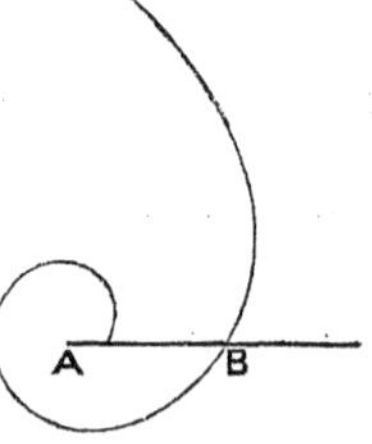

FIG. 177.

and the pole (see **119**), and we have the singular result that their entire length is $\sqrt{2}$. When $\theta = 2\pi = 6.2831853+$, we have $\log r = 6.2831853+$, and in the common system $r = 1919487.61+$ and $z = (M^2 + 1)^{\frac{1}{2}} \times 1919487.61+$, in which $(M^2 + 1)^{\frac{1}{2}} = [(.4243+)^2 + 1]^{\frac{1}{2}} = 1.086+$. Here we have another singular result, viz., that the whole of the infinite number of spires within the value $r = 1$ (**AB**), and the spire generated by the revolution from $\theta = 0$ to $\theta = 2\pi$, are together only a trifle longer than the radius vector after this revolution.

Ex. 2. Rectify the spiral of Archimedes, $r = \dfrac{\theta}{2\pi}$.

Result.—For convenience put $\dfrac{1}{2\pi} = a$, writing the equation $r = a\theta$. Then $\int dz = \dfrac{1}{a}\int (r^2 + a^2)^{\frac{1}{2}} dr$, and the process of integration is identical with that used in rectifying the common parabola (*Ex.* 2, **239**).

$$z = \frac{r(a^2 + r^2)^{\frac{1}{2}}}{2a} + \frac{a}{2} \log \left\{ \frac{r + (a^2 + r^2)^{\frac{1}{2}}}{a} \right\}.$$

Ex. 3. Rectify the cardioid, $r = a(1 + \cos\theta)$.

Solution.—This, as the name implies, is a heart-shaped curve. The student should first construct it. $dr^2 = a^2 \sin^2\theta d\theta^2$, and $r^2 = a^2 + 2a^2\cos\theta + a^2\cos^2\theta$; whence $dz = (a^2 + 2a^2\cos\theta + a^2\cos^2\theta + a^2\sin^2\theta)^{\frac{1}{2}} d\theta = a(2 + 2\cos\theta)^{\frac{1}{2}} d\theta = 2a\cos\frac{1}{2}\theta d\theta$, since $1 + \cos\theta = 2\cos^2\frac{1}{2}\theta$. (See Trigonometry **57**, P.) $\therefore z = 4a\sin\frac{1}{2}\theta + C$. Estimating the arc from $\theta = 0$, we have $z = 0$, whence $C = 0$; and the corrected integral is $z = 4a\sin\frac{1}{2}\theta$. Making $\theta = 180°$ we have $z = 4a$. This being $\frac{1}{2}$ the circumference, the entire length is $8a$.

SECTION X.

Quadrature of Plane Surfaces.

(*a*) BY RECTANGULAR CO-ORDINATES.

244. Def.—***The Quadrature*** of a surface is finding its area. The term *quadrature* comes from the conception that we find an equivalent square. Thus the quadrature of the circle consists in finding a square of the same area.

245. ***Prop.***—*The formula for the quadrature of plane surfaces is*

$$dA = ydx, \text{ or } xdy.$$

Dem.—Recurring to *Fig.* 172, it is proposed to find the area of the surface lying between **MN** and **AX**. Calling this area A, the trapezoid **PP′D′D** included by two consecutive ordinates is dA, a differential element of the area contemporaneous with dx, and dy, as heretofore considered. But area $\mathbf{PP'D'D} = \frac{\mathbf{PD} + \mathbf{P'D'}}{2} \times \mathbf{DD'}$, or $dA = \frac{y + y + dy}{2}dx = ydx + \frac{1}{2}dydx$. Since the last term is a differential of the second order with reference to the others, it must be dropped, and there results $dA = ydx$. In like manner an element of the area lying between a curve and the axis of ordinates may be shown to be $dA = xdy$. Q. E. D.

Ex. 1. Find the area of the common parabola.

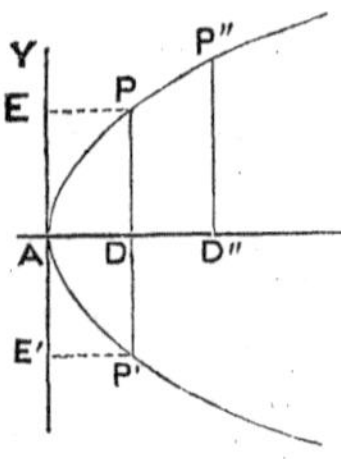

Solution.—From $y^2 = 2px$, we have $dx = \frac{1}{p}ydy$. Whence $dA = \frac{1}{p}y^2dy$, and integrating $A = \frac{y^3}{3p} + C$. Reckoning the area from the origin **A**, for $y = 0$, $A = 0$, whence $C = 0$, and the corrected integral is $A = \frac{y^3}{3p} = \frac{y^2 \cdot y}{3p} = \frac{2pxy}{3p} = \frac{2}{3}xy$; *i. e.*, the area **APD** is $\frac{2}{3}$**ADPE**. Consequently **PAP'** $= \frac{2}{3}$**EPP'E'**. The latter rectangle is sometimes called the circumscribed rectangle, and the area of the parabola is said to be $\frac{2}{3}$ *of the circumscribed rectangle.*

Fig. 178.

To find the area of any specified portion, as that between $x = \mathbf{AD} = a$, and $x = \mathbf{AD''} = b$, or of the surface **PP''D''D**, we resume the indefinite integral $A = \frac{y^3}{3p} + C = \frac{(2px)^{\frac{3}{2}}}{3p} + C$; substituting a and b successively for x, and subtracting the former result from the latter, *i. e.*, integrating between the limits $x = a$, $x = b$, we have

$$A = \frac{(2pb)^{\frac{3}{2}} - (2pa)^{\frac{3}{2}}}{3p} = \frac{(8)^{\frac{1}{2}}p^{\frac{3}{2}}}{3p}(b^{\frac{3}{2}} - a^{\frac{3}{2}}) = \tfrac{2}{3}(2p)^{\frac{1}{2}}(b^{\frac{3}{2}} - a^{\frac{3}{2}}).$$

Ex. 2. Find the area of $y = x - x^3$.

Solution.—Substituting the value of y in the general formula, it becomes $dA = xdx - x^3dx$ $\therefore A = \frac{1}{2}x^2 - \frac{1}{4}x^4 + C$. Sketching the curve, we observe that it will be natural to inquire for the area **A**m**B**, or **A**m'**B'**. Thus reckoning the area from the origin, $x = 0$, gives $A = 0$, and consequently $C = 0$. The corrected integral is $A = \frac{1}{2}x^2 - \frac{1}{4}x^4$. Hence area **A**$m$**B**, or **A**$m'$**B'** $= \frac{1}{4}$. To find the area of any other portion as **BCD**, we integrate between the limits $x = \mathbf{AB} = 1$, and $x = \mathbf{AD} = 2$. This gives $A = 2 - 4 - \frac{1}{4} = -2\frac{1}{4}$. Hence area **BCD** $= 2\frac{1}{4}$.

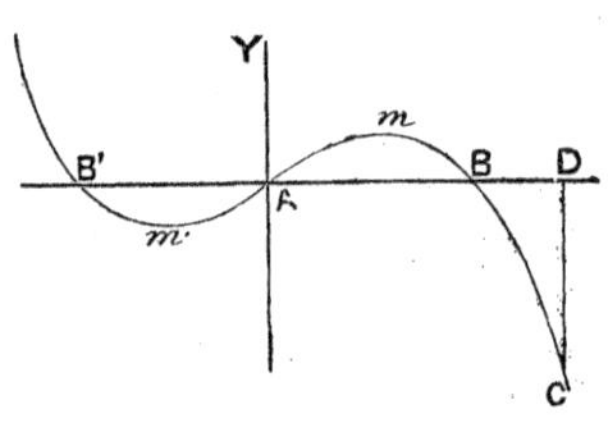

Fig. 179.

Ex. 3. Find the area of $y = x^3 - b^2x$

Area between $x = 0$ *and* $x = b$ *is* $\frac{1}{4}b^4$.

Ex. 4. Find the area of $y = x^3 + ax^2$, constructing the curve, and observing the natural limits of integration.

Result, Between $x = 0$, and $x = a$, $A = \frac{7}{12}a^4$; and between $x = 0$ and $x = -a$, $\int_{-a}^{0} dA = \frac{1}{12}a^4$.

Ex. 5. Construct and find the area of $a^2y^2 = x^2(a^2 - x^2)$.

Area $= \frac{4}{3}a^2$.

Ex. 6. Required the quadrature of $xy^2 = a^3$ between the limits $y = b$, and $y = c$. *Area* $= 2a^3\frac{b-c}{bc}$.

Ex. 7. Required the quadrature of the circle.

SOLUTION.—From $x^2 + y^2 = r^2$, we have $y = (r^2 - x^2)^{\frac{1}{2}}$, hence $dA = (r^2 - x^2)^{\frac{1}{2}}dx$. Applying formula **C** of reduction and integrating, we find $A = \frac{1}{2}x(r^2 - x^2)^{\frac{1}{2}} + \frac{1}{2}r^2 \sin^{-1}\frac{x}{r} + C$. Estimating the arc from the axis of ordinates, so that for $x = 0$, $A = 0$, C is also 0, and the corrected integral is $A = \frac{1}{2}x(r^2 - x^2)^{\frac{1}{2}} + \frac{1}{2}r^2 \sin^{-1}\frac{x}{r}$. Making $x = r$, we have *area of* $\frac{1}{4}$ *of the circle* $= \frac{1}{4}r^2\pi$; whence the entire area of a circle whose radius is r, is πr^2. Now π has been determined with sufficient accuracy in *Ex.* 3, ART. **239.**

Without assuming the value of π, we may expand $(r^2 - x^2)^{\frac{1}{2}}$ and integrate approximately. Thus $dA = (r^2 - x^2)^{\frac{1}{2}}dx = r\left(1 - \frac{x^2}{r^2}\right)^{\frac{1}{2}}dx$, and putting $\frac{x}{r} = x_1$, we have $dA = r^2(1 - x_1^2)^{\frac{1}{2}}dx_1 = r^2\left(1 - \frac{x_1^2}{2} - \frac{x_1^4}{8} - \frac{x_1^6}{16} - \frac{5x_1^8}{128} - \frac{7x_1^{10}}{256} - \frac{21x_1^{12}}{1024} - \text{etc.}\right)dx_1$;

whence $A = r^2[\int dx_1 - \frac{1}{2}\int x_1^2dx_1 - \frac{1}{8}\int x_1^4dx_1 - \frac{1}{16}\int x_1^6dx_1 - \frac{5}{128}\int x_1^8dx_1 - \frac{7}{256}\int x_1^{10}dx_1 - \frac{21}{1024}\int x_1^{12}dx_1 - \text{etc.}]$

$$= r^2\left[x_1 - \frac{x_1^3}{6} - \frac{x_1^5}{40} - \frac{x_1^7}{112} - \frac{5x_1^9}{1152} - \frac{7x_1^{11}}{2816} - \frac{21x_1^{13}}{13312} - \text{etc.}\right] + C,$$

in which C is 0 when the arc is estimated from the axis of y. Now, making $x = r$, makes $x_1 = 1$; whence the area of $\frac{1}{4}$ the circle is

$$A = r^2(1 - \tfrac{1}{6} - \tfrac{1}{40} - \tfrac{1}{112} - \tfrac{5}{1152} - \tfrac{7}{2816} - \tfrac{21}{13312} - \text{etc.}).$$

The sum of this series, thus extended is $1 - .208999 = .791001$. Hence

Area of circle $= 4 \times r^2 \times .791001 = 3.164005r^2$, approximately. This number 3.164005, if more accurately computed is found to be exactly the same as π, *Ex.* 3, ART. **239.** A nearer approximation can readily be made by extending the series.

246. COR. 1.—*The areas of circles are to each other as the squares of their radii.*

DEM.—Let r, and r_1 be the radii of two circles. Their areas are πr^2 and πr_1^2. But $\pi r^2 : \pi r_1^2 :: r^2 : r_1^2$.

247. COR. 2.—*The area of a circle whose radius is* 1 *is* π.

248. Cor. 3.—*The area of a circle is equal to its circumference into $\frac{1}{2}$ its radius.*

Dem. $r^2\pi = 2r\pi \times \frac{1}{2}r$, and $2r\pi$ is the circumference (**239**, *Ex.* 3).

249. Prob.—*To find the area of any segment of a circle.*

Solution.—Let the distances from the centre to the bases of the segment be b and a, so that $a - b =$ the height of the segment. We have but to resume the indefinite integral and take its value between these limits. Thus

$$\int_{x=b}^{x=a} dA = \tfrac{1}{2}a(r^2 - a^2)^{\frac{1}{2}} + \tfrac{1}{2}r^2 \sin^{-1}\frac{a}{r} - \tfrac{1}{2}b(r^2 - b^2)^{\frac{1}{2}} - \tfrac{1}{2}r^2 \sin^{-1}\frac{b}{r}.$$

If the segment is reckoned from the diameter, $b = 0$, and

$\int_{x=0}^{x=a} dA = \frac{1}{2}a(r^2 - a^2)^{\frac{1}{2}} + \frac{1}{2}r^2 \sin^{-1}\frac{a}{r}$. The significance of this is readily seen by inspecting *Fig.* 174. AD $= a$, and $\frac{1}{2}a(r^2 - a^2)^{\frac{1}{2}} =$ area of APD. Arc BP $= r \sin^{-1}\frac{a}{r}$; whence $\frac{1}{2}r^2 \sin^{-1}\frac{a}{r} =$ area sector ABP.

If the segment has but one base, $a = r$, and we have

$\int_{x=b}^{x=r} dA = \frac{1}{4}\pi r^2 - \frac{1}{2}b(r^2 - b^2)^{\frac{1}{2}} - \frac{1}{2}r^2 \sin^{-1}\frac{b}{r}$. From *Fig.* 174, we see that $\frac{1}{4}\pi r^2 = \frac{1}{4}$ the area of the circle, $\frac{1}{2}r^2 \sin^{-1}\frac{b}{r} =$ area BPA, and $\frac{1}{2}b(r^2 - b^2)^{\frac{1}{2}} =$ area APD.

In each case we have $\frac{1}{2}$ the segment.

Ex. 8. Find the area of the ellipse.

Sug's.— The area $= \frac{B}{A}\int_0^A (A^2 - x^2)^{\frac{1}{2}} dx$. But $\int_0^A (A^2 - x^2)^{\frac{1}{2}} dx$ is $\frac{1}{4}$ the area of a circle whose radius is A, which is πA^2. ∴ *Area of ellipse* $= \pi AB$.

250. Cor.—*The area of an ellipse is to the area of the circumscribed circle as the conjugate axis is to the transverse axis. The area of an ellipse is to the area of the inscribed circle as the transverse axis is to the conjugate axis. The area of an ellipse is a mean proportional between the inscribed and circumscribed circles.*

Ex. 9. Find the area of the cycloid.

Sug's.—We have $dx = \frac{ydy}{(2ry - y^2)^{\frac{1}{2}}}$; whence $dA = \frac{y^2dy}{(2ry - y^2)^{\frac{1}{2}}} = y^{\frac{3}{2}}(2r - y)^{-\frac{1}{2}}dy$. Integrating by applying formula **A** twice, we obtain $A = \int_0^{2r} y^{\frac{3}{2}}(2r - y)^{-\frac{1}{2}}dy = \frac{3}{2}r^2 \text{vers}^{-1}2 = \frac{3}{2}r^2\pi$, since $\text{vers}^{-1}2 = \pi$. ∴ The entire area is $3\pi r^2$, or *three times the generating circle.*

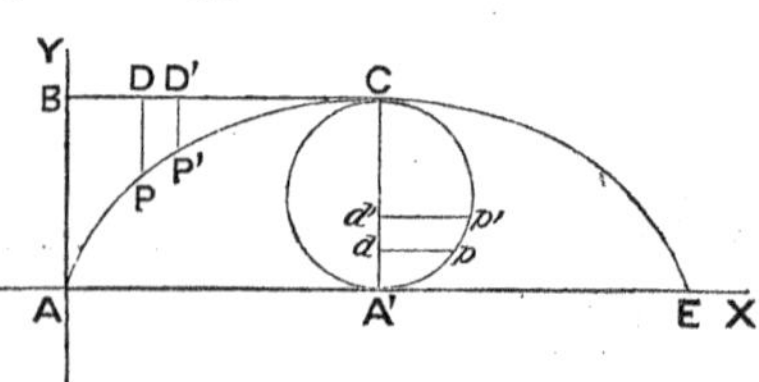

Fig. 180.

A somewhat indirect but simple method of quadrature is as follows:

1st. Find the area of APCB, the

element of which **DPP′D′** $= dx(2r - y) = (2r - y)\frac{ydy}{(2ry - y^2)^{\frac{1}{2}}} = (2ry - y^2)^{\frac{1}{2}}dy$. Now considering the circle **A′pC**, we observe that an element as $pdd'p' = dy(2ry - y^2)^{\frac{1}{2}}$. $\therefore$ **APCB** $= \frac{1}{2}$ *area of the generatrix*, $= \frac{1}{2}\pi r^2$. But **ABCA′** $= 2r \times$ **AA′** $= 2r \times \pi r = 2\pi r^2$. $\therefore$ **ACA′** $= \frac{3}{2}\pi r^2$, or the entire area of the cycloid $= 3\pi r^2$. Observe that both these integrals are to be taken between the same limits, viz., $y = 0$, and $y = 2r$.

Ex. 10. Find the area of the curve $a^4y^2 = a^2b^2x^2 - b^2x^4$.

Area $= \frac{4}{3}ab$.

(b) QUADRATURE OF POLAR CURVES.

251. Prop.—*An element of the area of a polar curve is* $\frac{1}{2}r^2d\theta$ *and the formula for the quadrature is* $dA = \frac{1}{2}r^2d\theta$.

DEM.—In *Fig.* 176 **PAP′** $= dA$. But area **PAP′** $=$ **AP′** $\times \frac{1}{2}$**PD** $= (r + dr)\frac{1}{2}rd\theta = \frac{1}{2}r^2d\theta$, omitting $\frac{1}{2}rdrd\theta$, and also remembering that **PD** $= rd\theta$. $\therefore dA = \frac{1}{2}r^2d\theta$. Q. E. D.

Ex. 1. Find the area of the spiral of Archimedes.

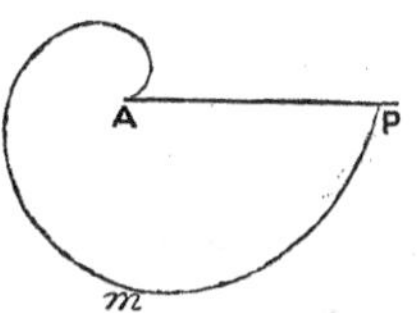

FIG. 181.

SOLUTION.—The equation is $r = \frac{1}{2\pi}\theta$; whence $dr = \frac{1}{2\pi}d\theta$, and $d\theta = 2\pi dr$. Substituting in the formula, we have $dA = \pi r^2dr$, and $A = \frac{1}{3}\pi r^3 + C$. If we estimate the area from the pole, we have $A = 0$, when $r = 0$, and consequently $C = 0$. The corrected integral is, therefore, $A = \frac{1}{3}\pi r^3$. This is the general expression for the area passed over by the radius vector in its revolution from its starting at 0 to any value, as r. At the end of one revolution $r =$ **AP** $= 1$, and $A =$ *area of the first spire* $=$ **AmP** $= \frac{1}{3}\pi = \frac{1}{3}$ the area of a circle described with **AP** as a radius. At the end of the second revolution $r = 2$, and the area traced by the radius vector, or $A_1 = \frac{8}{3}\pi$. This evidently includes twice the first spire + the second. The area of the *second spire* is, therefore $\frac{8}{3}\pi - \frac{2}{3}\pi = 2\pi$. The area of the first two spires is $\frac{7}{3}\pi$.

Ex. 2. Show that the area of the Napierian logarithmic spiral is $\frac{1}{4}$ the square described on the radius vector.

Ex. 3. Find the area of the Lemniscate of Bernouilli. ($r^2 = a^2\cos 2\theta$.)

SOLUTION.—From the formula we have $dA = \frac{1}{2}a^2\cos 2\theta\, d\theta$. Whence $A = \frac{1}{2}a^2\int_{0°}^{45°}\cos 2\theta\, d\theta = \frac{1}{4}a^2\sin 2\theta = \frac{1}{4}a^2$. $\therefore$ The entire area is a^2, *i. e.*, the square on the semi-axis.

Ex. 4. Trace the curve $r = a(\cos 2\theta + \sin 2\theta)$, and find its area.

SUG'S. $r^2 = a^2(\cos^2 2\theta + \sin^2 2\theta + 2\sin 2\theta \cos 2\theta) = a^2(1 + \sin 4\theta)$ $\therefore A = \int \frac{1}{2}a^2(1 + \sin 4\theta)d\theta = \frac{1}{2}a^2\{\int d\theta + \int \sin 4\theta\, d\theta\}$. The entire area, which is comprised between $\theta = 0$ and $\theta = 2\pi$, is πa^2.

SECTION XI.

Quadrature of Surfaces of Revolution.

252. DEF.—***A Surface of Revolution*** is a surface generated by a line (right or curved) revolving around a fixed right line as an axis, so that sections of the volume generated made by a plane perpendicular to the axis are circles.

ILL'S.—A right line parallel to and revolving around another right line, in the manner described in the definition, generates a cylindrical surface with a circular base. A semi-ellipse revolving around its transverse axis generates a *prolate spheroid*, around its conjugate axis an *oblate spheroid*. These two are varieties of ellipsoids. A paraboloid is generated by the revolution of a parabola around its axis. The number of these surfaces is, of course, infinite; and the specific character of each depends upon the nature of the generating curve.

253. ***Prop.***—*The differential element of a surface of revolution is*

$$dS = 2\pi y\sqrt{dy^2 + dx^2}.$$

DEM.—Let MN be the generatrix, AX the axis of revolution and PP′ two consecutive points on the generatrix. As MN revolves about AX, any point, as P, whose co-ordinates are x and y describes a circle whose circumference is $2\pi y$. The circumference described by the consecutive point P′ will be $2\pi(y + dy)$. PP′ describes the frustum of a cone whose surface is half the sum of the circumferences traced by P and P′ multiplied by PP′, or $= \frac{2\pi y + 2\pi y + 2\pi dy}{2}\sqrt{dy^2+dx^2}$. Whence omitting $2\pi dy$, it being an infinitesimal of a higher degree than the other terms, we have $dS = 2\pi y\sqrt{dy^2+dx^2}$. Q. E. D.

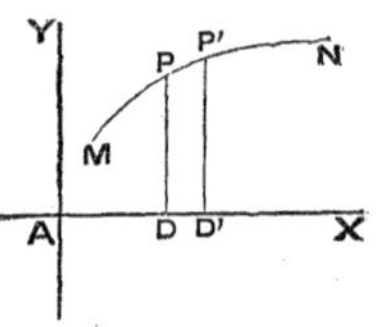

FIG. 182.

SCH.—To apply this formula, let $y = \varphi(x)$ be the equation of the generatrix, from which find dx or dy, and, having substituted in the formula, integrate between the proper limits.

Ex. 1. Find the area of the surface of a sphere.

SOLUTION.—From $y^2 + x^2 = r^2$ we have $dy^2 = \frac{x^2dx^2}{y^2}$. $\therefore dS = 2\pi y\sqrt{dx^2 + \frac{x^2dx^2}{y^2}} = 2\pi r dx$; and $S = \int_0^r 2\pi r dx = 2\pi r^2$. This being the surface of a hemisphere the entire surface of the sphere is $4\pi r^2$.

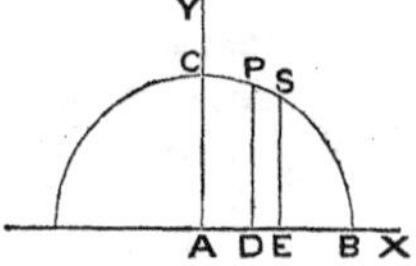

Fig. 183.

Cor. 1.—*Since* $\pi r^2 =$ *area of a great circle, the surface of a sphere* $= 4$ *great circles. Again, since* $4\pi r^2 = 2\pi r \times 2r$, *the surface of a sphere = the circumference of a great circle × the diameter.*

254. Cor. 2.—*The portion of the surface generated by any arc, as* **PS** *is a zone. To find the surface of a zone, let* **AD** $= a$ *and* **AE** $= b$ *and integrate* $dS = 2\pi r dx$ *between these limits; thus,* $S = \int_a^b 2\pi r dx = 2\pi r(b - a)$; *i. e., the surface of a zone = its altitude multiplied by the circumference of the great circle of the sphere. Hence, also, the surfaces of zones on the same sphere are to each other as their altitudes.*

Sch.—If a cylinder be circumscribed about a sphere, its convex surface is the circumference of its base $(2\pi r) \times$ by its altitude $(2r)$; *i. e.*, $4\pi r^2$, the same as the surface of the sphere. Add to this $2\pi r^2$, the upper and lower bases, and the *entire* surface of the cylinder is found to be $6\pi r^2$, The surface of the cylinder is, therefore, to the surface of the sphere as 3 to 2. This fact, together with the same relation between the volumes, was discovered by Archimedes, and the discovery was so much thought of by him that he expressed a wish that the device on his tombstone might be a sphere inscribed in a cylinder. The great geometer was murdered by the soldiers of Marcellus, B. C. 212, though contrary to the orders of that general. Marcellus executed the device of the sphere and cylinder upon the tomb, and buried the philosopher with honors. 140 years after, when Cicero was questor in Sicily, the place of the grave had been forgotten; but he, chancing to remember the device upon the stone, sought out and restored the tomb, which had become overrun with weeds and thorns.

Ex. 2. To find the area of the surface of the paraboloid.

Solution.—From $y^2 = 2px$, we have $dx^2 = \frac{y^2dy^2}{p^2}$. $\therefore dS = \frac{2\pi}{p}y(p^2 + y^2)^{\frac{1}{2}}dy$, and $S = \int \frac{2\pi y}{p}(p^2 + y^2)^{\frac{1}{2}}dy = \frac{2\pi}{3p}(p^2 + y^2)^{\frac{3}{2}} + C$. Reckoning S from $y = 0$, $C = -\frac{2}{3}\pi p^2$ and the corrected integral is $S = \frac{2\pi}{3p}[(p^2 + y^2)^{\frac{3}{2}} - p^3]$, which may be satisfied for any value of y.

Ex. 3. To find the area of the surface generated by the revolution of the hypocycloid $x^{\frac{2}{3}} + y^{\frac{2}{3}} = c^{\frac{2}{3}}$ about the axis of x.

$S = 2\pi c^{\frac{1}{3}} \int_0^c y^{\frac{2}{3}}\, dy = \frac{6}{5}\pi c^2$; *whence* $2S = \frac{12}{5}$, *the area of the fixed circle.*

Ex. 4. Find the area of the surface generated by the cycloid revolved about its base.

Sug's. $dS = 2\pi(2r)^{\frac{1}{2}}y(2r - y)^{-\frac{1}{2}}dy$, which integrated by formula **A** is $S =$

$-\frac{4}{3}\pi\sqrt{2ry}(2r-y)^{\frac{1}{2}} - \frac{16}{3}\pi r\sqrt{2r}(2r-y)^{\frac{1}{2}} + C$. This taken between $y=2r$, and $y=0$ gives for half the surface $\frac{32}{3}\pi r^2$, and for the whole surface $\frac{64}{3}$ times the generating circle.

Ex. 5. Find the area of the surface of the prolate spheroid.

SOLUTION.—From $a^2y^2 + b^2x^2 = a^2b^2$ we get $dy^2 = \frac{b^4x^2}{a^4y^2}dx^2$. $\therefore dS = 2\pi y\left(\frac{a^4y^2+b^4x^2}{a^4y^2}\right)^{\frac{1}{2}}dx = \frac{2\pi}{a^2}[a^2(a^2b^2-b^2x^2)+b^4x^2]^{\frac{1}{2}}dx = \frac{2\pi b}{a^2}(a^4-a^2e^2x^2)^{\frac{1}{2}}dx = 2\pi\frac{b}{a}(a^2-e^2x^2)^{\frac{1}{2}}dx$. Whence half the surface of the ellipsoid $=2\pi\frac{b}{a}\int_0^a(a^2-e^2x^2)^{\frac{1}{2}}dx$. To integrate $(a^2-e^2x^2)^{\frac{1}{2}}dx$ apply formula 𝕮, which gives $\int(a^2-e^2x^2)^{\frac{1}{2}}dx = \frac{x(a^2-e^2x^2)^{\frac{1}{2}} + a^2\int(a^2-e^2x^2)^{-\frac{1}{2}}dx}{2} = \frac{1}{2}x(a^2-e^2x^2)^{\frac{1}{2}} + \frac{1}{2}a^2\int\frac{dx}{(a^2-e^2x^2)^{\frac{1}{2}}} = \frac{1}{2}x(a^2-e^2x^2)^{\frac{1}{2}} + \frac{1}{2}\frac{a^2}{e}\sin^{-1}\frac{ex}{a} + C$. $\therefore S = \pi\frac{b}{a}\left[x(a^2-e^2x^2)^{\frac{1}{2}} + \frac{a^2}{e}\sin^{-1}\frac{ex}{a}+C\right]$. Reckoning $S=0$ for $x=0$, $C=0$; and then putting $x=a$ we have for half the surface $\pi ab[(1-e^2)^{\frac{1}{2}} + \frac{1}{e}\sin^{-1}e]$, or $\pi b^2 + \frac{\pi ab}{e}\sin^{-1}e$.

Ex. 6. Apply the above formula to find the area of the surface of the prolate spheroid whose generatrix is $25y^2 + 16x^2 = 400$.

Area, 235.41 nearly.

SECTION XII.

Cubature of Volumes of Revolution.

255. Prop.—*The differential element of volume of a solid of revolution is* $dV = \pi y^2 dx$.

DEM.—Using *Fig.* 182, with the same notation, the volume generated by the revolution of $DPP'D'$ about DD', is a frustum of a cone, and is therefore equal to three cones having for their common altitude the altitude of the frustum (dx), and for bases the upper base (πy^2), the lower base $[\pi(y+dy)^2]$, and a mean proportional between the two bases $[\pi y(y+dy)]$. Hence $dV = \frac{\pi(y^2+y^2+2ydy+dy^2+y^2+ydy)dx}{3} =$ (omitting terms having infinitesimals of a higher order) πy^2dx. Q. E. D.

Ex. 1. To find the volume of a sphere.

SOLUTION. $y^2 = r^2 - x^2$. $\therefore dV = \pi(r^2-x^2)dx = \pi r^2dx - \pi x^2dx$, $V = \int_0^r(\pi r^2dx - \pi x^2dx) = \frac{2}{3}\pi r^3$; or, letting $D=$ the diameter, and doubling for the entire volume $V = \frac{1}{6}\pi D^3$

256. Cor. 1.—*Since* $\frac{4}{3}\pi r^3 = \frac{1}{3}r \times 4\pi r^2$, *the volume of a sphere = the surface* $\times \frac{1}{3}$ *the radius.*

257. Cor. 2.—*The volumes of spheres are to each other as the cubes of their radii or their diameters.*

258. Cor. 3.—*The portion of a sphere included between two parallel planes is called a segment. If* a *and* b *represent the distances of these planes from the centre, the volume of the segment is*

$$V = \pi[r^2(b - a) - \tfrac{1}{3}(b^3 - a^3)].$$

259. Cor. 4. $\frac{4}{3}\pi r^3 = \frac{2}{3}$ *of* $\pi r^2 \times 2r = \frac{2}{3}$ *of the circumscribed cylinder. See* Scholium under Art. **254.**

Ex. 2. Find the volume of the Prolate Spheroid. Also of the oblate Spheroid. Show that each is $\frac{2}{3}$ of the circumscribed cylinder. Deduce from each the volume of the sphere.

Ex. 3. Find the volume of the paraboloid.

Volume = one half the circumscribed cylinder.

Ex. 4. Find the volume of the cissoid revolved about its asymptote.

Solution.—Let the curve AM revolve about BT, then will PDP′D′ be an elemental section of the volume if $DD' = dy$. $PD = 2a - x$, and the circle traced by PD in its revolution is $\pi(2a - x)^2$. Therefore the element of volume or $dV = \pi(2a - x)^2 dy$. From the equation of the curve $y^2 = \dfrac{x^3}{2a - x}$, we have $dy = \dfrac{3ax^2 - x^3}{y(2a - x)^2}dx = \dfrac{(3a - x)(2ax - x^2)^{\frac{1}{2}}}{(2a - x)^2}dx$. Whence substituting, $dV = \pi(3a - x)(2ax - x^2)^{\frac{1}{2}}dx, = 3a\pi x^{\frac{1}{2}}(2a - x)^{\frac{1}{2}}dx - \pi x^{\frac{3}{2}}(2a - x)^{\frac{1}{2}}dx$. These terms may be integrated by the use of formulas $\mathfrak{A}$ and $\mathfrak{C}$, and we find

$$V = \int_0^{2a} 3a\pi x^{\frac{1}{2}}(2a - x)^{\frac{1}{2}}dx - \int_0^{2a} \pi x^{\frac{3}{2}}(2a - x)^{\frac{1}{2}}dx = \pi^2 a^3.$$

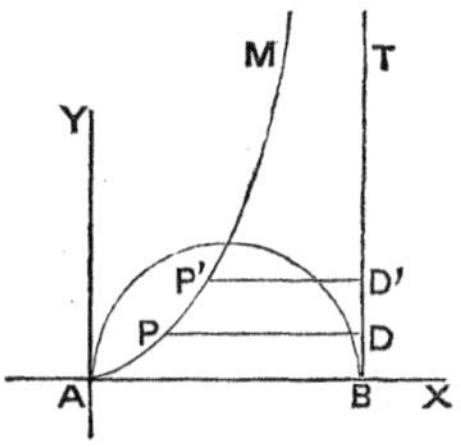

Fig. 184.

SECTION XIII.

Equations of Curves deduced by the aid of the Calculus.

260. Def.—***The Tractrix*** *or Equitangential Curve* is generated by the motion of a weight drawn upon a plane by a cord of constant length, the extremity of the cord moving along a straight line not in the direction of the cord itself. The portion of the tangent intercepted between the curve and the fixed line, is a constant quantity (the length of the string), and hence the second appellation.

ILLUS.—Let a weight be placed at **B**, *Fig.* 185, with a string of the length **AB** attached. As the extremity of the string, **A**, is carried along the line **AX**, **B** will trace **BM**, the *tractrix*. (This conception supposes friction to exist but not momentum.) When the extremity of the string is at any point in the line, as **T**, the weight will be at **P**, and it is evident that **PT**, the tangent, will be constant.

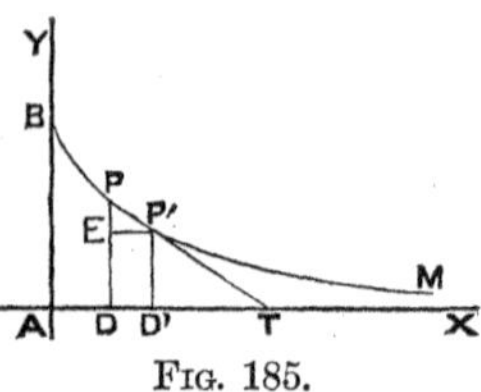

FIG. 185.

261. Prob.—*To find the equation of the* Tractrix.

SOLUTION.—Let **P** and **P**$'$ be consecutive points on the curve, **PD** $= y$, **PE** $= -dy$, (minus since y decreases as x increases), **AD** $= x$ and **DD**$' = dx$. Let **AB** $=$ **PT** $= a$. Then **DT** $= \sqrt{a^2 - y^2}$, and **PE** : **EP**$'$ :: **PD** : **DT**, or by the notation $-dy : dx :: y : (a^2 - y^2)^{\frac{1}{2}}$. $\therefore \frac{dy}{dx} + \frac{y}{(a^2 - y^2)^{\frac{1}{2}}} = 0$.

Or this differential equation may be obtained from the general differential value of the tangent; thus $\left(1 + \frac{dx^2}{dy^2}\right)^{\frac{1}{2}} y = a$, whence $\frac{dy}{dx} = \pm \frac{y}{(a^2 - y^2)^{\frac{1}{2}}}$. The $+$ sign indicates that the curve is generated by motion to the left, as then $\tan^{-1}\frac{dy}{dx}$ is $+$, and the $-$ sign is to be taken when the curve is generated as in the illustration above. In order to have the equation in finite terms it remains to integrate this differential between proper limits. The inferior limit of x is evidently 0, to which $y = a$ corresponds. The superior limit must be left general, as the curve extends to infinity. Putting the equation in the form $dx = -\frac{(a^2 - y^2)^{\frac{1}{2}}}{y} dy$ we have $\int_0^x dx = -\int_a^y \frac{(a^2 - y^2)^{\frac{1}{2}}}{y} dy = -\int_a^y \left(\frac{a^2}{y(a^2 - y^2)^{\frac{1}{2}}} - \frac{y}{(a^2 - y^2)^{\frac{1}{2}}}\right) dy = a \log \frac{a + (a^2 - y^2)^{\frac{1}{2}}}{y} - (a^2 - y^2)^{\frac{1}{2}}$. $\therefore$ The equation is $x = a \log \frac{a + (a^2 - y^2)^{\frac{1}{2}}}{y} - (a^2 - y^2)^{\frac{1}{2}}$.

---

262. Prob.—*To find the equation of a curve whose subnormal is constant.*

SOLUTION.—The general differential value of the subnormal is $y\frac{dy}{dx}$. $\therefore$ Letting p be the constant value of the subnormal, we have $y\frac{dy}{dx} = p$, or $y\,dy = p\,dx$. Integrating, $y^2 = 2px + C$. This is the equation of the parabola, as it should be, since the constant subnormal characterizes that curve. C is not determined by the problem. If the condition "which passes through the origin," were added to the problem, C would be 0.

263. Prob.—*To find the equation of the curve whose normal is of constant length.*

SOLUTION.—The general differential value of the normal is $y\left(1+\frac{dy^2}{dx^2}\right)^{\frac{1}{2}}$. Putting this equal to the constant r, we have $y\left(1+\frac{dy^2}{dx^2}\right)^{\frac{1}{2}}=r$, or $\frac{dy^2}{dx^2}=\frac{r^2}{y^2}-1$, or $dx=\frac{y\,dy}{(r^2-y^2)^{\frac{1}{2}}}=y(r^2-y^2)^{-\frac{1}{2}}dy$. Placing the origin on the curve, so that the inferior limit of the integral shall be $x=0$, $y=0$ and we have $\int_0^x dx=\int_0^y y(r^2-y^2)^{-\frac{1}{2}}dy$, or $x=r-(r^2-y^2)^{\frac{1}{2}}$ whence $y^2=2rx-x^2$. This is the well known equation of the circle, as it should be, the normal of the circle being its radius, and hence constant:

264. Prob.—*To determine the equation of the curve whose subtangent is constant* (m).

$m\log y=x+C$. The logarithmic curve.

265. Prob.—*To determine the equation of the curve whose subnormal varies as the square of the abscissa.*

$y^2=\frac{2}{3}x^3+C$. The semicubical parabola.

---

266. Prob.—*To determine the equation of a curve such that the area shall equal twice the product of its co-ordinates.*

SOLUTION. $\int ydx=2xy$, or $ydx=2xdy+2ydx$ or $\frac{dy}{y}=-\frac{1}{2}\frac{dx}{x}$. Integrating, $\log y=-\frac{1}{2}\log x+C=-\log x^{\frac{1}{2}}+C$. $\therefore$ $C=\log x^{\frac{1}{2}}y$. As $\log x^{\frac{1}{2}}y$ is constant $x^{\frac{1}{2}}y$ must be constant; therefore we may put $x^{\frac{1}{2}}y=m$, or $xy^2=m^2$ is the equation sought.

---

267. Prob.—*Find the equation of the curve whose arc varies as the square root of the third power of the abscissa.*

SOLUTION. $\int(dx^2+dy^2)^{\frac{1}{2}}=mx^{\frac{3}{2}}$, using m for any constant factor. Removing the sign of integration by differentiation and squaring both sides we have $dx^2+dy^2=\frac{9}{4}m^2xdx^2$, or $dy=(\frac{9}{4}m^2x-1)^{\frac{1}{2}}dx$. Whence integrating, $y=\frac{8}{27m^2}(\frac{9}{4}m^2x-1)^{\frac{3}{2}}+C$, the semi-cubical parabola.

SECTION XIV.

Of Tangents and Normals.

[WITHOUT THE AID OF THE CALCULUS.]

[NOTE.—Students taking a shorter course, without the Calculus, will omit the thirteen preceding sections of this chapter, and conclude the course with this and the following section. Such as take the fuller course will not need to read this section, but should read Sec. XV.]

268 Prob.—*To produce the equation of a tangent to a plane curve.*

SOLUTION.—Let **MN** be any plane curve whose equation is $y = f(x)$*, or $f(x, y) = 0$. Let **P'** be a point in the curve whose co-ordinates are (x', y'), and **P''** a point whose co-ordinates are (x'', y''). Then the equation of the *secant* line **RT** is $y - y' = \frac{y' - y''}{x' - x''}(x - x')$ (*31*). Now as **P'** and **P''** are points in the curve their co-ordinates will satisfy the equation of the curve, and we have

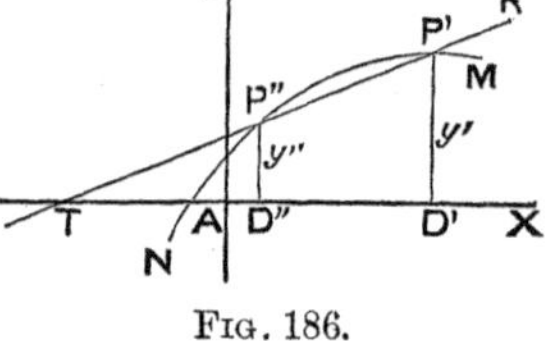

FIG. 186.

(1) $y' = f(x')$ and $y'' = f(x'')$; or

(2) $f(x', y') = 0$ and $f(x'', y'') = 0$.

If now we can find the value of $\frac{y' - y''}{x' - x''}$ from the two equations (1) or (2), in such a form that it will take a determinate, finite form when $y' = y''$, and $x' = x''$, that is when **P'** and **P''** coincide, this value substituted in the equation of the secant will transform it into the equation of a tangent, since it is evident that as **P'** and **P''** approach each other the line **RT** passing through them approaches a tangent, and becomes a tangent when the points coincide. Q. E. D.

Ex. 1. Produce the equation of a tangent to the parabola.

SOLUTION.—Letting (x', y') and $(x'' y'')$ be two points on the curve, the equation of a line passing through them is $y - y' = \frac{y' - y''}{x' - x''}(x - x')$, which is the equation of a secant, since the two points are in the curve. Moreover as (x', y') and (x'', y'') are in the curve they satisfy the equation of the curve $y^2 = 2px$, giving (1) $y'^2 = 2px'$, and (2) $y''^2 = 2px''$. Subtracting (2) from (1) $y'^2 - y''^2 = 2p(x' - x'')$, or dividing by $x' - x''$, and $y' + y''$, we have

$\frac{y' - y''}{x' - x''}, = \frac{2p}{y' + y''}, = \frac{p}{y'}$ when the point (x'', y'') is made to coincide with (x', y'). Substituting this value in the equation of the secant, we have for the tangent $y - y' = \frac{p}{y'}(x - x')$, or $yy' - y'^2 = px - px'$, or $yy' = y'^2 + px - px'$.

* This is read "y equals a function of x," and is simply a general form comprehending every equation containing x and y, supposed solved in respect to y. The second form, read "function of x and $y = 0$," indicates the same thing except that the equation is not supposed to be solved.

But as (x', y') is a point in the curve $y'^2 = 2px'$. Substituting this, and uniting terms we have $yy' = p(x + x')$.

Ex. 2. Produce the equation of a tangent to the parabola whose parameter is 9, at $x = 4$. Construct the tangent from its equation, and then construct the parabola, thus observing that the right line is tangent to the curve at the given point.

SOLUTION.—[We might proceed as in the general method, but it will be better to use the equation of the tangent to the parabola, both because it will require less work, and because it is important that this form should be made familiar.]

The equation of this parabola is $y^2 = 9x$. For $x = 4$, $y = \pm 6$. Taking the point (4, 6) we have $x' = 4$, and $y' = 6$. Hence $yy' = p(x + x')$ becomes $6y = 4\frac{1}{2}(x + 4)$, or $y = \frac{3}{4}x + 3$, as the equation of a straight line which is tangent to the parabola $y^2 = 9x$ at $x = 4$, $y = 6$.

Construction.—1st. Constructing the line $y = \frac{3}{4}x + 3$, we find that it cuts the axis of y at (0, 3), that is at **C**, and the axis of x at (— 4, 0), that is at **T**. Whence we draw **RT**. 2nd. Constructing the parabola $y^2 = 9x$, we find that the line **RT** touches it at **P**, whose co-ordinates are (4, 6).

As $x = 4$ gives $y = \pm 6$, we have another point (4, —6) which is embraced by the example. The equation of a tangent at this point is $y = -\frac{3}{4}x - 3$. [Let the student produce and illustrate it.]

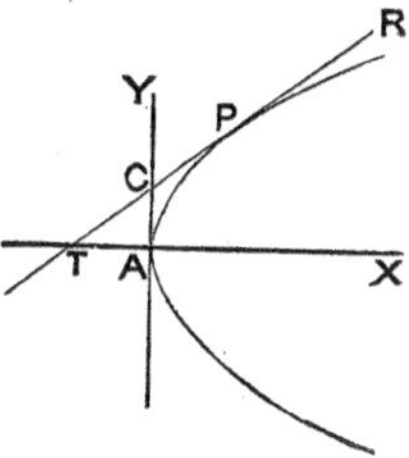

FIG. 187.

Ex. 3. Produce the equation of a tangent to the parabola whose parameter is 10, at $x = 10$, and construct and illustrate as above.

Ex. 4. Produce the equation of the tangent to the ellipse referred to its axes.

SUG'S.—Equation of secant, $y - y' = \dfrac{y' - y''}{x' - x''}(x - x')$. The equation of the ellipse $A^2y^2 + B^2x^2 = A^2B^2$, satisfied for the points (x', y') and (x'', y''), gives

(1) $A^2y'^2 + B^2x'^2 = A^2B^2$ and

(2) $A^2y''^2 + B^2x''^2 = A^2B^2$,

Subtracting, $A^2(y'^2 - y''^2) + B^2(x'^2 - x''^2) = 0$; whence

$$\frac{y' - y''}{x' - x''} = -\frac{B^2}{A^2}\left(\frac{x' + x''}{y' + y''}\right) = -\frac{B^2x'}{A^2y'},$$ when (x'', y'') becomes coincident with (x', y'). Hence the equation of the tangent is

$$y - y' = -\frac{B^2x'}{A^2y'}(x - x');$$ or, clearing of fractions and transposing,

$A^2yy' + B^2xx' = A^2y'^2 + B^2x'^2$. But since (x', y') is in the curve $A^2y'^2 + B^2x'^2 = A^2B^2$, which value substituted gives for the equation of the tangent to the ellipse $A^2yy' + B^2xx' = A^2B^2$.

Ex. 5. What is the equation of a tangent to the ellipse whose axes are 6 and 4, at $x = 2$? Construct as above.

Ans., Calling the point of tangency (2, 1.49) the equation is $y = -.5966x + 2.66$. Calling the point of tangency (2, —1.49) the equation is $y = .5966x - 2.66$.

Ex. 6. Produce the equation of a tangent to $3y^2 + x^2 = 5$, at $x = 1$. Construct and illustrate as before.

Sug.—See *Ex.* 2, page 97, second solution.

Ex. 7. Show that the form of the equation of a tangent to a circle at a given point (x', y') in the circumference is $yy' + xx' = R^2$.

Ex. 8. What is the equation of a tangent to a circle whose radius is 5 at $x = 3$? *Equation*, $y = \mp \frac{3}{4}x \pm 6\frac{1}{4}$.

Ex. 9. What is the equation of a tangent to a circle whose radius is 10 at $x = -3$? At $y = -4$? At $y = 1$?

One equation is $y = \pm \frac{1}{2}\sqrt{21}x - 25$.

Ex. 10. Produce the equation of the tangent to the hyperbola.

Equation, $A^2yy' - B^2xx' = -A^2B^2$.

Ex. 11. Produce the equation of a tangent to the hyperbola whose axes are 4 and 2, at $x = 4$. Construct and illustrate.

The points of tangency are $(4, \sqrt{3})$, and $(4, -\sqrt{3})$.

Equation of tangent at $(4, \sqrt{3})$, $y = \frac{1}{3}\sqrt{3}x - \frac{1}{3}\sqrt{3}$.

" " $(4, -\sqrt{3})$, $y = -\frac{1}{3}\sqrt{3}x + \frac{1}{3}\sqrt{3}$.

Ex. 12. Produce the equation of a tangent to $3y^2 - 2x^2 = 10$, at $x = 4$. Is there more than one tangent? Construct as above.

Equation, $y = \pm .7127x \pm .8909$.

Ex. 13. Produce the equation of a tangent to $y^2 = \frac{x^3}{4 - x}$, at $x = 2$.

Sug's.—Reasoning as on *Ex.* 1, we have $y'^2 = \frac{x'^3}{4 - x'}$, and $y''^2 = \frac{x''^3}{4 - x''}$.

Subtracting, $y'^2 - y''^2 = \frac{x'^3(4 - x'') - x''^3(4 - x')}{(4 - x')(4 - x'')} = \frac{4x'^3 - x'^3x'' - 4x''^3 + x''^3x'}{(4 - x')(4 - x'')} = \frac{4(x'^3 - x''^3) - x'x''(x'^2 - x''^2)}{(4 - x')(4 - x'')}$. Whence dividing by $y' + y''$, and $x' - x''$, we have $\frac{y' - y''}{x' - x''} = \frac{4(x'^2 + x'x'' + x''^2) - x'x''(x' + x'')}{(y' + y'')(4 - x')(4 - x'')} = \frac{6x'^2 - x'^3}{y'(4 - x')^2}$ when $x'' = x'$, and $y'' = y'$. Substituting this value of $\frac{y' - y''}{x' - x''}$ in the general equation for the secant, we have for the general equation of a tangent to this locus $y - y' = \frac{6x'^2 - x'^3}{y'(4 - x')^2}(x - x')$. For the point on the curve $x = 2$, this gives for the *two* tangents $y = 2x - 2$, and $y = -2x + 2$.

---

269. Cor.—*The tangent of the angle which a tangent to an ellipse at* (x, y) *makes with the axis of abscissas is* $-\frac{\mathrm{B}^2\mathrm{x}}{\mathrm{A}^2\mathrm{y}}$, *to an hyperbola* $\frac{\mathrm{B}^2\mathrm{x}}{\mathrm{A}^2\mathrm{y}}$, *and to a parabola* $\frac{\mathrm{p}}{\mathrm{y}}$.

Dem.—These are the values of the coefficient $\frac{y' - y''}{x' - x''}$ as obtained in the several cases in *Ex.* 4, 10, and 1. But this coefficient is the tangent of the angle which the line makes with the axis of x (**31**, Cor. 1).

Ex. 1. What angle does a tangent to $y^2 = 8x$ make with the axis of x, when the point of tangency is at $x = 2$? At $x = 5$? At $x = 10$?

Answers, $\text{Tan}^{-1}1$ or 45°, $\tan^{-1}(-1)$ or 135°; $\tan^{-1}(\pm .63245)$ or 32° 18′ 30″ and 147° 41′ 30″; $\tan^{-1}(\pm .447214)$ or 24° 5′ 41″ and 155° 54′ 19″.

Ex. 2. What angle does the focal tangent to the parabola make with the axis of x?

Ex. 3. In an ellipse whose axes are 10 and 6 what angle does a tangent at the point $x = 2$ make with the axis of x?

Ans., 165° 19′ 33″ and 14° 40′ 27″.

270. Sch.—To determine at what point on a given curve a tangent must be drawn to make a given angle with the axis of x; or, what is the same thing, to find a point at which a curve has a given direction, put the general value of the co-efficient of x in the equation of the tangent to the particular curve equal to the tangent of the proposed angle. The equation thus formed together with the equation of the locus will enable us to find the values of x and y which locate the point sought. In case the tangent is to be parallel, the co-efficient is put equal to 0; if perpendicular, equal to ∞.

Ex. 4. At what point on an ellipse whose axes are 16 and 8 must a tangent be drawn to make an angle with the axis of x whose tangent is 2? At what point to make an angle of 45°? At what point to make an angle of 135°? At what point is the tangent parallel? At what point perpendicular?

Sug's.—The *general* value of the tangent of the angle which a tangent to an ellipse makes with the axis of x is $-\frac{B^2x}{A^2y}$. Hence for the 1st inquiry we have $-\frac{16x}{64y} = 2$, and $64y^2 + 16x^2 = 1024$, from which to find x and y.

Ex. 5. At what point on an hyperbola whose axes are 8 and 6 must a tangent be drawn to make an angle of 45° with the axis of x? What angle does the focal tangent make with the axis of x?

Ex. 6. On an elliptical track whose transverse axis is 1 mile, and whose conjugate is $\frac{3}{4}$ of a mile in length and coincides with the meridian, in what direction is a man travelling when he is on the north-west quarter of the track, travelling around from left to right, and is at 40 rods from the transverse axis?

Answer, North 25° 14′ 21″ east.

SUBTANGENTS.

271. Def.—***A Subtangent*** is the portion of the axis of abscissas intercepted between the foot of the ordinate from the point of tangency, and the intersection of the tangent with this axis; or it may be defined as the projection of the corresponding portion of the tangent upon the axis of x. **DT** is the subtangent corresponding to the point **P**.

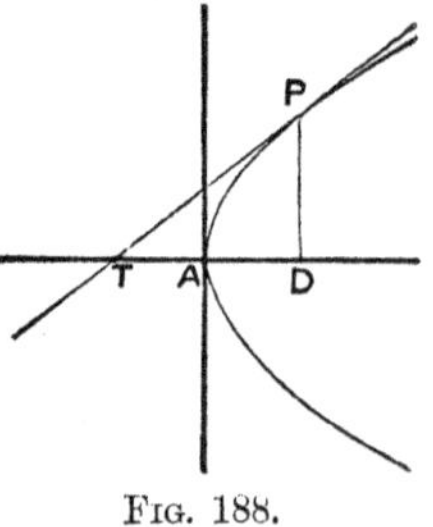

Fig. 188.

272. Prob.—*To find the length of the subtangent.*

Solution.—Letting α represent the angle which a tangent to the curve at the specified point makes with the axis of x, find $\tan \alpha$ as in (**268**). Whence from the triangle **PDT**, *Fig.* 188, we have $\frac{PD}{DT} = \tan \alpha$, or **DT** (the subtangent) $= \frac{y}{\tan \alpha}$. A slightly different method of solution is to produce the equation of the tangent to the curve, and then find where it intersects the axis of x. This intercept, as **AT**, *Fig.* 188, and the abscissa **AD** make known the subtangent, which is always their *algebraic* difference.

Ex. 1. Find the value of the subtangent of the common parabola at (x', y').

Solution.—We find in (**268** *Ex.* 1) that $\tan \alpha = \frac{p}{y'}$. Whence $Subt. = \frac{y'}{\tan \alpha} = \frac{y'^2}{p} = \frac{2px'}{p} = 2x'$.

273. Sch.—From this example we learn that the subtangent in a parabola is always equal to twice the abscissa of the point of tangency. Hence to draw a tangent to a parabola at any point as **P**, *Fig.* 188, let fall the ordinate **PD**, take **AT** = **AD**, and draw **PT**.

Ex. 2. What is the subtangent to $y^2 = 10x$, at $x = 6$? At $y = 8$? At $y = 12$? Draw these tangents according to the scholium.

Ex. 3. Find the value of the subtangent to the ellipse at (x', y').

SOLUTION.—We have $\tan\alpha = -\frac{B^2x'}{A^2y'}$ by *Ex.* 4 (**268**), whence *Subt.* $= \frac{y'}{\tan\alpha} = -\frac{A^2y'^2}{B^2x'} = \frac{A^2 - x'^2}{x'}$ neglecting the — sign as we are inquiring simply for a *value*, not a direction.

274. SCH. 1.—From this example we learn that the subtangent in the ellipse does not depend upon the conjugate axis, but only on the transverse axis and the abscissa of the point of tangency. Hence *if several ellipses be drawn on the same transverse axis the subtangents corresponding to the same abscissa are equal.* Thus in *Fig.* 189 let APB, AP′B, AP″B, and AP‴B be several ellipses on the same transverse axis CB, then AD being any abscissa (x'), the subtangent corresponding to each point of tangency P, P′, P″, P‴, is $\frac{A^2 - x'^2}{x'}$. Hence the tangents at these several points cut the axis of x at the same point T.

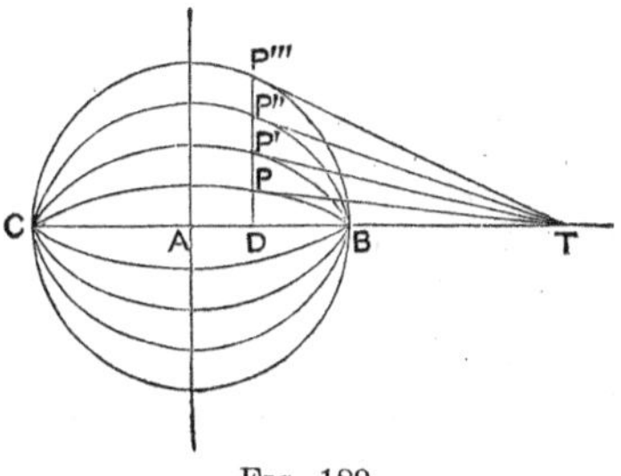

FIG. 189.

275. SCH. 2.—This property affords a convenient method of drawing a tangent to an ellipse at a given point in the curve. Thus let P be the point, *Fig.* 189. Draw the circle upon the transverse axis CB, draw the ordinate PD and produce it till it meets the circumference of the circle in P‴, and then draw a tangent to the circle at P‴. T being the point at which the latter intersects the axis of x, is also the point at which all corresponding tangents to ellipses on the same axis intersect. Hence drawing PT it is the tangent sought.

276. COR. —*The expression for the subtangent being independent of* B, *the property is the same in the hyperbola as in the ellipse* (**60**). *However as* x $>$ A *in the hyperbola, we may write* Subt. $= \frac{x'^2 - A^2}{x'}$, *in order to have it positive.*

Ex. 4. ***Prop.***—In an ellipse or hyperbola if from any point in the curve a tangent and an ordinate be drawn to the transverse axis, half this axis is a mean proportional between the distances of the intersections from the centre.

DEM.—Let **P** be any point in the curve and **PD**, **PT** the ordinate and tangent. Then **AT** : **AB** :: **AB** : **AD**. For we have the equation of a tangent at **P** (x', y'), $A^2yy' \pm B^2xx' = \pm A^2B^2$. Making $y = 0$, we get $x = \frac{A^2}{x'}$. But x is **AT** and x' is **AD**. Hence $x : A :: A : x'$, or **AT** : **AB** :: **AB** : **AD**. The demonstration being the same in each case.

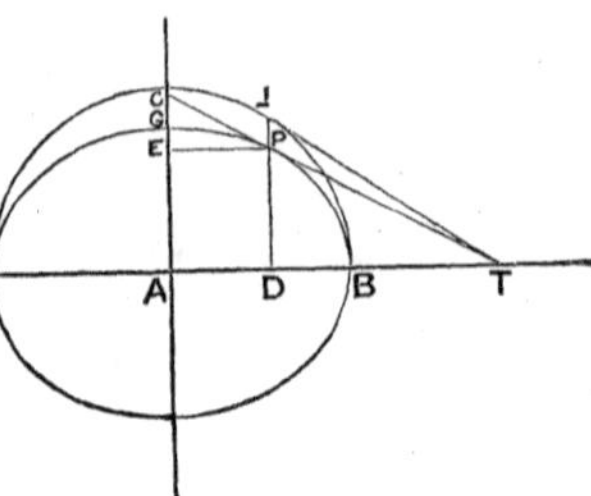

FIG. 190.

277. SCH. 1.—*To draw a tangent to a hyperbola at a given point.* From the given point of tangency **P**, *Fig.* 191, let fall the ordinate **PD**; and upon the transverse axis **HB**, and the abscissa **AD**, draw semi-circumferences. From their intersection let fall **LT** a perpendicular upon the axis of x. Draw a line through **P** and **T** and it is the tangent sought. *Proof.* Drawing **AL** and **LD**, we have **AD** (or x') : **AL** (or A) :: **AL** (or A) : **AT**. Whence **AT** $= \frac{A^2}{x'}$ and is the intercept made by a tangent at **P**.

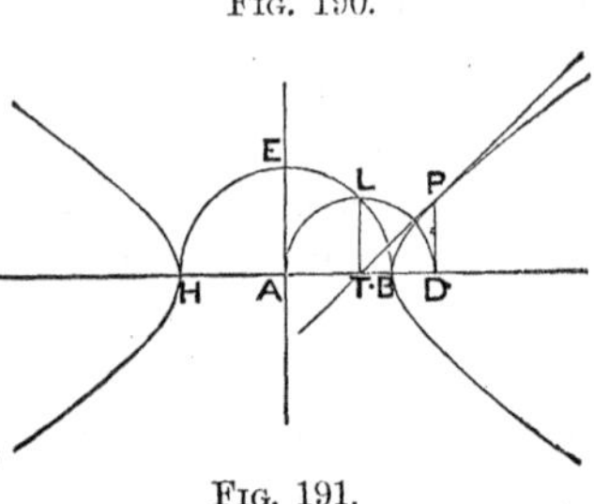

FIG. 191.

278. SCH. 2.—The pupil can scarcely fail to notice the close analogy between the forms of the equations of the conic sections and the equations of their tangents; *by simply dropping the accents the latter return to the former.* Thus dropping the accents

$A^2yy' + B^2xx' = A^2B^2$ ....becomes $A^2y^2 + B^2x^2 = A^2B^2$,
$A^2yy' - B^2xx' = - A^2B^2$ ......becomes $A^2y^2 - B^2x^2 = - A^2B^2$,
$yy' = p(x + x')$ ....becomes $y^2 = 2px$, and
$yy' + xx' = R^2$ ....becomes $y^2 + x^2 = R^2$.

NORMALS.

279. DEF.—*A* ***Normal*** to a plane curve is a perpendicular to a tangent at the point of tangency.

280. ***Prob.***—*To produce the equation of a normal to a plane curve.*

DEM.—Let **PE** be a normal to the curve **MN** at the point **P**, the co-ordinates of which are (x', y'). The equation of a tangent at the point **P** is $y - y' = a(x - x')$, in which a is the tangent of the angle which the tangent at **P** makes with the axis of x, that is tan **STX**; and is to be determined from the equation of the curve as in the preceding part of this section. Now the equation of a line passing

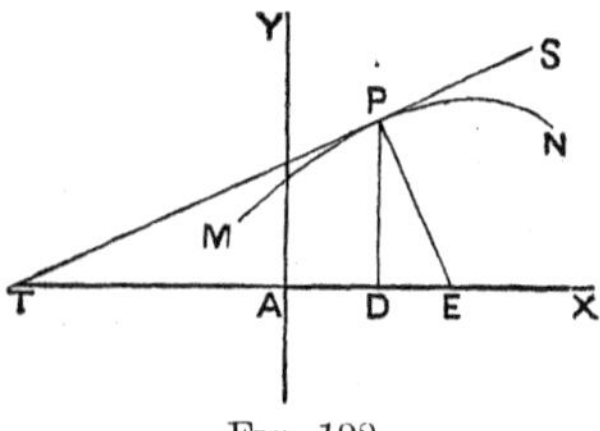

FIG. 192.

through (x', y') and perpendicular to the line $y - y' = a(x - x')$, is $y - y' = -\frac{1}{a}(x - x')$ (**39**), the coefficient $-\frac{1}{a}$ being the negative reciprocal of the tangent of the angle which the tangent to the curve makes with the axis of x.

281. Cor.—*The tangent of the angle which a normal to a curve at a particular point makes with the axis of* x *is the negative reciprocal of the tangent of the angle which is made by a tangent to the curve at the same point.*

Ex. 1. To find the equation of a normal to an ellipse.

Solution.—The equation of the tangent to the ellipse is $A^2yy' + B^2xx' = A^2B^2$, or $y = -\frac{B^2x'}{A^2y'}x + \frac{B^2}{y'}$. Now the equation of *any* line passing through (x', y') is $y - y' = a(x - x')$; but in order that this should be a normal to the ellipse we must have $a = \frac{A^2y'}{B^2x'}$, the negative reciprocal of the tangent of the angle which the tangent makes with the axis of x. Hence $y - y' = \frac{A^2y'}{B^2x'}(x - x')$ is the equation of a normal to the ellipse.

Ex. 2. Show that the equation of a normal to an hyperbola is $y - y' = -\frac{A^2y'}{B^2x'}(x - x')$.

Ex. 3. Show that the equation of a normal to the parabola is $y - y' = -\frac{y'}{p}(x - x')$.

Ex. 4. Show that the equation of a normal to the circle is $y = \frac{y'}{x'}x$, and hence is the radius.

Ex. 5. What is the equation of a normal to the ellipse whose axes are 8 and 4 at $x = 1$? At $x = -1$? At $x = 3$?

One equation is $y = \pm \frac{2}{3}\sqrt{7}x \mp \frac{3}{2}\sqrt{7}$.

Ex. 6. What is the equation of a normal to the parabola whose parameter is 9, at $x = 4$? At $x = 9$? At $x = 5\frac{4}{9}$?

One equation is $y = \mp 2x \pm 27$.

282. Cor.—*The general expressions for the tangents of the angles which normals to the conic sections make with the axis of* x *are:*

For the ellipse $\frac{A^2y'}{B^2x'}$, *for the hyperbola* $-\frac{A^2y'}{B^2x'}$,

For the circle $\frac{y'}{x'}$, *for the parabola* $-\frac{y'}{p}$.

SUBNORMALS.

283. Def.—***The Subnormal*** is the projection of the normal upon the axis of x; or it is the distance from the foot of the ordinate let fall from the point in the curve to which the normal is drawn, to the intersection of the normal with the axis of x, as **DE**, *Fig.* 193.

Ex. 1. Show that the subnormal in the ellipse is equal to $\frac{B^2x'}{A^2}$, and has the same numerical value in the hyperbola.

Sug's.—In case of the ellipse, let **PE** be the normal at **P** and **ED** the subnormal. Now $\frac{PD}{ED} = \tan PED$, or $\frac{y'}{Subnor} = \frac{A^2y'}{B^2x'}$. Whence $Subnor = \frac{B^2x'}{A^2}$.

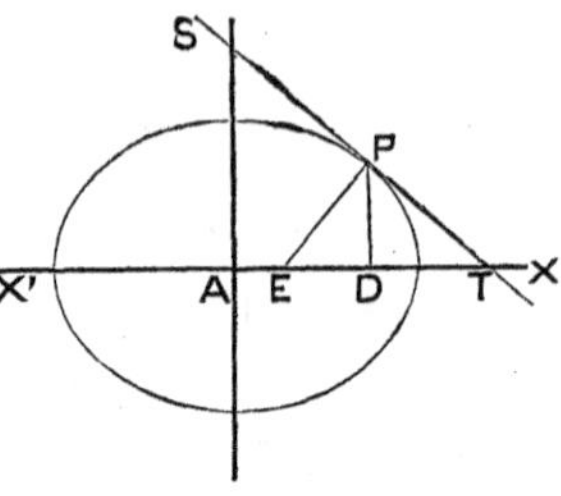

Fig. 193.

Ex. 2. Show that in the parabola the subnormal is constant and equals the semi-parameter. Show how this property may be used to draw a tangent to a parabola when the focus is known.

---

284. ***Prop.***—*In the parabola a line joining the focus and the intersection of a tangent with the axis of* y *(a tangent at the vertex), is perpendicular to the tangent.*

Dem.—Let **F** be the focus, **PT** a tangent, and **PE** a normal. Join the intersection **L** with **F**. Then is **FL** perpendicular to **PT**. For, since **AT** = **AD** (*Ex.* 1, **272**), **TL** = **LP**. Again **TF** = **AT** + **AF** = **AD** + $\frac{1}{2}p$. Also **FE** = **AD** + **DE** − **AF** = **AD** + $p - \frac{1}{2}p$ = **AD** + $\frac{1}{2}p$. Whence as **TP** and **TE** are bisected by **FL** it is parallel to the normal **PE** and hence perpendicular to the tangent **PT**. Q. E. D.

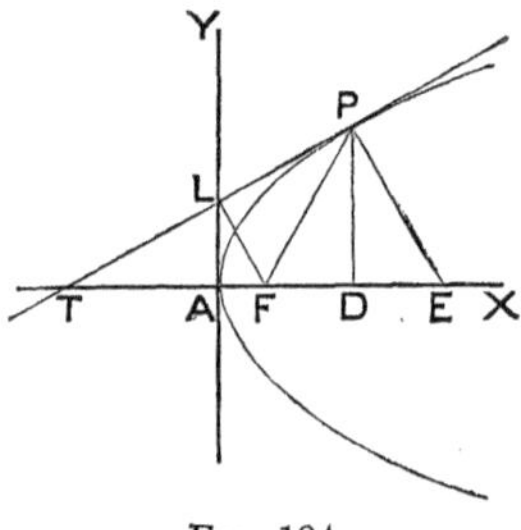

Fig. 194.

285. Cor. **PF** = **TF** = **FE**. *Also angle* **PTF** = **TPF**, *and* **FPE** = **FEP**. *Again* **PFX** = 2**PTX**.

286. Sch. 1.—Having given the curve and its axis, to find the focus draw a tangent to any point, as **P**, by (273), and then erect a perpendicular to it where it intersects the tangent at the vertex. The intersection of this perpendicular with the axis of x, will be the focus

287. Sch. 2.—Having given the axis and focus, to draw a tangent and a normal at **P**, take **FT** = **FP** = **FE** and draw **PT** and **PE**.

288. Sch. 3.—Having the axis and focus, a tangent may be drawn making any given angle with the axis of x, by making **PFX** = twice the given angle, and drawing a tangent from the point where **PF** intersects the curve.

SECTION XV.

Special Properties of the Conic Sections.

[NOTE.—The importance of the Conic Sections renders it necessary that their properties should be more fully developed than is found expedient in a compendious presentation of the subject of the General Geometry, and hence this section. Similar sections might be added on other curves, as of the cycloid, the catenary ; or sections discussing the loci embraced by equations of the 3rd degree, or the 4th degree, etc. But these subjects are not of sufficient importance to require treatment in an elementary course, nor capable of being epitomized so as to be brought within proper limits for such a course. Those who wish to pursue the subject farther will find Salmon's Conic Sections and Higher Plane Curves in two volumes, or Price's Infinitesimal Calculus in four large volumes, the best English resources. Todhunter's four volumes, two on the Calculus, and two on the Co-ordinate Geometry, are also among the most valuable recent treatises. The author of this volume proposes to prepare a second volume on loci in space, and a more extended course in the Calculus, as soon as he is able.]

(a) RADII VECTORES AND THE ANGLES WHICH THEY MAKE WITH A TANGENT.

289. DEF.—***A Radius Vector*** of a conic section is a line drawn from a focus to a point in the curve.

290. Prop.—*In an ellipse the sum of the radii vectores to any point in the curve is constant and equal to the transverse axis, and in the hyperbola the difference is constant and equal to the transverse axis.*

DEM. **P** being any point in the curve, let **PF** $= r$ and **PF′** $= r'$. We have $r = \sqrt{\overline{PD}^2 + \overline{DF}^2} = \sqrt{y^2 + (Ae - x)^2}$; also $r' = \sqrt{\overline{PD}^2 + \overline{DF'}^2} = \sqrt{y^2 + (Ae + x)^2}$. But $y^2 = (A^2 - x^2)(1 - e^2)$ (***52***). Substituting this value of y^2, we have $r = \sqrt{(A^2 - x^2)(1 - e^2) + (Ae - x)^2} = \sqrt{A^2 - 2Aex + e^2x^2} = A - ex$; and

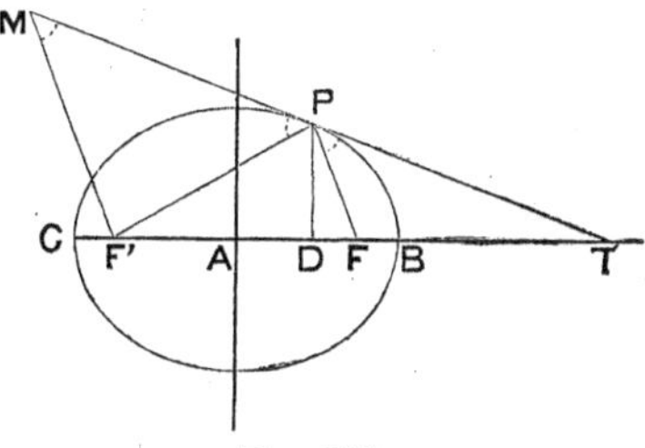

FIG. 195.

$r' = \sqrt{(A^2 - x^2)(1 - e^2) + (Ae + x)^2} = \sqrt{A^2 + 2Aex + e^2x^2} = A + ex$. Adding, $r' + r = 2A$, for the ellipse.

In the hyperbola $A - ex$, is negative, since $ex > A$, and we write $r' = A + ex$, and $r = ex - A$. Whence, subtracting, $r' - r = 2A$. Q. E. D.

291. COR.—*The length of a radius vector drawn to the nearer focus is* $r = A - ex$, *and to the more remote* $r' = A + ex$.

292. SCH.—The principles enunciated in this proposition afford very simple means for constructing the loci mechanically. For the ellipse take a string equal in length to the transverse axis, and fastening its ends at the foci, put a pencil against the string and move it around the perimeter of the curve, keeping the string tense. Thus **F′PF** *Fig.* 195, represents the string, and **P** the pencil when the point **P** is located.

To construct an hyperbola, take a ruler **AB**, and a string **BPF**; making the string shorter than the ruler by the length of the transverse axis of the required hyperbola; fasten one end of the string to one end of the ruler, as at **B**, fasten the other end of the string at one focus, as at **F**, and the other end of the ruler at the other focus, as at **F′**. Place a pencil against the string and bear it against the side of the ruler, as at **P**, and keeping the string tense move the pencil around the curve. It is evident that **F′P** — **PF** $= 2A$ in all positions of **P**. Hence **P** traces the curve. To trace the other branch the attachments have to be changed, so that the free end of the string shall be attached at **F′**, and the end of the ruler at **F**.

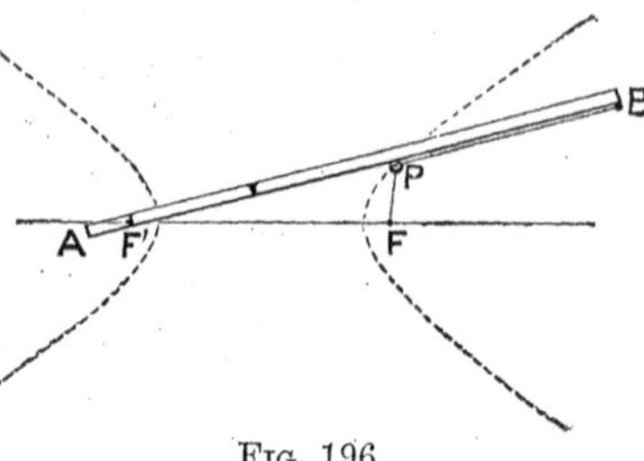

FIG. 196.

---

293. Prop.—*The radii vectores drawn to any point in an ellipse or hyperbola make equal angles with the tangent at that point.*

DEM.—Let **PF** and **PF′**, be the radii vectores, and **MT** the tangent, *Fig's.* 195, 197. Then **FPT** = **F′PM**. For, **AT** $= \frac{A^2}{x}$ (***137***, *Ex.* 1, or ***276***, *Ex.* 4), and **AF** = **AF′** $= Ae$. Hence, in the ellipse, **FT** $= \frac{A}{x}(A - ex)$, and **F′T** $= \frac{A}{x}(A + ex)$; and in the hyperbola **FT** $= \frac{A}{x}(ex - A)$, and **F′T** $= \frac{A}{x}(ex + A)$. Wherefore in either case we have **FT** : **F′T** :: $r : r'$ (***291***). Now drawing **F′M** parallel to **PF** we have **FT** : **F′T** :: **PF** : **F′M**, or $r : r' :: r :$ **F′M**. ∴ **F′M** $= r' =$ **F′P**, and **F′MP** = **FPT** = **F′PM**. Q. E. D.

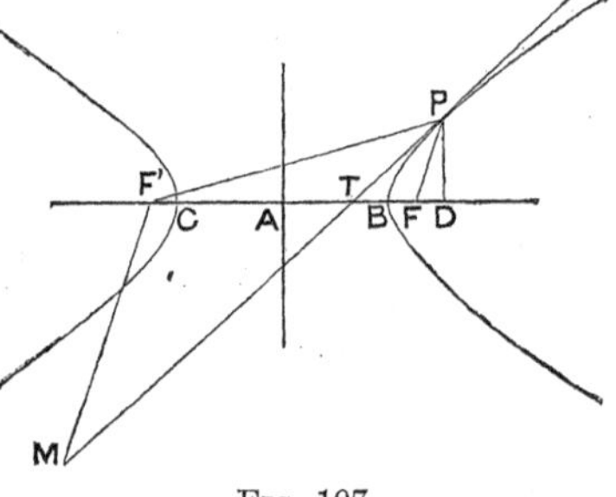

FIG. 197.

294. COR.—*In the ellipse the normal bisects the angle included by the radii vectores to the same point; and in the hyperbola it bisects the angle included by one radius vector and the other produced.*

295. SCH.—This principle affords one of the most convenient methods of drawing tangents to these curves.

1st. *To draw a tangent through a given point in the curve.* Let P, *Fig's.* 195, 197, be the point. Draw the radii vectores to the point and bisect the included angle for the hyperbola, or the angle included by one radius vector and the other produced in the case of the ellipse.

2nd. *To draw a tangent from a point without the curve.* Let P be the point. Join P with the nearer focus, and from P as a centre pass an arc of a circle through that focus. From the other focus, with a radius equal to the transverse axis, strike an arc cutting the former as at D and D′. Join D and D′ with F′, and T and T′ are the points of tangency. To prove this for T, join D and F, and F and T. Now F′T + TF = F′D, since each = $2A$. Hence TD = TF. Moreover DP = PF. Hence TP is perpendicular to DF, and angle DTP = MTF′ = PTF. Whence we know that PM is tangent at T. In a similar manner PM′ can be shown perpendicular to FD′, and hence tangent at T′. [The student should complete the figure and give the demonstration in the case of the hyperbola.]

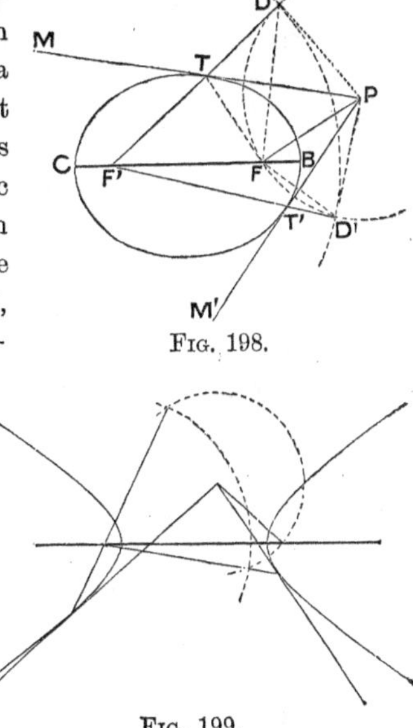

FIG. 198.

FIG. 199.

3rd. If a circle be described on PF and another on CB, the lines passing from P through their intersections are tangents to the curve, and this whether P is in or without the curve. [Why? The student should be able to answer after having read in this section through the next two propositions.] This method, however, is impracticable when P is without the curve, as it does not indicate the precise point of tangency.

---

296. *Prop.*—*In the parabola the radius vector drawn to the point of tangency makes the same angle with the tangent as a diameter through the same point, or as the tangent does with the axis of abscissas.*

DEM.—Let PF be the radius vector, PF′ the diameter, and MT the tangent. Then FPT = F′PM. For, AT = x (**140**, or **272**, *Ex.* 1), and AF = $\frac{1}{2}p$. Hence FT = $x + \frac{1}{2}p$. But PF = PD = BE = BA + AE = $\frac{1}{2}p + x$. ∴ FT = FP, and angle FPT = FTP = F′PM. Q.E.D.

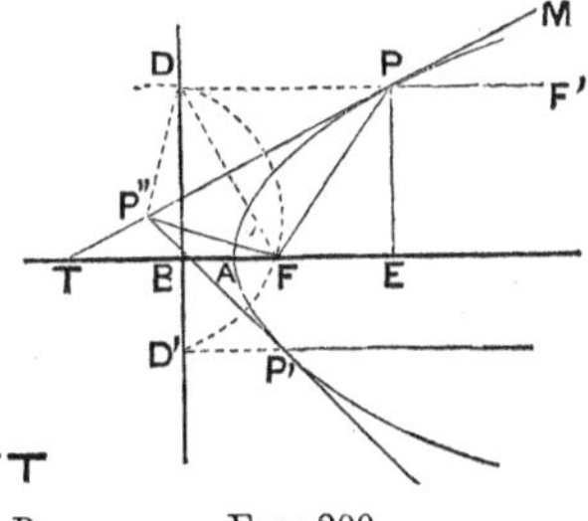

FIG. 200.

297. COR.—*In the parabola the normal bisects the angle included by the radius vector and a diameter at the same point in the curve.*

298. SCH.—*To draw a tangent to the point* P *in the parabola,* draw the radius vector and the diameter to the point, produce one of them (as PD) and bisect the angle thus formed.

To draw a tangent from a point without as P'', join the point with the focus, from P'' as a centre, pass an arc of a circle through the focus, and through its intersections with the directrix draw diameters. The vertices of these diameters are the points of tangency on the curve, as P and P'. To prove this, observe that as PF = PD, and P''F = P''D, P''P is perpendicular to DF and angle FPP'' = DPP'' = MPF'. [Let the student give the proof for the point P'.]

[NOTE.—The properties demonstrated in these propositions give elliptic, hyperbolic, and parabolic reflectors their peculiar properties. Thus, rays of light, sound, or heat diverging from one focus of an elliptic reflector are converged at the other; diverging from one focus of an hyperbolic reflector, they diverge after reflection as though they proceeded directly from the other focus; and diverging from the focus of a parabolic reflector, they are reflected parallel. Conversely to the last, incident rays parallel to the axis are concentrated at the focus of a parabolic reflector.]

---

299. Prop.—*The rectangle of the perpendiculars from the foci upon the tangent of the ellipse or hyperbola is constant, and equal to the square of the semi-conjugate axis.*

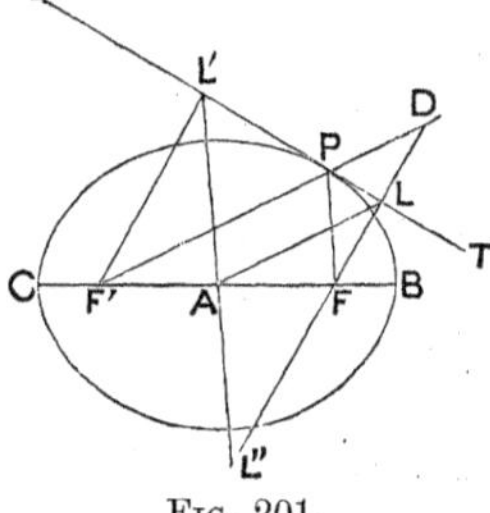

FIG. 201.

DEM.—Let L'T be a tangent at any point P, and FL a perpendicular from the focus upon it. Produce FL till it meets F'P, produced if necessary, in D, and draw AL. Since AF = AF' and FL = LD, AL is parallel to F'D and equal to $\frac{1}{2}$F'D = AB. *Hence the foot of a perpendicular from the focus upon a tangent lies in the circumference of a circle described on the transverse axis.* Now let F'L' be the perpendicular from the second focus upon the tangent LT, we are to show that FL × F'L' = B^2. Join A and L' and produce the line till it meets LF produced in L''. Then the triangles AL'F' and AL''F are equal, and AL'' = AL', and L'' is in the circumference of the circle described on CB. Finally, FL × F'L' = FL × FL'' = FB × FC (CB and LL'' being chords in the same circle). But FB × FC = $(A + Ae)(A - Ae) = A^2(1 - e^2) = B^2$ (**49**, 3rd and 7th). ∴ FL × F'L' = B^2. Q. E. D.

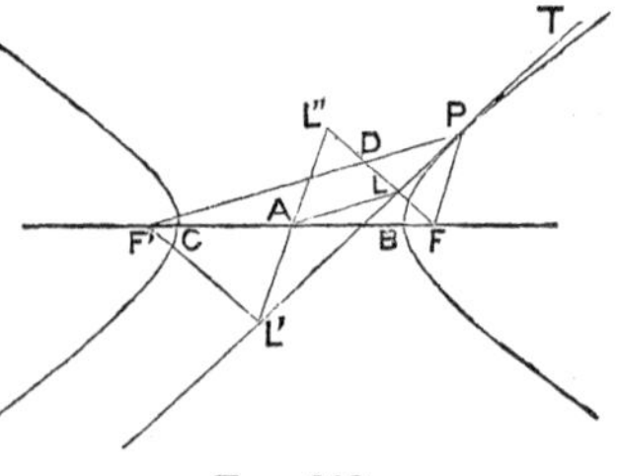

FIG. 202.

300. COR.—*The semi-conjugate axis is a mean proportional between the focal distances.*

(*b*) SUPPLEMENTARY CHORDS AND CONJUGATE DIAMETERS.

301. DEF.—The term ***Ordinate,*** as used in connection with the conic sections, may mean any line drawn from a point in the curve to any diameter, and parallel to a tangent at the extremity of that diameter.

302. DEF.—***Supplementary Chords*** are chords drawn from any point in the curve to the extremities of any diameter.

303. DEF.—One diameter is said to be ***conjugate*** to another when it is parallel to a tangent at the extremity of the latter.

---

304. Prop.—*In an ellipse the rectangle of the tangents of the angles which a pair of supplementary chords make with the transverse axis is equal to* $-\frac{B^2}{A^2}$.

DEM.—Let **PC** and **PB** be supplementary chords drawn to the axis. Let the angles **PCX** and **PBX** be represented by α and α', and $\tan\alpha = a$, and $\tan\alpha' = a'$. Then $aa' = -\frac{B^2}{A^2}$, A and B being the semi-axes. The equation of **PC** is $y = a(x + A)$, and of **PB** $y = a'(x - A)$, disregarding the signs of a and a', since **PC** and **PB** are, as yet, any lines passing through **C**, and **B**. For the intersection of these lines these equations are simultaneous, and we may have $y^2 = aa'(x^2 - A^2)$. Again, when the point of intersection is in the ellipse, this result is simultaneous with the equation of the curve, $A^2y^2 + B^2x^2 = A^2B^2$, or $y^2 = -\frac{B^2}{A^2}(x^2 - A^2)$; whence combining the two, we have $aa' = -\frac{B^2}{A^2}$. Q. E. D.

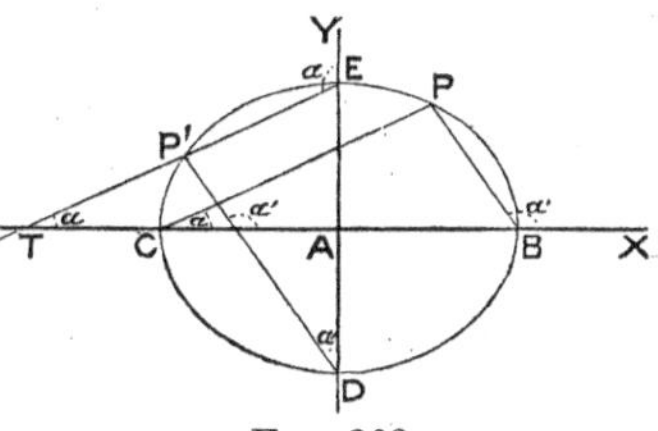

FIG. 203.

305. COR. 1.—*By a similar course of reasoning, or by simply changing the sign of* B², *we have for the hyperbola* $aa' = \frac{B^2}{A^2}$.

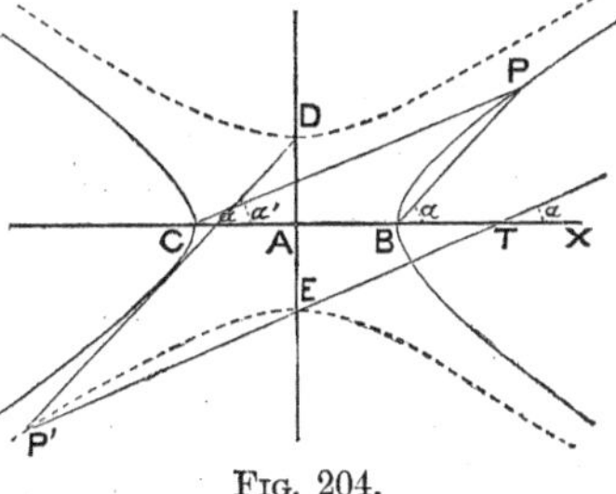

FIG. 204.

306. COR. 2.—*In the ellipse, if supplementary chords are drawn to the extremities of the conjugate axis* $aa' = -\frac{B^2}{A^2}$ (*the same as before*) *if the angles are measured from the axis of* x, *but* $aa' = -\frac{A^2}{B^2}$ *if they are measured from the axis of* y.

DEM.—The equations of **P′E** and **P′D**, *Fig.* 203, are respectively $y - B = ax$ and $y + B = a'x$; whence $y^2 - B^2 = aa'x^2$. From $A^2y^2 + B^2x^2 = A^2B^2$ we have $y^2 - B^2 = -\frac{B^2}{A^2}x^2$. $\therefore\ aa' = -\frac{B^2}{A^2}$. If the angles are reckoned from the axis of y we observe that for a we shall have $-\frac{1}{a}$, and for $-a'$, $\frac{1}{a'}$. $\therefore\ aa' = -\frac{A^2}{B^2}$.

307. COR. 3.—*In the hyperbola, if supplementary chords are drawn from any point in the conjugate curve to the extremities of the conjugate axis* $aa' = \frac{B^2}{A^2}$ *if the angles are reckoned from the axis of* x, *but* $aa' = \frac{A^2}{B^2}$ *if they are reckoned from the axis of* y.

DEM.—[Let the student supply the demonstration, and also show that if the chords be drawn from a point in the x hyperbola to the extremities of the conjugate axis, aa' is not constant.]

308. COR. 4.—*In the circle this relation becomes* $aa' = -1$, *or* $1 + aa' = 0$; *which shows that the chords are perpendicular to each other, which is a well known property of the circle. In the equilateral hyperbola the relation is* $aa' = 1$, or $a = \frac{1}{a'}$, *signifying that the angles are complementary.*

309. COR. 5.—*If one of two supplementary chords to either axis is parallel to one of two drawn to the other axis the other two are parallel.*

DEM.—In either *Fig.* 203 or 204 if **P′E** is parallel to **PC**, **P′D** is parallel to **PB**. For we have in case of each set of chords $aa' = \mp\frac{B^2}{A^2}$. $\therefore$ If a is the same in each, a' is also.

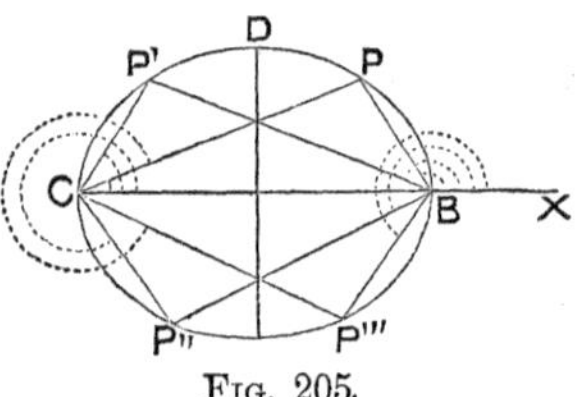

FIG. 205.

310. SCH.—The $-$ sign in the formula $aa' = -\frac{B^2}{A^2}$ indicates that a and a' have opposite signs in the ellipse. Thus **P** being the point from which the chords are drawn, **PBX** $> 90°$ and $< 180°$ gives $-a'$, **PCX** $< 90°$, gives $+a$. Again **P″BX** being an angle between 180° and 270°, tan **P″BX** $= +a'$, but **P″CX** being between 270° and 360°, its tangent is $-a$. In a similar manner the $+$ sign in the formula $aa' = +\frac{B^2}{A^2}$ signifies that a and a' have always the same sign, as may be readily observed from a figure.

---

311. Prob.—*To discuss the angle included between supplementary chords to the transverse axis of an ellipse.*

Solution.—Let **V** be the included angle **CPB**; then $\tan \mathbf{V} = \dfrac{a' - a}{1 + aa'} = \dfrac{a' + \dfrac{B^2}{A^2 a'}}{1 - \dfrac{B^2}{A^2}}$. Limiting the discussion to the upper segment **CPB**, *Fig.* 205, a' is always —. Hence $\tan \mathbf{V}$ is —, and **V** is always an obtuse angle.

*Again, as **V** is a variable angle, we may inquire when it is a maximum. Differentiating $a' + \dfrac{B^2}{A^2 a'}$ with respect to a', and putting the result $= 0$, we have $a' = \pm \dfrac{B}{A}$. The ambiguous sign is explained by the fact that the result applies equally well to either angle **PBX**, or **PCX**. Hence **V** is a maximum when $\tan \mathbf{PBX} = -\dfrac{B}{A}$, or $\tan \mathbf{PCX} = \dfrac{B}{A}$, *i. e.*, when **P** is at **D**.

To find the value of **V** when it is a maximum, we have simply to substitute $-\dfrac{B}{A}$ for a' in $\tan \mathbf{V}$; whence $\tan \mathbf{V} = -\dfrac{2AB}{A^2 - B^2}$.

Sch.—The angle included by supplementary chords to the conjugate axis of an ellipse, and also the corresponding cases in the hyperbola, may be discussed in a similar manner; but the results are not important.

---

312. Prop.—*If one of two supplementary chords to the transverse axis of an ellipse is parallel to a tangent, the other is parallel to the diameter drawn through the point of tangency, and conversely.*

Dem.—Let **PB** be parallel to **MT**, then is **PC** parallel to **DD′**. Let $\tan \mathbf{PCX} = a$, $\tan \mathbf{PBX} = a'$, $\tan \mathbf{DAX} = a_1$, and $\tan \mathbf{MTX} = a_1'$. Now the equation of **DD′** is $y = a_1 x$; whence $a_1 = \dfrac{y}{x}$. Also $\tan \mathbf{MTX} = -\dfrac{B^2 x}{A^2 y}$ (***136***, *Ex.* 1, or ***269***). Hence, $a_1 a_1' = -\dfrac{B^2}{A^2}$. But $aa' = -\dfrac{B^2}{A^2}$. $\therefore aa' = a_1 a_1'$, and if $a' = a_1'$, $a = a_1$. Conversely, if $a = a_1$, $a' = a_1'$. Q. E. D.

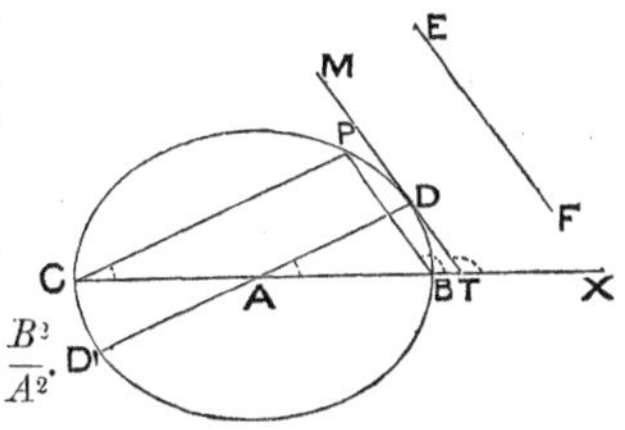

Fig. 206.

313. Cor. 1.—*The same property exists in the hyperbola and is demonstrated in the same way.*

314. Sch.—This property affords a convenient method of drawing tangents. Thus to draw a tangent at **D** *Fig's.* 206, 207, draw the diameter **DD′**, the chord **PC** parallel to it, the supplementary chord **PB**, and through **D** draw **MT** parallel to **PB**.

* Students who have not read the Calculus will omit this paragraph.

To draw a tangent parallel to a given line, or what is the same thing, making a given angle with the axis of x, draw a chord **PB** parallel to the given line **EF**, or making the given angle **PBX**, draw the supplementary chord **PC**, and the diameter **DD′** parallel to **PC**. Through **D** draw **MT** parallel to **PB** and it is the tangent sought.

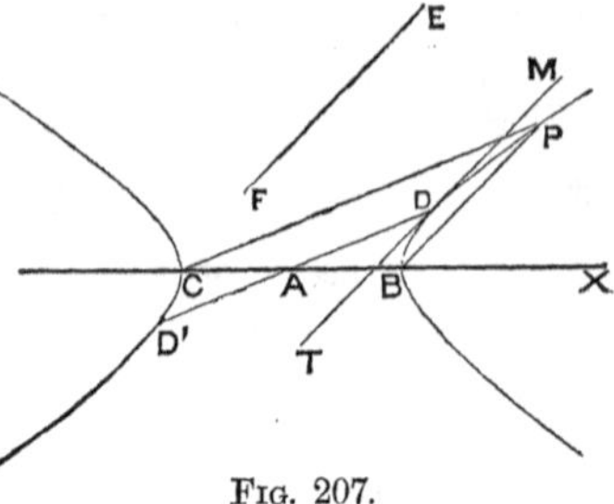

FIG. 207.

315. COR. 2.—*If one of two supplementary chords to the transverse axis is parallel to a diameter, the other chord is parallel to the conjugate diameter, for the latter diameter is parallel to a tangent at the vertex of the former* (**303**). *Hence, also, if one diameter is conjugate to another, reciprocally, the latter is conjugate to the former.*

Ex. 1. In an ellipse whose axes are 8 and 6, one supplementary chord to the transverse axis makes an angle with that axis whose tangent is 2; what angle does the other make?

SOLUTIONS.—*Arithmetically.* Using the formula $aa' = -\frac{B^2}{A^2}$, $a = 2$, $B = 3$, and $A = 4$; whence $a' = -\frac{9}{32}$; or the angle is 164° 18′ nearly. *Geometrically.* Construct the ellipse with **CB** = 8 and **DE** = 6. Make **PCB** = $\tan^{-1} 2$ and draw **PB**. The angle **PBI** is the one required, whose tangent is found by measurement to be about $-\frac{9}{32}$.

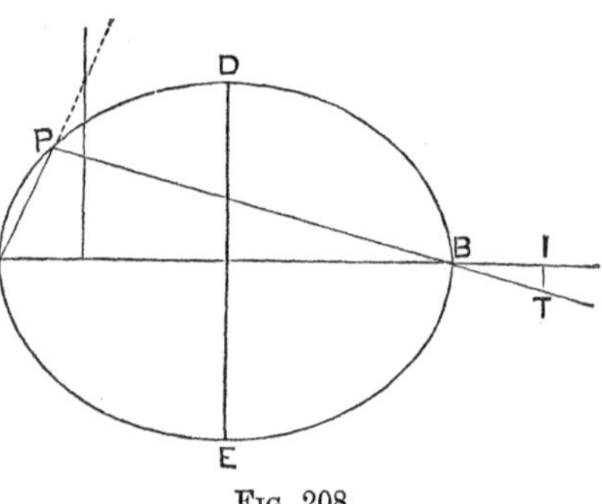

FIG. 208.

Ex. 2. The same as above, the tangent of the angle being —5, and the axes 10 and 6.

Ex. 3. The same numbers as in *Ex.* 1, applied to the hyperbola. What are the co-ordinates of the point in the curve from which the chords are drawn?

SUG.—For the solution of the last question we have $aa' = \frac{B^2}{A^2}$ or $a = \frac{9}{32}$, a' being 2; $y = 2(x - 4)$ and $y = \frac{9}{32}(x + 4)$ to find x and y, which are nearly 5.3 and 2.6.

Ex. 4. In an ellipse whose axes are 8 and 6 find the angle included by the supplementary chords to the transverse axis, from the point $x = 1$.

316. Prob.—*To draw, geometrically, a pair of supplementary chords to the transverse axis of a given ellipse so that the chords shall include a given angle.*

SOLUTION.—Upon the transverse axis describe a segment of a circle **BPMA** which shall contain the given angle—in this case **ABC**. From the intersection of this circumference with the ellipse, as from **P** or **P'**, draw supplementary chords. The pupil may give the reason, and also show how the construction will indicate the impossibility in case the angle given is larger or smaller than can be included by chords in the given ellipse.

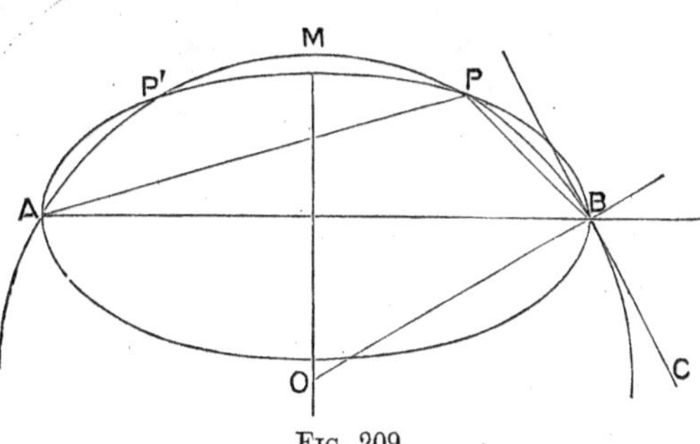

FIG. 209.

317. GENERAL SCHOLIUM.—From the preceding articles in this section it is evident:

1st. That to draw a diameter conjugate to a given one, we may draw a tangent through the extremity of the given diameter and then draw a diameter parallel to this tangent; or we may draw one of two supplementary chords parallel to the given diameter, and drawing the other supplementary chord, draw the second diameter parallel to the last chord.

2nd. To draw a pair of conjugate diameters which shall include a given angle, draw a pair of supplementary chords which shall include the angle, and parallel to these draw a pair of diameters.

3rd. If a be the tangent of the angle which one diameter makes with the transverse axis and a' the tangent of the angle which the other makes, $aa' = \frac{B^2}{A^2}$ in the ellipse, and $aa' = \frac{B^2}{A^2}$ in the hyperbola. Letting $\tan^{-1} a$ and $\tan^{-1} a'$ be respectively α and α', and we have $a = \frac{\sin \alpha}{\cos \alpha}$ and $\alpha' = \frac{\sin \alpha'}{\cos \alpha'}$. Substituting these values, there results in the case of the ellipse $A^2 \sin \alpha \sin \alpha' + B^2 \cos \alpha \cos \alpha' = 0$, and of the hyperbola $A^2 \sin \alpha \sin \alpha' - B^2 \cos \alpha \cos \alpha' = 0$, *formulæ* which are sometimes referred to as the *equations of condition of conjugate diameters.*

4th. From the relation $aa' = \mp \frac{B^2}{A^2}$ we may also solve the 2nd above. Thus, if β be the given included angle, $\beta = \alpha' - \alpha$, and $\tan \beta = \frac{a' - a}{1 + aa'} = \frac{a' \pm \frac{B^2}{A^2 a'}}{1 \mp \frac{B^2}{A^2}}$; whence a' can be determined, as all the other quantities are supposed known.

318. Prob.—*To investigate the relation between conjugate diameters and the axes.*

DEM.—The equation of the ellipse and hyperbola referred to the conjugate diameters $2A_2$ and $2B_2$ is $A_2{}^2y^2 \pm B_2{}^2x^2 = \pm A_2{}^2B_2{}^2$ (***127,*** *Ex's.* 10 and 11); the $+$ sign applying to the ellipse and the $-$ sign to the hyperbola. Transforming this equation so that the reference shall be to the axes, by means of the *formulæ*

$$y_2 = \frac{y\cos\alpha - x\sin\alpha}{\sin(\alpha' - \alpha)},\quad x_2 = \frac{x\sin\alpha' - y\cos\alpha'}{\sin(\alpha' - \alpha)}\quad (\textbf{127}, \text{ Sch.}),$$

we have

$$\left.\begin{array}{l} A_2{}^2y^2\cos^2\alpha - 2A_2{}^2xy\cos\alpha\sin\alpha + A_2{}^2x^2\sin^2\alpha \\ \pm B_2{}^2y^2\cos^2\alpha' \mp 2B_2{}^2xy\cos\alpha'\sin\alpha' \pm B_2{}^2x^2\sin^2\alpha' \end{array}\right\} = \pm A_2{}^2B_2{}^2\sin^2(\alpha' - \alpha).$$

Comparing this with $A^2y^2 \pm B^2x^2 = \pm A^2B^2$, we have

(1) $A_2{}^2\cos^2\alpha \pm B_2{}^2\cos^2\alpha' = A^2$,

(2) $A_2{}^2\sin^2\alpha \pm B_2{}^2\sin^2\alpha' = \pm B^2$,

(3) $A_2{}^2\cos\alpha\sin\alpha \pm B_2{}^2\cos\alpha'\sin\alpha' = 0$, and

(4) $A_2{}^2B_2{}^2\sin^2(\alpha' - \alpha) = A^2B^2$.

1. Adding (1) and (2) and remembering that $\sin^2 + \cos^2 = 1$, we have $A_2{}^2 \pm B_2{}^2 = A^2 \pm B^2$, that is

(a) *In the ellipse the sum of the squares of conjugate diameters is constant, and equal to the sum of the squares on the axes. In the hyperbola the difference of the squares is constant and equal to the difference of the squares on the axes.*

2. Making $A_2{}^2 = B_2{}^2$ (3) gives $\dfrac{\cos\alpha\sin\alpha}{\cos\alpha'\sin\alpha'} = \mp 1$; but as A_2 and B_2 are conjugate $\tan\alpha\tan\alpha' = \dfrac{\sin\alpha\sin\alpha'}{\cos\alpha\cos\alpha'} = \mp\dfrac{B^2}{A^2}$. Multiplying the members of these equations and rejecting common factors we have $\dfrac{\sin^2\alpha}{\cos^2\alpha'} = \dfrac{B^2}{A^2}$, or $\dfrac{\sin\alpha}{\cos\alpha'} = \pm\dfrac{B}{A}$, the $-$ sign characterizing the ellipse since α' is obtuse in the ellipse, and the $+$ sign characterizing the hyperbola, as in it α and α' are acute. Hence $-\dfrac{A}{\cos\alpha'} = \dfrac{B}{\sin\alpha}$ indicates the position of equal conjugate diameters in the ellipse, and $\dfrac{A}{\cos\alpha'} = \dfrac{B}{\sin\alpha}$ in the hyperbola. From *Fig.* 210 we see that **AD** and **AF** meet this condition in the ellipse; for $\mathbf{AK} = \dfrac{\mathbf{AC}}{\cos\mathbf{KAC}} = \dfrac{A}{-\cos\alpha'}$, and $\mathbf{AH} = \dfrac{\mathbf{HB}}{\sin\mathbf{HAB}} = \dfrac{B}{\sin\alpha'}$. Hence

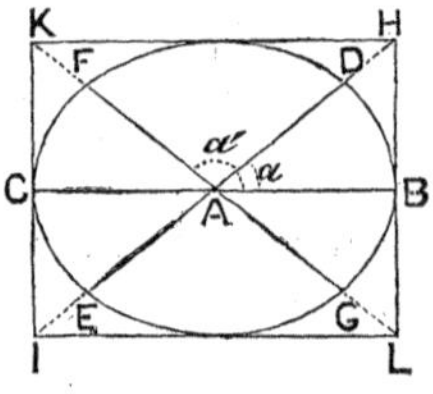

FIG. 210.

(b) *In the ellipse the conjugate diameters which fall upon the diagonals of the rectangle on the axes are equal, and α and α' are supplementary.*

3. In the hyperbola in general, the condition $\dfrac{A}{\cos\alpha'} = \dfrac{B}{\sin\alpha}$ is met only when the two diameters, as **GF** and **DE** *Fig.* 212, coincide and fall on the asymptote. Hence, in the hyperbola, asymptotes are the analogues of the equal conjugate diameters of the ellipse. But from $A_2{}^2 - B_2{}^2 = A^2 - B^2$, we observe that if $A = B$, $A_2 = B_2$, independently of α and α'. Hence

(c) *In the hyperbola, in general, there are no equal (finite) conjugate diameters; but, in the equilateral hyperbola, any pair of conjugate diameters are equal each to each.* The conjugate diameters of the equilateral hyperbola find their analogues in the diameters of a circle.

4. Making $A_2 = B_2$ in $A_2^2 + B_2^2 = A^2 + B^2$, we find that

(d) *The length of one of the equal conjugate diameters of an ellipse is* $\sqrt{2}\sqrt{A^2+B^2}$.

$\therefore$ *The semi-conjugate diameter : the semi-diagonal on the axes* $:: 1 : \sqrt{2}$.

5. Extracting the square root of both members of (4), we have $A_2B_2 \sin(\alpha' - \alpha) = AB$; which signifies that

(e) *The parallelogram formed by tangents drawn through the vertices of any pair of conjugate diameters is constant, and equal to the rectangle on the axes.* This will be more apparent from *Fig's.* 211, 212. **DAF** $= (\alpha' - \alpha)$, and **DA** $= B_2$; whence **AO** $= B_2 \sin(\alpha' - \alpha)$. Hence $A_2B_2 \sin(\alpha' - \alpha)$ = area **AFHD** = $\frac{1}{4}$**LKIH** $= AB = \frac{1}{4}$ the rectangle on the axes. $\therefore$ **LKIH** = the rectangle on the axes.

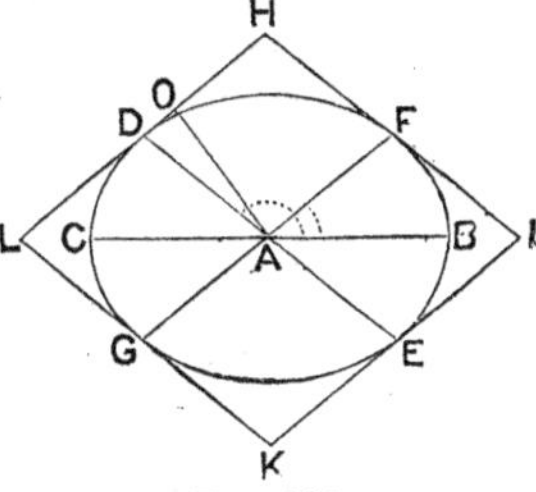

FIG. 211.

Ex. 1. Write the equation of an ellipse referred to a pair of conjugate diameters whose lengths are 8 and 6, and the included angle $\tan^{-1}(-2)$. Having written the equation construct it as in Chapter I., Sec. II.

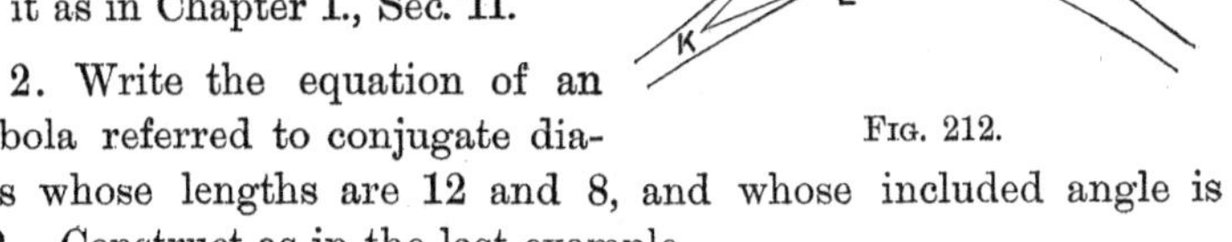
FIG. 212.

Ex. 2. Write the equation of an hyperbola referred to conjugate diameters whose lengths are 12 and 8, and whose included angle is $\tan^{-1}2$. Construct as in the last example.

Ex. 3. In an ellipse whose axes are 8 and 6 what is the length of a diameter which makes an angle of 45° with the axis of x? What is the length of its conjugate?

SUG'S.—From the relation $aa' = -\frac{B^2}{A^2}$, we learn that the conjugate diameter makes with the axis of x an angle of 150° 39′ nearly. Hence $A_2B_2 \sin(\alpha' - \alpha) = AB$, becomes $A_2B_2 \sin 105° 39' = 12$, or $A_2B_2 = \frac{12}{.96293}$. Also $A_2^2 + B_2^2 = A^2 + B^2$ gives $A_2^2 + B_2^2 = 25$. These two equations will give the values of A_2 and B_2.

Ex. 4. In an ellipse whose axes are 8 and 6, what are the sides of the circumscribed parallelogram whose sides are parallel to the equal conjugate diameters? What is the altitude of this parallelogram?

Altitude, 6.79 *nearly.*

(c) PROPERTIES OF ORDINATES.

319. Prop.—*The squares of ordinates to the transverse axis of an ellipse are to each other as the rectangles of the segments into which they respectively divide the axis.*

DEM.—Let **PD** $= y$, **P′D′** $= y'$. **AD** $= x$, **AD′** $= x'$, then $A^2y^2 + B^2x^2 = A^2B^2$, and $A^2y'^2 + B^2x'^2 = A^2B^2$; whence $y^2 = \frac{B^2}{A^2}(A^2 - x^2)$ and $y'^2 = \frac{B^2}{A^2}(A^2 - x'^2)$. Dividing and rejecting the common factor we have $\frac{y'^2}{y^2} = \frac{A^2 - x'^2}{A^2 - x^2} = \frac{(A + x')(A - x')}{(A + x)(A - x)}$; or $y^2 : y'^2 :: (A + x)(A - x) : (A + x')(A - x')$ or :: **CD** $\times$ **DB** : **CD′** $\times$ **D′B.** Q. E. D.

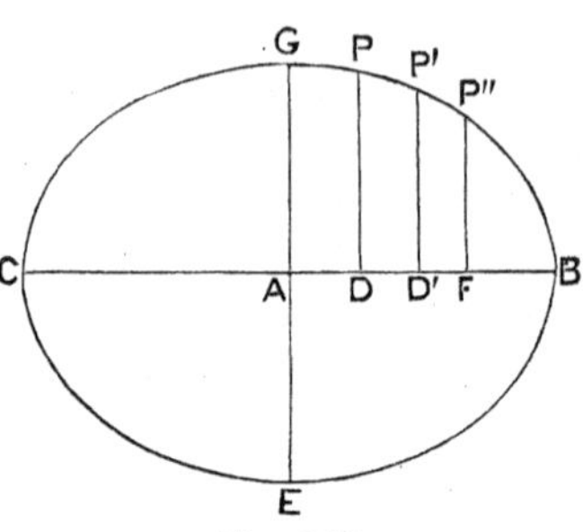

FIG. 213.

320. COR. 1.—*The square of any ordinate to the transverse axis of an ellipse is to the rectangle of the segments into which it divides that axis, as the square of the conjugate axis is to the square of the transverse.*

DEM.—In the above proportion if $y' =$ **GA** $= B$, $(A + x')(A - x') = A^2$, and we have $y^2 : B^2 ::$ **CD** $\times$ **DB** $: A^2$. $\therefore\ y^2 :$ **CD** $\times$ **DB** $:: 4B^2 : 4A^2$. Q. E. D.

321. COR. 2.—*The latus rectum is a third proportional to the transverse and conjugate axes.*

DEM.—In the last proportion let y become the focal ordinate **P′F**, which call p, and **CD** $\times$ **DB** becomes **CF** $\times$ **FB** $= (A + c)(A - c)$, c being **AF**. Now $(A + c)(A - c) = A^2 - c^2 = B^2$, hence $p^2 : B^2 :: B^2 : A^2$, or $2A : 2B :: 2B : 2p$. Q. E. D.

322. SCH.—The properties demonstrated in this proposition, and in the 1st and 2nd corollaries, are equally true for the hyperbola, and can be proved in the same way. In the case of the hyperbola, however, the statement should be, *The rectangles of the distances from the feet of the ordinates to the vertices*, instead of "the rectangles of the segments, etc.," as, in this case the ordinates do not

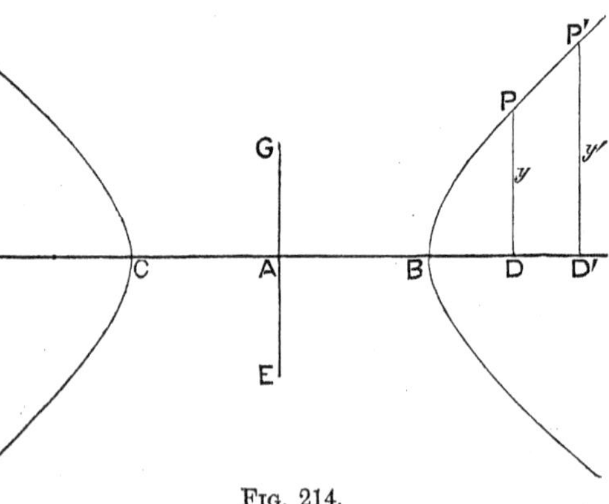

FIG. 214.

divide the axis, but fall upon its prolongation; so that, in *Fig.* 214, we have $y^2 : y'^2 :: \mathsf{CD} \times \mathsf{BD} : \mathsf{CD'} \times \mathsf{BD'}$.

323. COR. 3.—*In the case of the circle* COR. 1st *shows that the square of the ordinate equals the rectangle of the segments into which it divides the diameter—a well known property.*

324. COR. 4.—*This proposition and* COR. 1st *may be asserted of ordinates to the conjugate axis.* [Let the student give the proof and a figure to illustrate it.]

325. COR. 5.—*This proposition and* COR. 1*st may also be asserted of ordinates to* ANY *diameter of an ellipse or an hyperbola.*

DEM.—The corollary can be proved in the same way as the proposition, by using the equation of the curves referred to conjugate diameters (***127,*** *Ex's.* 10 and 11), since these equations are of the same form as those used above. In the annexed figures, therefore, $\overline{\mathsf{PD}}^2 : \overline{\mathsf{P'D'}}^2 :: \mathsf{CD} \times \mathsf{DB} : \mathsf{CD'} \times \mathsf{D'B}$. Also $\overline{\mathsf{PD}}^2 : \mathsf{CD} \times \mathsf{DB} :: B'^2 : A'^2$.

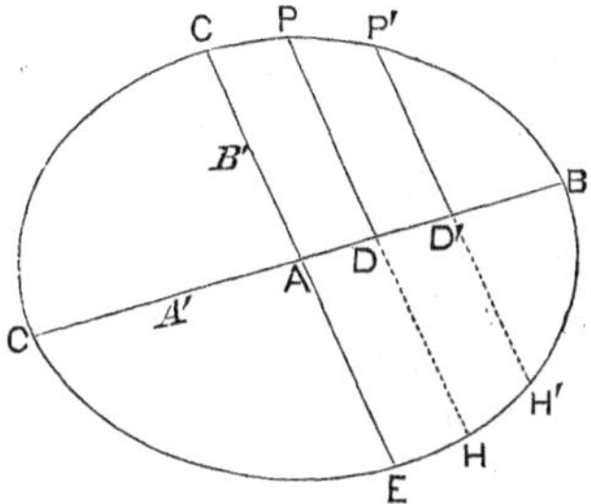

FIG. 215.

326. COR. 6.—*From the last relation, it follows that chords parallel to any diameter are bisected by its conjugate, i. e.* $\mathsf{PD} = \mathsf{DH}, \mathsf{P'D'} = \mathsf{D'H'}$, *etc.; and hence that these curves are symmetrical with respect to any diameter.*

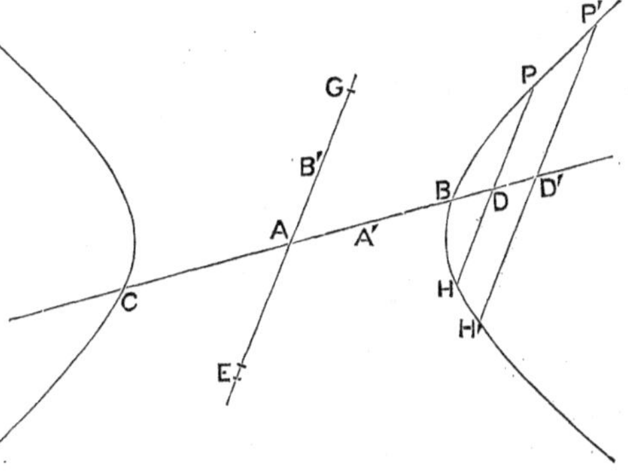

FIG. 216.

327. SCH.—These principles, together with others already known, enable us to find the centre, axes, and foci of the curves, geometrically, when the perimeters alone are given. Thus, in the case of the ellipse, let the curve NHIM be given, to find the centre, axes, and foci. Draw any two parallel chords as DE and BC, bisect them at K and L, and draw FG; it will be a diameter by COR. 6. Bisect this diameter and A will be the centre. From A with

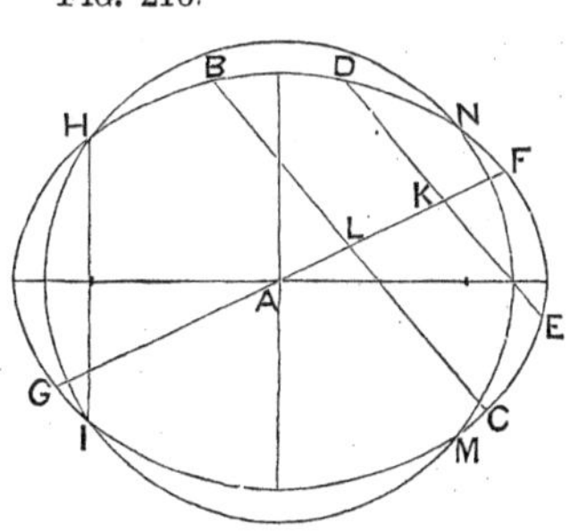

FIG. 217.

a radius sufficiently long to cut the curve, construct the circle **HIMN**, join two of the intersections, as **I** and **H**, and perpendicular to this chord pass a line through the centre; it will be the axis. [The student can readily finish the problem.]

The construction is the same for the hyperbola except in finding the conjugate axis when the conjugate hyperbola is not given.

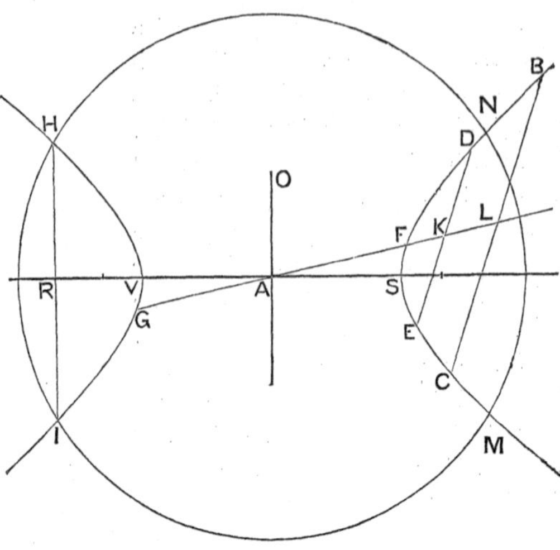

FIG. 218.

For this purpose use the proposition in COR. 1. In the figure, take $SR \times VR : \overline{HR}^2 :: \overline{AV}^2 : \overline{AO}^2$; whence **AO** can be constructed.

---

328. Prop.—*In different ellipses upon the same transverse axis, the corresponding ordinates to the transverse axis are to each other as the conjugate axes of the respective curves.*

DEM.—We have $\overline{PG}^2 : CG \times GB :: \overline{AD}^2 : \overline{AB}^2$, also $\overline{P'G}^2 : CG \times GB :: \overline{AD'}^2 : \overline{AB}^2$ and $\overline{P''G}^2 : CG \times GB :: \overline{AD''}^2 : \overline{AB}^2$. $\therefore PG : P'G : P''G :: AD : AD' : AD''$, etc. Q. E. D.

329. COR.—*Any ordinate to the transverse axis of an ellipse is to the corresponding ordinate of the circumscribed circle as the conjugate axis of the ellipse is to the transverse.*

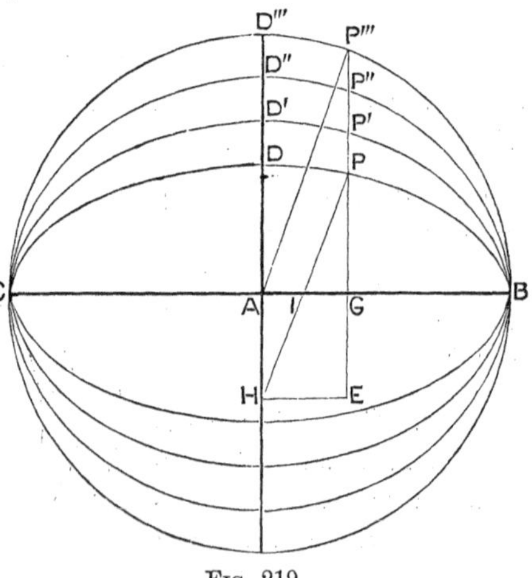

FIG. 219.

DEM.—Let **CD'''B** be the circumscribed circle, then as it may be considered as an ellipse with equal axes, we have $PG : P'''G :: AD : AD'''(= AB)$. Q. E. D.

330. SCH.—An instrument called a *Trammel* is constructed upon the principle enunciated in this corollary. It consists of two grooved bars X'X, YY', fastened together at right angles, and an adjustable arm PH. H and I are pins which can be fastened anywhere on PH, and have heads on the under side which run in the grooves of the bars. Any point in the movable bar, as P, traces an ellipse as H and I slide back and forth in the grooves. To prove that P is a point in an ellipse of which PH is the semi-transverse axis and PI the semi-conjugate, draw AP''' and PH parallel to it, *Fig.* 219. Produce PG till it meets HE drawn parallel to AB, in E. Then AP''' $= A =$ PH. Again, P'' G : PG :: P'''A : PI, or ordinate of circle : ordinate of ellipse :: A : PI. And as this is true for all positions, PI being made = B, and PH $= A$, P is always in the curve.

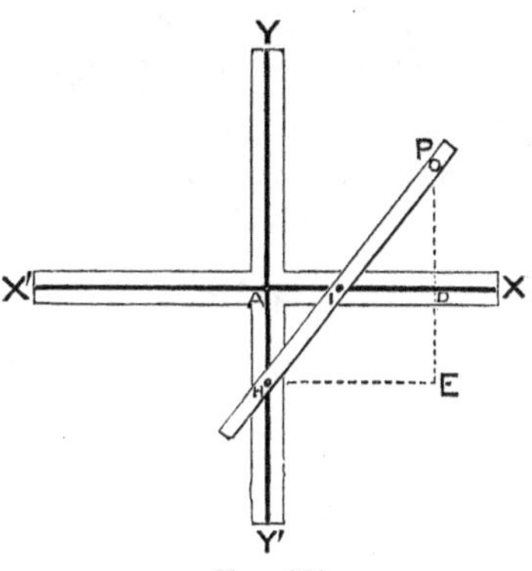

FIG. 220.

We may also demonstrate directly that the locus of P is an ellipse. From *Fig.* 220, using the common notation PI : PH :: PD : PE, gives $B : A :: y : \sqrt{A^2 - x^2}$, or, squaring, $B^2 : A^2 :: y^2 : A^2 - x^2$; whence $A^2y^2 + B^2x^2 = A^2B^2$.

---

331. Prop.—*In different ellipses on the same conjugate axis, corresponding ordinates to this axis are to each other as the transverse axes of the respective curves.*

DEM.—We have $\overline{PG}^2 : DG \times GE :: \overline{AB}^2 : \overline{AD}^2$, and $\overline{P'G}^2 : DG \times GE :: \overline{AB'}^2 : \overline{AD}^2$. $\therefore$ PG : P'G :: AB : AB'. Q. E. D.

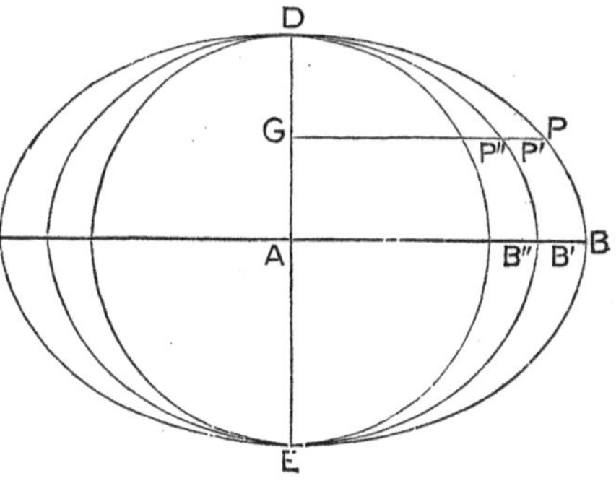

FIG. 221.

332. COR.—*Any ordinate to an ellipse is to the corresponding ordinate of the inscribed circle, as the transverse axis of the ellipse is to the conjugate.* [The student may make the deduction from the proposition.]

333. Prop.—*The squares of ordinates to any diameter of a parabola are to each other as their corresponding abscissas.*

DEM.—Referred to any diameter, as **AX** or **A_1X_1**, the equation of the parabola is $y^2 = 2px$ (***127,*** *Ex.* 12). Whence, letting y and y' represent any two ordinates, as **PD** and **P'D**, or **P_1D_1** and **$P_1'D_1'$**, and x and x' the corresponding abscissas we have $y^2 = 2px$ and $y'^2 = 2px'$. Dividing, $\frac{y^2}{y'^2} = \frac{x}{x'}$. Q. E. D.

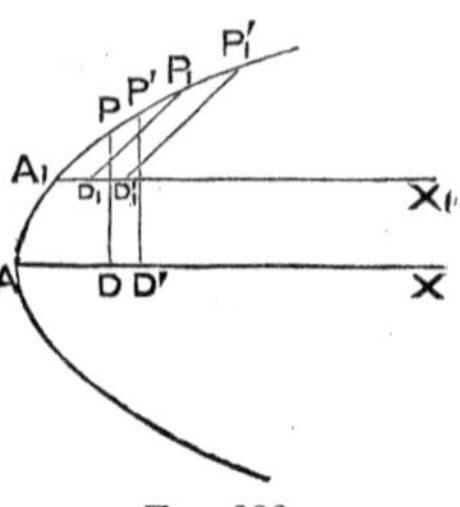

FIG. 222.

334. COR.—*All chords drawn parallel to a tangent at the extremity of a diameter of a parabola are bisected by that diameter.*

335. SCH.—Having the curve to find the axis and focus of a parabola, we draw any pair of parallel chords, and bisect them by a right line. This line is a diameter. Draw two other parallel chords perpendicular to the diameter thus found, bisect these chords by a right line, and it will be the axis. Find the focus by (***164***, or ***284***).

(*d*) ECCENTRIC ANGLE.

336. DEF.—***The Eccentric Angle*** in an ellipse is the angle formed with the axis of abscissas by a line drawn from the centre to a point in the circumference of the circumscribed circle where a produced ordinate meets it, that is **P'''AB** *Fig.* 219.

337. Prop.—*The abscissa of any point in the ellipse equals the semi-transverse axis into the cosine of the eccentric angle, and the corresponding ordinate equals the semi-conjugate axis into the sine of the same angle.* That is, letting φ represent the eccentric angle,

$$x = A\cos\varphi, \text{ and } y = B\sin\varphi.$$

DEM.—In *Fig.* 219, **AG** $= x =$ **P'''A** cos **P'''AB** $= A\cos\varphi$. Also **PG** $= y$ $=$ **PI** sin **PIG** $= B\sin\varphi$.

SCH.—The introduction of this angle is a recent device to facilitate the deduction of certain properties of the ellipse. It enables us to transform an equation in terms of rectangular co-ordinates (x, y) into one containing but one variable, φ, which is sometimes of much advantage. We will give a few specimens of its use.

338. Prop.—*The equation of a tangent to the ellipse in terms of the eccentric angle is*

$$A\sin\varphi \cdot y + B\cos\varphi \cdot x = AB.$$

DEM.—The equation of a tangent to an ellipse is $A^2y'y + B^2x'x = A^2B^2$. As (x', y') is a point in the ellipse, we have $x' = A\cos\varphi$, and $y' = B\sin\varphi$. Substituting these values and dividing by AB, we have $A\sin\varphi \cdot y + B\cos\varphi \cdot x = AB$. Q. E. D.

339. Prop.—*The eccentric angles of the vertices of conjugate diameters differ by* 90°.

Dem.—Let **DAB** $= \varphi$, and **D′AB** $= \varphi'$, be the eccentric angles of the vertices of the conjugate diameters **PC** and **P′C′**. Letting (x, y) be **P**, and (x_1, y_1) be **P′**, the equations of **AP** and **AP′** are, respectively $y = ax$, or $a = \frac{y}{x}$, and $y_1 = a'x_1$, or $a' = \frac{y_1}{x_1}$. Whence $aa' = -\frac{B^2}{A^2} = \frac{yy_1}{xx_1} = \frac{B\sin\varphi \times B\sin\varphi'}{A\cos\varphi \times A\cos\varphi'}$, or $\tan\varphi\tan\varphi'$ $= -1$. Hence **AD** and **AD′** are perpendicular to each other, and $\varphi' = \varphi + 90°$. Q. E. D.

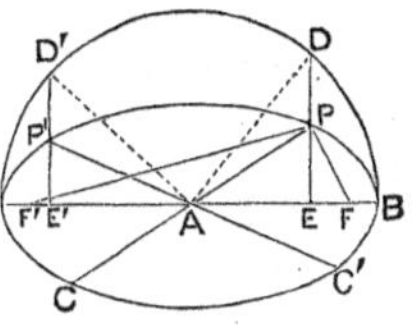

Fig. 223.

340. Sch.—This proposition affords a ready method of drawing a diameter conjugate to a given diameter. Thus let **PC**, *Fig.* 223, be the given diameter. Circumscribe the circle, produce the ordinate **PE** to **D**, draw **DA**, and **D′A** perpendicular to it. From **D′** let fall the perpendicular **D′E′**, and **P′** is the vertex of the conjugate diameter required.

341. Prop.—*The rectangle of the radii vectores drawn to the extremity of any diameter equals the square of the semi-conjugate diameter.*

Dem.—Let **F′P** $= r'$, and **PF** $= r$, *Fig.* 223, and the other notation remain as before. Then from **F′PE** we have $r' = \sqrt{y^2 + (Ae + x)^2} = \sqrt{(A^2 - x^2)(1 - e^2) + A^2e^2 + 2Aex + x^2} = \sqrt{A^2 + 2Aex + e^2x^2} = A + ex$. In like manner from **PEF**, $r = A - ex$. Whence $rr' = A^2 - e^2x^2$. Again $\overline{\mathbf{PA}}^2 = y_1{}^2 + x_1{}^2 = B^2\sin^2\varphi' + A^2\cos^2\varphi' = (A^2 - A^2e^2)\sin^2\varphi' + A^2\cos^2\varphi' = A^2 - A^2e^2\sin^2\varphi'$. But $\varphi' = 90° + \varphi$; whence $\sin\varphi' = \cos\varphi$, and **P′A** $= A^2 - e^2 \cdot A^2\cos^2\varphi = A^2 - e^2x^2$. ∴ $rr' =$ **P′A**. Q. E. D.

342. Prop.—*The sum of the squares of any pair of conjugate diameters is constant and equal to the sum of the squares of the axes.*

Dem.—In *Fig.* 223 we have $\overline{\mathbf{P'A}}^2 = x_1{}^2 + y_1{}^2 = B^2\sin^2\varphi' + A^2\cos^2\varphi'$; or since $\varphi' = 90° + \alpha$, $\sin\varphi' = \cos\varphi$, and $\cos\varphi' = -\sin\varphi$,

$$\overline{\mathbf{P'A}}^2 = A^2\sin^2\varphi + B^2\cos^2\varphi; \text{ and in like manner,}$$

$$\overline{\mathbf{PA}}^2 = A^2\cos^2\varphi + B^2\sin^2\varphi.$$

Adding $\overline{\mathbf{PA}}^2 + \overline{\mathbf{P'A}}^2 = A^2 + B^2$. Multiplying by 4,

$$4\overline{\mathbf{PA}}^2 + 4\overline{\mathbf{P'A}}^2 = 4A^2 + 4B^2. \quad \text{Q. E. D.}$$

Sch.—This proposition has been demonstrated before (***318,*** *a*), but is inserted here as its demonstration affords an example of the utility of the eccentric angle.

Ex. 1. What is the eccentric angle of the extremity of the transverse axis? What of the extremity of the latus rectum? What of the extremity of the conjugate axis?

$$Ans., \varphi = 0°, \varphi = \cos^{-1} e = \sin^{-1}\frac{B}{A}, \varphi = 90°.$$

Ex. 2. In an ellipse whose axes are 8 and 6, what is the eccentric angle at $x = 1$? What are the co-ordinates of the point of which the eccentric angle is 60°? 45°? 30°?

Ex. 3. In an ellipse whose axes are 12 and 8 what is the length of the diameter from the point whose eccentric angle is 60°?

SUG.—Calling the semi-diameter A_2 we have $A_2{}^2 = A^2\cos^2\varphi + B^2\sin^2\varphi = 36 \times (\frac{1}{2})^2 + 16 \times (\frac{1}{2}\sqrt{3})^2 = 21$, and $A_2 = \sqrt{21}$.

---

343. Prop.—*The intercepts of a secant between the hyperbola and its asymptotes are equal.*

DEM.—Let **DD**′ be any secant, and **P** and **P**′, the points in which it cuts the curve, be designated respectively as (x', y') and (x'', y''). Since **DD**′ is a line passing through the two points (x', y'), and (x'', y''), we have for its equation $y - y' = \frac{y' - y''}{x' - x''}(x - x')$. And since (x', y') and (x'', y'') are points in the curve $x'y' = x''y'' = m$. If in the equation of **DD**′ we make $y = 0$, $x =$ **AD**, and $x - x' =$ **CD**. Hence we have **CD** $= x - x' = \frac{x'y' - y'x''}{y'' - y'} = \frac{x''y'' - y'x''}{y'' - y'} = x'' =$ **C′P′**. Now as **PCD** and **P′C′D′** are equiangular and have **CD** = **C′P′**, the triangles are equal, and **PD** = **P′D′**. Q. E. D.

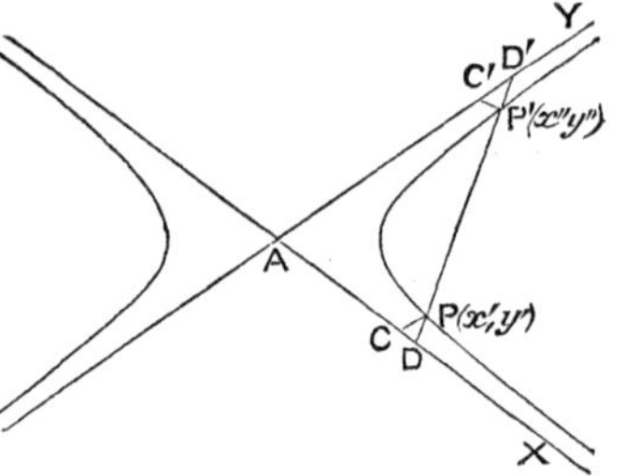

FIG. 224.

SCH.—This proposition affords an elegant and convenient method of constructing the hyperbola. If the axes are given, put them in position and draw the asymptotes, which are the diagonals of the rectangle on the axes. Then, through the extremities of the transverse axis, draw a convenient number of radiant lines, as aa', bb', cc', dd', and make the intercepts $1a'$, $2b'$, $3c'$, $4d'$ respectively equal to **B**a, **B**b, **B**c, **B**d. Then are 1, 2, 3, 4, points in the curve.

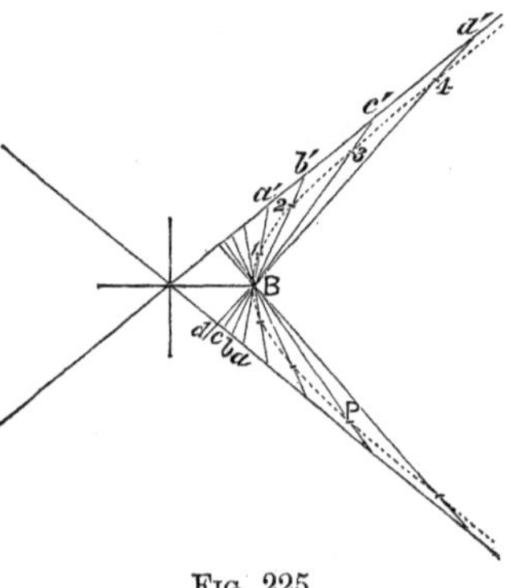

FIG. 225.

If the asymptotes are given, or the

angle included, and *any* point in the curve as **P**, the asymptotes can be drawn; and then radiant lines through **P** will be secants whose intercepts will make known points in the curve.

PARAMETER TO ANY DIAMETER.

344. *A Parameter to any Diameter* of an Ellipse or Hyperbola, is a third proportional to that diameter and its conjugate. In the Parabola it is a third proportional to any abscissa and its corresponding ordinate.

345. *Prop.*—*The distance from any point in a Parabola to the focus is one fourth the parameter to the diameter from that point.*

DEM.—Let $\mathbf{A}_2\mathbf{F} = f$; there is $y_2{}^2 = 4fx_2$. From *Ex.* 12, page 88, we have $y_2{}^2 = \frac{2p}{\sin^2\alpha'}x_2$; and also $2n\sin\alpha' - 2p\cos\alpha' = 0$. From the latter, $n^2\sin^2\alpha' = p^2\cos^2\alpha' = p^2 - p^2\sin^2\alpha'$; whence $\sin^2\alpha' = \frac{p^2}{n^2+p^2}$. Hence $\frac{2p}{\sin^2\alpha'} = \frac{2(n^2+p^2)}{p} = \frac{2(2pm+p^2)}{p} = 4(m+\frac{1}{2}p)$, since $n^2 = 2pm$. But $m + \frac{1}{2}p = \mathbf{TF} = \mathbf{FA}_2 + f$. Therefore $y_2{}^2 = 4fx_2$, or $x_2 : y_2 :: y_2 : 4f$; and $4f$ is the parameter to the diameter $\mathbf{A}_2x_2$ Q. E. D.

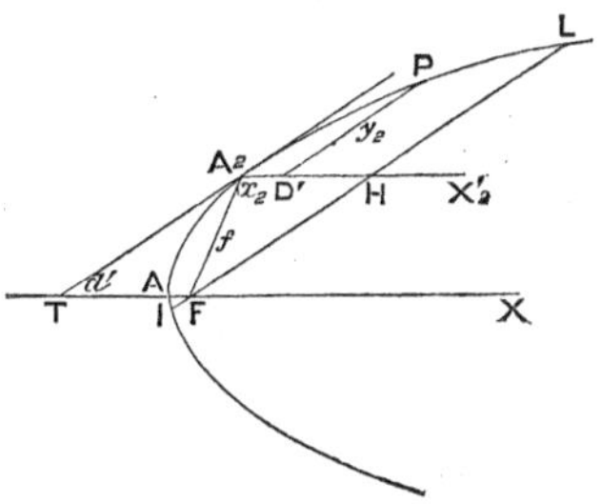

FIG. 226.

346. COR. 1.—*The parameter to any diameter of a Parabola is four times the distance from the vertex of that diameter to the directrix.*

347. COR. 2.—*The double ordinate to any diameter of a Parabola, which (ordinate) passes through the focus, is the parameter to that diameter.*

DEM.—Let $\mathbf{A}_2\mathbf{H} = x_2$, and $\mathbf{LH} = y_2$; whence $y_2{}^2 = 4fx_2$. Now $\mathbf{A}_2\mathbf{H} = x_2 = \mathbf{TF} = \mathbf{A}_2\mathbf{F} = f$. Wherefore $y_2{}^2 = 4f^2$; and $y_2 = 2f$. But $\mathbf{IL} = 2\mathbf{LH} = 2y_2 = 4f$. $\therefore$ $\mathbf{IL}$ is the parameter to $\mathbf{A}_2x_2$.

---

348. *Prop.*—*Any chord which passes through the focus of an Ellipse is a third proportional to the transverse axis and a diameter parallel to the chord.*

DEM.—Let $\mathbf{PF} = r$, $\mathbf{PFB} = \alpha$, and $\mathbf{P'F} = r'$. Then $r = \frac{p}{1 - e\cos\alpha}$, and $r' = \frac{p}{1 + e\cos\alpha}$ (***107***); whence $r + r' = \mathbf{PP'} = \frac{2p}{1 - e^2\cos^2\alpha}$. But from *Ex.* 10, page 86, $B_1{}^2 = \frac{A^2B^2}{A^2\sin^2\alpha + B^2\cos^2\alpha} = \frac{A^2(1-e^2)}{1 - e^2\cos^2\alpha} =$

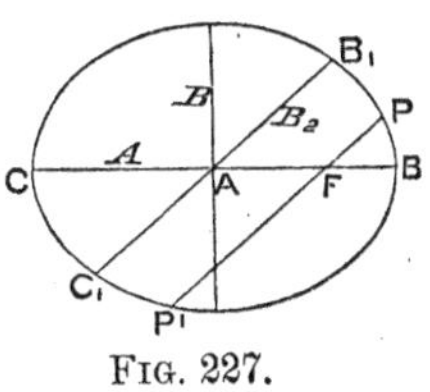

FIG. 227.

$\dfrac{Ap}{1 - e^2 \cos^2\alpha}$; by substituting $A^2(1 - e^2)$ for B^2 and reducing. Therefore $\dfrac{\mathbf{PP'}}{B_1{}^2} = \dfrac{2}{A}$, or $2A : \mathbf{B_1C_1} :: \mathbf{B_1C_1} : \mathbf{PP'}$. Q. E. D.

349. SCH.—The statement in (***347***) is not true in case of the ellipse, as will appear from this proposition.

CHORD OF CURVATURE.

[NOTE.—The following proposition is designed to be read by those who have taken the Differential Calculus, and have studied Section VI, Chapter IV, or have some knowledge of the subject of *radius of curvature.*]

350. A Chord of Curvature is a chord of the *Osculatory Circle,* drawn from the point of contact.

351. Prop.*—In the parabola, the chord of curvature which passes through the focus is the parameter to the diameter passing through the point of contact.*

DEM.—**O** being the centre of the osculatory circle at **P**, in the parabola whose focus is **F**, **PM** is the chord of curvature passing through the focus, and is the parameter to **PD**, the diameter through **P**. For, draw **FL** perpendicular to the tangent through P, and we have from the similar triangles **PRM** and **FPL**,

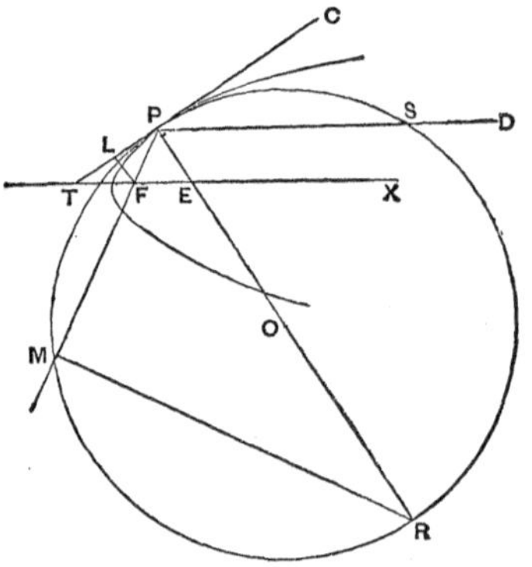

$$\mathbf{PR} : \mathbf{PM} :: \mathbf{PF} : \mathbf{FL}, \text{ or,}$$

$$\mathbf{PM} = \frac{\mathbf{PR} \times \mathbf{FL}}{\mathbf{PF}}.$$

FIG. 228.

But $\mathbf{PR} = \dfrac{2N^3}{p^2}$, N being the normal, and p the *semi-latusrectum* of the curve (***211***); $\mathbf{FL} = \frac{1}{2}\mathbf{PE} = \frac{1}{2}N$ (***164*** or ***284***), and $\mathbf{PF} = \sqrt{\overline{\mathbf{FL}}^2 + \overline{\mathbf{LP}}^2} = \sqrt{\dfrac{1}{4}N^2 + \dfrac{y^2}{4p^2}N^2} = \dfrac{1}{2}N\sqrt{1 + \dfrac{y^2}{p^2}} = \dfrac{1}{2}N\sqrt{\dfrac{p^2 + y^2}{p^2}} = \dfrac{N^2}{2p}$ (***143***, Ex. 2), remembering that $p^2 + y^2 = N^2$. Substituting these values, we have, $\mathbf{PM} = \dfrac{2N^3}{p^2} \times \dfrac{N}{2} \times \dfrac{2p}{N^2} = \dfrac{2N^2}{p} = 4\mathbf{PF}$, and hence is the parameter to **PD** (***346***).

352. COR.—*The chord* **PS** *intercepted on the diameter* **PD** *is equal to the chord of curvature passing through the focus,* since angle **CPD** = **TPM**.

END OF PART FIRST.

GENERAL SCHOLIUM.

Book Second, treating of Loci in Space, is reserved for a second volume. The present volume is deemed sufficient for the use of all students in our colleges, except such as pursue mathematical studies as a specialty. Volume II. will contain Loci in Space, and a more extended course in the Calculus.

THE

INFINITESIMAL CALCULUS.

INTRODUCTION.

[NOTE.—The four following chapters on the Differential Calculus are to be read immediately after the first three chapters of the General Geometry, that is, the first 92 pages of this volume.]

1. *Quantity* is the amount or extent of that which may be measured; it comprehends number and magnitude. (See ART. 4, General Geometry, and the two Scholiums under it on pages 1 and 2.)

2. *Number* is quantity conceived as made up of parts, and answers to the question, "How many?" (See ART. 5, Illustration, General Geometry.)

3. Number is of two kinds, ***Discontinuous*** and ***Continuous.***

4. *Discontinuous Number* is number conceived as made up of finite parts; or it is number which passes from one state of aggregation to another by the successive additions of finite units, *i. e.*, units of appreciable magnitude.

5. *Continuous Number* is number which is conceived as composed of infinitesimal parts; or it is number which passes from one state of value to another by passing through all intermediate values, or states.

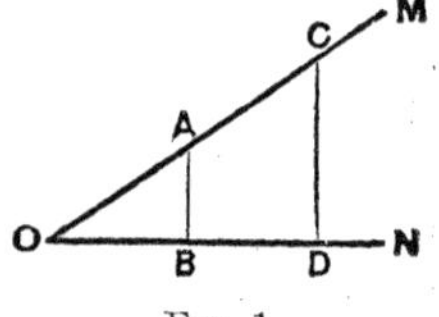

FIG. 1.

ILL'S.—The method of conceiving number with which the pupil has become familiar in arithmetic and algebra, characterizes *discontinuous* number. Thus the number 13 is conceived as produced from 5 by the successive additions of finite units, either integral or fractional. In either case we advance by successive steps of *finite* length. If we say 5, 6, 7, etc., till we reach 13, we pass by one kind of steps; and, if we say 5.1, 5.2, 5.3, etc., till we reach 13, we pass by another sort of steps (*tenths*), but as really by *finite* ones. If, however, we call the line **AB**, *Fig.* 1, x, and **CD**, x', and conceive **AB** to slide to the position **CD**, increasing in length as it moves so as to keep its extremities in the lines **OM** and

ON, it will pass by *infinitesimal* elements of growth from the value x, to the value x'; or, it will pass from one value to the other by passing through *all* intermediate values, and thus becomes an illustration of *continuous* number.

Again, if the line **AB**, *Fig.* 2, be considered as generated by a point moving from **A** to **B**, and we call the portion generated when the point has reached **C**, x, and the whole line x', x will pass to x', by receiving infinitesimal increments, or by passing through all states of value between x and x'.

A C B

FIG. 2.

A surface may be considered as generated by the motion of a line, and thus afford another illustration of continuous number. Thus let the parallelogram **AF** be conceived as generated by the right line **AB** moving from **AB** to **EF**. When **AB** has reached the position **CD**, call the surface traced, namely **ABCD**, x, and the entire surface **ABEF**, x'; then will x pass to x' by receiving infinitesimal increments, or by passing through all intermediate values.

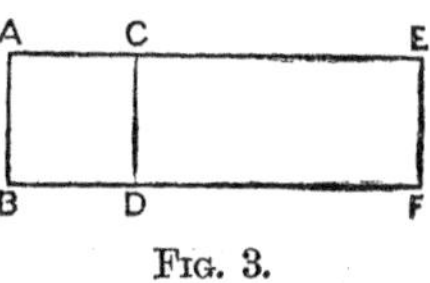

FIG. 3.

Finally, as volumes may be conceived as generated by the motion of planes, all geometrical magnitudes afford illustrations of continuous number.

We usually conceive of *time* as discontinuous number, as when we think of it as made up of hours, days, weeks, etc. But it is easy to see that such is not the way in which time actually grows. A period of one day does not grow to be a period of one week by taking on a whole day at a time, or a whole hour, or even a whole second. It grows by imperceptible increments (additions). These inconceivably small parts of which continuous number is made up are called *Infinitesimals*.

Motion and force afford other illustrations of continuous number. In fact, the conception which regards number as continuous, will be seen to be less artificial—more true to nature—than the conception of it as discontinuous.

6. *An Infinite Quantity* is a quantity conceived under such a form, or law, as to be necessarily *greater* than any assignable quantity.

7. *An Infinitesimal* is a quantity conceived under such a form, or law, as to be necessarily *less* than any assignable quantity.

8. SCH.—By an infinite quantity is not meant one larger than any other, or the largest possible quantity. It simply means a quantity larger than any assignable quantity; *i. e.*, larger than any one which has limits. The mathematical notion concerns rather the manner of conceiving the quantity, than its absolute value. Thus, a series of 1s, as 1 1 1, etc., repeated without stopping, represents an infinite quantity, because, from the method of conceiving the quantity, it is necessarily greater than any quantity which we can assign or mention. If we assign a row of 9s reaching around the world, though it is an inconceivably great number, it is not as great as a series of 1s extending *without limit.* Moreover, one infinite may be larger than another; for a series of 2s extending without limit, as 2 2 2 2, etc., is *twice as large* as a series of 1s conceived in the same way. It is never of any use to try to comprehend the *magnitude* of an infinite quantity; we cannot do it; although we can *compare* infinites just as well as finites.

Again, and what is more to our purpose, an infinitesimal quantity is not a quantity so small that there can be no smaller. There would be but *one* such quantity and hence no comparison of infinitesimals. All that is meant by the term as used in mathematics is, *a quantity which is to be treated in the argument* as less than any assignable quantity. Whether we can or cannot comprehend its absolute magnitude is of no manner of consequence. Nor is *absolute value* usually of any importance in pure mathematical reasoning. Thus 2 times 5 is 10 whether 5 be mites or mountains. In order to free himself from needless embarrassment in the use of infinitesimals, the student needs to keep constantly in mind the fact that, ***In pure mathematics, it is the relation of quantities, rather than their absolute values, with which we are concerned.***

---

9. Prop.—*Any finite quantity divided by an infinite is an infinitesimal; and any finite quantity divided by an infinitesimal is an infinite.*

DEM.—Let a represent any finite quantity and x any infinite. Then $\frac{a}{x}$ is an infinitesimal; for the value of a fraction depends upon the *relative* values of its numerator and denominator, and is less as the ratio of numerator to denominator is less. Now, in this case, a is *infinitely* less than x, by the definition of an infinite. Hence $\frac{a}{x}$ is an infinitesimal. Again $\frac{a}{x}$ is infinite if x is infinitesimal, since a is infinitely greater than x.

10. COR.—*The reciprocal of an infinite is infinitesimal, and the reciprocal of an infinitesimal is infinite.*

11. The products of infinites by infinites, and of infinitesimals by infinitesimals are denominated ***Orders:*** thus, if x and y are infinites, x^2, y^2, and xy are infinites of the ***Second Order;*** if x, y, and z are infinites, x^3, z^3, xyz, x^2y, xy^2, etc., are infinites of the ***Third Order.*** The corresponding expressions are used with reference to infinitesimals, the product of two infinitesimals being called an infinitesimal of the second order, of three, the third, etc.

12. SCH.—An infinite of a lower order sustains a relation to the next higher similar to that which a finite sustains to an infinite. Thus if x and y are infinites, x^2, xy, and y^2 are infinitely greater than x and y. On the other hand if x and y are infinitesimals, x^2, xy, and y^2 are infinitely less, and sustain a relation to x and y, similar to that which infinitesimals sustain to finites.

AXIOMS.

13. From expressions containing the sum or difference of finites and infinites, the finites may be dropped without affecting the ratio.

14. From expressions containing the sum or difference of infinites of different orders, the terms containing the lower orders may be dropped without affecting the ratio.

15. The order of an infinite is not altered by multiplying or dividing it by a finite.

16. From expressions containing the sum or difference of finites and infinitesimals, the infinitesimal terms may be dropped without affecting the ratio.

17. From expressions containing the sum or difference of infinitesimals of different orders the terms containing the higher orders may be dropped without affecting the ratio.

18. The order of an infinitesimal is not changed by multiplying or dividing it by a finite.

ILL'S.—Although the above are conceived to be axioms in the strictest sense, that is truths to which the mind at once assents as soon as the terms used are clearly comprehended, the true notion of infinites and infinitesimals is so removed from common thought that a familiar illustration or two may aid the comprehension. Suppose, then, that the quantities under consideration were the masses of matter in the earth and in the sun. If a grain of sand were added to or subtracted from each or either it would not *appreciably* affect the ratio of these masses. But in this instance the grain of sand is by no means infinitesimal with reference to either mass; it is a *finite*, though *very small* part, of either mass.

Again, let x and y be two infinite quantities, and a and b two finite ones. There can be no difference between $\frac{x \pm a}{y \pm b}$ and $\frac{x}{y}$; since to assume such a difference would be to assign some values to a and b, as respects x and y. But by hypothesis, the former have no assignable values in relation to the latter.

Once more, if a and b are finite quantities, and x and y infinitesimal, $\frac{a \pm x}{b \pm y} = \frac{a}{b}$, since x and y have no assignable values as compared with a and b. So also, x and y still being infinitesimal, $\frac{x \pm x^2}{y \pm y^2} = \frac{x}{y}$, as x^2 and y^2 are infinitesimals, (have no assignable values) with respect to x and y.

EVALUATION OF EXPRESSIONS CONTAINING FINITES AND INFINITESIMALS, AND FINITES AND INFINITES.

Ex. 1. What is the value of the fraction $\frac{2x - a}{3x + b}$ if x is infinite and a and b finite?

SOLUTION.—Since a and b have no assignable values in relation to x they must be dropped, and we have $\frac{2x}{3x}$. Now dividing both terms by x, we have $\frac{2}{3}$ as the value of $\frac{2x - a}{3x + b}$ when x is infinite and a and b finite.

Ex. 2. What is the value of the fraction in the last example if x is infinitesimal and a and b finite?

SOLUTION.—As x is infinitesimal $2x$ and $3x$ are also infinitesimal, and hence have no value in relation to a and b, and must be dropped. Hence the value of the fraction is $-\frac{a}{b}$.

Ex. 3. What is the value of $\frac{12x^2 - 3x}{2x^2 - x}$ when x is infinite? When x is infinitesimal?

Ans., When x is infinite, 6; when infinitesimal, 3.

Ex. 4. What is the value of y in the equation $y = \frac{\frac{a}{x} - 5x}{\frac{b}{x} + x}$ when x is infinite? When x is infinitesimal?

Ans., When x is infinite, -5; when infinitesimal, $\frac{a}{b}$.

Ex. 5. What is the value of y in the expression $y = \frac{1}{1 + x}$ when x is infinite? When x is infinitesimal?

Ans., When x is infinite, 0; when infinitesimal, 1.

Ex. 6. What is the value of $\frac{ax^3 + bx^2 + cx + d}{mx^3 + nx^2 + px + q}$ when x is infinite? When x is infinitesimal?

Ans., When x is infinite, $\frac{a}{m}$; when infinitesimal, $\frac{d}{q}$.

Ex. 7. What is the value of y in the expression $y = \frac{2x^2 - 5m^2x}{3x^3 - mx}$ when x is infinite? When x is infinitesimal?

Ans., When x is infinite, 0; when infinitesimal, $5m$.

Ex. 8. What is the value of y in the equation $y = \dfrac{ax^5 + 2x^2 - 1}{mx^3 - 3x + 2}$ when x is infinite? When x is infinitesimal?

Ans., When x is infinite, $y = \infty$; when infinitesimal, $y = -\frac{1}{2}$.

Ex. 9. When x and y are infinitesimals what is the value of $\dfrac{3x}{5y}$?

Ans., We cannot tell; as we know nothing about the relation between x and y.

Ex. 10. What is the value of $\dfrac{3x}{5y}$ when $y^2 = 9x$ and x and y are infinite? *Ans.*, ∞.

Ex. 11. Same as *Ex.* 10, only x and y infinitesimal? *Ans.*, 0.

Ex. 12. What is the value of y in the equation $y^2 = \dfrac{x^3}{2a - x}$, when x is infinitesimal? *Ans.*, 0.

CONSTANTS AND VARIABLES.

19. A Constant quantity is one which maintains the same value throughout the same discussion, and is represented in the notation by one of the leading letters of the alphabet.

20. Variable quantities are such as may assume in the same discussion any value, within certain limits determined by the nature of the problem, and are represented by the final letters of the alphabet.

21. Cor.—*Any expression containing a variable is, when taken as a whole, a variable. Thus the value of the* ENTIRE *expression* $(4a - 3x^2 + 5)^{\frac{1}{2}}$ *varies if* X *varies; so that* TAKEN AS A WHOLE *it is a variable.*

[Note.—These notions should be already familiar from General Geometry, page 9, and are introduced here only to give completeness, and for review.]

22. Variables are distinguished as *Independent* and *Dependent.*

23. An Independent Variable is one to which we assign arbitrary values, or upon whose law of variation we make some arbitrary hypothesis.

24. A Dependent Variable is one which varies in value in consequence of the variation of the independent variable or variables.

Ill.—Thus, in the equation of the parabola, $y^2 = 2px$, if we assign arbitrary values to x and find the corresponding values of y, we make x the independent variable, and y the dependent variable. Again, and what is more to our present purpose, if we assume x to vary in some particular way, as by taking on *equal* increments, as DD′, D′D″, D″D‴, etc., y will evidently vary in *some other* way, but still in a way depending upon the way in which x varies, and upon the nature of the curve, or, what is the same thing, upon the form of the equation of the curve. In this case also, x is the independent and y the dependent variable.

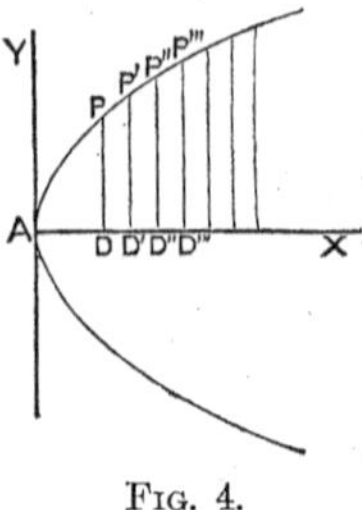

FIG. 4.

SCH.—This distinction is made simply for convenience, and is not founded in any difference in the nature of the variables; either variable may be treated as the independent variable.

25. *An Equicrescent* variable is one which is *assumed* to increase or decrease by *equal* increments or decrements, as x in the last illustration.

26. *Contemporaneous Increments* are such as are generated at the same time.

ILL.—Thus let $y = f(x)$ represent the equation of AM in the figure. Suppose we contemplate the values of x and y at the point P′. Now if x takes the increment D′D″, y takes the *contemporaneous* increment P″E′. So also we see that DD′, P′E, and PP′ are contemporaneous increments of the abscissa, ordinate, and arc, respectively.

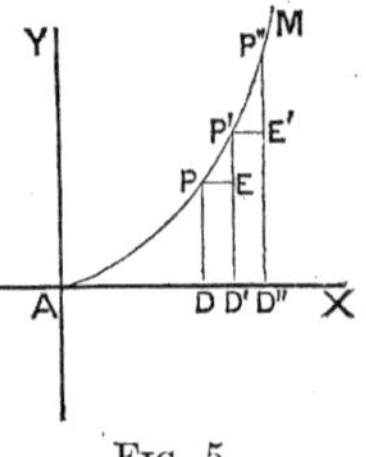

FIG. 5.

FUNCTIONS AND THEIR FORMS.

27. *A Function* is a quantity, or a mathematical expression, conceived as depending for its value upon some other quantity or quantities.

ILL.—A man's wages *for a given time* is a function of the amount received per day; or, in general, his wages is a function of *both* the *time* of service and the *amount* received per day. Again, in the expressions $y = 2ax^2$, $y = x^3 - 2bx + 5$, $y = 2\log ax$, $y = a^x$, y is a function of x; since, the numbers 2, 5, a and b being considered constant, the value of y depends upon the value we assign to x. For a like reason $\sqrt{a^2 - x^2}$, and $3ax^2 - 2\sqrt{x}$ may be spoken of as functions of x. Once more, the ordinate of a curve is a function of the abscissa.

SCH.—There is a sense in which the dependent variable (or function) is a function of the *constants* as well as of the variable or variables which enter into its value. So also it is a function of the *form* of the expression, that is, its value depends in part upon the *form* of the expression as well as upon the value of the independent variable. Thus if we have $y = a\log x$

$+ b$, and $y = x^3 - cx$, though in each case y is a function of x, speaking according to the definition, nevertheless it is not the *same* function in both cases. Its value depends upon the value of x, upon the constants, and upon the *form* of the expression involving these quantities. But the conception expressed in the definition is the ordinary one.

28. Functions are classified by their *forms* as ***Algebraic*** and ***Transcendental,*** and the latter are subdivided into ***Trigonometrical*** and ***Circular, Logarithmic*** and ***Exponential.***

29. An Algebraic Function is one which involves only the elementary methods of combination, viz., addition, subtraction, multiplication, division, involution and evolution. Thus in $y = ax^2 - 3x^3$, y is an algebraic function of x.

30. A Trigonometrical Function is one which involves sines, cosines, tangents, cotangents, etc., as variables; thus $y = \sin x$, $y = \sin x \tan x$, etc.

31. A Circular Function is one in which the concept is a variable arc (in the trigonometrical the concept is a right line). These are written thus: $y = \sin^{-1}x$, read "y equals *the arc* whose sine is x"; $y = \tan^{-1}x$, read "y equals *the arc* whose tangent is x."

ILL.—Notice that in the expression $y = \tan^{-1}x$, it is the *arc* which we are to think of, while in the expression $x = \tan y$ it is the *tangent*, which is a right line. Trigonometrical functions are right lines; circular functions are arcs. These functions are mutually convertible into each other; thus $y = \sin^{-1}x$ is equivalent to $x = \sin y$, the only difference being that in the former we think of the *arc*, the sine being given to tell *what* arc, and in the latter, we think of its *sine*, the arc being given to tell *what* sine.

The circular functions $y = \sin^{-1}x$, $y = \cos^{-1}x$, $y = \sec^{-1}x$, etc., are often called *Inverse Trigonometrical Functions.*

32. A Logarithmic Function is one which involves logarithms of the variable; as $y = \log x$, $\log^2 y = 3 \log ax$, etc.

33. An Exponential Function is one in which the variable occurs as an exponent; as $y = a^x$, $z = x^y$, etc.

---

34. Functions are further distinguished as ***Explicit*** and ***Implicit.***

35. An Explicit* Function is a variable whose value is expressed in terms of another variable or other variables and constants. Thus in $y = 2ax^3 - 3x^{\frac{1}{2}}$, y is an explicit function of x.

* From *explicitum*, unfolded. The function is disentangled from the other quantities.

36. *An Implicit* Function* is a variable involved in an equation which is not solved. Thus in $x^2 - 3xy + 2y = 16$, y is an implicit function of x, or x is an implicit function of y. When we can solve the equation, an implicit function may always be expressed as explicit.

37. *Notation.* When we wish to write that y is an explicit function of x, and do not care to say precisely what the *form* of the function is, we write $y = f(x)$, read "y = a function of x." If we wish to indicate several different forms of dependence in the same discussion, we use other letters, as $y = f(x)$, $y = F(x)$, $y = \varphi(x)$, etc., or use subscripts or accents as $y = f_1(x)$, $y = f'(x)$, etc. Such symbols are read "y = the f, large F, φ, f sub-one, f prime, etc., function of x," as the case may be.

When we wish to write that x and y are functions of each other, or that y is an implicit function of x, or x an implicit function of y, without being more specific, we write $F(x, y) = 0$, or $f(x, y) = 0$, or $\varphi(x, y) = 0$, etc.; and read "function x and $y = 0$," the F function x and $y = 0$, etc. This form symbolizes *any* equation between two variables with all the terms transposed to the first member.

---

38. Again, functions are distinguished as ***Increasing*** and ***Decreasing.***

39. *An Increasing Function* is a function that increases as its variable increases, and decreases as its variable decreases.

40. *A Decreasing Function* is a function which decreases as its variable increases, and increases as its variable decreases.

Ill.—In the expressions $y^2 = 2px$, $y = \log x$, $y = a^x$, y is an increasing function of x. In the expressions $y = \frac{1}{x^2}$, $y^2 + x^2 = R^2$, $y = \log\frac{1}{x}$, y is a decreasing function of x. For what values of x is y an increasing function of its variable, and for what a decreasing, in the following: $y^3 = ax^3 - x^2$, $y = \sin x$, $y = \cos x$?

---

41. *The Infinitesimal Calculus* treats of *Continuous Number*, and is chiefly occupied in deducing the relations of the contemporaneous infinitesimal elements of such number from given relations between finite values, and the converse process, and also in pointing out the nature of such infinitesimals and the methods of using them in mathematical investigation.

* From *implicitum*, infolded, entangled.

ILL.—Let $y^2 = 8x$ be the equation of the parabola in the figure. Here we have the relation between finite values of y and x expressed. Now suppose x takes an infinitesimal increment as DD′*, what increment does y take? The calculus shows us that the increment which y takes is $\frac{4}{y}$ times as large as the increment which x takes; that is, it shows us the relation between the elements of the variables y and x, when we know the relation between finite values. This is the province of the *Differential Calculus.* The converse of this problem is, What is the equation of the curve whose ordinate varies $\frac{4}{y}$ times as fast as its abscissa? that is, having given the relation between the infinitesimal elements of y and x, to find the relation between finite values. This is the province of *The Integral Calculus.*

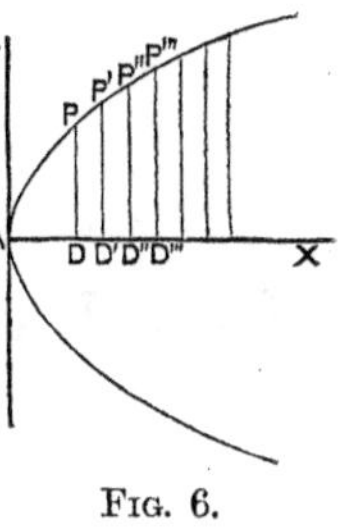

FIG. 6.

42. There are two branches of the *Calculus*, viz., ***The Differential Calculus,*** and ***The Integral Calculus.***

* Of course all such attempts to represent infinitesimals to the eye, are egregious exaggerations; nevertheless they are of great service to the mind.

THE

INFINITESIMAL CALCULUS.

CHAPTER I.

THE DIFFERENTIAL CALCULUS.

SECTION I.

Differentiation of Algebraic Functions.

43. ***The Differential Calculus*** is that branch of the Infinitesimal Calculus which treats of the methods of deducing the relations between the contemporaneous infinitesimal elements of variables, from given relations between finite values.

44. ***A Differential*** is the difference between two consecutive states of a function, or variable. It is the same as an infinitesimal.

45. ***Consecutive Values*** of a function or variable are values which differ from each other by *less than any assignable quantity.*

Consecutive Points on a line are points nearer to each other than any assignable distance.

Ill.—Suppose $y = 2x^2 - 3x$. Now let x be supposed to increase infinitesimally, y will also change infinitesimally. Call the new value of y, y'. Then $y' = 2x'^2 - 3x'$. In such a case x and x' are *consecutive* values of the variable, and y and y' are consecutive values of the function. *But by this we do not mean that* x *and* x' *(or* y *and* y'*) are so nearly equal that there can be no intermediate value,* for this would be to make an infinitesimal mean a quantity so small that there can be no smaller, which is not its meaning as used in mathematics (**7**). All that is meant by saying that y and y' are consecutive values is that they are *to be reasoned upon* as having no assignable difference.

So also in speaking of consecutive points on a line, as D and D', or P and P', *Fig.* 6, we do not conceive them as actually in juxtaposition; but we mean simply that *we are to reason upon them* as nearer each other than any assignable distance.

46. ***Notation.*** The differential of a variable (one of its infinitesimal elements) is represented by writing the letter d before it.

Thus, dx, read, "differential x." Of course the letter d is not to be confounded with a factor ; it is simply an abbreviation for *differential.*

[CAUTION.—The student should be careful and not allow himself to read such expressions as dy, dx, etc., by merely naming the letters as he would ay, ax, etc. The former should always be read "differential y," "differential x," etc.]

RULES FOR DIFFERENTIATING ALGEBRAIC FUNCTIONS.

47. *RULE* 1.—TO DIFFERENTIATE A SINGLE VARIABLE SIMPLY WRITE THE LETTER d BEFORE IT.

DEM.—Let us take the function $y = x$. The consecutive state of the variable is $x + dx$. Now representing the change in y which is produced by this change in x by dy (dx and dy being the contemporaneous increments of the variable and the function), we have

1st state of the function, ... $y = x$,

2nd, or consecutive state, ... $y + dy = x + dx$.

Subtracting the 1st from the 2nd, ... $dy = dx$, which being the difference between two consecutive states of the function is its differential (**44**). Q. E. D.

SCH.—This rule is evidently only the same thing as the notation requires, and its formal demonstration would be unnecessary except for the purpose of uniformity in treating the several cases of differentiation.

ILL.—Let **MN** be a line passing through the origin and making an angle of 45° with the axis of x. Its equation is $y = x$. Let **P** be *any* point in the line, **AD** $= x$, and **PD** $= y$. Let x take the infinitesimal increment **DD**$'$ (dx), then y becomes **P$'$D$'$**. Now the first state of the function is

PD $=$ **AD**, or $y = x$,

The second or consecutive state is **PD** $+$ **P$'$E** $=$ **AD** $+$ **DD$'$**, or $y + dy = x + dx$.

Subtracting we have **P$'$E** $=$ **DD$'$**, or $dy = dx$.

Now that the increment of y (or dy) is equal to the increment of x (or dx) in this case is readily seen from the figure ; for, as **P$'$PE** $= 45°$, **P$'$E** $=$ **PE**, or **DD$'$**. $dy = dx$, then, means that the contemporaneous increments of x and y are equal, or that x and y increase at the same rate.

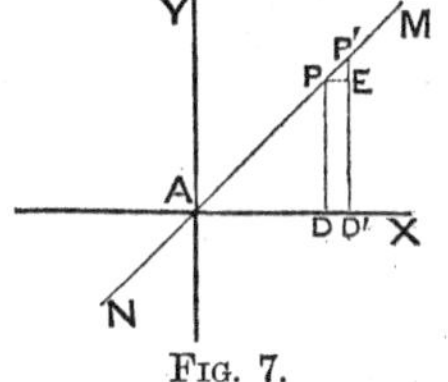

FIG. 7.

48. *RULE* 2.—CONSTANT FACTORS OR DIVISORS APPEAR IN THE DIFFERENTIAL THE SAME AS IN THE FUNCTION.

DEM.—Let us take the function $y = ax$, in which a is any constant, integral or fractional. Let x take an infinitesimal increment and become $x + dx$; and let dy be the contemporaneous increment of y, so that when x becomes $x + dx$, y becomes $y + dy$. We then have

1st state of the function, ... $y = ax$;
2nd, or consecutive state, ... $y + dy = a(x + dx) = ax + adx.$
Subtracting the 1st from the 2nd, $dy = adx$,

which being the difference between two consecutive states of the function is its differential (***44***). Now the factor a appears in this differential just as it was in the function. Q. E. D.

ILL.—Let $y = ax$ be the equation of the line MN. PD and P′D′ representing consecutive ordinates, DD′ represents dx, and P′E represents dy. Here it is evident that P′E $= a \times$ DD′; for from the triangle PP′E we have P′E = tan P′PE × PE. But tan P′PE = tan MAX $= a$. The meaning in this case is, therefore, that the ordinate increases a times as fast as the abscissa.

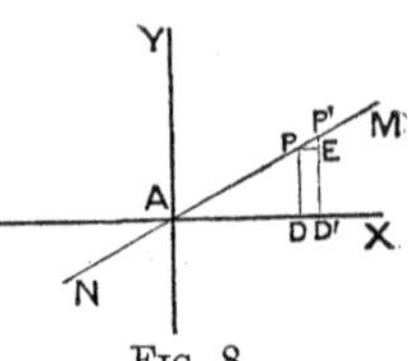

FIG. 8.

If $a = 1$, or tan 45°, the ordinate and abscissa increase at equal rates; if $a < 1$, *i. e.*, if the angle is less than 45° the ordinate increases more slowly than the abscissa; if $a > 1$, the ordinate increases more rapidly than the abscissa.

49. *RULE* 3.—CONSTANT TERMS DISAPPEAR IN DIFFERENTIATING: OR THE DIFFERENTIAL OF A CONSTANT IS 0.

DEM.—Let us take the function $y = ax \pm b$, in which a and b are constants. Let x take an infinitesimal increment and become $x + dx$; and let dy be the contemporaneous increment of y, so that when x becomes $x + dx$, y becomes $y + dy$. We then have

1st state of the function, ... $y = ax \pm b$;
2nd, or consecutive state, ... $y + dy = a(x + dx) \pm b$,
or . . . . $y + dy = ax + adx \pm b.$
Subtracting the 1st state from the 2nd, $dy = adx$, which

being the difference between two consecutive states of the function is its differential (***44***). Now from this differential the constant term $\pm b$ has disappeared. We may also say that as a constant retains the same value there is no difference between its consecutive states (properly it has no consecutive states). Hence the differential of a constant may be spoken of (though with some latitude) as 0. Q. E. D.

ILL.—Let $y = ax + b$ be the equation of the line MN. Now the *relative* rates of increase of the abscissa and ordinate, that is the relation of dy to dx, is evidently not affected by b which is AB; for, if we were to draw a line through the origin parallel to MN, the contemporaneous increments of *its* co-ordinates would be the same as those of MN. Again, we can see that the constant term does not affect the differential, *i. e.*, the difference between the consecutive states of y, by observing that these two states are represented by PD and P′D′, each of which contains the constant as a part of it, whence the *difference between them* is not affected by it.

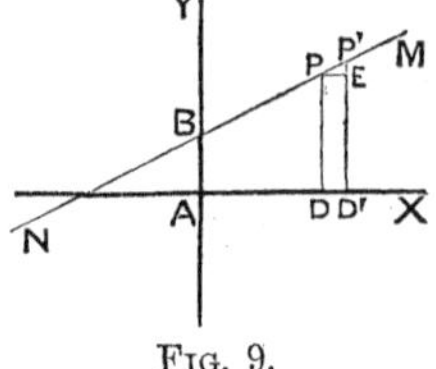

FIG. 9.

50. COR.—*An infinite variety of functions differing from each other only in their constant terms still have the same differential.*

51. *RULE* 4.—TO DIFFERENTIATE THE ALGEBRAIC SUM OF SEVERAL VARIABLES, DIFFERENTIATE EACH TERM SEPARATELY AND CONNECT THE DIFFERENTIALS WITH THE SAME SIGNS AS THE TERMS.

DEM.—Let $u = x + y - z$, u representing the algebraic sum of the variables x, y, and $-z$. Then is the differential of this sum or $du = dx + dy - dz$. For let dx, dy, and dz be infinitesimal increments of x, y, and z; and let du be the increment which u takes in consequence of the infinitesimal changes in x, y, and z. We then have

1st state of the function,........ $u = x + y - z$;

2nd, or consecutive state, ... $u + du = x + dx + y + dy - (z + dz)$,

or .... $u + du = x + dx + y + dy - z - dz$.

Subtracting the 1st state from the 2nd, $du = dx + dy - dz$. Q. E. D.

ILL.—We may illustrate this by conceiving x and y to be forces acting to *raise* a weight, and z a force acting to prevent the raising, u being the *aggregate* effect of all, *i. e.* their algebraic sum (Complete Algebra, 65). Now if x, y, and z each receive an infinitesimal increment, which we will call respectively dx, dy, and dz, it is evident that the increment of *lifting* force is $dx + dy$, and as the increment of the depressing force is dz, the combined effect of the change is $dx + dy - dz$, which is the change in u. Moreover, since this quantity $dx + dy - dz$ is the aggregate of a finite number of infinitesimals, it must be itself infinitesimal. Hence the change in u is infinitesimal, or du.

SCH.—It is important to notice that the above reasoning is entirely independent of the relative values of the infinitesimals dx, dy, and dz. These may be conceived as equal, or as sustaining any finite ratio whatever to each other, only so that they remain infinitesimal.

52. *RULE* 5.—THE DIFFERENTIAL OF THE PRODUCT OF TWO VARIABLES IS THE DIFFERENTIAL OF THE FIRST INTO THE SECOND, PLUS THE DIFFERENTIAL OF THE SECOND INTO THE FIRST.

DEM.—Let $u = xy$ be the first state. The consecutive state is $u + du = (x + dx)(y + dy) = xy + ydx + xdy + dxdy$. Subtracting the 1st state from the 2nd, or consecutive state, we have $du = ydx + xdy + dx \cdot dy$. Now ydx and xdy are infinitesimals of the 1st order, and $dx \cdot dy$, being the product of two infinitesimals, is of the 2nd order and must be dropped (***17***). Therefore $du = ydx + xdy$. Q. E. D.

ILL.—Let u represent the area of the rectangle **ABCD**, $x =$ ***AB**, and $y =$ ***BC**. Then $u = xy$. Let **B**b represent dx, and **C**c'', dy. Whence **B**b**C**c' = *ydx, **D**d**C**c'' = *xdy, **C**$c'cc''$ =* $dx \cdot dy$, and du = ***B**b**C**c' + **D**d**C**c'' + **C**$c'cc''$. Now since cc' is infinitesimal and $c'b$ is finite, **C**$c'cc''$ is infinitesimal with reference to **B**b**C**c', as for a like reason it is with reference to **D**d**C**c''; hence it is to be omitted as having no assignable value with reference to them.

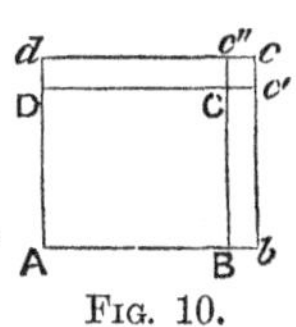

FIG. 10.

Another view which may be taken of this is to consider that it is *the rate at which*

* In such cases = signifies "is represented by," and is used for brevity.

the rectangle is increasing when $x = $ **AB** and $y = $ **BC**, not the *amount* of change in the area after x and y shall have increased more or less : in other words, we seek for the difference between *consecutive* values of the area. Now it is easy to see that the *rate* at which the rectangle **ABCD** *starts* to increase, depends upon the *length* of the side **BC** (y) and the *rate* at which it *starts to move* to the right, + the length of **DC** (x) and the rate at which it *starts* to move upward. Letting dx represent the rate at which **AB** starts to increase (by being the amount which it would increase in an infinitesimal of time), and dy represent in like manner the rate at which y *starts* to increase, we readily see that $du = ydx + xdy$ is the rate at which the area *starts* to increase. Moreover, we see that this is equally true whether $dy = dx$, or whether one is any finite multiple of the other ; all that is necessary being that both be infinitesimals of the same order.

53. *RULE* 6.—THE DIFFERENTIAL OF THE PRODUCT OF SEVERAL VARIABLES IS THE SUM OF THE PRODUCTS OF THE DIFFERENTIAL OF EACH INTO THE PRODUCT OF ALL THE OTHERS.

DEM.—Let $u = xyz$; then $du = yzdx + xzdy + xydz$.
For the 1st state of function is .... $u = xyz$,
2nd, or consecutive state,................ $u + du = (x + dx)(y + dy)(z + dz)$,
or.......... $u + du = xyz + yzdx + xzdy + xydz + xdydz + ydxdz + zdxdy + dxdydz$.
Subtracting and dropping infinitesimals of higher orders than the first we have $du = yzdx + xzdy + xydz$.

In a similar manner the rule can be demonstrated for any number of variables. Q. E. D.

54. *RULE* 7.—THE DIFFERENTIAL OF A FRACTION HAVING A VARIABLE NUMERATOR AND DENOMINATOR IS THE DIFFERENTIAL OF THE NUMERATOR MULTIPLIED BY THE DENOMINATOR, MINUS THE DIFFERENTIAL OF THE DENOMINATOR MULTIPLIED BY THE NUMERATOR, DIVIDED BY THE SQUARE OF THE DENOMINATOR.

DEM.—Let $u = \frac{x}{y}$; then is $du = \frac{ydx - xdy}{y^2}$. For clearing of fractions $yu = x$. Differentiating this by Rule 5, $udy + ydu = dx$. Substituting for u its value, we have $\frac{xdy}{y} + ydu = dx$. Finding the value of du, we have $du = \frac{ydx - xdy}{y^2}$. Q. E. D.

55. COR.—*The differential of a fraction having a constant numerator and a variable denominator is the product of the numerator with its sign changed into the differential of the denominator, divided by the square of the denominator.*

DEM.—Let $u = \frac{a}{y}$. Differentiating this by the rule and calling the differential of the constant (a), 0, we have $du = \frac{0 - ady}{y^2} = \frac{-ady}{y^2}$. Q. E. D.

SCH.—If the numerator is variable and the denominator constant it falls under *Rule* 2.

56. *RULE* 8.—THE DIFFERENTIAL OF A VARIABLE AFFECTED WITH AN EXPONENT IS THE CONTINUED PRODUCT OF THE EXPONENT, THE VARIABLE WITH ITS EXPONENT DIMINISHED BY 1, AND THE DIFFERENTIAL OF THE VARIABLE.

DEM.—1st. *When the exponent is a positive integer.*—Let $y = x^m$, m being a positive integer; then $dy = mx^{m-1}dx$. For $y = x^m = x \cdot x \cdot x \cdot x$ to m factors. Now differentiating this by *Rule* 6, we have
$dy = (xxx$ to $m - 1$ factors$)\,dx + (xxx$ to $m - 1$ factors$)\,dx +$ etc., to m terms,
or $dy = x^{m-1}dx + x^{m-1}dx + x^{m-1}dx +$ etc., to m terms.
$\therefore\ dy = mx^{m-1}dx.$

2nd. *When the exponent is a positive fraction.*—Let $y = x^{\frac{m}{n}}$, $\frac{m}{n}$ being a positive fraction; then $dy = \frac{m}{n}x^{\frac{m}{n}-1}dx$. For involving both members to the nth power we have $y^n = x^m$. Differentiating as just shown, $ny^{n-1}dy = mx^{m-1}dx$. Now from $y = x^{\frac{m}{n}}$, we have $y^{n-1} = x^{\frac{mn-m}{n}}$. Substituting this in the last form, we have $nx^{\frac{mn-m}{n}}dy = mx^{m-1}dx$, or $dy = \frac{m}{n}x^{m-1-\frac{mn-m}{n}}dx = \frac{m}{n}x^{\frac{m}{n}-1}dx.$

3rd. *When the exponent is negative.*—Let $y = x^{-n}$, n being integral or fractional; then $dy = -nx^{-n-1}dx$. For $y = x^{-n} = \frac{1}{x^n}$, which differentiated by *Rule* 7, *Cor.*, gives $dy = -\frac{nx^{n-1}dx}{x^{2n}} = -nx^{-n-1}dx$. All three of which forms agree with the enunciation of the rule. Q. E. D.

57. COR.—*The differential of the square root of a variable is the differential of the variable divided by twice the square root of the variable.*

DEM.—Let $y = \sqrt{x} = x^{\frac{1}{2}}$. Differentiating by the rule we have $dy = \frac{1}{2}x^{\frac{1}{2}-1}dx = \frac{1}{2}x^{-\frac{1}{2}}dx = \frac{dx}{2\sqrt{x}}$. Q. E. D.

SCH.—Special rules can be readily made for other roots, but it is unnecessary. The square root is of such frequent occurrence as to make the special process expedient. Of course the general rule can always be used, if desired.

EXERCISES.

[NOTE.—The following examples are designed to give practical skill in applying the rules for differentiating algebraic functions. The student should not advance beyond these, till he has the rules firmly fixed in memory, and can apply them with facility to all forms of algebraic functions.]

Ex. 1. Differentiate $y = 6x - 4$. $dy = 6dx$.

QUERY.—What three rules apply? Be careful to repeat the rules in applying them to the solution of the examples, and thus render them familiar.

Ex. 2. Differentiate $y = a^6 + 3a^4x^2 + 3a^2x^4 + x^6$.

SOLUTION.—The differential of y is dy. [Repeat *Rule* 1.] To differentiate the second member we notice 1st, that it consists of several terms, and hence proceed to differentiate each term separately. [Repeat *Rule* 4.] a^6 being a constant *term*, disappears. [Repeat *Rule* 3.] To differentiate $3a^4x^2$, we notice 1st that the constant factor $3a^4$ will be a factor in the differential. [Repeat *Rule* 2.] The differential of x^2 is $2xdx$. [Repeat *Rule* 8.] Hence the differential of $3a^4x^2$ is $6a^4xdx$. [In like manner proceed with the other terms, *giving the reason for each step by repeating the appropriate rule.*]

Ex. 3. Differentiate $u = 2ax - 3x^2 + abx^3 - 5$.

Result, $du = (2a - 6x + 3abx^2)dx$.

Ex. 4. Differentiate $y = 3x^4 - 2x - 5m$.

Ex. 5. Differentiate $u = ab - 6x^3 + 2ax$.

Ex. 6. Differentiate $u = ax^2y^3$.

QUERIES.—What is the most general feature of the function ax^2y^3? What rule applies first? Rule 5. What other rule applies?

Result, $du = 2axy^3dx + 3ax^2y^2dy$.

Ex. 7. Differentiate $u = 6ax^{\frac{2}{3}}y^3$.

Result, $du = 4ax^{-\frac{1}{3}}y^3dx + 18ax^{\frac{2}{3}}y^2dy$.

Ex. 8. Differentiate $y = 2bz^{-2} + 3ax^{\frac{5}{3}}z^{\frac{1}{2}}$.

Result, $dy = 5ax^{\frac{2}{3}}z^{\frac{1}{2}}dx + \dfrac{3ax^{\frac{5}{3}}dz}{2\sqrt{z}} - \dfrac{4bdz}{z^3}$.

Ex. 9. Differentiate $u = x^{\frac{1}{2}}y^{\frac{1}{2}}$. *Result*, $\dfrac{xdy + ydx}{2x^{\frac{1}{2}}y^{\frac{1}{2}}}$.

Ex. 10. From $y^2 = 2px$ find the value of dy. $dy = \dfrac{p}{y}dx$.

Ex. 11. From $A^2y^2 + B^2x^2 = A^2B^2$ find the value of dy.

$$dy = -\frac{B^2x}{A^2y}dx.$$

Ex. 12. From $A^2y^2 - B^2x^2 = - A^2B^2$ find the value of dy.

$$dy = \frac{B^2x}{A^2y}dx.$$

Ex. 13. From $x^2 + y^2 = R^2$ find the value of dy. $dy = -\dfrac{x}{y}dx$.

Ex. 14. From $2xy^2 - ay^2 = x^3$ find the value of dy.

$$dy = \frac{3x^2 - 2y^2}{4xy - 2ay}dx.$$

Ex. 15. Differentiate $u = \frac{x^2}{3y^3}$. *Result*, $du = \frac{2xydx - 3x^2dy}{3y^4}$.

Ex. 16. Differentiate $y = \frac{1}{x}$. $dy = -\frac{dx}{x^2}$.

Ex. 17. Differentiate $u = \frac{a}{b - 2y^2}$. $du = \frac{4aydy}{(b - 2y^2)^2}$.

Ex. 18. Differentiate $y = \frac{3ax^2}{5}$. $dy = \frac{3a}{5} \times 2xdx = \frac{6ax}{5}dx$.

Sug.—Do not treat this as a fraction under Rule 7.

Ex. 19. Differentiate $u = x^2y^3z$.

Ex. 20. Differentiate $u = \frac{2x^2 - 3}{4x + x^2}$.

Operation. $du = \frac{d(2x^2 - 3)(4x + x^2) - d(4x + x^2)(2x^2 - 3)}{(4x + x^2)^2} =$

$\frac{4xdx(4x + x^2) - (4dx + 2xdx)(2x^2 - 3)}{(4x + x^2)^2} = \frac{\{4x(4x + x^2) - (4 + 2x)(2x^2 - 3)\}dx}{(4x + x^2)^2}$

$= \frac{(8x^2 + 6x + 12)dx}{(4x + x^2)^2}$.

Sug's.—The first step is the application of the rule for fractions, since the function is a fraction with a variable numerator and a variable denominator. The second step is to *perform* the differentiation of $2x^2 - 3$, and $4x + x^2$. This step involves the rules for constant factors, variables affected with exponents, constant terms, and the sum of variables. The remainder of the work is reduction and addition of terms.

Ex. 21. Differentiate $u = \frac{2x^4}{a^2 - x^2}$. $du = \frac{8a^2x^3 - 4x^5}{(a^2 - x^2)^2}dx$.

Ex. 22. Differentiate $y = \frac{a - x}{x}$. $dy = -\frac{a}{x^2}dx$.

Ex. 23. Differentiate $y = \frac{1 + x}{1 + x^2}$. $dy = \frac{(1 - 2x - x^2)dx}{(1 + x^2)^2}$.

Ex. 24. Differentiate $y = \frac{1 + x^2}{1 - x^2}$.

Ex. 25. Differentiate $y = \frac{x}{1 - x}$.

Ex. 26. Differentiate $y = 3x^m - 4$. $dy = 3mx^{m-1}dx$.

Ex. 27. Differentiate $y = 2mx^{\frac{m}{n}}$. $dy = \frac{2m^2}{n}x^{\frac{m-n}{n}}dx$.

Ex. 28. Differentiate $u = 2nx^{\frac{m}{n}}y^{\frac{1}{n}}$.

$$du = 2mx^{\frac{m-n}{n}}y^{\frac{1}{n}}dx + 2x^{\frac{m}{n}}y^{\frac{1-n}{n}}dy.$$

Ex. 29. Differentiate $y = \frac{1}{x^n}$. $\qquad dy = -\frac{n}{x^{n+1}}dx.$

Ex. 30. Differentiate $y = \sqrt{x^3 - a^2}$.

Operation.—By the special rule for the square root (**57**), we have $dy = \frac{d(x^3 - a^2)}{2\sqrt{x^3 - a^2}} = \frac{3x^2dx}{2\sqrt{x^3 - a^2}}$.

Ex. 31. Differentiate $y = \sqrt{ax} + \sqrt{c^2x^3}$.

$$dy = \frac{adx}{2\sqrt{ax}} + \frac{3c^2x^2dx}{2\sqrt{c^2x^3}} = (\tfrac{1}{2}a^{\frac{1}{2}}x^{-\frac{1}{2}} + \frac{3c}{2}x^{\frac{1}{2}})dx, \text{ or } \frac{a^{\frac{1}{2}} + 3cx}{2\sqrt{x}}dx.$$

Ex. 32. Differentiate $y = a\sqrt{x} - \frac{x}{3}$.

Ex. 33. Differentiate $y = \sqrt{ax + bx^2 + cx^3}$.

Ex. 34. Differentiate $y = (ax^2 - x^3)^4$.

Solution.—Regarding $ax^2 - x^3$ as a variable, it is affected with the exponent 4; hence we have $dy = 4(ax^2 - x^3)^3 \times d(ax^2 - x^3)$, the operation of differentiating the variable $ax^2 - x^3$ being as yet unperformed. Performing this operation and reducing, we have $dy = 4(ax^2 - x^3)^3 \times (2ax - 3x^2)dx =$

$$8ax^7(a - x)^3dx - 12x^8(a - x)^3dx.$$

Ex. 35. Differentiate $y = (a + bx^2)^{\frac{5}{3}}$. $\qquad dy = \frac{10}{3}(a + bx^2)^{\frac{2}{3}}bxdx.$

Ex. 36. Differentiate $y = (a^2 + x^2)^3$. $\qquad dy = 6x(a^2 + x^2)^2dx.$

Ex. 37. Differentiate $y = \frac{a}{(b^2 + x^2)^3}$. $\qquad dy = -\frac{6ax}{(b^2 + x^2)^4}dx.$

Ex. 38. Differentiate $y = (1 + 2x^2)(1 + 4x^3)$.

Solution.—Regarding this function as the product of the two variables $1 + 2x^2$ and $1 + 4x^3$, we have $dy = d(1 + 2x^2) \times (1 + 4x^3) + d(1 + 4x^3) \times (1 + 2x^2)$. Performing the operation of differentiating $1 + 2x^2$ and $1 + 4x^3$, we have $dy = 4x(1 + 4x^3)dx + 12x^2(1 + 2x^2)dx = 4x(1 + 3x + 10x^3)dx$.

Ex. 39. Differentiate $y = (x^3 + a)(3x^2 + b)$.

$$dy = (15x^4 + 3bx^2 + 6ax)dx.$$

Ex. 40. Differentiate $y = \frac{x^3}{(1 + x)^2}$. $\qquad dy = \frac{3x^2 + x^3}{(1 + x)^3}dx.$

Ex. 41. Differentiate $y = \frac{a}{(a - x)^3}$. $\qquad dy = \frac{3adx}{(a - x)^4}.$

Ex. 42. Differentiate $y = \frac{ax^2}{(ab - x^2)^3}$. $\quad dy = \frac{2ax(ab + 2x^2)dx}{(ab - x^2)^4}$.

Ex. 43. Differentiate, without first expanding, $y = (1 + x)^4(1 + x^2)^2$.

$$dy = 4(1 + x)^3(1 + x^2)(1 + x + 2x^2)dx.$$

Ex. 44. Differentiate $y = x^2 - \sqrt{1 - x^3}$.

$$dy = 2xdx + \frac{3x^2dx}{2\sqrt{1 - x^3}}.$$

Ex. 45. Differentiate $u = \sqrt{2ax - x^2}$. $\quad du = \frac{(a - x)dx}{\sqrt{2ax - x^2}}$.

Ex. 46. Differentiate $u = \sqrt{a^2 + x^2} \times \sqrt{b^2 + y^2}$.

$$du = \frac{(b^2 + y^2)xdx + (a^2 + x^2)ydy}{\sqrt{a^2 + x^2} \times \sqrt{b^2 + y^2}}.$$

Ex. 47. Differentiate $y = \frac{x}{\sqrt{a^2 - x^2}}$. $\quad dy = \frac{a^2dx}{\sqrt{(a^2 - x^2)^3}}$.

Ex. 48. Differentiate $y = \frac{x}{\sqrt{1 + x^2}}$.

SUG'S. $y = x(1 + x^2)^{-\frac{1}{2}}$. $\therefore dy = dx(1 + x^2)^{-\frac{1}{2}} + x \cdot d(1 + x^2)^{-\frac{1}{2}} = dx(1 + x^2)^{-\frac{1}{2}} - x^2(1 + x^2)^{-\frac{3}{2}}dx = \frac{dx(1 + x^2) - x^2dx}{(1 + x^2)^{\frac{3}{2}}} = \frac{dx}{(1 + x^2)^{\frac{3}{2}}}$. Or, we may apply the rules for a fraction and a square root, thus $dy = \frac{dx\sqrt{1 + x^2} - xd\sqrt{1 + x^2}}{1 + x^2} = \frac{dx\sqrt{1 + x^2} - x\frac{xdx}{\sqrt{1 + x^2}}}{1 + x^2} = \frac{dx(1 + x^2) - x^2dx}{(1 + x^2)^{\frac{3}{2}}} = \frac{dx}{(1 + x^2)^{\frac{3}{2}}}$.

Ex. 49. Differentiate $u = (1 + x)\sqrt{1 - x}$. $\quad du = \frac{(1 - 3x)dx}{2\sqrt{1 - x}}$.

Ex. 50. Differentiate $u = \frac{x^3}{\sqrt{(1 - x^2)^3}}$. $\quad du = \frac{3x^2dx}{(1 - x^2)^{\frac{5}{2}}}$.

Ex. 51. Differentiate $u = \frac{a^4}{2\sqrt{a^2x^2 - x^4}}$.

$$du = \frac{-a^4(a^2 - 2x^2)dx}{2x^2(a^2 - x^2)^{\frac{3}{2}}}.$$

Ex. 52. Differentiate $u = \sqrt{x + \sqrt{1 + x^2}}$.

SUG'S.—Squaring $u^2 = x + \sqrt{1 + x^2}$. $\quad 2udu = dx + \frac{xdx}{\sqrt{1 + x^2}}$. $\quad du =$

$$\frac{dx + \dfrac{xdx}{\sqrt{1+x^2}}}{2u} = \frac{\dfrac{(x+\sqrt{1+x^2})dx}{\sqrt{1+x^2}}}{2\sqrt{x+\sqrt{1+x^2}}} = \frac{(x+\sqrt{1+x^2})dx}{2\sqrt{1+x^2}\sqrt{x+\sqrt{1+x^2}}} =$$

$\dfrac{\sqrt{x+\sqrt{1+x^2}}dx}{2\sqrt{1+x^2}}$. Or, we may differentiate without squaring, thus $du =$

$$\frac{dx + \dfrac{xdx}{\sqrt{1+x^2}}}{2\sqrt{x+\sqrt{1+x^2}}} = \frac{(x+\sqrt{1+x^2})dx}{2\sqrt{1+x^2}\sqrt{x+\sqrt{1+x^2}}} = \frac{\sqrt{x+\sqrt{1+x^2}}dx}{2\sqrt{1+x^2}}.$$

Ex. 53. Differentiate $u = \dfrac{x}{\sqrt{a^2+x^2}-x}$.

SUG'S.—As the denominator is more involved in the differential of a fraction than the numerator, it is expedient to reduce the fraction to a form having as simple a denominator as possible. Rationalizing this denominator, we have $u = \dfrac{x\sqrt{a^2+x^2}+x^2}{a^2} = \dfrac{1}{a^2}x\sqrt{a^2+x^2} + \dfrac{1}{a^2}x^2$. $du = \dfrac{(2x^2+a^2)dx}{a^2\sqrt{x^2+a^2}} + \dfrac{2xdx}{a^2}$.

Ex. 54. Differentiate $u = \dfrac{\sqrt{x^2+1}-x}{\sqrt{x^2+1}+x}$.

$$du = 2\left\{2x - \frac{2x^2+1}{\sqrt{x^2+1}}\right\}dx.$$

Ex. 55. Differentiate $u = \sqrt{\dfrac{1-\sqrt{x}}{1+\sqrt{x}}}$.

SUG'S. $u = \sqrt{\dfrac{1-\sqrt{x}}{1+\sqrt{x}}} = \dfrac{\sqrt{1-\sqrt{x}}}{\sqrt{1+\sqrt{x}}} = \dfrac{\sqrt{1-x}}{1+\sqrt{x}}$. $du = -\dfrac{dx}{2(1+\sqrt{x})\sqrt{x-x^2}}$.

Ex. 56. Differentiate $u = \left(\dfrac{x}{1+x}\right)^n$. $du = \dfrac{nx^{n-1}dx}{(1+x)^{n+1}}$.

Ex. 57. Differentiate $u = \dfrac{\sqrt{1+x}+\sqrt{1-x}}{\sqrt{1+x}-\sqrt{1-x}}$.

$$du = -\frac{1+\sqrt{1-x^2}}{x^2\sqrt{1-x^2}}dx.$$

Ex. 58. Differentiate $u = \sqrt[3]{x} \cdot \sqrt{\sqrt{x}+1}$.

$$du = \frac{7x^{\frac{1}{2}}+4}{12\sqrt[3]{x^2}\cdot\sqrt{x^{\frac{1}{2}}+1}}dx.$$

Ex. 59. Differentiate $u = \sqrt{2x - 1 - \sqrt{2x - 1 - \sqrt{2x - 1 -}}}$, etc., to infinity.

SUG'S.—We have $u = \sqrt{2x - 1 - u}$; whence $u^2 = 2x - 1 - u$, and $u = -\frac{1}{2} \pm \frac{1}{2}\sqrt{8x - 3}$. $\therefore du = \pm \frac{2dx}{\sqrt{8x - 3}}$.

Ex. 60. Differentiate $u = \sqrt[4]{\left[a - \frac{b}{\sqrt{x}} + \sqrt[3]{(c^2 - x^2)^2}\right]^3}$.

$$du = \frac{\frac{3b}{2x\sqrt{x}} - \frac{4x}{\sqrt[3]{(c^2 - x^2)}}}{4\sqrt[4]{a - \frac{b}{\sqrt{x}} + \sqrt[3]{(c^2 - x^2)^2}}}dx.$$

ILLUSTRATIVE EXAMPLES.

[NOTE.—The following examples are designed to illustrate more fully the significance of the process of differentiation.]

Ex. 1. In a parabola whose parameter is 12, which is increasing the faster at $x = 2$, the ordinate or the abscissa, and how much? At $x = 3$? At $x = 8$? At $x = 24$? How does the relative rate of change vary as we recede from the vertex? At what point are ordinate and abscissa varying equally?

SOLUTION.—The equation of this parabola is $y^2 = 12x$. Differentiating, we have $dy = \frac{6}{y}dx$. Now, as differentiating is the process of finding the difference between two consecutive states of a function, dx represents one of the infinitesimal increments of x, as **D D'**, and dy the *contemporaneous, infinitesimal* increment of y, as **P'E.** We, therefore, learn from $dy = \frac{6}{y}dx$ that in general dy is $\frac{6}{y}$ times as great as dx; or, in other words, that y changes $\frac{6}{y}$ times as fast as x. At **P** where $x = 2$, $y = \sqrt{24}$. Hence, at this point, $dy = \frac{6}{\sqrt{24}}dx = \frac{1}{2}\sqrt{6}dx$; that is, y is increasing nearly $1\frac{1}{4}$ times as fast as x. At **P''** where $x = 3$, $y = 6$, and $dy = dx$; that is, x and y are increasing equally. In general, at the focus the ordinate and abscissa of a parabola are increasing equally, since at this point $y = p$. At $\mathbf{P^{IV}}$ where $x = 8$, y is increasing only about .6 as fast as x. At $\mathbf{P^{VI}}$ where $x = 24$, y is increasing at the still slower rate of about .35 as fast as x. Finally, it is evident, from a slight inspection of the figure,

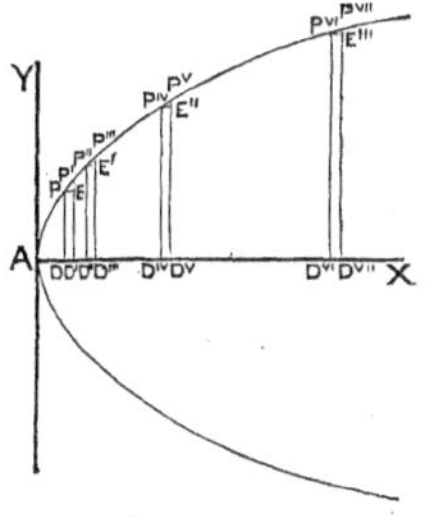

FIG. 11.

that y increases less and less rapidly as x becomes larger, x continuing to increase at a uniform rate. At $x = \infty$, y ceases to increase, *i. e.* the branches become parallel to the axis of x.

Ex. 2. Examine the relative rates of change of the ordinate and abscissa in the ellipse.

SOLUTION.—Differentiating $A^2y^2 + B^2x^2 = A^2B^2$, we find $dy = -\frac{B^2x}{A^2y}dx = -\frac{Bx}{A\sqrt{A^2 - x^2}}dx$. On this we observe 1st, That the — sign shows that x and y are decreasing functions of each other; that is, that as x takes an increment y takes a decrement. This is evident from a consideration of the curve. 2nd, That in general terms y diminishes $\frac{Bx}{A\sqrt{A^2 - x^2}}$ times as fast as x increases. 3rd, At $x = 0$, *i. e.* at the extremity of the conjugate axis, y is not increasing or decreasing, since here $\frac{Bx}{A\sqrt{A^2 - x^2}} = 0$, and $dy = 0 \cdot dx = 0$. At the extremity of the transverse axis $dy = -\infty \cdot dx$, *i. e.* y is decreasing infinitely faster than x increases. There are, therefore, all relative rates of change between x and y from 0 to ∞. Moreover as x begins to increase from 0, y commences to decrease (at first slowly, as the fraction $\frac{Bx}{A\sqrt{A^2 - x^2}}$ is small when x is small), and then more and more rapidly as x increases, till it reaches an infinitely rapid rate of decrease at $x = A$. This, it is easy to see, is the law of change in the fraction $\frac{Bx}{A\sqrt{A^2 - x^2}}$ as x increases. The same law is also rendered probable from an inspection of the curve. Finally, we may inquire at what point the relative rates of change sustain any given relation to each other, as, for example, when y decreases twice as fast as x increases, or just as fast, or 10 times as fast. Thus when y decreases twice as fast as x increases, we must have $dy = -2dx$, *i. e.* $\frac{Bx}{A\sqrt{A^2 - x^2}} = 2$. From this we find $x = \pm \frac{2A^2}{\sqrt{4A^2 + B^2}}$; hence at these points, y is diminishing twice as fast as x is increasing.

Ex. 3. A boy is running on a horizontal plane directly toward the foot of a tower 100 feet in height. How much faster is he nearing the foot than the top of the tower? How far is he from the foot of the tower when he is approaching the base twice as fast as he approaches the top? How far off must he be to be approaching both base and top equally? Where is he when he is not approaching the top at all, or is making infinitely more progress toward the base than towards the top? When he is at 200 feet from the base of the tower how much faster is he approaching the base than the top?

SUG'S.—Let **AB** represent the tower, and **AX** the line in the plane of the base in which the boy is approaching the base. Suppose the boy at any point, as **P** and let **AP** $= x$, and **PB** $= y$. Then $y^2 - x^2 = 10000$. Whence $dy = \frac{x}{y}dx$. Hence we see that in general he is only approaching the top an $\frac{x}{y}$th part as fast as he is the base; *i. e.*, letting **PP′** represent an infinitesimal element of the distance to the foot of the tower, **PF** represents a contemporaneous, infinitesimal element of the distance to the top; and also, that **PF** is an $\frac{x}{y}$th part of **PP′**. Secondly, when he is approaching the foot of the tower twice as fast as he is the top; we have $dy = \frac{1}{2}dx$, or $\frac{x}{y} = \frac{1}{2}$; whence $y = 2x$. But $y^2 - x^2 = 10000$; and, substituting, $3x^2 = 10000$, or $x = \frac{100}{\sqrt{3}} = 58$ nearly. Lastly, when he is at 200 feet from the base $y = \sqrt{50000} = 224$ nearly, and $\frac{x}{y} = \frac{200}{224} = \frac{25}{28}$ nearly. Hence $dy = \frac{25}{28}dx$, or he is approaching the top $\frac{25}{28}$ as fast as he is the base. [Let the pupil decide the other points himself.]

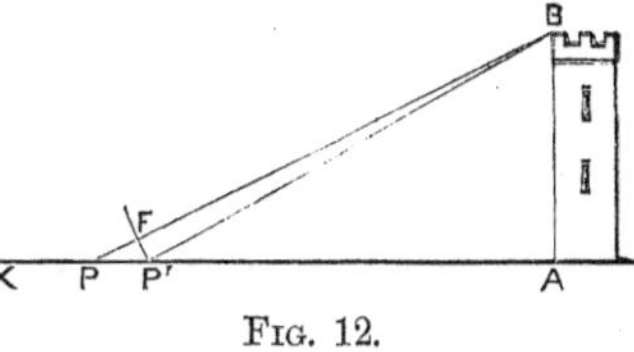

Fig. 12.

Ex. 4. A ship is sailing northwest at 15 miles an hour. At what rate is she making north latitude?

Ans., At 10.605+ miles an hour.

Sug's.—Let y represent any distance run in the northwest course, and x the corresponding northing. Then as the course is northwest there is made in the same time x westing, and we have $y^2 = 2x^2$. From this $dy = \frac{2x}{y}dx$, and the ship is running $\frac{2x}{y}$ times as fast as she is making northing. But $y = x\sqrt{2}$, whence $\frac{2x}{y} = \frac{2x}{x\sqrt{2}} = \sqrt{2}$, and $dy = \sqrt{2}dx$, or $dx = \frac{1}{2}\sqrt{2}dy$; *i. e.*, she is making northing .707+ as fast as she is running.

Ex. 5. In the function $y = 27x + 3x^2$, required the value of x when y is increasing 45 times as fast as x. *Result*, $x = 3$.

Ex. 6. What is the relative rate of variation of the side and altitude of an equilateral triangle? *i. e.*, if the side takes an infinitesimal increment, what is the contemporaneous infinitesimal increment of the altitude? When the side is increasing at the rate of 2 inches per second how rapidly is the altitude increasing? Is the relative rate of increase constant or variable; that is, does the altitude increase more or less rapidly in comparison with the side when the side is small than it does when it is large, or is the relative rate of increase always the same?

Sug's.—Let $y =$ the altitude and x one of the sides of the triangle. Then

$y^2 = \frac{3}{4}x^2$, and $dy = \frac{3x}{4y}dx = \frac{3x}{2\sqrt{3}x}dx = \frac{1}{2}\sqrt{3}dx$. Hence we see that the infinitesimal increment of y is *always* $\frac{1}{2}\sqrt{3}$ times as much as the contemporaneous infinitesimal increment of x. When x is increasing at 2 inches per second y is increasing $\frac{1}{2}\sqrt{3}$ times 2 inches, or $\sqrt{3}$ inches per second.

SCH.—The student should now be able to comprehend with considerable clearness the object of the Differential Calculus; viz., having given the relation between finite values of variables, to find the relation between the contemporaneous infinitesimal increments of those variables, or their relative rate of change. Thus, in the last example, the relation between the altitude and one side of an equilateral triangle, $y^2 = \frac{3}{4}x^2$, is the relation between finite values of the variables, from which we find the relation between the contemporaneous infinitesimal increments dy and dx, by the Differential Calculus.

SECTION II.

Differentiation of Logarithmic and Exponential Functions.*

58. *The Modulus* of a system of logarithms is a constant factor which depends upon the base of the system and characterizes the system.

59. *Prop.*—*The differential of the logarithm of a variable is the differential of the variable multiplied by the modulus of the system, divided by the variable; or, in the Napierian system the modulus being* 1, *the differential of the logarithm is the differential of the variable divided by the variable.*

DEM.—Let $y = x^n$, n being constant. Then $\log y = n \log x$. Differentiating $y = x^n$, we have $dy = nx^{n-1}dx$, or $n = \frac{dy}{x^{n-1}dx} = \frac{dy}{\frac{y}{x}dx} = \frac{\frac{dy}{y}}{\frac{dx}{x}}$, since $x^{n-1} = \frac{y}{x}$. Again, whatever the differentials of $\log y$ and $\log x$ are, we have $d(\log y) = n \cdot d(\log x)$, or $n = \frac{d(\log y)}{d(\log x)}$. Placing these values of n equal to each other, we obtain $\frac{d(\log y)}{d(\log x)} = \frac{\frac{dy}{y}}{\frac{dx}{x}}$. Now let m be the factor by which $\frac{dy}{y}$ must be multiplied to make it equal to $d(\log y)$, then is $d(\log x) = \frac{mdx}{x}$.

---

* See **32, 33.**

We are now to show that m is a constant depending upon the base of the system. To do this take $y = z^{n'}$, from which we find as before $n' = \frac{d(\log y)}{d(\log z)} = \frac{\frac{dy}{y}}{\frac{dz}{z}}$. But m is the ratio of $d(\log y)$ to $\frac{dy}{y}$; hence $d(\log z) = \frac{mdz}{z}$. Thus we see that in any case the same ratio exists between the differential of the log. of a number, and the differential of the number divided by the number. Therefore m is a constant factor. To show that m depends upon the base of the system we have but to recur to the definition of a logarithm to see that the only quantities involved are *the number*, its *logarithm*, and the *base* of the system. Of these the two former are variable, whence, as the base is the only constant in the scheme, m is a function of the base.*

Finally, as m depends upon the base of the system, the base may be so taken that $m = 1$. The system of logarithms founded on this base is called the Napierian system. Q. E. D.

---

60. Prop.—*The differential of an exponential function with a constant base is the function itself, into the logarithm of the base, into the differential of the exponent, divided by the modulus.*

DEM.—Let $y = a^x$. Taking the logarithms of both members $\log y = x \log a$. Differentiating $\frac{mdy}{y} = \log a dx$, or $dy = \frac{a^x \log a dx}{m}$, remembering that $y = a^x$, and that $\log a$ is constant. Q. E. D.

61. COR. 1.—*The differential of an exponential function with a constant base, taken with reference to the Napierian system, is the function itself, into the logarithm of the base, into the differential of the exponent. Thus if* $y = a^x$, $dy = a^x \log a dx$.

62. COR. 2.—*If the base of the exponential is the base of the system of logarithms in reference to which the differentiation is made, we have, in general,* $dy = \frac{a^x dx}{m}$, *or in the Napierian system* $dy = e^x dx$, *since the logarithm of the base of a system, taken in that system, is* 1, *and in the Napierian system* e *is used to represent the base and* $m = 1$.

---

63. Prop.—*The differential of an exponential with a variable base is best obtained by passing to logarithms, and then differentiating.*

ILL.—Let $u = y^x$. Passing to logarithms, $\log u = x \log y$. Differentiating, we have *in general*, $\frac{mdu}{u} = \log y\, dx + \frac{mxdy}{y}$, whence $du = \frac{u \log y\, dx}{m} + \frac{u\, x\, dy}{y} = \frac{y^x \log y dx}{m} + \frac{y^x x dy}{y}$. If the logarithms are taken in the Napierian system, $m = 1$,

---

* What this relation is, it does not concern us at present to know. It will be determined hereafter.

and $du = y^x \log y\, dx + y^{x-1}x dy$. If in addition $y = x$, so that $u = x^x$, $du = x^x(\log x + 1)dx$.

EXERCISES.

[NOTE.—The following exercises are designed to familiarize the rules for differentiating logarithmic and exponential functions, and give the needed facility in applying them.]

Ex. 1. Differentiate $u = x \log x$.

$$du = \log x dx + m dx, \text{ or } (\log x + 1)dx.$$

Ex. 2. Differentiate $u = \log x^2$. $\quad du = 2m\frac{dx}{x}$, or $\frac{2dx}{x}$.

Ex. 3. Differentiate $u = \log^2 x$. $\quad du = 2m \log x\frac{dx}{x}$, or $2 \log x\frac{dx}{x}$.

Ex. 4. Differentiate $u = x^{x^x}$.

$$du = x^{x^x}x^x \left\{ \log x(\log x + 1) + \frac{1}{x} \right\} dx.*$$

Ex. 5. Differentiate $u = a^{\log x}$. $\quad du = \frac{a^{\log x} \log a}{x}dx.$

Ex. 6. Differentiate $u = \log \sqrt{1 - x^2}$. $\quad du = -\frac{xdx}{1 - x^2}$.

Ex. 7. Differentiate $u = \log (3x^2 + x)$. $\quad du = \frac{6x + 1}{3x^2 + x}dx$.

Ex. 8. Differentiate $u = \log (x + \sqrt{1 + x^2})$. $\quad du = \frac{dx}{\sqrt{1 + x^2}}$.

Ex. 9. What is the differential of $u = a^{x^2}$ in the common system when a is the base of the system? $\quad du = \frac{2}{m}a^{x^2}xdx$.

Ex. 10. Differentiate $u = e^{\log x}$, in the common system, e being the base of the Napierian system. $\quad du = \frac{udx}{x} \log e = \frac{mudx}{x}$.

SUG.—If the student has studied the subject of logarithms as usually presented in our higher Algebras, he has learned that the common logarithm of the Napierian base is the modulus of the common system; *i. e.*, in this example $\log e = m$. This fact will also appear hereafter.

Ex. 11. Differentiate $u = \log \frac{\sqrt{x^2 + 1} - x}{\sqrt{x^2 + 1} + x}$.

SUG.—First rationalize the denominator of the fraction, obtaining $u = \log (\sqrt{x^2+1} - x)^2 = 2 \log(\sqrt{x^2+1} - x)$, and then differentiate. $\quad du = -\frac{2dx}{\sqrt{x^2 + 1}}$.

---

* The student will observe for himself whether common or Napierian logarithms are used.

64. SCH.—The differentiation of algebraic functions is often performed with greater facility by first passing to logarithms.

Ex. 12. Differentiate $u = \frac{1+x^2}{1-x^2}$.

SUG.—Passing to logarithms we have $\log u = \log(1+x^2) - \log(1-x^2)$. Differentiating, $\frac{du}{u} = \frac{2xdx}{1+x^2} - \frac{-2xdx}{1-x^2} = \frac{4xdx}{(1+x^2)(1-x^2)}$. $\therefore du = \frac{4xdx}{(1+x^2)(1-x^2)}u$ $= \frac{4xdx}{(1+x^2)(1-x^2)} \times \frac{1+x^2}{1-x^2} = \frac{4xdx}{(1-x^2)^2}$. This example illustrates the method referred to in the scholium, although the student will find the direct method quite as expeditious.

Ex. 13. Differentiate $u = x(a^2+x^2)\sqrt{a^2-x^2}$, by first passing to logarithms. $du = \frac{a^4+a^2x^2-4x^4}{\sqrt{a^2-x^2}}dx.$

Ex. 14. Differentiate $u = (a^x+1)^2$. $du = 2a^x(a^x+1)\log a dx.$

Ex. 15. Differentiate $u = \frac{a^x-1}{a^x+1}$. $du = \frac{2a^x \log a dx}{(a^x+1)^2}.$

65. COR.—*The ordinary rule* (**56**) *for differentiating a variable affected with an exponent applies when the exponent is imaginary.*

DEM.—Let $u = x^{a\sqrt{-1}}$. Passing to logarithms, $\log u = a\sqrt{-1}\log x$. Differentiating, $\frac{du}{u} = a\sqrt{-1}\frac{dx}{x}$. $\therefore du = a\sqrt{-1}\frac{udx}{x} = a\sqrt{-1}\,x^{a\sqrt{-1}-1}dx$. Q. E. D.

ILLUSTRATIVE EXAMPLES.

Ex. 1. Which increases the faster, a number or its logarithm?

SOLUTION.—Let x represent any number and y its logarithm, so that $y = \log x$. We now wish to find the relation between the contemporaneous, infinitesimal increments of x and y; *i. e.*, if the *number* (x) changes how does the *logarithm* (y) change? Hence we differentiate, and have $dy = \frac{m}{x}dx$. From this we see that the increment of the logarithm (dy) is $\frac{m}{x}$ times the increment of the number (dx). Therefore when $x < m$ the logarithm increases faster than the number; when $x > m$ the logarithm increases slower than the number; and when $x = m$ they increase equally.

[NOTE.—The student should not fail to see in every such example the real object of the Differential Calculus (**42**). In the last example the relation between finite values of the variables x and y is $y = \log. x$. The relation between the contemporaneous, infinitesimal elements of these variables is found by differentiating, this being the object of the Differential Calculus.]

Ex. 2. When the number is 2124 and is conceived as passing on to larger values by the law of growth of continuous number, *i. e.* by taking on infinitesimal increments, how much faster is the number increasing than its common logarithm? If this relative rate of change continued uniform (which it does not) while the number passed to 2125, *i. e.* increased by 1, how much would the logarithm have increased?

SOLUTION.—Letting x be any number and y its logarithm, we have found that $dy = \frac{m}{x}dx$. But m, the modulus of the common system $= .43429448$. Hence when $x = 2124$, we have $dy = \frac{.43429448}{2124}dx = .000204dx$, or the increment of the logarithm is .000204 part of the increment of the number. The number is, therefore increasing $\frac{1000000}{204}$, or about 4902 times as fast as the logarithm. Secondly, If this relative rate of change continued the same while the number passed from 2124 to 2125, *i. e.* increased by 1, the logarithm would increase once .000204, or .000204. Hence the logarithm of 2125 would be .000204 larger than the logarithm of 2124.

GEOMETRICAL ILLUSTRATION.—Let **MN** be the curve whose equation is $y = \log x$. Take **AD** $= 2124$; then will **PD** represent its logarithm. Let **DD′** represent dx; then will **P′E** represent dy.*

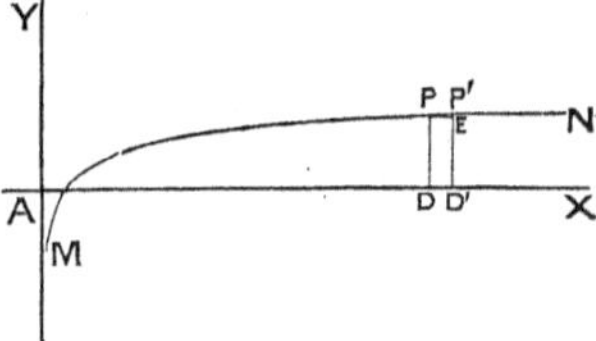

FIG. 13.

Ex. 3. The common logarithm of 327 is 2.514548. What is the logarithm of 327.12, on the hypothesis that the relative rate of change of the number and its logarithm continues uniformly the same from 327 to 327.12 that it is at 327?

SUG.—At 327 $dy = \frac{.43429448}{327}dx = .001328dx$. Now as the number 327 increases .12 to become 327.12; according to the hypothesis the logarithm increases .12 times .001328 or .000159. Hence the logarithm of 327.12 is 2.514707.

SCH.—The hypothesis that the relative rate of change of a number and its logarithm continues constant for comparatively small changes in the number, is sufficiently accurate for practical purposes, and is the assumption made in using the *tabular difference* in the table of logarithms, as explained in THE COMPLETE SCHOOL ALGEBRA (***125***), and in the introduction to the table of logarithms (***14***) in the volume on Geometry and Trigonometry.

Ex. 4. What should be the tabular difference in the table of logarithms for numbers between 2688 and 2689? *Ans.*, .00016156+.

QUERY.—How is it that the tabular difference found in the table of logarithms for

* The figure is necessarily out of proportion, as the true relation of y and x requires that **AD** be nearly 700 times as long as **PD**.

numbers between 2688 and 2689, is 161? Show how the method of using this tabular difference makes the result agree substantially with the method of interpolating now being presented.

Ex. 5. According to the arrangement of our common tables, show that the tabular difference corresponding to 7487 is 58.

SECTION III.

Differentiation of Trigonometrical and Circular Functions.

TRIGONOMETRICAL FUNCTIONS.

66. Prop.—*The differential of the sine of an arc (or angle) is the cosine of the same arc into the differential of the arc.*

DEM.—Let x be any arc (or angle) and y its sine, *i. e.* let $y = \sin x$. If x takes an infinitesimal increment (dx), let dy represent the contemporaneous infinitesimal increment of y. Then the consecutive state of the function is

$$y + dy = \sin(x + dx) = \sin x \cos dx + \sin dx \cos x.$$

Now $\cos dx = 1$, since as an angle grows less its cosine approaches the radius in value, and *at the limit, is* radius. Moreover, as an angle grows less the sine and the corresponding arc approach equality, and *at the limit* we have $\sin dx = dx$.* The consecutive state may therefore be written $y + dy = \sin x + \cos x\, dx.$

From this subtract $y = \sin x$

and we have $dy = \cos x\, dx,$

which, being the difference between two consecutive states of the function, is the differential. Q. E. D.

---

67. Prop.—*The differential of the cosine of an arc (or angle) has the opposite algebraic sign from the function, and is numerically equal to the sine of the same arc into the differential of the arc.*

DEM.—Let x be any arc (or angle) and y its cosine, so that $y = \cos x$. Since $\cos x = \sin(90° - x)$ we have $y = \sin(90° - x)$. Differentiating this by the preceding proposition, we obtain

$$dy = \cos(90° - x) \times d(90° - x) = \cos(90° - x)(-dx) = -\sin x\, dx,$$

since $d(90° - x) = -dx$, and $\cos(90° - x) = \sin x$. Q. E. D.

68. SCH.—The opposition in the signs of the differential of the cosine, and of the corresponding arc, signifies that they are decreasing functions of each other (***40***); *i. e.*, if one takes an *increment* the other suffers a *decrement.*

---

* The student may be inclined to say that *at the limit* $\sin dx = 0$. This is true, and no error would follow from the assumption; but the statement in the demonstration is equally true, and we consider $\sin dx = dx$ instead of $= 0$, simply because we do not wish to have dx vanish from the formula, our object being to find the relation between dy and dx.

69. Prop.—*The differential of the tangent of an arc (or angle) is the square of the secant of the same arc into the differential of the arc; or for the square of the secant we may write the reciprocal of the square of the cosine.*

Dem.—Let $y = \tan x$. Now $\tan x = \frac{\sin x}{\cos x}$, whence $y = \frac{\sin x}{\cos x}$. Differentiating this, observing that $\frac{\sin x}{\cos x}$ is a fraction with a variable numerator and denominator, and hence can be differentiated by the rule for fractions (**54**), and the two propositions (**66, 67**), we have $dy = \frac{\cos x\, d(\sin x) - \sin x\, d(\cos x)}{\cos^2 x} = \frac{\cos^2 x\, dx + \sin^2 x\, dx}{\cos^2 x}$ $= \frac{\cos^2 x + \sin^2 x}{\cos^2 x}dx = \frac{1}{\cos^2 x}dx = \sec^2 x\, dx$. Q. E. D.

Another Demonstration.—The consecutive state of the function $y = \tan x$, being $y + dy = \tan(x + dx) = \frac{\tan x + \tan dx}{1 - \tan x \tan dx} = \frac{\tan x + dx}{1 - \tan x\, dx}$, the difference between the two states, *i. e.* the differential is $dy = \frac{\tan x + dx}{1 - \tan x\, dx} - \tan x =$ $\frac{1 + \tan^2 x}{1 - \tan x\, dx}dx = (1 + \tan^2 x)dx = \sec^2 x\, dx$. [Let the student give the detailed explanation of the process.]

---

70. Prop.—*The differential of the cotangent of an arc (or angle) has the opposite algebraic sign from the function, and is numerically equal to the square of the cosecant of the same arc into the differential of the arc; or for the square of the cosecant we may write the reciprocal of the square of the sine.*

Dem.—Let $y = \cot x = \tan(90° - x)$. Differentiating by the last proposition $dy = \sec^2(90° - x) \times d(90° - x) = \operatorname{cosec}^2 x(-dx) = -\operatorname{cosec}^2 x\, dx$, or $-\frac{1}{\sin^2 x}dx$. Q. E. D.

[Let the student demonstrate this rule by remembering that $y = \cot x = \frac{\cos x}{\sin x}$, and also by taking the difference between the consecutive states $y = \cot x$, and $y + dy = \cot(x + dx)$, developing and reducing as in the second demonstration under (**69**).]

Query.—What is the significance of the opposition in signs?

---

71. Prop.—*The differential of the secant of an arc (or angle) is the tangent of the same arc into its secant into the differential of the arc.*

Dem.—Let $y = \sec x = \frac{1}{\cos x}$. Differentiating by (**55, 67**) we have $dy =$ $\frac{\sin x\, dx}{\cos^2 x} = \frac{\sin x}{\cos x} \times \frac{1}{\cos x} \times dx = \tan x \sec x dx$. Q. E. D.

72. Prop.—*The differential of the cosecant of an arc (or angle) has the opposite algebraic sign from the function, and is numerically equal to the cotangent of the same arc into its cosecant into the differential of the arc.*

DEM.—Let $y = \operatorname{cosec} x = \sec(90° - x)$. Differentiating by the last proposition, $dy = \tan(90° - x)\sec(90° - x)d(90° - x) = \cot x \operatorname{cosec} x(-dx) = -\cot x \operatorname{cosec} x dx$. Q. E. D.

[Let the student demonstrate this proposition from the relation $y = \operatorname{cosec} x = \frac{1}{\sin x}$.]

QUERY.—What is the significance of the opposition in signs?

---

73. Prop.—*The differential of the versed-sine of an arc (or angle) is the sine of the same arc into the differential of the arc.*

DEM.—Let $y = \operatorname{vers} x = 1 - \cos x$. Differentiating by (**67**), $dy = \sin x dx$. Q. E. D.

QUERY.—Why should the differential of the versed-sine be numerically the same as the differential of the cosine, but have an opposite sign? Illustrate geometrically.

---

74. Prop.—*The differential of the coversed-sine of an arc (or angle) has the opposite algebraic sign from the function, and is numerically equal to the cosine of the same arc into the differential of the arc.*

DEM.—(Similar to the preceding.)

QUERY.—Why should the coversed-sine have the same differential as the sine, but with an opposite algebraic sign? Illustrate geometrically.

EXERCISES.

Ex. 1. Differentiate $u = \sin x \cos x$.

SUG.—Observe that we have here the product of two variables, viz., $\sin x$ and $\cos x$. Hence $du = \cos x\, d(\sin x) + \sin x\, d(\cos x) = \cos^2 x\, dx - \sin^2 x\, dx = (\cos^2 x - \sin^2 x)dx$, or $(2\cos^2 x - 1)dx$, or $(1 - 2\sin^2 x)dx$, or $\cos 2x\, dx$.

Ex. 2. Differentiate $u = \cos^3 x$.

SUG.—Observe that this is the cube of the variable $\cos x$. Hence apply (**56**) and we have $du = 3\cos^2 x\, d(\cos x) = -3\cos^2 x \sin x\, dx = 3(\sin^3 x - \sin x)dx$.

Ex. 3. Differentiate $u = \tan 5x$.

SUG. $du = \sec^2 5x d(5x) = 5\sec^2 5x\, dx$.

Ex. 4. Differentiate $u = \cot^2 x^3$. $du = -6x^2 \cot x^3 \operatorname{cosec}^2 x^3\, dx$.

Ex. 5. Differentiate $u = \sin^3 x \cos x$.

$du = \sin^2 x(3 - 4\sin^2 x)dx$.

Ex. 6. Differentiate $u = 3 \sin^4 x$. $du = 12 \sin^3 x \cos x \, dx$.

Ex. 7. Differentiate $u = \cos mx$. $du = - m \sin mx \, dx$.

Ex. 8. Differentiate $u = \sin 3x \cos 2x$.

$$du = (3 \cos 3x \cos 2x - 2 \sin 3x \sin 2x)dx.$$

Ex. 9. Differentiate $u = \sec^2 5x$. $du = 10 \sec^2 5x \tan 5x \, dx$.

Ex. 10. Differentiate $u = \tan^n nx$. $du = n^2 \tan^{n-1} nx \sec^2 nx \, dx$.

Ex. 11. Differentiate $u = \log \sin x$.

SOLUTION.—We have here a logarithm to differentiate, *i. e.* the logarithm of $\sin x$. Hence the differential is the differential of $\sin x$, divided by $\sin x$, in the Napierian system, or m times this, in the common system. Therefore $du = \frac{md(\sin x)}{\sin x} = \frac{m \cos x \, dx}{\sin x} = m \cot x \, dx$, or in the Napierian system, $\cot x \, dx$.

Ex. 12. Differentiate $u = \log \cos x$.

$$du = - m \tan x \, dx, \text{ or } - \tan x \, dx.$$

Ex. 13. Differentiate $u = \log \tan x$.

$$du = \frac{\sec^2 x}{\tan x} dx = \frac{dx}{\sin x \cos x} = \frac{2dx}{\sin 2x}.$$

Ex. 14. Differentiate $u = \log \cot x$.

Ex. 15. Differentiate $u = \log \sec x$.

Ex. 16. Differentiate $u = \log \operatorname{cosec} x$. $du = - \cot x \, dx$.

Ex. 17. Differentiate $u = e^x \cos x$, e being the Napierian base.

SUG'S. $du = e^x d(\cos x) + \cos x \, d(e^x) = - e^x \sin x \, dx + e^x \cos x \, dx = e^x(\cos x - \sin x)dx$.

Ex. 18. Differentiate $u = xe^{\cos x}$.

SUG'S. $du = e^{\cos x}dx + xe^{\cos x}d(\cos x) = e^{\cos x}dx - xe^{\cos x} \sin x \, dx = e^{\cos x}(1 - x \sin x)dx$.

Ex. 19. Differentiate $u = \frac{e^{ax}(a \sin x - \cos x)}{a^2 + 1}$.

$$du = e^{ax} \sin x \, dx.$$

Ex. 20. Differentiate $u = \log \sqrt{\sin x} + \log \sqrt{\cos x}$.

SUG'S. $u = \frac{1}{2} \log \sin x + \frac{1}{2} \log \cos x$. $\therefore du = \frac{1}{2}(\cot x - \tan x)dx = \frac{dx}{\tan 2x}$.

ILLUSTRATIVE EXAMPLES.

[NOTE.—The object of these examples is to still farther illustrate the meaning of the process of differentiation.]

Ex. 1. Which changes the faster an arc or its sine? What is the relative rate of change? When is the disparity greatest and when least? What is the relative rate of change when the arc is 60°? When 20°? When 80°?

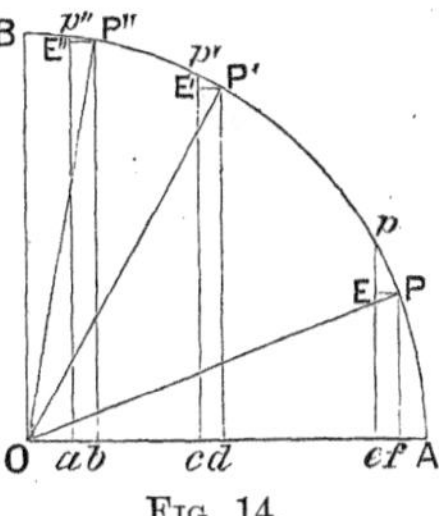

FIG. 14.

SOLUTION.—From $y = \sin x$, we have by differentiating, $dy = \cos x dx$. The meaning of this is, that if the arc (x, **AP**) takes an infinitesimal increment (dx, **P**p) the sine (y, **P**f) takes an infinitesimal increment (dy, p**E**) which is $\cos x$ times the increment of the arc. Now $\cos x$ is, in general, less than unity, so that the increment of the sine is, in general, less than the contemporaneous increment of the arc. But as x grows less $\cos x$ becomes greater and approaches unity as x approaches 0. So, also, $\cos x$ approaches 0 as x approaches 90°. Hence the disparity between the contemporaneous increments of an arc and its sine is less as the arc is less, disappears when the arc is 0, and becomes infinite when the arc is 90°. For $x = 0$ we may, therefore, consider the arc and its sine to be increasing at equal rates. For $x = 90°$, the arc is increasing infinitely faster than its sine. When $x = 60°, \cos x = \frac{1}{2}$. Hence at 60° the sine is increasing just $\frac{1}{2}$ as fast as the arc. In the figure, letting **P**$'p'$ represent dx, p'**E**$'$ represents dy and p'**E**$' = \frac{1}{2}$**P**$'p'$. When $x = 20°$ $\cos x = .94$ nearly. Hence at 20° the sine is increasing .94 as fast as the arc. At 80°, the sine is increasing only about .17 as fast as the arc. These facts are illustrated in the figure.

Ex. 2. Assuming that the relative rate of increase remains constantly the same as at 40°, how much does the sine increase when the arc increases from 40° to 40° 10′? What when the arc increases to 41°?

SUG.—Since the arc of $10' = \frac{3.14159}{180 \times 6} = .0029088$; we find the increase of the sine, on the above hypothesis, to be .002228, which is slightly in excess of the real increase, as will be found by examining a table of natural sines in which the decimals are extended to 7 places. The table gives .0022156.

At the same rate of increase the sine of 41° should be .013369 above the sine of 40°; whereas from a table the increase is found to be .0132714.

[The student will observe that the cause of this disagreement is that the relative rate of increase of the sine as compared with its arc, is greater at 40° than at any point between 40° and 40° 10′, or at any point after 40°.]

Ex. 3. The natural tangent of 27° 20′ is .5168755. Assuming that the relative rate of increase of the tangent as compared with its arc remains the same as at this point, for the next 25″ increase of the arc, what is the natural tangent of 27° 20′ 25″? *Ans.*, .517029.

Ex. 4. Which increases faster, the arc or its tangent? When is this difference greatest? When least? What is the value of the arc when the tangent is increasing just twice as fast as the arc?

Answer to the last, 45°.

Ex. 5. The natural cosine of 5° 31′ is .995368. Assuming that the relative rate of change of the cosine as compared with the arc remains the same as at 5° 31′, while the arc increases to 5° 32′, what is the cosine of 5° 32′? *Ans.*, .995340.

Ex. 6. At 36° what is the relative rate of increase of the arc and the logarithm of its tangent?

SUG.—From $u = \log \tan x$, we have $du = m(\tan x + \cot x)dx$. When $x = 36°$ this becomes $du = .43429 \times 2.102925dx$; or the logarithm of the tangent increases about .91 times as fast as the arc.

Ex. 7. The logarithmic cosine of 67° 30′ is 9.582840. Assuming that the relative rate of change of the logarithmic cosine and the arc remains the same as at this point while the arc passes to 67° 31′, what is the logarithmic cosine of the latter arc? *Ans.*, 9.582535.

Ex. 8. The log cot 58°21′ = 9.789868. On the same assumption as above, what is the decrease of this logarithm for 1 second increase in the arc? *Ans.*, .00000471.

Ex. 9. The log cos 42°14′ = 9.869474. What is the corresponding tabular difference?

Ex. 10. At what rate relative to its velocity, is a point in the circumference of a wheel revolving in a vertical plane, ascending, when it is 60° above the horizontal plane through the centre of motion?

Ans., One half as fast.

CIRCULAR FUNCTIONS.

75. Prop.—*The differential of an arc in terms of its sine is the differential of the sine divided by the square root of* 1 *minus the square of the sine; or the differential of the sine divided by the cosine.*

DEM.—Let $y = \sin^{-1}x$*, whence $x = \sin y$. Differentiating and finding the value of dy, we have $dy = \dfrac{dx}{\cos y}$. But $\cos y = \sqrt{1 - \sin^2 y} = \sqrt{1 - x^2}$. $\therefore dy = \dfrac{dx}{\sqrt{1 - x^2}}$. Q. E. D.

* This notation is explained in the Trigonometry of this series. It means simply "y = the arc whose sine is x, and hence $y = \sin^{-1}x$ is equivalent to $x = \sin y$."

76. SCH.—The student should not fail to observe the essential identity of this proposition with (**66**). Thus, when we differentiate $u = \sin x$, we get $du = \cos x dx$, which expresses the differential of the *sine* (u) in terms of its arc (x). From this we have $dx = \frac{du}{\cos x} = \frac{du}{\sqrt{1-u^2}}$, which expresses the differential of the *arc* (x) in terms of the *sine* (u). The one conception is the converse of the other.

---

77. Prop.—*The differential of an arc in terms of its cosine has the opposite sign from the function, and is numerically equal to the differential of the cosine divided by the square root of* 1 *minus the square of the cosine; or the differential of the cosine divided by the sine.*

DEM.—Let $y = \cos^{-1} x$, whence $x = \cos y$. Differentiating, and finding the value of dy, we have $dy = -\frac{dx}{\sin y} = -\frac{dx}{\sqrt{1-x^2}}$. Q. E. D.

78. SCH.—Compare this and the following propositions with their equivalents in Trigonometrical functions, as was done in the case of the preceding proposition.

---

79. Prop.—*The differential of an arc in terms of its tangent is the differential of the tangent divided by* 1 *plus the square of the tangent.*

DEM.—Let $y = \tan^{-1} x$, whence $x = \tan y$. Differentiating and finding the value of dy, we have $dy = \frac{dx}{\sec^2 y} = \frac{dx}{1+x^2}$, since $\sec^2 y = 1 + \tan^2 y = 1 + x^2$. Q. E. D.

---

80. Prop.—*The differential of an arc in terms of its cotangent has the opposite sign from the function, and is numerically equal to the differential of the cotangent divided by* 1 *plus the square of the cotangent.*

DEM.—Let $y = \cot^{-1} x$, whence $x = \cot y$. Differentiating, and finding the value of dy, we have $dy = -\frac{dx}{\operatorname{cosec}^2 y} = -\frac{dx}{1+x^2}$. Q. E. D.

---

81. Prop.—*The differential of an arc in terms of its secant is the differential of the secant divided by the secant into the square root of the square of the secant minus* 1.

DEM.—Let $y = \sec^{-1} x$, whence $x = \sec y$. Differentiating and finding the value of dy we have $dy = \frac{dx}{\sec y \tan y} = \frac{dx}{x\sqrt{x^2-1}}$, since $\tan y = \sqrt{\sec^2 y - 1} = \sqrt{x^2-1}$. Q. E. D.

82. Prop.—*The differential of an arc in terms of its cosecant has the opposite sign from the function, and is numerically equal to the differential of the cosecant divided by the cosecant into the square root of the square of the cosecant, minus* 1.

DEM.—Let $y = \text{cosec}^{-1}x$, whence $x = \text{cosec}\, y$. Differentiating, and finding the value of dy, we have $dy = -\frac{dx}{\text{cosec}\, y \cot y} = -\frac{dx}{x\sqrt{x^2-1}}$, since $\cot y = \sqrt{\text{cosec}^2 y - 1}$ $= \sqrt{x^{\cdot} - 1}$. Q. E. D.

---

83. Prop.—*The differential of an arc in terms of its versed-sine is the differential of the versed-sine divided by the square root of twice the versed-sine minus the square of the versed-sine.*

DEM.—Let $y = \text{vers}^{-1}x$, whence $x = \text{vers}\, y$. Differentiating and finding the value of dy we have $dy = \frac{dx}{\sin y}$. But $\sin y = \sqrt{1-\cos^2 y} = \sqrt{1-(1-\text{vers}\, y)^2} =$ $\sqrt{1-(1-x)^2} = \sqrt{2x-x^2}$. Therefore, substituting, $dy = \frac{dx}{\sqrt{2x-x^2}}$. Q. E. D.

84. Prop.—*The differential of an arc in terms of its coversed-sine has the opposite sign from the function, and is numerically equal to the differential of the coversed-sine divided by the square root of twice the coversed-sine minus the square of the coversed-sine.*

DEM.—Let $y = \text{covers}^{-1}x$, whence $x = \text{covers}\, y$. Differentiating, and finding the value of dy, we have $dy = -\frac{dx}{\cos y} = -\frac{dx}{\sqrt{1-\sin^2 y}} = -\frac{dx}{\sqrt{1-(1-\text{covers}\, y)^2}} =$ $-\frac{dx}{\sqrt{1-(1-x)^2}} = -\frac{dx}{\sqrt{2x-x^2}}$. Q. E. D.

EXERCISES.

Ex. 1. Differentiate $y = \sin^{-1}\frac{x}{r}$; $y = \cos^{-1}\frac{x}{r}$; $y = \tan^{-1}\frac{x}{r}$; $y =$ $\cot^{-1}\frac{x}{r}$; $y = \sec^{-1}\frac{x}{r}$; $y = \text{cosec}^{-1}\frac{x}{r}$; $y = \text{vers}^{-1}\frac{x}{r}$; $y = \text{covers}^{-1}\frac{x}{r}$.

SUG.—We have $dy = \frac{d\left(\frac{x}{r}\right)}{\sqrt{1-\left(\frac{x}{r}\right)^2}}$, by (**75**). Now since $d\left(\frac{x}{r}\right) = \frac{dx}{r}$, we have $dy = \frac{\frac{dx}{r}}{\sqrt{1-\frac{x^2}{r^2}}} = \frac{dx}{\sqrt{r^2-x^2}}$. In like manner $dy = d\left(\cos^{-1}\frac{x}{r}\right) = -\frac{dx}{\sqrt{r^2-x^2}}$;

$dy = d\left(\tan^{-1}\frac{x}{r}\right) = \frac{rdx}{r^2 + x^2}$; $dy = d\left(\cot^{-1}\frac{x}{r}\right) = -\frac{rdx}{r^2 + x^2}$; $dy = d\left(\sec^{-1}\frac{x}{r}\right) = \frac{rdx}{x\sqrt{x^2 - r^2}}$; $dy = d\left(\text{cosec}^{-1}\frac{x}{r}\right) = -\frac{rdx}{x\sqrt{x^2 - r^2}}$; $dy = d\left(\text{vers}^{-1}\frac{x}{r}\right) = \frac{dx}{\sqrt{2rx - x^2}}$; and $dy = d\left(\text{covers}^{-1}\frac{x}{r}\right) = -\frac{dx}{\sqrt{2rx - x^2}}$.

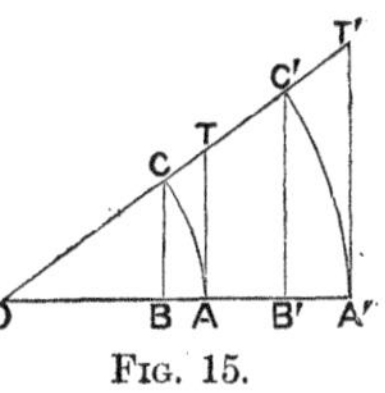

Fig. 15.

Geometrical Illustration.—Let $OA = 1$, and $OA' = r$. Let $y =$ arc CA (to radius 1), and $x = C'B'$ the sine of the same number of degrees as y, but to a radius r. Now $CB = \frac{C'B'}{r} = \frac{x}{r}$, and we have y (or CA) $= \sin^{-1} CB = \sin^{-1}\frac{C'B'}{r} = \sin^{-1}\frac{x}{r}$, the arc ($y$) being taken to radius 1 while the sine x is taken to the radius r.

Ex. 2. Differentiate $\frac{y}{r} = \sin^{-1}\frac{x}{r}$; $\frac{y}{r} = \cos^{-1}\frac{x}{r}$; $\frac{y}{r} = \tan^{-1}\frac{x}{r}$; $\frac{y}{r} = \cot^{-1}\frac{x}{r}$; $\frac{y}{r} = \sec^{-1}\frac{x}{r}$; $\frac{y}{r} = \text{cosec}^{-1}\frac{x}{r}$; $\frac{y}{r} = \text{vers}^{-1}\frac{x}{r}$; and $\frac{y}{r} = \text{covers}^{-1}\frac{x}{r}$.

Results in order: $dy = \frac{rdx}{\sqrt{r^2 - x^2}}$; $dy = -\frac{rdx}{\sqrt{r^2 - x^2}}$; $dy = \frac{r^2dx}{r^2 + x^2}$; $dy = -\frac{r^2dx}{r^2 + x^2}$; $dy = \frac{r^2dx}{x\sqrt{x^2 - r^2}}$; $dy = -\frac{r^2dx}{x\sqrt{x^2 - r^2}}$; $dy = \frac{rdx}{\sqrt{2rx - x^2}}$; and $dy = -\frac{rdx}{\sqrt{2rx - x^2}}$.

Geometrical Illustration.—In *Fig.* 15 let $OA' = r$, $C'A' = y$, and $C'B' = x$. Now if $OA = 1$, we have $CA = \frac{y}{r}$, and $CB = \frac{x}{r}$. Hence $dy = d(\sin^{-1} x)$ to radius $r = \frac{rdx}{\sqrt{r^2 - x^2}}$, etc.

85. Sch.—The results in the last example will be seen to correspond with those in *Ex.* 1, by noticing that in *Ex.* 1 y represents CA, whereas in *Ex.* 2 it represents $C'A'$. Now an increment of $C'B'$ (which is x in both cases) which makes an increment in CA, will make r times as great an increment in $C'A'$. Hence we have but to multiply the increments of CA (the dy's) as found in *Ex.* 1, by r to get the corresponding increments in $C'A'$, which are the dy's in *Ex.* 2.

Ex. 3. Differentiate $u = \tan^{-1}\frac{x}{y}$. $\qquad du = \frac{ydx - xdy}{y^2 + x^2}$.

Ex. 4. Differentiate $u = \sin^{-1}(2x\sqrt{1 - x^2})$. $\qquad du = \frac{2dx}{\sqrt{1 - x^2}}$.

Ex. 5. Differentiate $u = \cos^{-1}(x\sqrt{1-x^2})$.

SUG'S.—By the rule the differential of the arc u is negative, and numerically equal to the differential of its cosine, $x\sqrt{1-x^2}$, divided by the square root of 1 minus the square of its cosine. The differential of $x\sqrt{1-x^2}$ is $dx\sqrt{1-x^2} - \frac{x^2dx}{\sqrt{1-x^2}}$, and $1-(x\sqrt{1-x^2})^2 = 1-x^2+x^4$ $\therefore$ $du = -\frac{(1-2x^2)dx}{\sqrt{(1-x^2+x^4)(1-x^2)}}$.

Ex. 6. Differentiate $u = \sin^{-1}\frac{x}{\sqrt{1+x^2}}$. $\quad du = \frac{dx}{1+x^2}$.

Ex. 7. Differentiate $u = \sin^{-1}(3x - 4x^3)$. $\quad du = \frac{3dx}{\sqrt{1-x^2}}$.

Ex. 8. Differentiate $u = \text{vers}^{-1}y - \sqrt{2ry-y^2}$, understanding that $\text{vers}^{-1}y$ is taken to radius r.

SUG. $du = \frac{rdy}{\sqrt{2ry-y^2}} - \frac{rdy - ydy}{\sqrt{2ry-y^2}} = \frac{ydy}{\sqrt{2ry-y^2}}$.

Ex. 9. Differentiate $u = \tan^{-1}(\sqrt{1+x^2} - x)$.

$$du = -\frac{dx}{2(1+x^2)}.$$

Ex. 10. Differentiate $u = \log\sqrt[4]{\frac{1+x}{1-x}} + \frac{1}{2}\tan^{-1}x$.

SUG. $u = \frac{1}{4}\log(1+x) - \frac{1}{4}\log(1-x) + \frac{1}{2}\tan^{-1}x$. $\quad du = \frac{dx}{1-x^4}$.

Ex. 11. Differentiate $y = \sin^{-1}mx$. $\quad dy = \frac{mdx}{\sqrt{1-m^2x^2}}$.

Ex. 12. Differentiate $y = e^{\tan^{-1}x}$. $\quad dy = e^{\tan^{-1}x}\frac{dx}{1+x^2}$.

Ex. 13. Differentiate $y = \tan^{-1}\frac{2x}{1-x^2}$. $\quad dy = \frac{2dx}{1+x^2}$.

Ex. 14. Differentiate $y = x^{\sin^{-1}x}$.

$$dy = x^{\sin^{-1}x}\left\{\frac{x\log x + (1-x^2)^{\frac{1}{2}}\sin^{-1}x}{x(1-x^2)^{\frac{1}{2}}}\right\}dx.$$

GENERAL SCHOLIUM.

86. The preceding sections comprise the fundamental rules of the differential calculus; and it only remains to extend and apply them, in order to complete this portion of our subject.

SECTION IV.

Successive Differentiation and Differential Coefficients.

SUCCESSIVE DIFFERENTIATION.

87. Def.—***Successive Differentials*** are differentials of differentials; or a successive differential is the difference between two consecutive states of a differential.

Ill.—Let **MN** *Fig.* 16, be a straight line whose equation is $y = ax + b$; whence $dy = adx$. Now suppose x to be considered equicrescent, and let **DD′**, **D′D″**, **D″D‴**, and **D‴D^{IV}** represent the successive equal increments. **P′E**, **P″E′**, **P‴E″**, and **P^{IV}E‴** represent the contemporaneous increments of y, *i. e.* the dy's. But *in this case* the dy's are all *equal*. Hence there being no difference between two successive states of dy, as between **P′E** and **P″E′**, there is no successive differential, or the differential of dy is 0, since dy is constant. This fact appears also from the relation $dy = adx$, in which, if we conceive dx to be constant (*i. e.*, x equicrescent), adx is constant; whence dy, which equals adx, is constant.

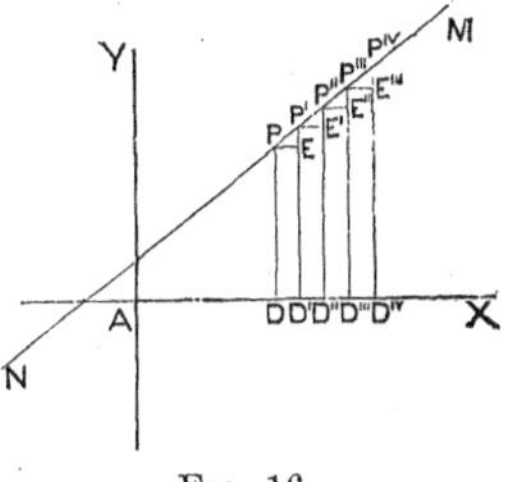

Fig. 16.

But consider in a similar manner the parabola in *Fig.* 17, whose equation is $y^2 = 2px$; whence $dy = \frac{pdx}{y}$. Still considering dx as constant, *i. e.* **DD′** $=$ **D′D″** $=$ **D″D‴** $=$ **D‴D^{IV}**, etc., it is evident that the dy's, which are represented by **P′E**, **P″E′**, **P‴E″**, etc., are *not* constant. Now the difference between any two successive values of dy, as between **P′E** and **P″E′**, is a successive differential, *i. e.* a differential of a differential. The fact that dy is a variable in this case when dx is constant is also readily seen from its value $dy = \frac{pdx}{y}$. In this pdx is constant, but y is variable. Hence dy varies inversely as y.

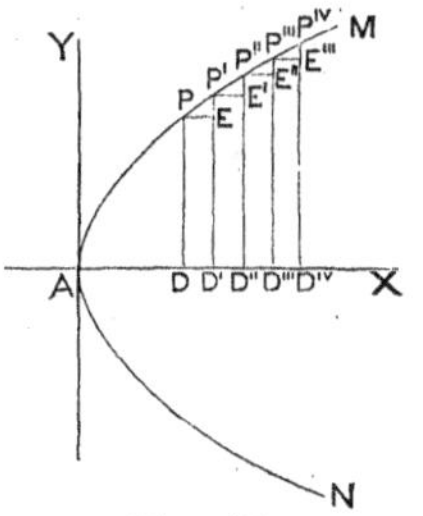

Fig. 17.

88. Def.—***A Second Differential*** is a differential of a first differential, is represented by d^2y, and read "Second differential y." ***A Third Differential*** is a differential of a second differential, is represented by d^3y, and read "Third differential y." In like manner we have fourth, fifth, etc., differentials.

Sch.—The student should be careful not to confound d^2y with dy^2. The latter is the square of dy. Nor should the superior 2 in d^2y be mistaken for an exponent: it has no analogy to an exponent. Observe the signification of the several expressions d^2y, dy^2 and $d(y^2)$. The latter is equivalent to $2ydy$.

89. Prop.—*Second differentials are formed by differentiating first differentials, third differentials by differentiating second differentials, etc., according to the rules already given.*

This proposition is self-evident, since the differentials are expressed as algebraic, trigonometric, logarithmic, or exponential functions, the rules for differentiating which are those heretofore given.

Ex. 1. Produce the several successive differentials of $y = ax^4$.

SOLUTION.—Differentiating $y = ax^4$, we have $dy = 4ax^3dx$. Differentiating this differential remembering that $d(dy)$, *i. e.* the differential of dy is written d^2y, and that dx is constant, we have $d^2y = 12ax^2dx\ dx$, or $12ax^2\ dx^2$. In like manner differentiating $d^2y = 12ax^2dx^2$, we have $d^3y = 24ax\ dx^3$. And again $d^4y = 24adx^4$. Here the operation terminates, since d^4y being equal to $24adx^4$ is constant.

Ex. 2. Produce the several successive differentials of $y = 8x^4 - 3x^3 - 5x$.

$$\textit{Results,}\begin{cases} dy = (32x^3 - 9x^2 - 5)dx, \\ d^2y = (96x^2 - 18x)dx^2, \\ d^3y = (192x - 18)dx^3, \\ d^4y = 192dx^4. \end{cases}$$

Ex. 3. Produce the first six successive differentials of $y = \sin x$.

$$\textit{Results,}\begin{cases} dy = \cos x\ dx,\ d^2y = -\sin x\ dx^2, \\ d^3y = -\cos x\ dx^3,\ d^4y = \sin x\ dx^4, \\ d^5y = \cos x\ dx^5,\ d^6y = -\sin x\ dx^6. \end{cases}$$

QUERY.—Does the above process ever terminate?

Ex. 4. What is the 3rd differential of $y = x^n$?

$$d^3y = n(n-1)(n-2)x^{n-3}dx^3.$$

Ex. 5. Produce the 4th differential of $y = ax^{\frac{1}{2}}$.

$$d^4y = -\frac{15a\ dx^4}{16\sqrt{x^7}}.$$

Ex. 6. Produce the first six successive differentials of $y = \cos x$.

Ex. 7. Produce the first four successive differentials of $y = \log x$, in the common system.

Results, $dy = \frac{m\ dx}{x}$, $d^2y = -\frac{m\ dx^2}{x^2}$, $d^3y = \frac{2m\ xdx^3}{x^4} = \frac{2m\ dx^3}{x^3}$, $d^4y = -\frac{6m\ dx^4}{x^4}$.

Ex. 8. Produce the first four successive differentials of $y = \log(1+x)$, in the common system.

Results, $dy = \frac{m\,dx}{1+x}$, $d^2y = -\frac{m\,dx^2}{(1+x)^2}$, $d^3y = \frac{2m(1+x)\,dx^3}{(1+x)^4} = \frac{2m\,dx^3}{(1+x)^3}$,

$d^4y = -\frac{6m\,dx^4}{(1+x)^4}$.

Ex. 9. Produce the fourth differential of $y = e^x$.

$$d^4y = e^x dx^4.$$

Ex. 10. Produce the fourth differential of $y = a^x$, in the common system.

$$d^4y = \frac{\log^4 a}{m^4} a^x dx^4.$$

DIFFERENTIAL COEFFICIENTS.

90. Defs.—***A First Differential Coefficient*** is the ratio of the differential of a function to the differential of its variable, and is represented thus, $\frac{dy}{dx}$, y being a function of the variable x.

A Second Differential Coefficient is the ratio of the second differential of a function to the square of the differential of its variable, and is expressed thus, $\frac{d^2y}{dx^2}$.

A Third Differential Coefficient is the ratio of the third differential of the function to the cube of the differential of its variable, and is represented thus, $\frac{d^3y}{dx^3}$. In like manner the nth differential coefficient is $\frac{d^ny}{dx^n}$.

Ill.—Having $y = ax^5$, we obtain $\frac{dy}{dx} = 5ax^4$. In strict propriety $\frac{dy}{dx}$ is a symbol representing the general conception of the ratio of an infinitesimal increment of the function to the contemporaneous infinitesimal increment of its variable; and $5ax^4$ is, in this case, its value. But it is customary to speak of either as the differential coefficient. The appropriateness of the term differential coefficient arises from the fact that the $5ax^4$ is the coefficient by which the differential of the variable has to be multiplied in order to produce the differential of the function. Strictly, therefore, the differential coefficient is the coefficient of the differential of the variable; but it is customary to speak of it as the differential coefficient of* the function.

* The "of" meaning, perhaps, "derived from," or "appertaining to."

Ex. 1. Given $y = ax^3 - x^2$, to find the 1st, 2nd, and 3rd differential coefficients.

Results, $\frac{dy}{dx} = 3ax^2 - 2x$, $\frac{d^2y}{dx^2} = 6ax - 2$, $\frac{d^3y}{dx^3} = 6a$.

Ex. 2. Given $y = \frac{1+x}{1-x}$, to find the 5th differential coefficient.

$$\frac{d^5y}{dx^5} = \frac{240}{(1-x)^6}.$$

Ex. 3. Given $y = e^{ax}$, to find the 4th differential coefficient.

$$\frac{d^4y}{dx^4} = a^4e^{ax}.$$

Ex. 4. Given $y = a^x$, to find the 5th differential coefficient.

$$\frac{d^5y}{dx^5} = \log^5 a\ a^x.$$

Ex. 5. Produce the first three successive differential coefficients of $y = \tan x$.

Results, $\frac{dy}{dx} = \sec^2 x$, $\frac{d^2y}{dx^2} = 2\sec^2 x \tan x$, $\frac{d^3y}{dx^3} = 4\sec^2 x \tan^2 x + 2\sec^4 x$.

[NOTE.—The 10 examples in the former part of this section may be used for further illustration of this subject, if desired.]

Ex. 6. Given $y^2 = 2px$, to form the third differential coefficient.

SOLUTION.—Differentiating and finding the coefficient, we have $\frac{dy}{dx} = \frac{p}{y}$. To differentiate $\frac{dy}{dx}$ we have but to remember that dy is a variable, that its differential is d^2y, and that dx is constant and hence remains the same; whence $d\left(\frac{dy}{dx}\right) = \frac{d^2y}{dx}$. Differentiating $\frac{p}{y}$, we have $d\left(\frac{p}{y}\right) = -\frac{pdy}{y^2}$. Hence $\frac{d^2y}{dx} = -\frac{pdy}{y^2}$. But the *second* differential coefficient is the ratio of the second differential of the function to the *square* of the differential of its variable. Hence dividing by dx, we have $\frac{d^2y}{dx^2} = -\frac{p\frac{dy}{dx}}{y^2}$. But $\frac{dy}{dx} = \frac{p}{y}$. Substituting this value, $\frac{d^2y}{dx^2} = -\frac{p\frac{p}{y}}{y^2} = -\frac{p^2}{y^3}$. In a similar manner we find $\frac{d^3y}{dx^3} = \frac{3p^3}{y^5}$.

Ex. 7. From $A^2y^2 + B^2x^2 = A^2B^2$, find $\frac{d^2y}{dx^2}$. $\qquad \frac{d^2y}{dx^2} = -\frac{B^4}{A^2y^3}$.

Ex. 8. From $xy = c^2$ show that $\frac{d^2y}{dx^2} = \frac{2c^2}{x^3}$.

Ex. 9. From $y = \frac{x^3}{1-x}$ show that $\frac{d^4y}{dx^4} = \frac{24}{(1-x)^5}$.

Ex. 10. From $y^2 = \sec 2x$ show that $y + \frac{d^2y}{dx^2} = 3y^5$.

[NOTE.—The first differential coefficient expresses the relative rate of increase of the function and its variable, and its significance is illustrated in the examples on pages 34, 35, which it will be well to review for this purpose.]

Ex. 11. What is the third differential of $u = \frac{dy}{dx}$, dx being constant?

$$d^3u = \frac{d^4y}{dx}.$$

Ex. 12. What is the differential of $\frac{dy}{dx}$, dx not being considered constant?

Ans., $\frac{d^2ydx - d^2xdy}{dx^2}$.

SCH.—Differential coefficients are *generally* variables and hence can be differentiated. In differentiating them it is important to observe whether dx is conceived as constant or variable. To say that the first differential coefficient is *generally* variable is equivalent to saying, in the case of a curve, that the relative rate of change of the ordinate and abscissa varies for different points, as we have seen heretofore.

SECTION V.

Functions of Several Variables, Partial Differentiation, and Differentiation of Implicit and of Compound Functions.

91. When a quantity is a function of two or more variables, as $u = f(x, y)$, these variables may be independent of each other, or dependent upon each other. This distinction is marked by saying that u is a function of two or more independent,* or two or more dependent* variables as the case may be.

ILL.—Let x and y be the sides of a rectangle and u its area. Then $u = xy$, in which u is a function of the two *independent* variables x and y. That is, x may vary without causing any variation in y, and y may vary without causing a variation in x.

But suppose u to represent a rectangle which is to remain similar to a given rectangle. Now when x or y changes, the other must change also, and we have u a function of *two dependent* variables.

* These terms are here used in a slightly different sense from that hitherto assigned them.

92. DEF.—***A Partial Differential*** of a function of two or more variables is a differential taken under the hypothesis that one or more of the variables remains constant; usually we consider the partial differential under the hypothesis that *but one* of the variables changes, the others remaining constant.

93. DEF.—***A Total Differential*** of a function of two or more variables is a differential taken under the hypothesis that all the variables upon which it depends vary.

ILL.—If $u = 3ax^2y - 2y^2 + 3bx^3 - 5$, and x and y are conceived as entirely independent of each other,* it is evidently legitimate to inquire what effect upon u will be produced by a change of x or y alone. Thus, if we suppose x to take an infinitesimal increment and y to remain unchanged, the contemporaneous increment of u is $du = (6axy + 9bx^2)dx$. If, on the other hand, y takes an infinitesimal increment and x remains constant, the contemporaneous increment of u is $du = (3ax^2 - 4y)dy$.

But how is it when x and y are dependent upon each other? Evidently, if we wish to obtain the total change in u due to a change in *both* x and y, we may still *conceive* them to change *in succession*. Only, in this case, we must remember that the change in y (for example) is not independent of the change in x. That is, that dy is a function of dx, which is not the case under the former hypothesis.

94. DEF.—***A Partial Differential Coefficient*** is the ratio of a partial differential of a function of several variables, to the differential of the quantity supposed to vary.

95. DEF.—***A Total Differential Coefficient*** of a function of two or more variables is the ratio of the total differential of the function to the differential of some one of the variables; and there may be as many such coefficients as there are variables.

96. SCH.—When several *independent* variables enter into a function, we might, if we chose, consider each of them equicrescent, and its differential constant; but when the variables are mutually interdependent, it is evident that in general but *one* can be considered equicrescent.

---

97. ***Prop.***—*The total differential of a function of several variables is equal to the sum of the partial differentials.*

DEM.—Let $u = f(x, y)$. Represent the partial differential of u with respect to x, by d_xu and with respect to y by d_yu, while du represents the total differential. Now $du = f(x + dx, y + dy) - f(x, y)$; *i. e.* it is the difference between two consecutive states of the function when both x and y have taken increments. Now subtract and add $f(x + dx, y)$, which will not change the value, and we have

$$du = f(x + dx, y + dy) - f(x + dx, y) + f(x + dx, y) - f(x, y).$$

But $f(x + dx, y + dy)$† $- f(x + dx, y)$† $= d_yu$, since it is the difference between

* Something in the nature of the problem always determines whether x and y are dependent or independent.

† The st dent will notice that, however x and y are involved, the only difference between $f(x + dx,$ [illegible]$+ dy)$ and $f(x + dx, y)$ is what arises from y passing to $y + dy$.

two consecutive states of the function due to a change in y alone. So, also, $f(x + dx, y) - f(x, y) = d_x u$, for a like reason. $\therefore\ du = d_x u + d_y u$; and it is evident that a similar course of reasoning can be applied when u is a function of any number of variables. Q. E. D.

ILL.—This proposition may be perplexing to a thoughtful student, even after he has learned the demonstration. The following illustration, for the substance of which I am indebted to Price, may help him to realize its truth. Consider the amount of grain grown upon a piece of land. This is evidently a function of the *area*, the *soil* (quality of), and the *tillage*, to say nothing more. Let G represent the total amount of grain, A the area, S the soil (quality of), and T the tillage. In mathematical notation we have $G = f(A, S, T)$. Now consider A, S, and T as variable.

1st. We may inquire what effect upon G will be produced by an increment (for our present purpose infinitesimal) of A, while S and T remain constant. This will give a *partial differential* of G with respect to A. In like manner we may inquire what effect upon G, an infinitesimal increment of S, or infinitesimal increments of A and S, will produce. And again what effect will be produced by a change in any one, any two, or in *all three* of the variables, A, S, and T. The latter would be the *total* differential of G.

2nd. It is evident that A is independent of S, and T, while S and T are (probably) dependent upon each other, *i. e.* good tillage may have a larger proportionate effect upon the crop on good soil, than upon poor.

3rd. If we consider the effect upon G, produced by an infinitesimal increment of A, while S and T are constant, calling this $d_A G$; and then (not considering A as having increased) consider the effect on G of an infinitesimal increment of S, calling it $d_S G$; and then (not considering either A or S as having increased) consider the effect upon G produced by an infinitesimal increment of T, calling it $d_T G$; it may seem that

$$dG = d_A G + d_S G + d_T G$$

will not represent the total change in G when A, S, and T change together. For example, we lose the effect of the improvement in soil and tillage upon the increment of the area, and also the effect of the increased effect of better tillage upon the improvement of the soil. But it is easy to see that these effects are infinitesimals of infinitesimals, and that the effects we seek to trace are infinitesimals of the first order, hence the former must be omitted.

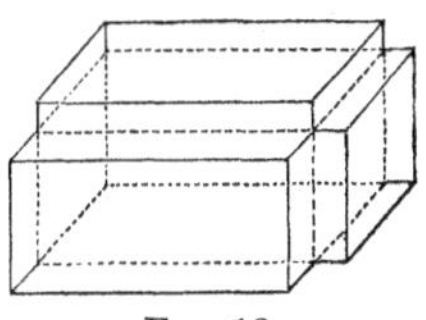

FIG. 18.

ANOTHER ILLUSTRATION to the same purpose is furnished by the parallelopipedon. Let u represent the volume of the parallelopipedon indicated by the dotted lines, of which x, y, and z are the length, breadth, and height, respectively. Then $u = f(x, y, z) = xyz$. The additions represented upon the end, side, and top respectively, are $d_x u$, $d_y u$, and $d_z u$, but it does not appear that the sum of these make up the total increment of u due to an increase of *all three* of the variables. But it *does* appear that the wanting parts will be infinitesimals of the second and third orders, when the increments of x, y, and z are infinitesimal, and as the increments represented in the figure are infinitesimals of the *first* order, the others must be omitted, in relation to the latter.

Ex. 1. On the principle of the above proposition differentiate $u =$

$3x^2y^2 - 5x^3 - 2y - 10$. Observe also that the result agrees with that obtained by the method before learned.

SOLUTION.—Differentiating with respect to x, we have $d_xu = 6xy^2dx - 15x^2dx$. Again, differentiating with respect to y, we have $d_yu = 6x^2ydy - 2dy$. Adding, $du = d_xu + d_yu = 6xy^2dx - 15x^2dx + 6x^2ydy - 2dy$. Finally, differentiating by the elementary methods, $du = 6xy^2dx + 6x^2ydy - 15x^2dx - 2dy$, a result identical with the preceding.

Ex. 2. Differentiate $u = x^y$ both by the above principle and by passing to logarithms and using the elementary methods, and compare the results.

SUG'S. $d_xu = yx^{y-1}dx$. $d_yu = x^y \log x\, dy$. $\therefore du = yx^{y-1}dx + x^y \log x\, dy$. Again $\log u = y \log x$. Hence $\frac{du}{u} = y\frac{dx}{x} + \log x\, dy$, or $du = \frac{uydx}{x} + u \log x\, dy = \frac{x^y ydx}{x} + x^y \log x\, dy = x^{y-1}y\, dx + x^y \log x\, dy$.

Ex. 3. Differentiate $u = \tan^{-1}\frac{y}{x}$, both by the method of partial differentials and by the elementary method.

SUG'S.—By partial differentiation, $d_xu = \frac{-\frac{ydx}{x^2}}{1+\frac{y^2}{x^2}} = -\frac{ydx}{x^2+y^2}$. $d_yu = \frac{\frac{dy}{x}}{1+\frac{y^2}{x^2}} = \frac{xdy}{x^2+y^2}$. $\therefore du = \frac{xdy - ydx}{x^2+y^2}$. By the elementary method, $du = \frac{d\left(\frac{y}{x}\right)}{1+\frac{y^2}{x^2}} = \frac{\frac{xdy - ydx}{x^2}}{1+\frac{y^2}{x^2}} = \frac{xdy - ydx}{x^2+y^2}$.

[NOTE.—The pupil will doubtless be led to inquire, Why use the method of partial differentiation when the elementary methods seem to be more expeditious? It is not for its use in such elementary processes that it is valuable. These examples are given only to illustrate the proposition; the *utility* of it will appear hereafter.]

Ex. 4. Differentiate as above $u = \frac{x+y}{x-y}$.

SUG'S. $d_xu = \frac{(x-y)dx - (x+y)dx}{(x-y)^2} = \frac{-2ydx}{(x-y)^2}$. $d_yu = \frac{(x-y)dy + (x+y)dy}{(x-y)^2} = \frac{2xdy}{(x-y)^2}$. $\therefore du = \frac{2(xdy - ydx)}{(x-y)^2}$.

Ex. 5. Differentiate as above $u = \sin(xy)$.

$$du = \cos(xy)[ydx + xdy].$$

Ex. 6. Differentiate as above $u = \log x^y$.

$$du = d_xu + d_yu = \frac{ydx}{x} + \log x\, dy.$$

Ex. 7. Differentiate as above $u = y^{\sin x}$.

$du = d_x u + d_y u = y^{\sin x} \log y \cos x \, dx + \sin x \, y^{\sin x - 1} dy = y^{\sin x} \log y \cos x \, dx + \sin x \dfrac{dy}{y^{\text{covers } x}}$.

Ex. 8. Differentiate as above $u = \text{vers}^{-1} \dfrac{x}{y}$.

$$du = \frac{dx}{\sqrt{2xy - x^2}} - \frac{\dfrac{x\,dy}{y}}{\sqrt{2xy - x^2}} = \frac{ydx - xdy}{y\sqrt{2xy - x^2}}.$$

Ex. 9. Differentiate as above $u = \sin(x + y)$.

$$du = \cos(x + y)[dx + dy].$$

Ex. 10. Differentiate as above $u = x^2y^2z^2$.

---

98. Notation.—Since when $u = f(x)$, $du =$ the first differential coefficient of u with respect to x, multiplied by dx, we may symbolize $d_x u$, by $\dfrac{du}{dx}dx$. So also $d_y u = \dfrac{du}{dy}dy$.

99. Prop.—*The formula for the total differential coefficient of* u *with respect to* x, *when* u = f(x, y) *is*

$$\left[\frac{du}{dx}\right] = \frac{du}{dx} + \frac{du}{dy}\frac{dy}{dx},$$

in which $\dfrac{\text{du}}{\text{dx}}$, *and* $\dfrac{\text{du}}{\text{dy}}$ *are partial differential coefficients, and the* [ ] *indicate the total coefficient.*

Dem.—We have $du = d_x u + d_y u = \dfrac{du}{dx}dx + \dfrac{du}{dy}dy$. Dividing by dx, and distinguishing the total differential coefficient by the [ ], we have $\left[\dfrac{du}{dx}\right] = \dfrac{du}{dx} + \dfrac{du}{dy}\dfrac{dy}{dx}$. Q. E. D.

Sch.—This formula has definite meaning only when y and x are mutually dependent, *i. e.* when $y = f(x)$, since otherwise $\dfrac{dy}{dx}$ is indeterminate. The formula, therefore, signifies that when $u = f(x, y)$, and $y = f_1(x)$, the total differential coefficient of u with respect to x is equal to the partial differential coefficient of u with respect to x, + the product of the partial differential coefficient of u with respect to y, multiplied by the differential coefficient of y with respect to x, obtained from the relation $y = f_1(x)$.

The following language is often used to express the relation of u to x and y in such a case; viz., u is *directly* a function of x, and *indirectly* a function of x through y.

It might seem that the relation $\left[\dfrac{du}{dx}\right] = \dfrac{du}{dx} + \dfrac{du}{dy}\dfrac{dy}{dx}$ is absurd, since by cancelling the dy, it reduces to $\left[\frac{du}{dx}\right] = 2\frac{du}{dx}$. But this is to misapprehend entirely the significance of the notation. It is to be observed that the du in $\frac{du}{dx}$, is by no means *necessarily* the same as the du in $\frac{du}{dy}$, or in $\left[\frac{du}{dx}\right]$. In $\frac{du}{dx}$, du is the increment of u due to an infinitesimal increment of x in the function $u = f(x, y)$, while y remains constant. In like manner the du in $\frac{du}{dy}$, is the increment of u due to an infinitesimal increment of y, while x remains constant. These will by no means be generally the same. Much less will either of these du's be the same as the du in $\left[\frac{du}{dx}\right]$, which is the change in u due to a change in both the variables x and y. Finally, there is nothing in the logic of the process by which the expression $\frac{du}{dy}\frac{dy}{dx}$ was arrived at, that makes the dy's in it equal to each other. The dy in $\frac{du}{dy}$ is simply an arbitrary infinitesimal increment assigned to y in $u = f(x, y)$, x remaining constant; while dy in $\frac{dy}{dx}$ is the increment which y takes on in the function $y = f_1(x)$, when x takes the increment dx.

100. Cor. 1.—*From* $u = f(x, y, z)$ *we have* $\left[\frac{du}{dx}\right] = \frac{du}{dx} + \frac{du}{dy}\frac{dy}{dx} + \frac{du}{dz}\frac{dz}{dx}$, *in which* y *and* z *are functions of* x.

Dem.—Since the total differential equals the sum of the partials, we have $du = \frac{du}{dx}dx + \frac{du}{dy}dy + \frac{du}{dz}dz$. Dividing by dx, $\left[\frac{du}{dx}\right] = \frac{du}{dx} + \frac{du}{dy}\frac{dy}{dx} + \frac{du}{dz}\frac{dz}{dx}$. Q. E. D.

Sch.—In the same manner the total differential of a function of any number of variables dependent or independent, may be found; and, when all are dependent upon some single variable, the differential coefficient with respect to that one may be formed.

101. Cor. 2.—*If* $u = f(y, z, w)$, *and* $y = \varphi(x)$, $z = \varphi_1(x)$, *and* $w = \phi_2(x)$, $\left[\frac{du}{dx}\right] = \frac{du}{dy}\frac{dy}{dx} + \frac{du}{dz}\frac{dz}{dx} + \frac{du}{dw}\frac{dw}{dx}$.

[Let the student give the proof.]

Ex. 1. Given $u = \tan^{-1}\frac{x}{y}$, and $y^2 + x^2 = r^2$, to find $\left[\frac{du}{dx}\right]$.

Solution.—We have $\left[\frac{du}{dx}\right] = \frac{du}{dx} + \frac{du}{dy}\frac{dy}{dx}$. Remembering that $\frac{du}{dx}$ and $\frac{du}{dy}$ are partial differential coefficients, we have from $u = \tan^{-1}\frac{x}{y}$, $d_x u = \dfrac{d\left(\frac{x}{y}\right)}{1+\left(\frac{x}{y}\right)^2} = \dfrac{\frac{dx}{y}}{\frac{y^2+x^2}{y^2}}$ $= \frac{ydx}{r^2}$; whence $\frac{du}{dx} = \frac{y}{r^2}$. In like manner we find $\frac{du}{dy} = -\frac{x}{r^2}$. Also from $y^2+x^2=r^2$, we find $\frac{dy}{dx} = -\frac{x}{y}$. Substituting these values, we obtain $\left[\frac{du}{dx}\right] = \frac{y}{r^2} + \left(-\frac{x}{r^2}\right)\left(-\frac{x}{y}\right)$ $= \frac{y}{r^2} + \frac{x^2}{r^2y} = \frac{y^2+x^2}{r^2y} = \frac{r^2}{r^2y} = \frac{1}{y}$, or $\frac{1}{\sqrt{r^2-x^2}}$.

Ex. 2. Given $u = \tan^{-1}(xy)$, and $y = e^x$ to form $\left[\frac{du}{dx}\right]$.

Result, $\left[\frac{du}{dx}\right] = \frac{e^x(1+x)}{1+x^2e^{2x}}$.

Ex. 3. Given $u = z^2 + y^3 + zy$, and $z = \sin x$, $y = e^x$, to form $\left[\frac{du}{dx}\right]$.

SUG'S.—We have
$$\begin{aligned}\left[\frac{du}{dx}\right] &= \frac{du}{dy}\frac{dy}{dx} + \frac{du}{dz}\frac{dz}{dx} \\ &= (3y^2+z)e^x + (2z+y)(\cos x) \\ &= (3e^{2x} + \sin x)e^x + (2\sin x + e^x)(\cos x) \\ &= 3e^{3x} + e^x(\sin x + \cos x) + 2\sin x\cos x \\ &= 3e^{3x} + e^x(\sin x + \cos x) + \sin 2x.\end{aligned}$$

Ex. 4. Given $u = yz$, and $y = e^x$, $z = x^4 - 4x^3 + 12x^2 - 24x + 24$, to find $\left[\frac{du}{dx}\right]$. *Result*, $\left[\frac{du}{dx}\right] = e^x x^4$.

Ex. 5. Given $u = \sin^{-1}(p-q)$, and $p = 3x$, $q = 4x^3$, to find $\left[\frac{du}{dx}\right]$.

SUG'S.—We have $\frac{du}{dp} = \frac{1}{\sqrt{1-(p-q)^2}}$, $\frac{du}{dq} = \frac{-1}{\sqrt{1-(p-q)^2}}$, $\frac{dp}{dx} = 3$, and $\frac{dq}{dx} = 12x^2$. Whence $\left[\frac{du}{dx}\right] = \frac{3}{\sqrt{1-(p-q)^2}} - \frac{12x^2}{\sqrt{1-(p-q)^2}}$ $= \frac{3-12x^2}{\sqrt{1-9x^2+24x^4-16x^6}} = \frac{3}{\sqrt{1-x^2}}$.

Ex. 6. Given $u = \frac{x^4y^2}{4} - \frac{x^4y}{8} + \frac{x^4}{32}$ and $y = \log x$, to find that $\left[\frac{du}{dx}\right]$ $= x^3(\log x)^2$.

Ex. 7. Given $u = \frac{e^{ax}(p-q)}{a^2+1}$, where $p = a\sin x$, and $q = \cos x$, to find that $\left[\frac{du}{dx}\right] = e^{ax}\sin x$.

[NOTE.—In such examples as the above, it is of course possible to substitute in the function u, the values of the several variables on which it depends, in terms of the single variable upon which each of them depends, and then have $u =$ a function of a single variable, which can be differentiated by the elementary processes. But it is the chief design of these examples to familiarize the important *formulæ* used, and render their meaning clear. Their precise practical value cannot be appreciated until the student has made further progress.]

IMPLICIT FUNCTIONS.

102. Prop.—*Having* $f(x, y) = 0 = u$, $\frac{dy}{dx} = -\frac{\frac{du}{dx}}{\frac{du}{dy}}$; *in which* $\frac{du}{dx}$, *and* $\frac{du}{dy}$ *are the partial differential coefficients of the function taken with reference to* x *and* y *respectively.*

DEM.—From (**99**) we have $\left[\frac{du}{dx}\right] = \frac{du}{dx} + \frac{du}{dy}\frac{dy}{dx}$. But as u remains constantly equal to 0, for all simultaneous values of x and y, when both x and y have taken on contemporaneous changes, $du = 0$. Therefore $\left[\frac{du}{dx}\right] = 0$, and $\frac{du}{dx} + \frac{du}{dy}\frac{dy}{dx} = 0$; whence $\frac{dy}{dx} = -\frac{\frac{du}{dx}}{\frac{du}{dy}}$. Q. E. D.

SCH.—It is to be observed that though $\left[\frac{du}{dx}\right] = 0$, it by no means follows that $\frac{du}{dx}$, or $\frac{du}{dy} = 0$. For example, $y^2 + x^2 - r^2 = 0$ is of the form $f(x, y) = 0$. Now if y changes and x *does not*, the *function* changes and is no longer $= 0$. So also if x changes and y *does not*, the function is not 0. But if both change together according to the law of their mutual dependence, *i. e.* if the changes are what we have called contemporaneous, the function remains equal to 0; and its total differential is 0. [The student can observe the geometrical signification of these statements, by noticing that $y^2 + x^2 - r^2 = 0$ is the equation of a circle, and that the function is 0 when x and y vary together, according to their mutual dependence: but when one varies and the other does not, the function varies; *i. e.* the point falls *out of* the circumference. Illustrate in like manner from $y^2 - 2px = 0$.]

Ex. 1. Given $x^4 + 2ax^2y - ay^3 = 0$, to form $\frac{dy}{dx}$ upon the principle just demonstrated.

SOLUTION.—Putting $u = 0 = x^4 + 2ax^2y - ay^3$, we have $d_xu = 4x^3dx + 4axydx$, and $d_yu = 2ax^2dy - 3ay^2dy$; whence we have the partial differential coefficients $\frac{du}{dx} = 4x^3 + 4axy$, and $\frac{du}{dy} = 2ax^2 - 3ay^2$. $\therefore \frac{dy}{dx} = -\frac{\frac{du}{dx}}{\frac{du}{dy}} = -\frac{4x^3 + 4axy}{2ax^2 - 3ay^2}$.

Ex. 2. Given $ax^3 + x^3y - ay^3 = 0$, to form $\frac{dy}{dx}$ as above.

$$Result,\ \frac{dy}{dx} = -\frac{3ax^2 + 3x^2y}{x^3 - 3ay^2}.$$

Ex. 3. Given $y^2 - 2axy + x^2 - b^2 = 0$, to form $\frac{dy}{dx}$ as above.

$$Result,\ \frac{dy}{dx} = \frac{ay - x}{y - ax}.$$

Ex. 4. Given $y^3 - 3y + x = 0$, to form $\frac{dy}{dx}$ as above.

$$Result,\ \frac{dy}{dx} = \frac{1}{3(1 - y^2)}.$$

Ex. 5. Given $x^y - y^x = 0$, to form $\frac{dy}{dx}$ as above.

$$Result,\ \frac{dy}{dx} = \frac{y^2 - xy\log y}{x^2 - xy\log x}.$$

Ex. 6. Given $a^{x^y} + \sqrt{\sec(xy)} = 0$, to form $\frac{dy}{dx}$ as above.

$$Result,\ \frac{dy}{dx} = -\frac{y\sqrt{\sec(xy)}\tan(xy) + 2a^{x^y}yx^{y-1}\log a}{x\sqrt{\sec(xy)}\tan(xy) + 2a^{x^y}x^y\log a\log x}.$$

COMPOUND FUNCTIONS.

103. DEF.—***A Compound Function*** is a function of a function. Thus, if $u = f(y)$, and $y = \varphi(x)$, $u = f[\varphi(x)]$,* and u is said to be a compound function of x. This relation is often indicated by saying that "u is a function of x through y," or that "u is *indirectly* a function of x through y."

In case $u = f(x, y)$, and $y = \varphi(x)$, we say that u is *directly* a function of x, and also *indirectly* through y.

104. Prop.—*If* $u = f(y)$ *and* $y = \varphi(x)$, $\frac{du}{dx} = \frac{du}{dy}\frac{dy}{dx}$.

DEM.—From $u = f(y)$, we have $du = \frac{du}{dy}dy$. But from $y = \varphi(x)$, we have $dy = \frac{dy}{dx}dx$. Hence $du = \frac{du}{dy}\frac{dy}{dx}dx$, and $\frac{du}{dx} = \frac{du}{dy}\frac{dy}{dx}$. Q. E. D.

SCH.—It will be seen that this is only a particular case of the preceding; but it is of such frequent occurrence that it is thought best to give it prominence.

---

* Read, "u equals the f function of the φ function of x."

Ex. 1. Given $u = a^y$, and $y = b^x$, to form $\frac{du}{dx}$ on the principle just demonstrated.

SOLUTION. $\frac{du}{dy} = a^y \log a$, and $\frac{dy}{dx} = b^x \log b$. $\therefore \frac{du}{dx} = \frac{du}{dy}\frac{dy}{dx} = a^y \log a \times b^x \log b = a^y b^x \log a \log b$, or $a^{b^x} b^x \log a \log b$.

Ex. 2. Given $u = by^3$, and $y = ax^2$, to form $\frac{du}{dx}$ as above.

$$\frac{du}{dx} = 6a^3bx^5.$$

Ex. 3. Given $u = \log y$, and $y = \log x$ to form $\frac{du}{dx}$ as above.

$$\frac{du}{dx} = \frac{1}{y} \cdot \frac{1}{x} = \frac{1}{x \log x}.$$

105. Prop.—Having given $u = \varphi(z)$ and $z = f(x, y)$ to differentiate u with respect to x and y without previously eliminating z.

DEM.—Since u is a function of x and y, we have

$$du = \frac{du}{dx}dx + \frac{du}{dy}dy \quad (\textbf{97}).$$

Now u is a function of x through z (*i. e.* it is a compound function of x). Hence $\frac{du}{dx} = \frac{du}{dz}\frac{dz}{dx}$; and for a like reason $\frac{du}{dy} = \frac{du}{dz}\frac{dz}{dy}$ (***104***). Therefore, substituting, $du = \frac{du}{dz}\frac{dz}{dx}dx + \frac{du}{dz}\frac{dz}{dy}dy$. Q. E. D.

SCH.—The student should not fail to observe that all the truths developed in this section are but deductions from the proposition that the total differential of a function of several variables equals the sum of the partial differentials. With this key in hand, he can readily unlock the mysteries of the whole subject.

SECTION VI.

Successive Differentiation of Functions of Two Independent Variables, and of Implicit Functions.

106. Prop.—*In a function of two independent variables, both variables may be considered equicrescent;* i. e., *their differentials may be regarded as constant.*

ILL.—This proposition is an axiom, and it is only necessary that its import be clearly understood. Thus, if $u = f(x, y)$ and x and y are independent, any change which x may undergo does not affect y, and any change which y may undergo does not affect x, as this is what is meant by their being independent. We may therefore conceive *each* of them to change according to any law we please; and it is found convenient to conceive that x increases by *equal* infinitesimal increments, as heretofore, and that y also increases by equal infinitesimal increments. Thus dx and dy are constants; but it does not follow that we are to regard $dy = dx$. In fact this would be to establish a relation between x and y, and hence would be contrary to the hypothesis.

107. DEF.—When $u = f(x, y)$, and x and y are independent of each other, d_xu and d_yu are, in general, functions of x and y, and hence may be differentiated with respect to either variable, thus obtaining a class of ***Second Partial Differentials.*** In like manner these second partial differentials are in general functions of x and y, and may be differentiated with reference to either, giving rise to ***Third Partial Differentials;*** etc., etc.

108. ***Notation.***—Having $u = f(x, y)$, $\frac{d^2u}{dx^2}dx^2$, $\frac{d^2u}{dxdy}dxdy$ or $\frac{d^2u}{dydx}dydx$, and $\frac{d^2u}{dy^2}dy^2$ are the symbols for the second partial differentials. The third partial differentials are indicated thus, $\frac{d^3u}{dx^3}dx^3$, $\frac{d^3u}{dx^2dy}dx^2dy$ or $\frac{d^3u}{dydx^2}dydx^2$, $\frac{d^3u}{dxdy^2}dxdy^2$ or $\frac{d^3u}{dy^2dx}dy^2dx$, and $\frac{d^3u}{dy^3}dy^3$. In each case the form of the numerator indicates the number of differentiations, and the denominator the variable or variables with which the successive differentiations have been made, and the order. Thus $\frac{d^3u}{dy^2dx}dy^2dx$ signifies that $u = f(x, y)$ has been differentiated three times in succession, twice with reference to y and once with reference to x, and in this order. So, in general, $\frac{d^mu}{dx^{m-n}dy^n}dx^{m-n}dy^n$ signifies that $u = f(x, y)$ has been differentiated $m - n$ times in succession with reference to x, and then n times with reference to y.

Ex. 1. Given $u = x^2y^2$ to form the several successive partial differentials.

Results, $\frac{du}{dx}dx = 2y^2xdx$, $\frac{du}{dy}dy = 2x^2ydy$; $\frac{d^2u}{dx^2}dx^2 = 2y^2dx^2$,

$\frac{d^2u}{dxdy}dxdy = 4xydxdy$; $\frac{d^2u}{dy^2}dy^2 = 2x^2dy^2$; $\frac{d^3u}{dx^3}dx^3 = 0$,

$\frac{d^3u}{dx^2dy}dx^2dy = 4ydx^2dy$, $\frac{d^3u}{dxdy^2}dxdy^2 = 4xdxdy^2$; $\frac{d^3u}{dy^3}dy^3 = 0$;

$\frac{d^4u}{dx^2dy^2} = dx^2dy^2 = 4dx^2dy^2$, etc.

SUG'S.—Having $\frac{du}{dx}dx = 2y^2xdx$, to differentiate it with respect to x, we notice in the first member that it will give $\frac{d\left(\frac{du}{dx}dx\right)}{dx}dx$; the dx being written in the denominator, and as a factor to designate with respect to which variable the differentiation is made, and also in accordance with the principle that the differential coefficient multiplied by the differential of the variable, is the differential of the function. Now, the differential coefficient being the differential of the function divided by the differential of the variable, we have for the differential coefficient of $\frac{du}{dx}dx$ taken with respect to x $\frac{d\left(\frac{du}{dx}dx\right)}{dx}$; hence the differential is $\frac{d\left(\frac{du}{dx}dx\right)}{dx}dx$. Finally, observing that dx is constant, this becomes $\frac{\frac{d^2u}{dx}dx}{dx}dx$, or $\frac{d^2u}{dx^2}dx^2$. In like manner the student should analyze the other processes. *It is of the utmost importance that he fully comprehend the reasons for these processes; if he do not he will become hopelessly entangled in the subsequent operations.*

Ex. 2. Produce the successive partial differentials of $u = (x + y)^m$ with respect to x; also with respect to y.

Results, d_xu or $\frac{du}{dx}dx = m(x + y)^{m-1}dx$,

$\frac{d^2u}{dx^2}dx^2$, or $d_xd_xu = m(m - 1)(x + y)^{m-2}dx^2$,

$\frac{d^3u}{dx^3}dx^3$ or $d_xd_xd_xu = m(m - 1)(m - 2)(x + y)^{m-3}dx^3$, etc.

SCH.—When $u = f(x, y)$, the following forms are called ***Partial Differential Coefficients***; $\frac{d^2u}{dx^2}, \frac{d^2u}{dy^2}, \frac{d^2u}{dydx}, \frac{d^3u}{dx^3}, \frac{d^3u}{dx^2dy}, \frac{d^3u}{dxdy^2}, \text{- - - -} \frac{d^mu}{dx^{m-n}dy^n}$, etc.

Ex. 3. Form the successive partial differential coefficients of $u = \sin(x + y)$ with respect to y.

Results, $\frac{du}{dy} = \cos(x + y)$, $\frac{d^2u}{dy^2} = -\sin(x + y)$, $\frac{d^3u}{dy^3} = -\cos(x + y)$, $\frac{d^4u}{dy^4} = \sin(x + y)$, $\frac{d^5u}{dy^5} = \cos(x + y)$, etc.

Ex. 4. Form the successive partial differential coefficients of $u = \cos(x - y)$ with respect to x.

Results, $\frac{du}{dx} = -\sin(x - y)$, $\frac{d^2u}{dx^2} = -\cos(x - y)$, $\frac{d^3u}{dx^3} = \sin(x - y)$, etc.

Ex. 5. Form the successive partial differential coefficients of $u =$

log $(x + y)$ with respect to x, and also with respect to y in the common system of logarithms.

Results, $\frac{du}{dx} = \frac{m}{x+y}$, $\frac{d^2u}{dx^2} = -\frac{m}{(x+y)^2}$, $\frac{d^3u}{dx^3} = \frac{2m(x+y)}{(x+y)^4} = \frac{2m}{(x+y)^3}$, etc. The partial differential coefficients with respect to y are altogether similar.

---

109. Prop.—*If* u = f(x, y), *in which* x *and* y *are independent, and several differentiations be performed,* m *with reference to one variable and* n *with reference to the other, the result is the same whatever the order of the operations.*

DEM.—1st., To show that $\frac{d^2u}{dxdy}dxdy = \frac{d^2u}{dydx}dydx$.

$\frac{du}{dx}dx = f(x + dx, y) - f(x, y)$, and

$\frac{d^2u}{dxdy}dxdy = f(x + dx, y + dy) - f(x, y + dy) - [f(x + dx, y) - f(x, y)] =$

$f(x + dx, y + dy) - f(x, y + dy) - f(x + dx, y) + f(x, y)$.

Again, $\frac{du}{dy}dy = f(x, y + dy) - f(x, y)$ and

$\frac{d^2u}{dydx}dydx = f(x + dx, y + dy) - f(x + dx, y) - [f(x, y + dy) - f(x, y)] =$

$f(x + dx, y + dy) - f(x + dx, y) - f(x, y + dy) + f(x, y)$.

These two results being identical, we have $\frac{d^2u}{dxdy}dxdy = \frac{d^2u}{dydx}dydx$, (1).

2nd., To show that $\frac{d^3u}{dx^2dy}dx^2dy = \frac{d^3u}{dydx^2}dydx^2$.

For convenience of notation put d_xd_yu for $\frac{d^2u}{dxdy}dxdy$, and d_yd_xu for $\frac{d^2u}{dydx}dydx$. Then, as before shown $d_yd_xu = d_xd_yu = f(x, y)$, and hence may be differentiated with reference to x or y. Differentiating with reference to x, we have, $d_xd_yd_xu = d_xd_xd_yu$.* But by (1) $d_xd_y(d_xu) = d_yd_x(d_xu)$ or $d_yd_xd_xu$. Whence $\frac{d^3u}{dydx^2}dydx^2 = \frac{d^3u}{dx^2dy}dx^2dy$.

In like manner we may proceed to any extent desired.

[Let the student show that $\frac{d^5u}{dx^3dy^2}dx^3dy^2 = \frac{d^5u}{dy^2dx^3}dy^2dx^3$.]

Ex. 1. Given $u = x\log y$ in which x and y are independent, to form the several second and third partial differentials, and to show that $d_yd_xu = d_xd_yu$, and also that $d_yd_xd_xu = d_xd_yd_xu = d_xd_xd_yu$.*

Results, $\frac{d^2u}{dx^2}dx^2$ (*i. e.* d_xd_xu) $= 0$, $\frac{d^2u}{dxdy}dxdy$ (*i. e.* d_xd_yu) $= \frac{dxdy}{y}$,

---

* These operations are sometimes indicated thus: $d_{yx}u = d_{xy}u$, $d_{yxx}u = d_{xyx}u = d_{xxy}u$, etc.

and $\frac{d^2u}{dydx}dydx$ also $= \frac{dydx}{y}$, $\frac{d^2u}{dy^2}dy^2 = -\frac{xdy^2}{y^2}$, $\frac{d^3u}{dxdy^2}dxdy^2 = -\frac{dxdy^2}{y^2}$, and $\frac{d^3u}{dydx^2} = 0$.

Ex. 2. Given $u = x^3y + ay^2$, to form the second, third, and fourth partial differential coefficients, and show the convertibility of the independent differentiations.

Results, $\frac{du}{dx} = 3x^2y$, $\frac{du}{dy} = x^3 + 2ay$; $\frac{d^2u}{dx^2} = 6xy$, $\frac{d^2u}{dy^2} = 2a$, $\frac{d^2u}{dxdy} = 3x^2$ and $\frac{d^2u}{dydx} = 3x^2$ also; $\frac{d^3u}{dx^3} = 6y$, $\frac{d^3u}{dy^3} = 0$, $\frac{d^3u}{dx^2dy} = 6x = \frac{d^3u}{dydx^2}$, $\frac{d^3u}{dxdy^2} = 0 = \frac{d^3u}{dy^2dx}$; $\frac{d^4u}{dx^4} = 0$, etc., etc.

Ex. 3 to 7. As above form and compare the successive partial differential coefficients of the following: $u = \frac{x^2 - y^2}{x^2 + y^2}$; $u = \tan^{-1}\frac{x}{y}$; $u = \sin x \cos y$; $u = x^y$; and $u = (x + y)^4$.

---

110. Prob.—*To form the successive differentials of a function of two independent variables.*

DEM.—Let $u = f(x, y)$, in which x and y are independent variables. The total differential being equal to the sum of the partials, we have

$$du = \frac{du}{dx}dx + \frac{du}{dy}dy, \quad (1).$$

Now, remembering that as x and y are independent and hence may be treated as equicrescent, dx and dy may be considered constant, and remembering also that $\frac{du}{dx}$, and $\frac{du}{dy}$ are, in general, functions of x and y (***107***), we proceed to differentiate (1) again. Thus

$$d^2u = d\left(\frac{du}{dx}\right)dx + d\left(\frac{du}{dy}\right)dy;$$

but
$$d\left(\frac{du}{dx}\right) = \frac{d^2u}{dx^2}dx + \frac{d^2u}{dxdy}dy,^*$$

and
$$d\left(\frac{du}{dy}\right) = \frac{d^2u}{dydx}dx + \frac{d^2u}{dy^2}dy, \quad \text{which substituted give}$$

$$d^2u = \frac{d^2u}{dx^2}dx^2 + \frac{d^2u}{dxdy}dydx + \frac{d^2u}{dydx}dxdy + \frac{d^2u}{dy^2}dy^2 = \frac{d^2u}{dx^2}dx^2 + 2\frac{d^2u}{dxdy}dxdy + \frac{d^2u}{dy^2}dy^2,$$

since $\frac{d^2u}{dxdy}dxdy = \frac{d^2u}{dydx}dydx$ (***109***).

Again, differentiating this second differential, we have

---

* To perform this operation observe that $\frac{du}{dx}$ is treated as a function of x and y, and hence its total differential is equal to the sum of its partial differentials. The partial differential with respect to x is $\frac{d^2u}{dx^2}dx$, and with reference to y, $\frac{d^2u}{dxdy}dy$.

$$d^3u = d\left(\frac{d^2u}{dx^2}\right)dx^2 + 2d\left(\frac{d^2u}{dxdy}\right)dxdy + d\left(\frac{d^2u}{dy^2}\right)dy^2;$$

but
$$d\left(\frac{d^2u}{dx^2}\right) = \frac{d^3u}{dx^3}dx + \frac{d^3u}{dx^2dy}dy,$$
$$d\left(\frac{d^2u}{dxdy}\right) = \frac{d^3u}{dx^2dy}dx + \frac{d^3u}{dxdy^2}dy,$$
and
$$d\left(\frac{d^2u}{dy^2}\right) = \frac{d^3u}{dy^2dx}dx + \frac{d^3u}{dy^3}dy.$$
Substituting, we have

$$d^3u = \left(\frac{d^3u}{dx^3}dx + \frac{d^3u}{dx^2dy}dy\right)dx^2 + 2\left(\frac{d^3u}{dx^2dy}dx + \frac{d^3u}{dxdy^2}dy\right)dxdy + \left(\frac{d^3u}{dy^2dx}dx + \frac{d^3u}{dy^3}dy\right)dy^2$$
$$= \frac{d^3u}{dx^3}dx^3 + \frac{d^3u}{dx^2dy}dx^2dy + 2\frac{d^3u}{dx^2dy}dx^2dy + 2\frac{d^3u}{dxdy^2}dxdy^2 + \frac{d^3u}{dy^2dx}dy^2dx + \frac{d^3u}{dy^3}dy^3$$
$$= \frac{d^3u}{dx^3}dx^3 + 3\frac{d^3u}{dx^2dy}dx^2dy + 3\frac{d^3u}{dy^2dx}dy^2dx + \frac{d^3u}{dy^3}dy^3,$$

since $\frac{d^3u}{dx^2dy}dx^2dy = \frac{d^3u}{dxdydx}\,dxdydx = \frac{d^3u}{dydx^2}dydx^2$, etc.

In like manner we may proceed to differentiate as often as desired.

SCH.—A little observation will enable the student to write out any required differential of $u = f(x, y)$ by analogy from the above. He only needs to notice that every distinct form of the partial differential of the required order is involved, and making x the leading letter insert the coefficients as in the binomial formula. Thus $d^5u = \frac{d^5u}{dx^5}dx^5 + 5\frac{d^5u}{dx^4dy}dx^4dy + 10\frac{d^5u}{dx^3dy^2}dx^3dy^2 + 10\frac{d^5u}{dx^2dy^3}dx^2dy^3 + 5\frac{d^5u}{dxdy^4}dxdy^4 + \frac{d^5u}{dy^5}dy^5$.

---

111. Prob.—*To form the successive differential coefficients of an implicit function of a single variable.*

SOLUTION.—Let $u = f(x, y) = 0$, in which y is an implicit function. We are to form $\frac{d^2y}{dx^2}, \frac{d^3y}{dx^3}$, etc.

First we have $\frac{dy}{dx} = -\frac{\frac{du}{dx}}{\frac{du}{dy}}$ by (***102***). (1).

The form for differentiating this is

$$\frac{d^2y}{dx^2} = -\frac{d\left(\frac{du}{dx}\right)\frac{du}{dy} - d\left(\frac{du}{dy}\right)\frac{du}{dx}}{\left(\frac{du}{dy}\right)^2} \div dx,$$ since the second member is a fraction.

To perform the operations thus indicated we have to remember that $\frac{du}{dx}$ and $\frac{du}{dy}$ are functions of x and y.

Hence $d\left(\frac{du}{dx}\right) = \frac{d^2u}{dx^2}dx + \frac{d^2u}{dx\,dy}dy,$

and $d\left(\frac{du}{dy}\right) = \frac{d^2u}{dy\,dx}dx + \frac{d^2u}{dy^2}dy$. Dividing these values by dx and substituting, we have

$$\frac{d^2y}{dx^2} = -\frac{\frac{du}{dy}\left(\frac{d^2u}{dx^2} + \frac{d^2u}{dx\,dy}\frac{dy}{dx}\right) - \frac{du}{dx}\left(\frac{d^2u}{dy\,dx} + \frac{d^2u}{dy^2}\frac{dy}{dx}\right)}{\left(\frac{du}{dy}\right)^2}.$$

Finally, substituting in this the value of $\frac{dy}{dx}$ as given in (1) we have

$$\frac{d^2y}{dx^2} = -\frac{\frac{du}{dy}\left(\frac{d^2u}{dx^2} - \frac{d^2u}{dx\,dy}\frac{\frac{du}{dx}}{\frac{du}{dy}}\right) - \frac{du}{dx}\left(\frac{d^2u}{dy\,dx} - \frac{d^2u}{dy^2}\frac{\frac{du}{dx}}{\frac{du}{dy}}\right)}{\left(\frac{du}{dy}\right)^2}$$

$$= -\frac{\left(\frac{du}{dy}\right)^2\left(\frac{d^2u}{dx^2}\right) - \frac{d^2u}{dx\,dy}\frac{du}{dx}\frac{du}{dy} - \frac{d^2u}{dy\,dx}\frac{du}{dx}\frac{du}{dy} + \frac{d^2u}{dy^2}\left(\frac{du}{dx}\right)^2}{\left(\frac{du}{dy}\right)^3}$$

$$= -\frac{\frac{d^2u}{dx^2}\left(\frac{du}{dy}\right)^2 - 2\frac{d^2u}{dx\,dy}\frac{du}{dx}\frac{du}{dy} + \frac{d^2u}{dy^2}\left(\frac{du}{dx}\right)^2}{\left(\frac{du}{dy}\right)^3}. \quad (2).$$

In like manner the higher coefficients may be produced, but the forms are too complicated for elementary purposes.

Ex. 1. Form the first and second differential coefficients of y as a function of x, when $y^2 - 2axy + x^2 - b^2 = 0$.

SOLUTION. $\frac{dy}{dx} = -\frac{\frac{du}{dx}}{\frac{du}{dy}} = -\frac{-2ay + 2x}{2y - 2ax} = \frac{ay - x}{y - ax}$. For convenience of notation put $\frac{dy}{dx} = \frac{ay-x}{y-ax} = p$, whence p is a function of x and y. Hence $\left[\frac{dp}{dx}\right] = \frac{dp^*}{dx} + \frac{dp^*}{dy}\frac{dy}{dx}$, (**99**). But $\left[\frac{dp}{dx}\right] = \frac{d^2y}{dx^2}$, $\frac{dp^*}{dx} = d_x\left(\frac{ay - x}{y - ax}\right) \div dx = \frac{-(y-ax) + a(ay-x)}{(y-ax)^2}$, and $\frac{dp^*}{dy} = d_y\left(\frac{ay-x}{y-ax}\right) \div dy = \frac{a(y-ax) - (ay-x)}{(y-ax)^2}$.

Reducing, $\frac{dp}{dx} = \frac{(a^2-1)y}{(y-ax)^2}$, and $\frac{dp}{dy} = -\frac{(a^2-1)x}{(y-ax)^2}$. Substituting these values and also the value of $\frac{dy}{dx}$ as at first found, we have,

$$\frac{d^2y}{dx^2} = \frac{(a^2-1)y}{(y-ax)^2} - \frac{(a^2-1)x}{(y-ax)^2} \times \frac{ay-x}{y-ax} = \frac{(a^2-1)(y^2 - 2axy + x^2)}{(y-ax)^3}.$$

Ex. 2. Form the first and second differential coefficients of y as a function of x, when $y^2 + x^2 - r^2 = 0$.

Results, $\frac{dy}{dx} = -\frac{x}{y}$, $\frac{d^2y}{dx^2} = -\frac{r^2}{y^3}$.

* Remember that these are *partial* differential coefficients.

SUG.—Be particular to use the method now being illustrated.

Ex. 3. Given $x^3 + 3axy + y^3 = 0$, to form $\frac{dy}{dx}$, and $\frac{d^2y}{dx^2}$, by the method for differentiating implicit functions.

$$Results,\ \frac{dy}{dx} = -\frac{x^2 + ay}{y^2 + ax},\ \frac{d^2y}{dx^2} = \frac{2a^3xy}{(y^2 + ax)^3}.$$

Ex. 4. Given $y^2 - 2xy + a^2 = 0$, to form the first and second differential coefficients of y as a function of x, by substituting in (1) and (2) of the preceding demonstration.

SUG'S. $\frac{du}{dx} = -2y,\ \frac{du}{dy} = 2y - 2x,\ \frac{d^2u}{dx^2} = 0,\ \frac{d^2u}{dxdy} = -2,\ \frac{d^2u}{dy^2} = 2.\ \therefore \frac{dy}{dx} = -\frac{-2y}{2y - 2x} = \frac{y}{y - x}$, and $\frac{d^2y}{dx^2} = -\frac{\frac{d^2u}{dx^2}\left(\frac{du}{dy}\right)^2 - 2\frac{d^2u}{dxdy}\frac{du}{dx}\frac{du}{dy} + \frac{d^2u}{dy^2}\left(\frac{du}{dx}\right)^2}{\left(\frac{du}{dy}\right)^3} =$
$-\frac{0 \cdot (2y - 2x)^2 - 2(-2)(-2y)(2y - 2x) + 2(-2y)^2}{(2y - 2x)^3} = -\frac{-16y(y - x) + 8y^2}{(2y - 2x)^3} = \frac{y(y - 2x)}{(y - x)^3}.$

Ex. 5. Given $\cos(x + y) = 0$, to form $\frac{dy}{dx}$, and $\frac{d^2y}{dx^2}$ by substituting as above.

SUG'S. $\frac{du}{dx} = -\sin(x + y),\ \frac{du}{dy} = -\sin(x + y),\ \frac{d^2u}{dx^2} = -\cos(x + y),\ \frac{d^2u}{dxdy} = -\cos(x + y),\ \frac{d^2u}{dy^2} = -\cos(x + y)$. Substituting, $\frac{dy}{dx} = -1$, and $\frac{d^2y}{dx^2} = -\frac{-\cos(x+y)\sin^2(x+y) + 2\cos(x+y)\sin^2(x+y) - \cos(x+y)\sin^2(x+y)}{-\sin^3(x + y)} = 0.$

These results are as might have been anticipated, since for $\cos(x + y) = 0$, $x + y = 90°$; hence as one arc (x) increases, the other (y) decreases at the same rate. Therefore $\frac{dy}{dx} = -1$, and, consequently, $\frac{d^2y}{dx^2} = 0$.

Ex. 6. Solve *Ex's* 1—3 inclusive by substituting in the general *formulæ* (1) and (2).

DERIVED EQUATIONS.

112. From $u = 0 = f(x, y)$, we have

$$\frac{dy}{dx} = -\frac{\frac{du}{dx}}{\frac{du}{dy}},\quad (1),$$

and $$\frac{d^2y}{dx^2} = -\frac{\frac{du}{dy}\left(\frac{d^2u}{dx^2} + \frac{d^2u}{dxdy}\frac{dy}{dx}\right) - \frac{du}{dx}\left(\frac{d^2u}{dydx} + \frac{d^2u}{dy^2}\frac{dy}{dx}\right)}{\left(\frac{du}{dy}\right)^2}, \quad (2),$$

From (1), we have $\frac{du}{dy}\frac{dy}{dx} + \frac{du}{dx} = 0.$ (1_1), which is called ***The First Derived Equation,*** or ***The Differential Equation of the First Order.***

From (2) we obtain

$$\frac{du}{dy}\frac{d^2y}{dx^2} = -\frac{\frac{du}{dy}\left(\frac{d^2u}{dx^2} + \frac{d^2u}{dxdy}\frac{dy}{dx}\right) - \frac{du}{dx}\left(\frac{d^2u}{dydx} + \frac{d^2u}{dy^2}\frac{dy}{dx}\right)}{\frac{du}{dy}}$$

$$= -\frac{d^2u}{dx^2} - \frac{d^2u}{dxdy}\frac{dy}{dx} + \frac{\frac{du}{dx}}{\frac{du}{dy}}\left(\frac{d^2u}{dydx} + \frac{d^2u}{dy^2}\frac{dy}{dx}\right)$$

$$= -\frac{d^2u}{dx^2} - \frac{d^2u}{dxdy}\frac{dy}{dx} - \frac{dy}{dx}\left(\frac{d^2u}{dydx} + \frac{d^2u}{dy^2}\frac{dy}{dx}\right)$$

$$= -\frac{d^2u}{dx^2} - 2\frac{d^2u}{dxdy}\frac{dy}{dx} - \frac{d^2u}{dy^2}\left(\frac{dy}{dx}\right)^2.$$

Whence, transposing, we have,

$$\frac{du}{dy}\frac{d^2y}{dx^2} + 2\frac{d^2u}{dxdy}\frac{dy}{dx} + \frac{d^2u}{dy^2}\left(\frac{dy}{dx}\right)^2 + \frac{d^2u}{dx^2} = 0, \quad (2_1),$$

which is called ***The Second Derived Equation*** or ***The Differential Equation of the Second Order.***

In a similar manner the Third Derived Equation is found to be

$$\frac{du}{dy}\frac{d^3y}{dx^3} + 3\left\{\frac{d^2u}{dxdy} + \frac{d^2u}{dy^2}\frac{dy}{dx}\right\}\frac{d^2y}{dx^2} + \frac{d^3u}{dy^3}\left(\frac{dy}{dx}\right)^3 + 3\frac{d^3u}{dxdy^2}\left(\frac{dy}{dx}\right)^2 +$$
$$3\frac{d^3u}{dx^2dy}\frac{dy}{dx} + \frac{d^3u}{dx^3} = 0.$$

SECTION VII.

Change of the Independent Variable.

113. In considering functions of a single variable, as $y = f(x)$, the hypothesis which we usually make that x is equicrescent, and hence that dx is constant, gives to all the differentials and differential coefficients of the function *after the first*, a different form from what they would have had if such hypothesis had not been made. Thus

$d\left(\frac{dy}{dx}\right)=\frac{d^2y}{dx}$, when x is equicrescent, but $\frac{d^2y\,dx-d^2x\,dy}{dx^2}$, when neither variable is regarded as equicrescent (*i. e.* when dy and dx are both treated as variable). In the course of a discussion it sometimes becomes important to change the conception and regard y as the equicrescent, or independent variable, and x as the function. Or it may be desirable to introduce a new variable of which x is a function, and make it the equicrescent variable.

Either of these changes can be readily effected by first giving to the expression under consideration the form which it would have had if neither variable had been treated as equicrescent. Then, to make y equicrescent, remember that all its differentials above the first are 0, and drop out the terms affected by them. To introduce a new independent equicrescent variable, as θ, of which x is a function, simply substitute in the general form in which neither x nor y is equicrescent, the values of x, dx, d^2x, etc., in terms of the new equicrescent variable θ.

114. *Prob.*—*To find the forms which* $\frac{dy}{dx}$, $\frac{d^2y}{dx^2}$, $\frac{d^3y}{dx^3}$, etc., *take when neither variable is considered equicrescent.*

Dem.—Since the hypothesis of an equicrescent variable has not modified the form of $\frac{dy}{dx}$, in it x or y may be considered equicrescent, or neither, at pleasure.

Again $\frac{d^2y}{dx^2}=\frac{d\left(\frac{dy}{dx}\right)}{dx}$. Now differentiating the latter without regarding dx as constant we have $\frac{d\left(\frac{dy}{dx}\right)}{dx}=\frac{\frac{d^2y\,dx-d^2x\,dy}{dx^2}}{dx}=\frac{d^2y\,dx-d^2x\,dy}{dx^3}$, which is therefore the form which the second differential coefficient takes when neither variable is equicrescent.

Once more, $\frac{d^3y}{dx^3}=\frac{d\left(\frac{d^2y}{dx^2}\right)}{dx}=\frac{d\left(\frac{d^2y\,dx-d^2x\,dy}{dx^3}\right)}{dx}=$

$\frac{d^3y\,dx^4+d^2y\,d^2x\,dx^3-d^3x\,dy\,dx^3-d^2x\,d^2y\,dx^3-3(d^2y\,dx-d^2x\,dy)dx^2\,d^2x}{dx^7}=$

$\frac{(d^3y\,dx-d^3x\,dy)dx-3(d^2y\,dx-d^2x\,dy)d^2x}{dx^5}$, which latter is the form assumed by the third differential coefficient when neither variable is equicrescent.

Ex. 1. Transform $x\frac{d^2y}{dx^2}+\left(\frac{dy}{dx}\right)^3-\frac{dy}{dx}=0$ in which x is equicrescent, into its equivalent when y is equicrescent.

Solution.—When y is equicrescent $d^2y=0$, hence $\frac{d^2y}{dx^2}=\frac{d^2y\,dx-d^2x\,dy}{dx^3}=$

$-\frac{d^2x\,dy}{dx^3}$. Substituting, we have $-x\frac{d^2x\,dy}{dx^3}+\frac{dy^3}{dx^3}-\frac{dy}{dx}=0$. Dividing by dy^3 and multiplying by dx^3 to give the differential of the independent variable its proper position, and changing signs, we obtain, $x\frac{d^2x}{dy^2}+\left(\frac{dx}{dy}\right)^2-1=0$.

Ex. 2. Transform $(dy^2+dx^2)^{\frac{3}{2}}+adxd^2y=0$, in which x is equicrescent, into its equivalent when y is equicrescent.

Result, $\left(1+\frac{dx^2}{dy^2}\right)^{\frac{3}{2}}-a\frac{d^2x}{dy^2}=0$.

Ex. 3. Transform $\frac{d^2y}{dx^2}-\frac{x}{1-x^2}\frac{dy}{dx}+\frac{y}{1-x^2}=0$, in which x is equicrescent, into its equivalent in terms of θ as the equicrescent variable, when $x=\cos\theta$.

Solution.—Introducing the general form of the second differential coefficient, the given equation becomes, $\frac{d^2y\,dx-d^2x\,dy}{dx^3}-\frac{x}{1-x^2}\frac{dy}{dx}+\frac{y}{1-x^2}=0$.

Now, from $x=\cos\theta$, $dx=-\sin\theta\,d\theta$, $d^2x=-\cos\theta\,d\theta^2$, and $1-x^2=\sin^2\theta$. Substituting these values, we have, $\frac{-d^2y\sin\theta\,d\theta+\cos\theta\,d\theta^2\,dy}{-\sin^3\theta\,d\theta^3}+\frac{\cos\theta}{\sin^2\theta}\frac{dy}{\sin\theta\,d\theta}+\frac{y}{\sin^2\theta}=0$, or reducing, $\frac{d^2y}{d\theta^2}+y=0$.

Ex. 4. Transform $R=\frac{\left(1+\frac{dy^2}{dx^2}\right)^{\frac{3}{2}}}{\frac{d^2y}{dx^2}}$, in which x is equicrescent, into its equivalent in terms of the variables r and θ, θ being the equicrescent variable, when $y=r\sin\theta$, and $x=r\cos\theta$.

Sug's.—The formula in its more general form, in which neither variable is regarded as equicrescent is

$$R=\frac{\left(1+\frac{dy^2}{dx^2}\right)^{\frac{3}{2}}}{\frac{d^2y\,dx-d^2x\,dy}{dx^3}}=\frac{(dx^2+dy^2)^{\frac{3}{2}}}{d^2y\,dx-d^2x\,dy}.$$

From $y=r\sin\theta$, we have, $dy=\sin\theta\,dr+r\cos\theta\,d\theta$, and $d^2y=\sin\theta\,d^2r+2\cos\theta\,d\theta\,dr-r\sin\theta\,d\theta^2$. From $x=r\cos\theta$, we have, $dx=\cos\theta\,dr-r\sin\theta\,d\theta$, and $d^2x=\cos\theta\,d^2r-2\sin\theta\,d\theta\,dr-r\cos\theta\,d\theta^2$. Substituting these values and reducing, we have,

$$R=\frac{\left(\frac{dr^2}{d\theta^2}+r^2\right)^{\frac{3}{2}}}{2\frac{dr^2}{d\theta^2}-r\frac{d^2r}{d\theta^2}+r^2}.$$

---

115. The method just given is sufficient to resolve all cases of change of the equicrescent variable, but general *formulæ* are sometimes convenient. We proceed to deduce the most important.

116. Prop.—*In operations where* y *has been treated as a function of the equicrescent variable* x, *to change the conception so that* x *shall be a function of the equicrescent variable* y, *we substitute for* $\frac{dy}{dx}$, $\frac{1}{\frac{dx}{dy}}$ *and for*

$$\frac{d^2y}{dx^2}, \quad -\frac{\frac{d^2x}{dy^2}}{\frac{dx^3}{dy^3}}.$$

Dem.—As the hypothesis of the equicrescent variable does not affect the first differential, we have the identical relation $\frac{dy}{dx} = \frac{1}{\frac{dx}{dy}}$.

Again, if neither variable is equicrescent we have $\frac{d^2y}{dx^2} = \frac{d^2y\,dx - d^2x\,dy}{dx^3}$.

Now, making y equicrescent makes $d^2y = 0$; hence $\frac{d^2y}{dx^2} = -\frac{d^2x dy}{dx^3} = -\frac{\frac{d^2x}{dy^2}}{\frac{dx^3}{dy^3}}$.

Q. E. D.

Sch.—In a similar manner the corresponding forms for the higher coefficients can be deduced; but they are not often required.

---

117. Prop.—*In expressions where* y *has been treated as a function of the equicrescent variable* x, *to change the expression so that* y *shall be a function of some new equicrescent variable as* θ, *having given* $x = \varphi(\theta)$, *we substitute for* $\frac{dy}{dx}$, $\frac{\frac{dy}{d\theta}}{\frac{dx}{d\theta}}$, *and for* $\frac{d^2y}{dx^2}$, $\frac{\frac{d^2y}{d\theta^2}\frac{dx}{d\theta} - \frac{d^2x}{d\theta^2}\frac{dy}{d\theta}}{\frac{dx^3}{d\theta^3}}$

Dem.—Since $y = f(x)$ and $x = \varphi(\theta)$, we have (**104**), $\frac{dy}{d\theta} = \frac{dy}{dx}\frac{dx}{d\theta}$; whence

$$\frac{dy}{dx} = \frac{\frac{dy}{d\theta}}{\frac{dx}{d\theta}}.$$

Again, the general value of $\frac{d^2y}{dx^2}$ (when neither variable is treated as equicrescent) is

$$\frac{d^2y}{dx^2} = \frac{d^2y\,dx - d^2x\,dy}{dx^3}.$$

Now from $x = \varphi(\theta)$ we have $dx = \frac{dx}{d\theta}d\theta$. Differentiating this, remembering that

$d\theta$ is constant, we have, $d^2x = \frac{d^2x}{d\theta^2}d\theta^2$. Substituting these values of dx and d^2x, in the general value above, we obtain $\frac{d^2y}{dx^2} = \frac{d^2y\,dx - d^2x\,dy}{dx^3} =$

$$\frac{d^2y\frac{dx}{d\theta}d\theta - \frac{d^2x}{d\theta^2}d\theta^2\,dy}{\frac{dx^3}{d\theta^3}d\theta^3} = \frac{\frac{d^2y}{d\theta^2}\frac{dx}{d\theta} - \frac{d^2x}{d\theta^2}\frac{dy}{d\theta}}{\frac{dx^3}{d\theta^3}}. \quad \text{Q. E. D.}$$

SCH.—To apply these *formulæ* in practice, we make the requisite substitutions of $\frac{\frac{dy}{d\theta}}{\frac{dx}{d\theta}}$ for $\frac{dy}{dx}$, and $\frac{\frac{d^2y}{d\theta^2}\frac{dx}{d\theta} - \frac{d^2x}{d\theta^2}\frac{dy}{d\theta}}{\frac{dx^3}{d\theta^3}}$ for $\frac{d^2y}{dx^2}$ in the given expression, and then finding the value of $\frac{dx}{d\theta}$ and of $\frac{d^2x}{d\theta^2}$ from the relation $x = \varphi(\theta)$, substitute these values, and have an expression in terms of $\frac{dy}{d\theta}$ and $\frac{d^2y}{d\theta^2}$, as required.

Ex. 1. Given $\frac{d^2y}{dx^2} = \frac{y}{(e^x + e^{-x})^2}$, in which x is the equicrescent (independent) variable, to transform so that t shall be the equicrescent variable, knowing that $x = \log\frac{t}{\sqrt{1 - t^2}}$.

SUG'S.—Substituting the value of $\frac{d^2y}{dx^2}$ requisite for this transformation, we have

$$\frac{\frac{d^2y}{dt^2}\frac{dx}{dt} - \frac{d^2x}{dt^2}\frac{dy}{dt}}{\frac{dx^3}{dt^3}} = \frac{y}{(e^x + e^{-x})^2}.$$

From $x = \log\frac{t}{\sqrt{1 - t^2}}$, we obtain $\frac{dx}{dt} = \frac{1}{t(1 - t^2)}$, and $\frac{d^2x}{dt^2} = -\frac{1 - 3t^2}{t^2(1 - t^2)^2}$. Introducing these values, the expression becomes $\frac{d^2y}{dt^2}\frac{1}{t(1 - t^2)} + \frac{1 - 3t^2}{t^2(1 - t^2)^2}\frac{dy}{dt} = \frac{y}{(e^x + e^{-x})^2}\frac{1}{t^3(1 - t^2)^3}$. This result readily reduces to $(t - t^3)\frac{d^2y}{dt^2} + (1 - 3t^2)\frac{dy}{dt} = ty$, by observing that $e^x = \frac{t}{\sqrt{1 - t^2}}$, and $e^{-x} = \frac{\sqrt{1 - t^2}}{t}$; whence $(e^x + e^{-x})^2 = \frac{1}{t^2(1 - t^2)}$.

Ex. 2. Transform $\frac{d^2y}{dx^2} + \frac{1}{x}\frac{dy}{dx} + y = 0$, into a form in which t is the equicrescent (independent) variable, knowing that $x^2 = 4t$.

Result, $t\frac{d^2y}{dt^2} + \frac{dy}{dt} + y = 0$.

SCH.—We observe from the foregoing that a change of the equicrescent variable may greatly simplify the expression.

Ex. 3. Making $x = \cos t$, and t the equicrescent variable, show that $(1 - x^2)\frac{d^2y}{dx^2} - x\frac{dy}{dx} = 0$, becomes $\frac{d^2y}{dt^2} = 0$.

---

118. Prob.—*Having* $u = f(x, y)$, *where* $x = \varphi(r, \theta)$ *and* $y = \varphi_1(r, \theta)$, *to express the partial differential coefficients* $\frac{du}{dx}$ *and* $\frac{du}{dy}$ *in terms of* r, θ *and the partial differential coefficients* $\frac{du}{dr}$, *and* $\frac{du}{d\theta}$.

SOLUTION.—Since u is a function of x and y, each of which is a function of r, we have

$$\frac{du}{dr} = \frac{du}{dx}\frac{dx}{dr} + \frac{du}{dy}\frac{dy}{dr}, \quad (1).$$

And in like manner

$$\frac{du}{d\theta} = \frac{du}{dx}\frac{dx}{d\theta} + \frac{du}{dy}\frac{dy}{d\theta}, \quad (2).$$

From these equations eliminating $\frac{du}{dy}$, and finding the value of $\frac{du}{dx}$, we obtain

$$\frac{du}{dx} = \frac{\frac{du}{dr}\frac{dy}{d\theta} - \frac{du}{d\theta}\frac{dy}{dr}}{\frac{dx}{dr}\frac{dy}{d\theta} - \frac{dy}{dr}\frac{dx}{d\theta}}, \text{ and similarly, } \frac{du}{dy} = -\frac{\frac{du}{dr}\frac{dx}{d\theta} - \frac{du}{d\theta}\frac{dx}{dr}}{\frac{dx}{dr}\frac{dy}{d\theta} - \frac{dy}{dr}\frac{dx}{d\theta}}.$$

119. COR.—*If* $x = r\cos\theta$, *and* $y = r\sin\theta$, *the above formulæ become* $\frac{du}{dx} = \cos\theta\frac{du}{dr} - \frac{\sin\theta}{r}\frac{du}{d\theta}$, *and* $\frac{du}{dy} = \sin\theta\frac{du}{dr} + \frac{\cos\theta}{r}\frac{du}{d\theta}$.

SCH. 1.—It will be observed that the relations $x = r\cos\theta$ and $y = r\sin\theta$, are the common *formulæ* for passing from rectangular to polar co-ordinates (PART I., ***128***).

GENERAL SCHOLIUM.

We here conclude the subject of the Differential Calculus, having developed the theory as fully as the plan of our course requires. In the next chapter we shall give a few applications; but its transcendent efficiency is best seen in the General Geometry and in Physics.

CHAPTER II.

APPLICATIONS OF THE DIFFERENTIAL CALCULUS.

SECTION I.

Development of Functions.

120. Def.—A Function is said to be ***Developed*** when the indicated operations are performed; or, more properly, when it is transformed into an equivalent series of terms following some general law.

Ill's.—Division affords a method of developing some forms of functions. Thus $y = \frac{1}{1-x}$ when developed by division becomes $y = 1 + x + x^2 + x^3 +$, etc. The binomial formula (Complete School Algebra, ***195***) is a formula for developing a binomial. Thus $y = (a + x)^5$ when developed becomes $y = a^5 + 5a^4x + 10a^3x^2 + 10a^2x^3 + 5ax^4 + x^5$. The subject is one of great importance in mathematics.

MACLAURIN'S FORMULA.

121. Def.—***Maclaurin's Formula*** is a formula for developing a function of a single variable in terms of the ascending powers of that variable and finite coefficients which depend upon the form of the function and upon its constants.

122. ***Prob.***—*To produce Maclaurin's Formula.*

Solution.—Let $y = f(x)$ be the function to be developed. It is proposed to discover the law of the development, when the function can be developed in the form

$$y = f(x) = A + Bx + Cx^2 + Dx^3 + Ex^4 +, \text{ etc.},$$

in which A, B, C, D, etc., are independent of x and depend upon the form of the function, and its constants.

Producing the successive differential coefficients, remembering that A, B, C, D, etc., are constant, we have,

$$\frac{dy}{dx} = B + 2Cx + 3Dx^2 + 4Ex^3 +, \text{ etc.},$$

$$\frac{d^2y}{dx^2} = 2C + 2\cdot 3Dx + 3\cdot 4Ex^2 +, \text{ etc.},$$

$$\frac{d^3y}{dx^3} = 2\cdot 3D + 2\cdot 3\cdot 4Ex +, \text{ etc.},$$

$$\frac{d^4y}{dx^4} = 2\cdot 3\cdot 4E +, \text{ etc.}$$

Now as the coefficients A, B, C, D, etc., are independent of x, they are the same for all values of it, and if we can find what they should be for any one value of x we shall have their values in all cases. Now, if $x = 0$ we have $(y) = f(0) = A$, the expressions (y), and $f(0)$ signifying the value of the function when $x = 0$. Also $\left(\frac{dy}{dx}\right) = B$, $\left(\frac{d^2y}{dx^2}\right) = 2C$, $\left(\frac{d^3y}{dx^3}\right) = 2 \cdot 3D$, $\left(\frac{d^4y}{dx^4}\right) = 2 \cdot 3 \cdot 4E$, the ( ) signifying in each case the value of the particular function when $x = 0$. Hence we find $A = (y)$, $B = \left(\frac{dy}{dx}\right)$, $C = \left(\frac{d^2y}{dx^2}\right)\frac{1}{1 \cdot 2}$, $D = \left(\frac{d^3y}{dx^3}\right)\frac{1}{1.2.3}$, $E = \left(\frac{d^4y}{dx^4}\right)\frac{1}{1.2.3.4}$, etc.

Substituting these values, we have

$$y = f(x) = (y) + \left(\frac{dy}{dx}\right)\frac{x}{1} + \left(\frac{d^2y}{dx^2}\right)\frac{x^2}{1 \cdot 2} + \left(\frac{d^3y}{dx^3}\right)\frac{x^3}{1 \cdot 2 \cdot 3} + \left(\frac{d^4y}{dx^4}\right)\frac{x^4}{1 \cdot 2 \cdot 3 \cdot 4} +, \text{ etc.},$$

which is the formula required.

123. Sch. 1.—The student should become perfectly familiar with this important formula, and for this purpose it will be well to describe it thus: Maclaurin's Formula develops $y = f(x)$ into a series of terms, the first of which is the value of the function when $x = 0$; the second is the first differential coefficient of the function, x being made 0, into x; the third, the second differential coefficient, x being made 0, into $\frac{x^2}{2}$, etc.

124. Sch. 2.—This formula may also be written $y = f(x) = f(0) + f_1(0)\frac{x}{1} + f_2(0)\frac{x^2}{1 \cdot 2} + f_3(0)\frac{x^3}{1 \cdot 2 \cdot 3} + f_4(0)\frac{x^4}{1 \cdot 2 \cdot 3 \cdot 4} +$, etc., in which $f(0)$, $f_1(0)$, $f_2(0)$, etc., signify the same as (y), $\left(\frac{dy}{dx}\right)$, $\left(\frac{d^2y}{dx^2}\right)$, etc., respectively.

Ex. 1. To develop $y = (a + x)^7$, by Maclaurin's Formula.

Solution.—Differentiating successively, we have $\frac{dy}{dx} = 7(a + x)^6$, $\frac{d^2y}{dx^2} = 6 \cdot 7(a+x)^5$, $\frac{d^3y}{dx^3} = 5 \cdot 6 \cdot 7(a+x)^4$, $\frac{d^4y}{dx^4} = 4 \cdot 5 \cdot 6 \cdot 7(a+x)^3$, $\frac{d^5y}{dx^5} = 3 \cdot 4 \cdot 5 \cdot 6 \cdot 7(a+x)^2$, $\frac{d^6y}{dx^6} = 2 \cdot 3 \cdot 4 \cdot 5 \cdot 6 \cdot 7(a + x)$, $\frac{d^7y}{dx^7} = 1 \cdot 2 \cdot 3 \cdot 4 \cdot 5 \cdot 6 \cdot 7$. Here the differentiation terminates. Making $x = 0$, we have, $(y) = (a + 0)^7 = a^7$, $\left(\frac{dy}{dx}\right) = 7(a + 0)^6 = 7a^6$, $\left(\frac{d^2y}{dx^2}\right) = 6 \cdot 7a^5$, $\left(\frac{d^3y}{dx^3}\right) = 5 \cdot 6 \cdot 7a^4$, $\left(\frac{d^4y}{dx^4}\right) = 4 \cdot 5 \cdot 6 \cdot 7a^3$, $\frac{d^5y}{dx^5} = 3 \cdot 4 \cdot 5 \cdot 6 \cdot 7a^2$, $\frac{d^6y}{dx^6} = 2 \cdot 3 \cdot 4 \cdot 5 \cdot 6 \cdot 7a$, and $\left(\frac{d^7y}{dx^7}\right) = 1 \cdot 2 \cdot 3 \cdot 4 \cdot 5 \cdot 6 \cdot 7$.

Substituting in the formula, we obtain

$$y = (a + x)^7 = a^7 + 7a^6x + 6 \cdot 7a^5\frac{x^2}{2} + 5 \cdot 6 \cdot 7a^4\frac{x^3}{2 \cdot 3} + 4 \cdot 5 \cdot 6 \cdot 7a^3\frac{x^4}{2 \cdot 3 \cdot 4} + 3 \cdot 4 \cdot 5 \cdot 6 \cdot 7a^2\frac{x^5}{2 \cdot 3 \cdot 4 \cdot 5} + 2 \cdot 3 \cdot 4 \cdot 5 \cdot 6 \cdot 7a\frac{x^6}{2 \cdot 3 \cdot 4 \cdot 5 \cdot 6} + 2 \cdot 3 \cdot 4 \cdot 5 \cdot 6 \cdot 7\frac{x^7}{2 \cdot 3 \cdot 4 \cdot 5 \cdot 6 \cdot 7};$$

or, reducing, $y = (a + x)^7 = a^7 + 7a^6x + 21a^5x^2 + 35a^4x^3 + 35a^3x^4 + 21a^2x^5 + 7ax^6 + x^7$, a result identical with that given by actual multiplication, or by the Binomial Formula.

Ex. 2. To deduce the Binomial Formula from Maclaurin's Formula.

SOLUTION.—Let $y=(a+x)^m$, in which m is either integral or fractional, positive or negative. Then differentiating successively, and taking the values for $x=0$, we have $(y)=a^m$, $\left(\frac{dy}{dx}\right)=ma^{m-1}$, $\left(\frac{d^2y}{dx^2}\right)=m(m-1)a^{m-2}$, $\left(\frac{d^3y}{dx^3}\right)=m(m-1)(m-2)a^{m-3}$, $\left(\frac{d^4y}{dx^4}\right)=m(m-1)(m-2)(m-3)a^{m-4}+$, etc. Substituting in Maclaurin's Formula, we obtain $y=(a+x)^m=a^m+ma^{m-1}x+m(m-1)a^{m-2}\frac{x^2}{2}+m(m-1)(m-2)a^{m-3}\frac{x^3}{2\cdot3}+m(m-1)(m-2)(m-3)a^{m-4}\frac{x^4}{2\cdot3\cdot4}+$, etc., or we may write $y=(a+x)^m=a^m+ma^{m-1}x+\frac{m(m-1)}{1\cdot2}a^{m-2}x^2+\frac{m(m-1)(m-2)}{1\cdot2\cdot3}a^{m-3}x^3+\frac{m(m-1)(m-2)(m-3)}{1\cdot2\cdot3\cdot4}a^{m-4}x^4+$, etc., which is the Binomial Formula.

Ex. 3. Develop $y=\sin x$.

SUG'S. $(y)=0$, $\left(\frac{dy}{dx}\right)=1$, $\left(\frac{d^2y}{dx^2}\right)=0$, $\frac{d^3y}{dx^3}=-1$, etc. $\therefore\ y=\sin x=x-\frac{x^3}{1\cdot2\cdot3}+\frac{x^5}{1\cdot2\cdot3\cdot4\cdot5}-\frac{x^7}{1\cdot2\cdot3\cdot4\cdot5\cdot6\cdot7}+$, etc.

Ex. 4. Develop $y=\cos x$.

Result, $y=\cos x=1-\frac{x^2}{1\cdot2}+\frac{x^4}{1\cdot2\cdot3\cdot4}-\frac{x^6}{1\cdot2\cdot3\cdot4\cdot5\cdot6}+\frac{x^8}{1\cdot2\cdot3\cdot4\cdot5\cdot6\cdot7\cdot8}-$, etc.

125. SCH.—These *formulæ* enable us to compute the natural sine and cosine of any arc directly. Thus, to obtain the natural sine of 10°, we have $x=\frac{\pi}{18}=.174533$ nearly. This value substituted in the *formulæ*, will give the $\sin 10°=.17365$, and $\cos 10°=.98481$. The series converge so rapidly that but few terms are necessary.

Ex. 5. Develop $y=(a^2+bx)^{\frac{1}{2}}$ by Maclaurin's Formula.

SUG'S. $y=(a^2+bx)^{\frac{1}{2}}$, $\therefore\ (y)=a$,

$$\frac{dy}{dx}=\tfrac{1}{2}b(a^2+bx)^{-\frac{1}{2}}_{[x=0]*}=\frac{b}{2a}=\left(\frac{dy}{dx}\right),$$

$$\frac{d^2y}{dx^2}=-\tfrac{1}{4}b^2(a^2+bx)^{-\frac{3}{2}}_{[x=0]}=-\frac{b^2}{4a^3}=\left(\frac{d^2y}{dx^2}\right)\text{ etc.}$$

$$\therefore\ y=(a^2+bx)^{\frac{1}{2}}=a+\frac{bx}{2a}-\frac{b^2x^2}{8a^3}+\frac{b^3x^3}{16a^5}-\text{, etc.}$$

---

* This notation signifies "x being made equal to 0."

Ex. 6. Develop $y = \sqrt{1+x^2}$.

Result, $y = \sqrt{1+x^2} = (1+x^2)^{\frac{1}{2}} = 1 + \frac{x^2}{2} - \frac{x^4}{8} + \frac{x^6}{16} - \frac{5x^8}{128} +$, etc.

Ex. 7. To produce the logarithmic series.

SOLUTION.—This series is the development of $y = \log(1+x)$. Differentiating with reference to a system of logarithms whose modulus is m, we have, $\frac{dy}{dx} = \frac{m}{1+x}$, $\frac{d^2y}{dx^2} = -\frac{m}{(1+x)^2}$, $\frac{d^3y}{dx^3} = \frac{2m}{(1+x)^3}$; $\frac{d^4y}{dx^4} = -\frac{2\cdot 3m}{(1+x)^4}$, etc. Whence $(y) = \log 1 = 0$, $\left(\frac{dy}{dx}\right) = m$, $\left(\frac{d^2y}{dx^2}\right) = -m$, $\left(\frac{d^3y}{dx^3}\right) = 2m$, $\left(\frac{d^4y}{dx^4}\right) = -2\cdot 3m$, etc. Substituting in Maclaurin's Formula, we have

$$y = \log(1+x) = m(x - \tfrac{1}{2}x^2 + \tfrac{1}{3}x^3 - \tfrac{1}{4}x^4 +, \text{etc.}),$$

the law of the series being apparent.

126. COR. 1.—*Since in the Napierian system* m = 1, we have

$$y = \log(1+x) = x - \tfrac{1}{2}x^2 + \tfrac{1}{3}x^3 - \tfrac{1}{4}x^4 + \tfrac{1}{5}x^5 -, \text{etc.}$$

127. SCH. 1.—This formula is not adapted to the purpose of computing logarithms, since it is diverging for integral values of x. Thus, letting $x = 2$, we have $y = \log 3 = 2 - 2 + \frac{8}{3} - 4 + \frac{32}{5} -$, etc., in which each term after the first two is greater than the preceding, and hence extending the series does not approximate the value of $\log 3$.

From the series in the corollary, however, a converging series may be readily deduced. The following is a simple method:

Substituting $-x$ for x we have

$$\log(1-x) = -x - \tfrac{1}{2}x^2 - \tfrac{1}{3}x^3 - \tfrac{1}{4}x^4 - \tfrac{1}{5}x^5 -, \text{etc.}$$

Subtracting this from the former, we obtain

$$\log(1+x) - \log(1-x) = \log\left\{\frac{1+x}{1-x}\right\} = 2(x + \tfrac{1}{3}x^3 + \tfrac{1}{5}x^5 + \tfrac{1}{7}x^7 +, \text{etc.})$$

Now putting $x = \frac{1}{2z+1}$, whence $\frac{1+x}{1-x} = \frac{z+1}{z}$, we have $\log\frac{z+1}{z} = \log(z+1) - \log z = 2\left(\frac{1}{2z+1} + \frac{1}{3(2z+1)^3} + \frac{1}{5(2z+1)^5} + \frac{1}{7(2z+1)^7} +, \text{etc.}\right)$,

or $\log(z+1) =$

$$\log z + 2\left(\frac{1}{2z+1} + \frac{1}{3(2z+1)^3} + \frac{1}{5(2z+1)^5} + \frac{1}{7(2z+1)^7} + \frac{1}{9(2z+1)^9} +, \text{etc.}\right).$$

This series converges for all positive values of z, and more rapidly as z increases.

To apply this formula in computing a table of Napierian logarithms, first let $z = 1$, whence $\log 2 = 0 +$

$$2\left(\frac{1}{3} + \frac{1}{3\cdot 3^3} + \frac{1}{5\cdot 3^5} + \frac{1}{7\cdot 3^7} + \frac{1}{9\cdot 3^9} + \frac{1}{11\cdot 3^{11}} + \frac{1}{13\cdot 3^{13}} + \frac{1}{15\cdot 3^{15}} +, \text{etc.}\right)$$

The numerical operations are conveniently performed as follows:

3	2.00000000		
9	.66666667	1	.66666667
9	.07407407	3	.02469136
9	.00823045	5	.00164609
9	.00091449	7	.00013064
9	.00010161	9	.00001129
9	.00001129	11	.00000103
9	.00000125	13	.00000009
	.00000014	15	.00000001
		∴ log 2 =	.69314718

Second. To find log 3, make $z = 2$, whence $\log 3 = \log 2 +$

$$2\left(\frac{1}{5} + \frac{1}{3 \cdot 5^3} + \frac{1}{5 \cdot 5^5} + \frac{1}{7 \cdot 5^7} + \frac{1}{9 \cdot 5^9} +, \text{etc.}\right)$$

Computation.

5	2.00000000		
25	.40000000	1	.40000000
25	.01600000	3	.00533333
25	.00064000	5	.00012800
25	.00002560	7	.00000366
	.00000102	9	.00000011
			.40546510
		log 2 =	.69314718
		∴ log 3 =	1.09861228

Third. To find log 4. $\log 4 = 2 \log 2 = 2 \times .69314718 = 1.38629436$.

Fourth. To find log 5. Let $z = 4$, whence $\log 5 = \log 4 +$
$$2\left(\frac{1}{9} + \frac{1}{3 \cdot 9^3} + \frac{1}{5 \cdot 9^5} + \frac{1}{7 \cdot 9^7} +, \text{etc.}\right)$$

Computation.

9	2.00000000		
81	.22222222	1	.22222222
81	.00274348	3	.00091449
81	.00003387	5	.00000677
	.00000042	7	.00000006
			.22314354
		log 4 =	1.38629436
		∴ log 5 =	1.60943790

In like manner we may proceed to compute the logarithms of the *prime* numbers from the formula, and obtain those of the composite numbers, on the principle that the logarithm of the product equals the sum of the logarithms of the factors.

The Napierian logarithm of the base of the common system, 10, = $\log 5 + \log 2 = 2.30258508$.

128. Cor. 2.—*The logarithms of the same number in different systems are to each other as the moduli of those systems; and the logarithm of a number in any system equals the Napierian logarithm of the same number multiplied by the modulus of the proposed system.*

129. Sch. 2.—*To find the modulus of the common system of logarithms,* we have com. log. $x = m$ Nap. log. x, whence $m = \frac{\text{com. log. } x}{\text{Nap. log. } x}$. Now having computed the Napierian logarithm of 10, by the formula above, and found it to be 2.302585, we have $m = \frac{\text{com. log. } 10}{\text{Nap. log. } 10} = \frac{1}{2.302585} =$ .43429448+.

130. SCH. 3.—To compute a table of common logarithms, first compute the Napierian logarithms and then multiply by the modulus of the common system, .43429448.

Ex. 8. To ascertain the relation of the modulus of a system of logarithms to its base.

SOLUTION.—Developing $y = a^x$, by Maclaurin's Formula, we have

$$y = a^x = 1 + \frac{1}{m}x + \frac{1}{m^2}\frac{x^2}{2} + \frac{1}{m^3}\frac{x^3}{2 \cdot 3} + \frac{1}{m^4}\frac{x^4}{2 \cdot 3 \cdot 4} + \text{etc.} \quad (1).$$

Again, putting $a = 1 + b$, and developing by the Binomial Formula, we obtain

$$y = a^x = (1+b)^x = 1 + xb + \frac{x(x-1)}{2}b^2 + \frac{x(x-1)(x-2)}{2 \cdot 3}b^3 + \frac{x(x-1)(x-2)(x-3)}{2 \cdot 3 \cdot 4}b^4$$
$$+ \frac{x(x-1)(x-2)(x-3)(x-4)}{2 \cdot 3 \cdot 4 \cdot 5}b^5 +, \text{etc.} \quad (2).$$

Expanding and collecting the coefficients of the 1st power of x we find it to be $b - \frac{b^2}{2} + \frac{b^3}{3} - \frac{b^4}{4} + \frac{b^5}{5} - \frac{b^6}{6} +$, etc.

Finally, since series (1) and (2) are equal the coefficients of like powers of x are equal, and $\frac{1}{m} = b - \frac{b^2}{2} + \frac{b^3}{3} - \frac{b^4}{4} + \frac{b^5}{5} - \frac{b^6}{6} +$, etc.; or restoring a and finding the value of m, we have

$$m = \frac{1}{(a-1) - \frac{1}{2}(a-1)^2 + \frac{1}{3}(a-1)^3 - \frac{1}{4}(a-1)^4 + \frac{1}{5}(a-1)^5 - \frac{1}{6}(a-1)^6 +, \text{etc.}}$$

131. SCH.—To find e, the base of the Napierian system. Since the logarithms of the same number in different systems are to each other as the moduli of those systems, we have

com. log e : Nap. log $e (= 1)$ :: .43429448 : 1.

∴ com. log e = .43429448, and e from the table of common logarithms, which we have shown how to compute, is 2.718281+.

Ex. 9. To develop $y = a^x$, *i. e.* to produce the exponential series.

Result, $y = a^x = 1 + \log a\frac{x}{1} + \log^2 a\frac{x^2}{1 \cdot 2} + \log^3 a\frac{x^3}{1 \cdot 2 \cdot 3} + \log^4 a\frac{x^4}{1 \cdot 2 \cdot 3 \cdot 4} +$, etc.

132. SCH.—If $a = e$, the Napierian base, this becomes $y = e^x = 1 + \frac{x}{1} + \frac{x^2}{1 \cdot 2} + \frac{x^3}{1 \cdot 2 \cdot 3} + \frac{x^4}{1 \cdot 2 \cdot 3 \cdot 4} + \frac{x^5}{1 \cdot 2 \cdot 3 \cdot 4 \cdot 5} +$, etc.

If $x = 1$, we have

$y = e^x = e = 2 + \frac{1}{2} + \frac{1}{2 \cdot 3} + \frac{1}{2 \cdot 3 \cdot 4} + \frac{1}{2 \cdot 3 \cdot 4 \cdot 5} +$, etc., a formula for finding the Napierian base, although the series converges slowly.

Ex. 10. Develop $y = \tan^{-1}x$.

Solution.—Differentiating $\frac{dy}{dx} = \frac{1}{1 + x^2}$, which by division becomes $\frac{dy}{dx} = 1 - x^2 + x^4 - x^6 + x^8 - x^{10} +$, etc.

Differentiating successively, $\frac{d^2y}{dx^2} = -2x + 4x^3 - 6x^5 + 8x^7 - 10x^9 +$, etc.

$\frac{d^3y}{dx^3} = -2 + 3 \cdot 4x^2 - 5 \cdot 6x^4 + 7 \cdot 8x^6 - 9 \cdot 10x^8 +$, etc.

$\frac{d^4y}{dx^4} = 2 \cdot 3 \cdot 4x - 4 \cdot 5 \cdot 6x^3 + 6 \cdot 7 \cdot 8x^5 - 8 \cdot 9 \cdot 10x^7 +$, etc.

$\frac{d^5y}{dx^5} = 2 \cdot 3 \cdot 4 - 3 \cdot 4 \cdot 5 \cdot 6x^2 + 5 \cdot 6 \cdot 7 \cdot 8x^4 - 7 \cdot 8 \cdot 9 \cdot 10x^6 +$, etc.

$\frac{d^6y}{dx^6} = -2 \cdot 3 \cdot 4 \cdot 5 \cdot 6x + 4 \cdot 5 \cdot 6 \cdot 7 \cdot 8x^3 - 6 \cdot 7 \cdot 8 \cdot 9 \cdot 10x^5 +$ etc.

Whence $(y) = \tan^{-1}0 = 0$, $\left(\frac{dy}{dx}\right) = 1$, $\left(\frac{d^2y}{dx^2}\right) = 0$, $\left(\frac{d^3y}{dx^3}\right) = -2$, $\left(\frac{d^4y}{dx^4}\right) = 0$, $\left(\frac{d^5y}{dx^5}\right) = 2 \cdot 3 \cdot 4$, $\left(\frac{d^6y}{dx^6}\right) = 0$, etc. Introducing these values into Maclaurin's Formula, we have $y = \tan^{-1}x = x - \frac{1}{3}x^3 + \frac{1}{5}x^5 - \frac{1}{7}x^7 + \frac{1}{9}x^9 - \frac{1}{11}x^{11} +$, etc.

Sch.—By means of this development we are enabled to find the value of π. Thus let $y = 45° = \frac{\pi}{4}$, whence $x = 1$, and we have $y = \frac{\pi}{4} = \tan^{-1}1 = 1 - \frac{1}{3} + \frac{1}{5} - \frac{1}{7} + \frac{1}{9} - \frac{1}{11} + \frac{1}{13} - \frac{1}{15} + \frac{1}{17} - \frac{1}{19} +$, etc.

---

133. Prop.—*Though Maclaurin's Formula is applicable to a very great variety of forms of functions of a single variable, it will not develop* ALL *such functions.*

The truth of this theorem will be substantiated if we can present examples of functions of a single variable which the formula will not develop properly. This we proceed to do.

Ex. 1. Show that $y = x^{\frac{1}{2}}$ is not properly developed by Maclaurin's Formula.

Solution.—From $y = x^{\frac{1}{2}}$, we have $\frac{dy}{dx} = \frac{1}{2x^{\frac{1}{2}}}$, $\frac{d^2y}{dx^2} = -\frac{1}{4x^{\frac{3}{2}}}$, etc. Hence $(y) = 0$, $\left(\frac{dy}{dx}\right) = \frac{1}{0} = \infty$, $\left(\frac{d^2y}{dx^2}\right) = -\infty$, etc. Substituting these values in the formula we have $y = x^{\frac{1}{2}} = 0 + \infty x - \infty \frac{x^2}{2} +$, etc. Such results as these will be simply.

unintelligible to the learner at first. But in this case it is easy to see that the development will consist of pairs of terms of the same general form as $\infty x - \infty \frac{x^2}{2}$. To ascertain just what is to be understood by this binomial, let us restore the values of ∞ as they were before x was made equal to 0, only using x' to indicate the x that is 0. We then have $\frac{1}{2x'^{\frac{1}{2}}}\frac{x}{1} - \frac{1}{4x'^{\frac{3}{2}}}\frac{x^2}{2}$, or $\frac{4x\,x' - x^2}{8x'^{\frac{3}{2}}}$. Now as $x' = 0$, this becomes $-\frac{x^2}{0}$, which is ∞ for *all* values of x except 0, and indeterminate for that. In like manner, it may be shown that each succeeding pair of terms equals ∞. Hence we have the absurd result that $y = x^{\frac{1}{2}} = \infty$, for all values of x, since the development should be true for all values of the variable.

Ex. 2. Show that $y = \log x$ is not properly developed by Maclaurin's Formula.

SUG'S.—The result is similar to the preceding except that the *first* term is ∞ in this case. Each succeeding binomial may be seen to be ∞, as in the former case. Hence we have the absurd result that $y = \log x = \infty$ for *all* values of x.

Ex. 3. Show that $y = \cot x$, and $y = a^{\frac{1}{x}}$ are not properly developed by Maclaurin's Formula.

134. SCH.—The *occasion* of the inapplicability of Maclaurin's Formula, in such cases as just given, is the fact that the form of the function is such that the coefficients $\frac{dy}{dx}$, $\frac{d^2y}{dx^2}$, etc., or the function itself, or both, become infinite for $x = 0$, which is contrary to the hypothesis upon which the formula was produced. Whether, in such cases, the failure to develop correctly by this formula is due to the fact that the particular function is incapable of *any* development, or whether it is simply because it will not develop in the particular form assumed in this theorem, does not as yet appear, and our limits forbid our entering upon the question.

TAYLOR'S FORMULA.

135. DEF.—***Taylor's Formula*** is a formula for developing a function of the sum of two variables in terms of the ascending powers of one of the variables, and finite coefficients which depend upon the other variable, the form of the function, and its constants.

136. Lemma.—*If* $u = f(x + y)$ *the partial differential coefficients* $\frac{du}{dx}$ *and* $\frac{du}{dy}$ *are equal.*

DEM.—Having $u = f(x + y)$, if x take an increment, we have $u + d_xu = f(x + dx + y) = f[(x + y) + dx]$; whence $d_xu = f[(x + y) + dx] - f(x + y)$. Again, if y take an increment, we have $u + d_yu = f(x+y+dy) = f[(x+y)+dy]$;

whence $d_y u = f[(x+y)+dy] - f(x+y)$. Now the *form* of the values of $d_x u$. and $d_y u$, as regards the way in which x and y are involved, is the same; hence, if it were not for dx and dy, they would be absolutely equal. Passing to the differential coefficients by dividing the first by dx and the second by dy, we have $\frac{du}{dx} = \frac{f[(x+y)+dx] - f(x+y)}{dx}$, and $\frac{du}{dy} = \frac{f[(x+y)+dy] - f(x+y)}{dy}$. But, in differentiating, the differential of the variable enters into every term; hence $f[(x+y)+dx] - f(x+y)$, as it would appear in application, would have a dx in each term which would be cancelled by the dx in the denominator in the coefficient, and $\frac{du}{dx}$ would be independent of dx. In like manner $\frac{du}{dy}$ is independent of dy. Hence, finally, as these values of the partial differential coefficients are simply functions of $(x+y)$, of the same form, and not involving dx or dy, they are equal. Q. E. D.

SCH.—The substance of this demonstration is that the values of the differential coefficients depend upon the *form* of the function, and are independent of the increment of the variable. Therefore when the form of the function is such as to give to the partial differential coefficients, the same *form* with respect to the variables, the coefficients are equal. But suppose we have $u = f(xy)$. $\frac{du}{dx} = \frac{f[(x+dx)y] - f(xy)}{dx} = \frac{f(xy + ydx) - f(xy)}{dx}$; and $\frac{du}{dy} = \frac{f[x(y+dy)] - f(xy)}{dy} = \frac{f(xy + xdy) - f(xy)}{dy}$. In these coefficients we see that the form is *not* such as to involve x and y in the same way; hence they are not necessarily equal. A few examples will render the truth of the lemma more clear.

Ex. 1. Given $u = (x+y)^m$ to show that the partial differential coefficients are equal.

$$\textit{Results}, \ \frac{du}{dx} = m(x+y)^{m-1}, \text{ and } \frac{du}{dy} = m(x+y)^{m-1}.$$

Ex. 2. Given $u = \log(x+y)$ to show that the partial differential coefficients are equal. *Results*, $\frac{du}{dx} = \frac{1}{x+y}$, and $\frac{du}{dy} = \frac{1}{x+y}$.

Ex. 3. Given $u = \tan^{-1}(x+y)$ to show that the partial differential coefficients are equal.

$$\textit{Results}, \ \frac{du}{dx} = \frac{1}{1+(x+y)^2}, \text{ and } \frac{du}{dy} = \frac{1}{1+(x+y)^2}.$$

Ex. 4. Given $u = \left(\frac{x}{y}\right)^m$, show that the partial differential coefficients are not equal.

$$\textit{Results}, \ \frac{du}{dx} = \frac{mx^{m-1}}{y^m}, \text{ and } \frac{du}{dy} = -\frac{mx^m y^{m-1}}{y^{2m}} = -mx^m y^{-m-1}.$$

Ex. 5. Given $u = \log(xy)$, show that the partial differential coefficients are not equal. *Results*, $\frac{du}{dx} = \frac{1}{x}$, and $\frac{du}{dy} = \frac{1}{y}$.

137. Prob.—*To produce Taylor's Formula.*

SOLUTION.—Let $u = f(x + y)$ be the function to be developed. It is proposed to discover the law of the development when the function can be developed in the form

$$u = f(x + y) = A + By + Cy^2 + Dy^3 + Ey^4 +, \text{ etc.,} \quad (1),$$

in which A, B, C, etc., are independent of y, and dependent upon x, the form of the function, and its constants.

Differentiating with respect to y, remembering that as A is independent of y it will disappear, and that as the factors B, C, D, etc., are likewise independent of y, they are to be regarded constant, we have

$$\frac{du}{dy} = B + 2Cy + 3Dy^2 + 4Ey^3 +, \text{ etc.} \quad (2).$$

Again, differentiating with respect to x, we have

$$\frac{du}{dx} = \frac{dA^*}{dx} + \frac{dB}{dx}y + \frac{dC}{dx}y^2 + \frac{dD}{dx}y^3 +, \text{ etc.} \quad (3).$$

Hence by (***136***)

$$B + 2Cy + 3Dy^2 + 4Ey^3 +, \text{ etc.} = \frac{dA}{dx} + \frac{dB}{dx}y + \frac{dC}{dx}y^2 + \frac{dD}{dx}y^3 +, \text{ etc.}$$

Now, by the theory of development by indeterminate coefficients, the coefficients of like powers of y are equal, and we have

$$B = \frac{dA}{dx}, \quad 2C = \frac{dB}{dx}, \quad 3D = \frac{dC}{dx}, \quad 4E = \frac{dD}{dx}, \quad \text{etc.}$$

But as (1) is true for all values of y, we may make $y = 0$, whence $A = f(x) = u'$, letting u' represent the value of the function u when $y = 0$. Hence $B = \frac{du'}{dx}$,

$$C = \frac{1}{2}\frac{dB}{dx} = \frac{1}{2}\frac{d\left(\frac{du'}{dx}\right)}{dx} = \frac{d^2u'}{dx^2}\frac{1}{2}, \quad D = \frac{1}{3}\frac{dC}{dx} = \frac{1}{2}\cdot\frac{1}{3}\frac{d\left(\frac{d^2u'}{dx^2}\right)}{dx} = \frac{d^3u'}{dx^3}\frac{1}{2\cdot 3},$$ and in like manner $E = \frac{d^4u'}{dx^4}\frac{1}{2\cdot 3\cdot 4}$, etc.

Substituting these values of A, B, C, D, etc. in (1), we have

$$u = f(x + y) = u' + \frac{du'}{dx}\frac{y}{1} + \frac{d^2u'}{dx^2}\frac{y^2}{1\cdot 2} + \frac{d^3u'}{dx^3}\frac{y^3}{1\cdot 2\cdot 3} + \frac{d^4u'}{dx^4}\frac{y^4}{1\cdot 2\cdot 3\cdot 4} +, \text{ etc.,}$$

which is the formula sought.

138. SCH.—Taylor's Formula develops $u_1 = f(x + y)$ into a series in which the *first term* is the value of the function when $y = 0$: the *second term* is the first differential coefficient of the function when $y = 0$, into y; the *third term* is the second differential coefficient of the function when $y = 0$, into $\frac{y^2}{1\cdot 2}$; etc., etc.

Ex. 1. Develop $u = (x + y)^m$ by Taylor's Formula, and thus deduce the Binomial Formula.

SOLUTION.—Making $y = 0$ we have $u' = x^m$. Differentiating $u' = x^m$, successively, we obtain $\frac{du'}{dx} = mx^{m-1}$, $\frac{d^2u'}{dx^2} = m(m-1)x^{m-2}$, $\frac{d^3u'}{dx^3} = m(m-1)(m-2)x^{m-3}$,

* This is the proper form, since A, B, C, etc., are functions of x.

$\frac{d^4u'}{dx^4} = m(m-1)(m-2)(m-3)x^{m-4}$, etc. Substituting these results in Taylor's Formula, we have

$u = (x+y)^m = x^m + mx^{m-1}y + \frac{m(m-1)}{2}x^{m-2}y^2 + \frac{m(m-1)(m-2)}{2\cdot 3}x^{m-3}y^3 + \frac{m(m-1)(m-2)(m-3)}{2\cdot 3\cdot 4}x^{m-4}y^4 +$, etc., which is the Binomial Formula.

Ex. 2. Develop $u = \log(x+y)$.

Sug's.—This being a function of the sum of two variables, we apply Taylor's Formula, $u' = \log x$, $\frac{du'}{dx} = \frac{1}{x}$, $\frac{d^2u'}{dx^2} = -\frac{1}{x^2}$, $\frac{d^3u'}{dx^3} = \frac{2}{x^3}$, etc.

Hence $u = \log(x+y) = \log x + \frac{y}{x} - \frac{y^2}{2x^2} + \frac{y^3}{3x^3} - \frac{y^4}{4x^4} +$, etc.

Ex. 3. Develop $u = a^{x+y}$.

Result, $u = a^x(1 + \log a\, y + \frac{\log^2 a}{2}y^2 + \frac{\log^3 a}{2\cdot 3}y^3 +$, etc.).

Ex. 4. Develop $u = \sin(x+y)$.

Sug's. $u' = \sin x$, $\frac{du'}{dx} = \cos x$, $\frac{d^2u'}{dx^2} = -\sin x$, $\frac{d^3u'}{dx^3} = -\cos x$, etc. Hence $u = \sin(x+y) = \sin x + \cos x\frac{y}{1} - \sin x\frac{y^2}{1\cdot 2} - \cos x\frac{y^3}{1\cdot 2\cdot 3} + \sin x\frac{y^4}{1\cdot 2\cdot 3\cdot 4} + \cos x\frac{y^5}{1\cdot 2\cdot 3\cdot 4\cdot 5} - \sin x\frac{y^6}{1\cdot 2\cdot 3\cdot 4\cdot 5\cdot 6} - \cos x\frac{y^7}{1\cdot 2\cdot 3\cdot 4\cdot 5\cdot 6\cdot 7} +$, etc. $=$

$\sin x(1 - \frac{y^2}{1\cdot 2} + \frac{y^4}{1\cdot 2\cdot 3\cdot 4} - \frac{y^6}{1\cdot 2\cdot 3\cdot 4\cdot 5\cdot 6} +$, etc.$) +$

$\cos x(y - \frac{y^3}{1\cdot 2\cdot 3} + \frac{y^5}{1\cdot 2\cdot 3\cdot 4\cdot 5} - \frac{y^7}{1\cdot 2\cdot 3\cdot 4\cdot 5\cdot 6\cdot 7} +$, etc.$) = \sin x\cos y + \cos x\sin y$, since the series in the parentheses are equal respectively to $\cos y$ and $\sin y$ (**124**, *Ex's* 3, and 4).

Ex. 5. Develop $u = \cos(x+y)$.

Result, $u = \cos(x+y) = \cos x(1 - \frac{y^2}{1\cdot 2} + \frac{y^4}{1\cdot 2\cdot 3\cdot 4} - \frac{y^6}{1\cdot 2\cdot 3\cdot 4\cdot 5\cdot 6} +$, etc.$)$

$- \sin x(y - \frac{y^3}{1\cdot 2\cdot 3} + \frac{y^5}{1\cdot 2\cdot 3\cdot 4\cdot 5} - \frac{y^7}{1\cdot 2\cdot 3\cdot 4\cdot 5\cdot 6\cdot 7} +$, etc.$) =$ $\cos x\cos y - \sin x\sin y$.

Ex. 6. Develop $u = \sin(x-y)$, and also $u = \cos(x-y)$.

Results, $u = \sin(x-y) = \sin x\cos y - \cos x\sin y$, and $u = \cos(x-y) = \cos x\cos y + \sin x\sin y$.

Ex. 7. Develop $u = (x+y)^5$, also $u = (x-y)^{\frac{1}{2}}$, by Taylor's Formula.

Ex. 8. Develop $u = (x-y)^{-4}$.

139. Taylor's Formula is much used for developing a function of a single variable after the variable has taken an increment. When so used the increment may be conceived as finite or infinitesimal, only so that it be regarded as a variable.

Ex. 1. Given $y = \log x$, to find y' which represents the value of y after x has taken the increment h.

SOLUTION. $y' = \log (x + h)$, which developed by Taylor's Formula gives $y' = \log (x + h) = \log x + m\left(\frac{h}{x} - \frac{h^2}{2x^2} + \frac{h^3}{3x^3} - \frac{h^4}{4x^4} +, \text{etc.}\right)$, m being the modulus of the system of logarithms.

SCH.—If h be considered infinitesimal with respect to x, so that we have $h = dx$, we may drop all the terms within the parenthesis except the first, and write $y' = \log x + \frac{mdx}{x}$. This is the consecutive state of the function $y = \log x$. Hence subtracting the latter from the former we have $y' - y = dy = d\log x = \frac{mdx}{x}$. This result is as it should be, in accordance with the rule for differentiating a logarithm.

Ex. 2. Given $y = 3x - 2x^3 - 5$, to find y', which represents the value of the function after x has taken the increment h.

Result, $y' = 3x - 2x^3 - 5 + (3 - 6x^2)h - 12x\frac{h^2}{2} - 12\frac{h^3}{2\cdot 3} = 3x - 2x^3 - 5 + (3 - 6x^2)h - 6xh^2 - 2h^3$.

SCH.—This result may be easily verified by direct substitution. Thus, $y' = 3(x + h) - 2(x + h)^3 - 5$. Expanding, $y' = 3x + 3h - 2x^3 - 6x^2h - 6xh^2 - 2h^3 - 5 = 3x - 2x^3 - 5 + (3 - 6x^2)h - 6xh^2 - 2h^3$.

---

140. Prop.—*Though Taylor's Formula gives the general form of the development of a function of the sum of two variables, there are sometimes particular values of one or the other of the variables for which the development is not true.*

We will illustrate this proposition with a few examples.

Ex. 1. Develop $u = (x + y - a)^{\frac{1}{2}}$ by Taylor's Formula, and show that the development is false when $x = a$.

SOLUTION. $u' = (x - a)^{\frac{1}{2}}$, $\frac{du'}{dx} = \frac{1}{2}(x - a)^{-\frac{1}{2}} = \frac{1}{2(x - a)^{\frac{1}{2}}}$, $\frac{d^2u'}{dx^2} = -\frac{1}{4}(x - a)^{-\frac{3}{2}} = -\frac{1}{4(x - a)^{\frac{3}{2}}}$, $\frac{d^3u'}{dx^3} = \frac{3}{8(x - a)^{\frac{5}{2}}}$, etc. Hence substituting in Taylor's Formula, we have $u = (x + y - a)^{\frac{1}{2}} = (x - a)^{\frac{1}{2}} + \frac{y}{2(x-a)^{\frac{1}{2}}} - \frac{y^2}{8(x-a)^{\frac{3}{2}}} + \frac{3y^3}{48(x-a)^{\frac{5}{2}}} -$, etc.

Now, no absurdity appears in this series for general values of x, but for $x = a$ the series becomes ∞, while $(x + y - a)^{\frac{1}{2}} = y^{\frac{1}{2}}$, for the same value. But by hypothesis x and y are independent and the development should be true for any value of y irrespective of the value assigned to x. Hence the conclusion that for $x = a$, $y^{\frac{1}{2}} = \infty$ is contradictory to the hypothesis, and false.

SCH.—It is evident that any form of function which, when developed by this formula, gives a factor of the form $(x \mp a)^m$ in the denominator of any term in the development, will afford an instance similar to the above, and the development will not be true for $x = \pm a$, since for this value $(x \mp a)^m = 0$, and the terms in the denominators of which it occurs will reduce to ∞.

Ex. 2. For what value of x is the development of $u = (x + y + b)^{\frac{7}{2}}$ by Taylor's Formula, untrue? *Ans.*, $x = -b$.

Ex. 3. Required the value of the function after x has taken an increment h, when $y = b + (x + c)^2 + (x - a)^{\frac{3}{2}}$. For what value of x does the development fail?

Result, $y' = b + (x + c)^2 + (x - a)^{\frac{3}{2}} + [2(x + c) + \frac{3}{2}(x - a)^{\frac{1}{2}}]h + [2 + \frac{3}{4}(x - a)^{-\frac{1}{2}}]\frac{h^2}{2} - \frac{3}{8}(x - a)^{-\frac{3}{2}}\frac{h^3}{2 \cdot 3} +$, etc.

$y' = \infty$ when $x = a$, and hence the development fails for this value.

SCH. 1.—If $h = dx$ the above development is true for all values of x, for then we have $y' = b + (x + c)^2 + (x - a)^{\frac{3}{2}} + [2(x + c) + \frac{3}{2}(x - a)^{\frac{1}{2}}]h$, which is the same as would be obtained by substituting $x + h$ for x in the first state of the function and developing, and then making $h = dx$, and dropping the higher powers of h. For $x = a$ this becomes $y' = b + (a + c)^2 + 2(a + c)h$, which is as it should be, since for $x + h = a + h$, $y' = b + (a + h + c)^2 + (a + h - a)^{\frac{3}{2}} = b + a^2 + 2ah + 2ac + h^2 + 2hc + c^2 + h^{\frac{3}{2}} =$ (dropping higher powers of h) $b + a^2 + 2ah + 2ac + 2hc + c^2 = b + (a^2 + 2ac + c^2) + (2ah + 2ch) = b + (a + c)^2 + 2(a + c)h$.

SCH. 2.—It will be observed that when Maclaurin's Formula fails to give the true development of a function it fails for *all* values of the variable; but when Taylor's fails it is only for *particular* values, the *general* development being still true.

GENERAL SCHOLIUM.—There are many other important *formulæ* for the development of functions, but the prescribed limits of this volume preclude their presentation.

SECTION II.

Evaluation of Indeterminate Expressions.

141. The following forms are called *The Indeterminate Forms*, viz.,

$$\frac{0}{0}, \frac{\infty}{\infty}, 0 \times \infty, \infty - \infty, 0^0, \infty^0, 1^\infty.$$

Whenever an expression assumes any one of these forms, the important question to be determined is whether it is *really* indeterminate, for it often happens that the indetermination is only apparent.

Of these forms, $\frac{0}{0}$ is the fundamental one, to which all the others can be reduced.

ILL.—That $\frac{0}{0}$ is an indeterminate form, is readily seen when we observe that the divisor, 0, multiplied by any finite number, produces the dividend, 0.

We may show that each of the other forms can be reduced to the first, and hence that they are indeterminate forms. Thus, let a represent a finite quantity; then $\dfrac{\frac{a}{0}}{\frac{a}{0}} = \frac{\infty}{\infty}$. But $\dfrac{\frac{a}{0}}{\frac{a}{0}} = \frac{a}{0} \times \frac{0}{a} = \frac{0}{0}$. That $\frac{\infty}{\infty}$ is an indeterminate form may also be seen directly; since one infinity may be any number of times another, and the symbols ∞ do not mean that numerator and denominator are the *same* infinity. Again $0 \times \infty = \frac{0}{a} \times \frac{a}{0} = \frac{0}{0}$, a being any finite quantity. Also $\infty - \infty$ is indeterminate, since the difference between *two* infinities may be any quantity whatever. Taking 0^0 and passing to logarithms, we have $0 \log 0 = 0(-\infty) = -0 \times \infty$, which has been shown equal to $\frac{0}{0}$. Finally, applying logarithms to ∞^0, and 1^∞, the former becomes $0 \log \infty = 0 \times \infty$, and the latter $\infty \log 1 = \infty \times 0$.

142. The apparent indetermination often occurs from the introduction of some hypothesis which introduces a factor 0, into both terms of the fraction.

ILL.—What is the value of $\frac{a^3 - x^3}{a - x}$ when $x = a$? Making $x = a$ reduces the expression to $\frac{0}{0}$; whence it would appear that $\frac{a^3 - x^3}{a - x}$ is indeterminate for $x = a$. But that such is not the case is evident, since $\frac{a^3 - x^3}{a - x} = a^2 + ax + x^2$ which $= 3a^2$ when $x = a$. This apparent indetermination arises from the fact that the hypothesis $x = a$ introduces a factor 0 into numerator and denominator. This factor being divided out, the true value is seen. But it is not always easy to discover the *factor* which becomes 0, so as to be able to cancel it; hence the necessity of some general method of procedure.

---

143. Prob.—*To evaluate* $y = \dfrac{f(x)}{\varphi(x)}$ *for* $x = a$, *when for this value of the variable the function assumes the form* $\dfrac{0}{0}$.

SOLUTION.—Let y' be the function when x has taken an increment h, so that $y' = \dfrac{f(x+h)}{\varphi(x+h)}$. Developing $f(x+h)$ and $\varphi(x+h)$ by Taylor's Formula, and for simplicity using $f'(x), f''(x), ----\ \varphi'(x), \varphi''(x)$, etc., for the coefficients, we have

$$y' = \frac{f(x+h)}{\varphi(x+h)} = \frac{f(x) + f'(x)\frac{h}{1} + f''(x)\frac{h^2}{1 \cdot 2} +, \text{ etc.}}{\varphi(x) + \varphi'(x)\frac{h}{1} + \varphi''(x)\frac{h^2}{1 \cdot 2} +, \text{ etc.}} \quad A.$$

But by hypothesis, when $x = a$, $f(x)$ and $\varphi(x)$ each equals 0. Hence dropping these terms and dividing by h, we have

$$y' = \frac{f'(x) + f''(x)\frac{h}{1 \cdot 2} +, \text{ etc.}}{\varphi'(x) + \varphi''(x)\frac{h}{1 \cdot 2} +, \text{ etc.}}$$

Now making $h = 0$, whence y' becomes y, there results $y = \dfrac{f'(x)}{\varphi'(x)} = \dfrac{f'(a)}{\varphi'(a)}$, as the value of the function for $x = a$.

If, however, $\dfrac{f'(a)}{\varphi'(a)} = \dfrac{0}{0}$, we can drop the first *two* terms of A, and dividing by h^2, making $h = 0$, and $x = a$, we have $y = \dfrac{f''(a)}{\varphi''(a)}$.

Thus we can continue to replace $f(x)$, and $\varphi(x)$ by their successive differential coefficients until a pair is reached which do not *both* reduce to 0 for $x = a$. The last result will be the true value of the symbol.

Ex. 1. Given $y = \dfrac{\sin x}{x}$, to evaluate the expression for $x = 0$.

SUG. $f(x) = \sin x$, and $\varphi(x) = x$. $f'(x) = \cos x$, and $\varphi'(x) = 1$. $\therefore y = \dfrac{\cos x_{x=0}*}{1} = \dfrac{1}{1} = 1$.

Ex. 2. Given $y = \dfrac{\log x}{x-1}$, to evaluate for $x = 1$.

For $x = 1$, $y = 1$.

Ex. 3. Given $y = \dfrac{x^5 - 1}{x - 1}$, to evaluate for $x = 1$.

For $x = 1$, $y = 5$.

---

* This subscript signifies "x being = to 0."

Ex. 4. Show that if $x = 0$, $y = \dfrac{a^x - b^x}{x} = \log \dfrac{a}{b}$.

Ex. 5. Show that if $x = a$, $y = \dfrac{e^{nx} - e^{na}}{(x-a)^s} = \infty$.

SUG. $f'(x) = ne^{nx}$, and $\varphi'(x) = s(x-a)^{s-1}$. Hence $f'(a) = ne^{na}$, and $\varphi'(a) = 0$. $\therefore$ For $x = a$, $y = \dfrac{ne^{na}}{0} = \infty$.

Ex. 6. Show that if $x = 0$, $y = \dfrac{x - \sin x}{x^3} = \dfrac{1}{6}$.

SUG.—The first and second differential coefficients of both numerator and denominator reduce to 0; but $f'''(x) = \cos x$, and $\varphi'''(x) = 6$. Hence for $x = 0$, $y = \dfrac{\cos x}{6} = \dfrac{1}{6}$.

Ex. 7. Evaluate $y = \dfrac{e^x - e^{-x} - 2x}{x - \sin x}$ for $x = 0$. $y_{x=0} = 2$.

Ex. 8. Evaluate $y = \dfrac{\log x}{(1-x)^{\frac{1}{2}}}$ for $x = 1$. $y_{x=1} = 0$.

Ex. 9. Evaluate $y = \dfrac{1 - \cos x}{x^2}$ for $x = 0$. $y_{x=0} = \frac{1}{2}$.

Ex. 10. Evaluate $y = \dfrac{(a^2 - x^2)^{\frac{3}{2}}}{(a-x)^{\frac{3}{2}}}$ for $x = a$.

SUG.—For $x = a$, the first and all succeeding differential coefficients of both numerator and denominator reduce to ∞. Hence we see that for $x = a$ the development of these functions by Taylor's Formula is not true. Moreover, if it were true, we should but exchange the symbol $\frac{0}{0}$ for $\frac{\infty}{\infty}$. In this case, however, it is easy to see the factor which gives the expression the indeterminate form. It is $(a-x)^{\frac{3}{2}}$. Cancelling it, $y = (a+x)^{\frac{3}{2}}_{x=a} = (2a)^{\frac{3}{2}}$.

Ex. 11. Evaluate $y = \dfrac{\sqrt{x} - \sqrt{a} + \sqrt{x-a}}{\sqrt{x^2 - a^2}}$ for $x = a$.

SUG.—This example is like the preceding. But dividing by $\sqrt{x-a}$, we have

$$y = \frac{\dfrac{\sqrt{x} - \sqrt{a}}{\sqrt{x-a}} + 1}{\sqrt{x+a}} = \frac{\sqrt{\dfrac{\sqrt{x} - \sqrt{a}}{\sqrt{x} + \sqrt{a}}} + 1}{\sqrt{x+a}}_{x=a} = \frac{\sqrt{\dfrac{0}{2\sqrt{a}}} + 1}{\sqrt{2a}} = \frac{1}{\sqrt{2a}}.$$

144. Prob.—*To evaluate* $y = \frac{f(x)}{\varphi(x)}$ *for* $x = a$, *when for this value of the variable the function assumes the form* $\frac{\infty}{\infty}$.

SOLUTION. $y = \frac{f(x)}{\varphi(x)} = \frac{\frac{1}{\varphi(x)}}{\frac{1}{f(x)}} = \frac{0}{0}$, when $f(x)$ and $\varphi(x)$ are each ∞. Now applying to $y = \frac{\frac{1}{\varphi(x)}}{\frac{1}{f(x)}}$ the method of the preceding problem, we have

$$\frac{f(x)}{\varphi(x)} = \frac{\frac{1}{\varphi(x)}}{\frac{1}{f(x)}} = \frac{-\frac{\varphi'(x)}{[\varphi(x)]^2}}{-\frac{f'(x)}{[f(x)]^2}} = \frac{\varphi'(x)}{f'(x)} \times \frac{[f(x)]^2}{[\varphi(x)]^2}.$$

Dividing the first and last members by $\frac{f(x)}{\varphi(x)}$, we have

$$1 = \frac{\varphi'(x)}{f'(x)} \times \frac{f(x)}{\varphi(x)}; \text{ whence } y = \frac{f(x)}{\varphi(x)} = \frac{f'(x)}{\varphi'(x)}.$$

Therefore the process in this case is the same as in the preceding.

Ex. 1. Evaluate $y = \frac{\log x}{x^n}$ for $x = \infty$.

SUG. $y_{x=\infty} = \frac{\frac{1}{x}}{nx^{n-1}} = \frac{1}{nx^n} = \frac{1}{n\,\infty^n} = 0.$

Ex. 2. Evaluate $y = \frac{\log x}{\cot x}$ for $x = 0$.

SUG. $y_{x=0} = \frac{\frac{1}{x}}{\operatorname{cosec}^2 x} = \frac{\frac{1}{x}}{\frac{1}{\sin^2 x}} = \frac{\sin^2 x}{x} = \frac{0}{0}$. Therefore differentiating again,

$y_{x=0} = \frac{2 \sin x \cos x}{1} = \frac{0}{1} = 0.$

Ex. 3. Evaluate $y = \frac{1 - \log x}{e^x}$ for $x = \infty$. $y_{x=\infty} = 0.$

Ex. 4. Evaluate $y = \frac{x^n}{e^x}$ for $x = \infty$.

SUG.—As the successive differential coefficients continue to be ∞ for $x = \infty$ until we reach the nth, we differentiate n times and obtain $y = \frac{n(n-1)(n-2) \cdots 3 \cdot 2 \cdot 1}{e^x}$ $= \frac{n(n-1)(n-2) \cdots 3 \cdot 2 \cdot 1}{\infty} = 0$, when $x = \infty$.

Ex. 5. Evaluate $y = \dfrac{\frac{1}{4}\frac{\pi}{x}}{\cot\frac{\pi x}{2}}$ for $x = 0$. $y_{x=0} = \dfrac{\pi^2}{8}$.

Ex. 6. Evaluate $y = \dfrac{\log\tan(2x)}{\log\tan x}$ for $x = 0$. $y_{x=0} = 1$.

---

145. Prob.—*To evaluate* $y = f(x) \times \varphi(x)$ *for* $x = a$, *when for this value of the variable the function assumes the form* $0 \times \infty$.

SOLUTION.—Since the reciprocal of that function of x which becomes ∞, is 0, we may write $y = f(x) \times \varphi(x) = \dfrac{f(x)}{\frac{1}{\varphi(x)}} = \dfrac{0}{0}$, for $x = a$, $f(x)$ being 0, and $\varphi(x)$ being ∞. Therefore, putting the expression in the form $y = \dfrac{f(x)}{\frac{1}{\varphi(x)}}$, it may be treated as the first case (**143**).

Ex. 1. Evaluate $y = 2^x \sin\dfrac{a}{2^x}$ for $x = \infty$.

SUG. Since for $x = \infty$, $2^x = \infty$, and $\sin\dfrac{a}{2^x} = 0$, we write $y = 2^x \sin\dfrac{a}{2^x} = \dfrac{\sin\frac{a}{2^x}}{2^{-x}}$. Whence replacing the numerator and denominator by their differential coefficients, we have $y = \dfrac{-a2^{-x}\log 2\cos\frac{a}{2^x}}{-2^{-x}\log 2} = a\cos\dfrac{a}{2^x} = a$, when $x = \infty$.

Ex. 2. Evaluate $y = (1 - x)\tan\dfrac{\pi x}{2}$ for $x = 1$.

SUG.—Since $\tan\dfrac{\pi x}{2} = \infty$ when $x = 1$, we write $y = \dfrac{1-x}{\frac{1}{\tan\frac{\pi x}{2}}} = \dfrac{1-x}{\cot\frac{\pi x}{2}}$ (differentiating) $= \dfrac{-1}{-\frac{\pi}{2}\operatorname{cosec}^2\frac{\pi x}{2}} = \dfrac{2}{\pi\operatorname{cosec}^2\frac{\pi}{2}} = \dfrac{2}{\pi}$, when $x = 1$.

Ex. 3. Evaluate $y = e^{\frac{1}{x}}\sin x$ for $x = 0$.

SUG. $y = e^{\frac{1}{x}}\sin x = \dfrac{\sin x}{e^{-\frac{1}{x}}} = \dfrac{\cos x}{\frac{1}{x^2}e^{-\frac{1}{x}}} = x^2 e^{\frac{1}{x}}\cos x = x^2 e^{\frac{1}{x}} = 0 \times \infty$, when $x = 0$.

Were we to repeat the process upon $x^2 e^{\frac{1}{x}}$ we should find that its form would remain the same. But put $\frac{1}{x} = z$, whence $x^2 e^{\frac{1}{x}} = \frac{e^z}{z^2}$, and differentiating twice, we have $\frac{e^z}{2} = \frac{e^{\frac{1}{x}}}{2} = \infty$, when $x = 0$. $\therefore\ y = e^{\frac{1}{x}} \sin x = x^2 e^{\frac{1}{x}} = \frac{e^{\frac{1}{x}}}{2} = \infty$, when $x = 0$.

Ex. 4. Evaluate $y = x^m \log^n x$ for $x = 0$.

Sug. $y = \frac{\log^n x}{\frac{1}{x^m}} = \frac{\infty}{\infty}$, when $x = 0$. Now differentiating, $y_{x=0} = \frac{n \log^{n-1} x \cdot \frac{1}{x}}{-\frac{m}{x^{m+1}}} = \frac{n \log^{n-1} x}{-\frac{m}{x^m}}$ (differentiating again) $= \frac{n(n-1)\log^{n-2} x \cdot \frac{1}{x}}{\frac{m^2}{x^{m+1}}} = \frac{n(n-1)\log^{n-2} x}{\frac{m^2}{x^m}}$. After n differentiations, we have $y_{x=0} = \frac{[n(n-1)(n-2) \text{ - - - - - } 3 \cdot 2 \cdot 1] \log^0 x}{\mp \frac{m^n}{x^m}} = \frac{n(n-1)(n-2) \text{ - - - - - } 3 \cdot 2 \cdot 1}{\mp \infty} = 0.$

[It is sometimes expedient to put the function in the form $\frac{\infty}{\infty}$ rather than $\frac{0}{0}$. Experiment must decide which is preferable in any given case.]

---

146. Prob.—*To evaluate* y = f(x) — φ(x) *for* x = a, *when for this value of the variable the function assumes the form* ∞ — ∞.

Solution.—Since $f(x) = \infty$, and $\varphi(x) = \infty$, we may write $y = \frac{1}{\frac{1}{f(x)}} - \frac{1}{\frac{1}{\varphi(x)}} = \frac{\frac{1}{\varphi(x)} - \frac{1}{f(x)}}{\frac{1}{f(x) \cdot \varphi(x)}} = \frac{0}{0}$. Having put the function in the latter form it may be treated as in the first case (***143***).

Ex. 1. Evaluate $y = \frac{2}{x^2 - 1} - \frac{1}{x - 1}$ for $x = 1$.

Solution.—In this case $f(x) = \frac{2}{x^2 - 1}$, and $\varphi(x) = \frac{1}{x - 1}$. Hence $\frac{1}{f(x)} = \frac{x^2 - 1}{2}$, and $\frac{1}{\varphi(x)} = \frac{x-1}{1}$. We may therefore write $y = \frac{2}{x^2 - 1} - \frac{1}{x - 1} = \frac{\frac{x-1}{1} - \frac{x^2 - 1}{2}}{\frac{x^2-1}{2} \cdot \frac{x-1}{1}} = \frac{1 - \frac{x+1}{2}}{\frac{x^2-1}{2}} = \frac{1-x}{x^2 - 1}_{x=1}$ (differentiating) $= \frac{-1}{2x} = -\frac{1}{2}$.

[In this case the factor $1 - x$ can be divided out without differentiating. Moreover $\frac{2}{x^2 - 1} - \frac{1}{x - 1} = \frac{2 - x - 1}{x^2 - 1} = \frac{1 - x}{x^2 - 1} = -\frac{1}{x + 1}$.]

Ex. 2. Evaluate $y = \frac{x}{x-1} - \frac{1}{\log x}$ for $x = 1$. $y_{x=1} = \frac{1}{2}$.

Ex. 3. Evaluate $y = \sec x - \tan x$ for $x = \frac{\pi}{2}$.

SUG.—This may be treated exactly as the last; but the following is more elegant. $y = \sec x - \tan x = \frac{1}{\cos x} - \frac{\sin x}{\cos x} = \frac{1 - \sin x}{\cos x} = \frac{0}{0}$, when $x = \frac{\pi}{2}$. Whence, differentiating, $y_{x=\frac{\pi}{2}} = \frac{-\cos x}{-\sin x} = 0$. Therefore when $x = \frac{\pi}{2}$, $\sec x$ and $\tan x$ are equal, a fact not difficult to observe from a figure.

Ex. 4. Evaluate $y = \frac{1}{\log x} - \frac{x}{\log x}$ for $x = 1$. $y_{x=1} = -1$.

---

147. Prob.—*To evaluate* $y = \{f(x)\}^{\varphi(x)}$ *for* $x = a$, *when for this value of the variable the function assumes either of the forms* 0^0, ∞^0, *or* 1^∞.

SOLUTION.—Passing to logarithms we have $\log y = (\varphi x) \log f(x)$. When $f(x) = 0$ and $\varphi(x) = 0$, $\log y = \varphi(x) \log f(x)_{x=a} = 0 \times (-\infty)$; when $f(x) = \infty$ and $\varphi(x) = 0$, $\log y = \varphi(x) \log f(x)_{x=a} = 0 \times \infty$; when $f(x) = 1$ and $\varphi(x) = \infty$, $\log y = \varphi(x) \log f(x)_{x=a} = \infty \times 0$. Hence all these cases fall under (**145**).

Ex. 1. Evaluate $y = x^x$ for $x = 0$.

SOLUTION. $\log y = x \log x = \frac{\log x}{\frac{1}{x}} = \frac{\infty}{\infty}$, when $x = 0$. Whence, replacing numerator and denominator by their first differential coefficients, we have $\frac{\frac{1}{x}}{-\frac{1}{x^2}} = -\frac{x^2}{x} = -x = 0$, for $x = 0$. Hence, $\log y = 0$, and $y = 1$.

Ex. 2. Evaluate $y = x^{\sin x}$, and $y = (\sin x)^{\sin x}$ for $x = 0$.

SUG.—Since for $x = 0$, $\sin x = x$, these are each $= x^x$. $\therefore\ y_{x=0} = (\sin x)^{\sin x} = x^x = 1$, by *Ex.* 1.

These may also be solved directly. Thus $y = x^{\sin x}$, gives $\log y = \sin x \log x = \frac{\log x}{\operatorname{cosec} x} = -\frac{\infty}{\infty}$, when $x = 0$. Hence by (**144**), and differentiating twice, we have

$$\log y_{x=0} = \frac{\log x}{\frac{1}{\sin x}} = \frac{\frac{1}{x}}{-\frac{\cos x}{\sin^2 x}} = -\frac{\sin^2 x}{x \cos x} = -\frac{2 \sin x \cos x}{\cos x - x \sin x} = -\frac{2 \sin x \cos x}{\cos x} = 2 \sin x = 0. \quad \therefore\ y = 1.$$

Also $y = (\sin x)^{\sin x}$ gives $\log y = \sin x \log \sin x = \frac{\log \sin x}{\operatorname{cosec} x}_{x=0}$ (differentiating) $= -\frac{\cot x}{\operatorname{cosec} x \cot x} = -\frac{1}{\operatorname{cosec} x} = 0. \quad \therefore \; y = 1.$

Ex. 3. Evaluate $y = (\cot x)^{\sin x}$ for $x = 0$.

Sug.—Put this in the form $\frac{\infty}{\infty}$. Thus $\log y = \sin x \log \cot x = \frac{\log \cot x}{\operatorname{cosec} x_{x=0}}$ (differentiating) $= \frac{-\frac{\operatorname{cosec}^2 x}{\cot x}}{-\operatorname{cosec} x \cot x} = \frac{\operatorname{cosec} x}{\cot^2 x} = \frac{\sin x}{\cos^2 x} = \frac{0}{1} = 0. \quad \therefore \; y = 1.$

Ex. 4. Evaluate $y = (1 + nx)^{\frac{1}{x}}$ for $x = 0$.

Sug. $\log y = \frac{\log (1 + nx)}{x}\Big]_{x=0} = \frac{0}{0}$. Differentiating, $\log y_{x=0} = \frac{n}{1} = n$. $\therefore \; y = e^n$.

Ex. 5. Evaluate $y = \{\cos(ax)\}^{\operatorname{cosec}^2(cx)}$ for $x = 0$.

Sug. $y = \{\cos(ax)\}^{\operatorname{cosec}^2(cx)} = 1^\infty$, when $x = 0$. Passing to logarithms $\log y = \operatorname{cosec}^2 (cx) \log \cos (ax) = \frac{\log \cos (ax)}{\sin^2 (cx)} = \frac{0}{0}$, when $x = 0$. Differentiating twice $\log y_{x=0} = \frac{-a \tan (ax)}{2c \sin (cx) \cos (cx)} = -\frac{a \tan (ax)}{c \sin (2cx)} = \frac{-a^2 \sec^2 (ax)}{2c^2 \cos (2cx)} = -\frac{a^2}{2c^2}$.

$\therefore \; y = e^{-\frac{a^2}{2c^2}}$.

SECTION III.

Maxima and Minima of Functions of One Variable.

148. Def.—***A Maximum*** value of a function of a single variable is a value which is greater than the immediately preceding and the immediately succeeding values; *i. e.*, the value when the variable takes an infinitesimal decrement, and the value when the variable takes an infinitesimal increment.

Ill's.—Let $y = \sin x$. When $x = \frac{\pi}{2}$, y is a maximum, since it is greater than the immediately preceding and the immediately succeeding values. If x takes an increment h, making $y' = \sin\left(\frac{\pi}{2} + h\right)$, or a decrement, $-h$, so that $y'' = \sin\left(\frac{\pi}{2} - h\right)$, y is evidently greater than y' and y'', as at 90° the sine is greater than it is at a little more or a little less than 90°.

Again, constructing the equation $y^3 = 6x^2 - x^3$, we find the right hand branch to be as given in the figure. Here $y = f(x)$, and y is a maximum when $x = AD = 4$, since for x infinitesimally less or greater than 4, y is less than for $x = 4$. The maximum value of y is, therefore, $y = \sqrt[3]{6 \cdot 4^2 - 4^3} = 3\frac{1}{4}$ nearly.

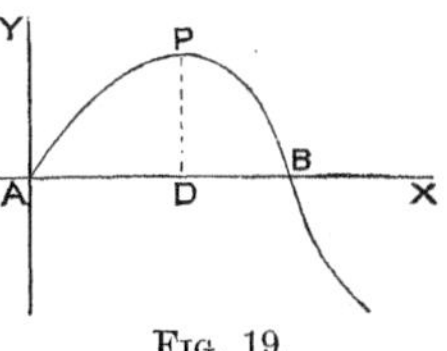

Fig. 19.

Once more, let $y = 8x - x^2$. If $x = 1$, $y = 7$; if $x = 2$, $y = 12$; if $x = 3$, $y = 15$; if $x = 4$, $y = 16$ (a maximum); if $x = 5$, $y = 15$; if $x = 6$, $y = 12$; and if $x = 7$, $y = 7$. Hence it appears that as x increases y increases till it has attained a certain value, when although x is made to continue its increase, y begins to diminish. The point at which the function ceases to increase and begins to decrease is its maximum. In this case it will be found that however little x varies from 4, either way, y becomes less than 16. Thus if $x = 3.9$, $y = 15.99$; and if $x = 4.1$, $y = 15.99$.

149. Def.—***A Minimum*** value of a function of a single variable is a value which is less than the immediately preceding and the immediately succeeding values; *i. e.*, the value when the variable takes an infinitesimal decrement, and the value when the variable takes an infinitesimal increment.

Ill's.—Let $y = \operatorname{cosec} x$. As x approaches $\frac{\pi}{2}$ y diminishes and approaches 1, reaching 1 at $x = \frac{\pi}{2}$. When x passes $\frac{\pi}{2}$, y begins to increase, so that $y = 1$, is a minimum value of the function $y = \operatorname{cosec} x$.

Again, $y = x^2 - 6x + 10$, has a minimum value for $x = 3$, at which value

$y = 1$. By substituting values of x a little greater than 3, as 3.01, and a little less as 2.09, y will be found to be greater than 1 in both cases. The locus of the function is given in *Fig.* 20, where **PD** represents the minimum value of y.

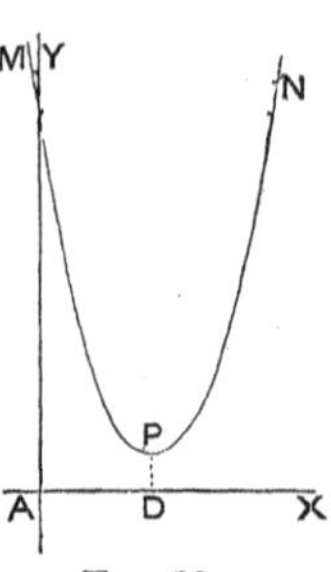

FIG. 20.

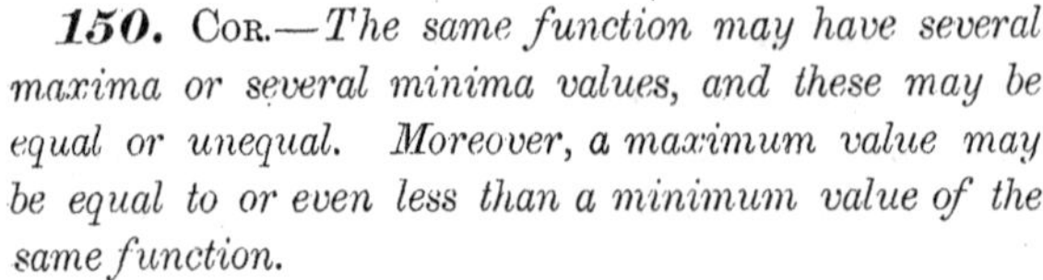
150. COR.—*The same function may have several maxima or several minima values, and these may be equal or unequal. Moreover, a maximum value may be equal to or even less than a minimum value of the same function.*

ILL'S.—The function $y = x^4 - 8x^3 + 22x^2 - 24x + 12$, has minima values for $x = 1$, and $x = 3$, which values are both $y = 3$; or two equal minima values, as illustrated by the ordinates at **P** and **P''** in the figure. For $x = 2$ $y = 4$, a maximum value, as illustrated by **P'D'**.

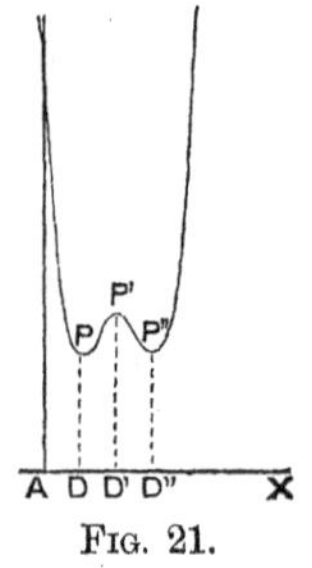

FIG. 21.

Again, let $y = f(x)$ be the equation of **MN** referred to **AX** and **AY** *Fig.* 22. Then **PD**, **P''D''**, and **P^IV D^IV** are maxima values of y; and **P'D'**, and **P'''D'''** are minima values. But the several maxima values are unequal and the minimum **P'D'** is greater than the maximum **P^IV D^IV**.

151. SCH.—It will be observed that the terms maximum and minimum, as here used, do not mean the greatest possible and least possible. Thus, if we ask for the maximum value of y in $y = x^3 - 3ax^2 - 5$, we do not inquire, what is the greatest possible value which y can have? but simply, whether if x vary continuously through all possible values, there is any point at which y will attain a greater value than it had immediately preceding that point, and than it will have immediately after passing that point; and, if there be such a value of y, what it is.

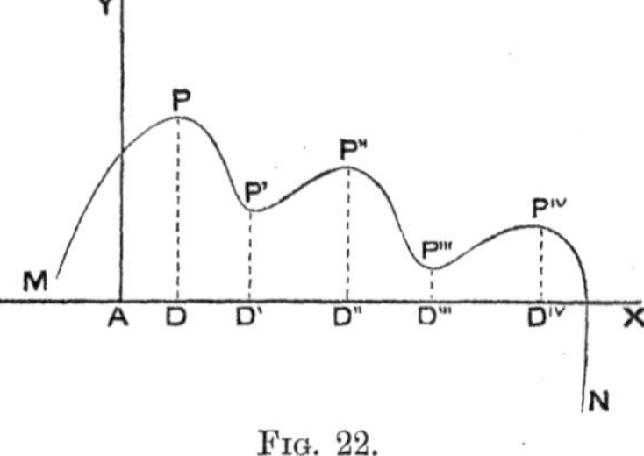

FIG. 22.

---

152. Prop.—*In an explicit function of a single variable,* $y = f(x)$, *the first differential coefficient,* $\frac{dy}{dx}$ *changes sign from + to —, for continuously increasing values of the variable, where the function is at a maximum, and from — to + where the function is at a minimum. Hence for such values the first differential coefficient* $= 0$ *or* ∞.

DEM.—Let $y = f(x)$ be the function. *First,* For $x = x'$, suppose y becomes y', a maximum. Then $y' = f(x')$ is at a maximum. Now the immediately preceding state of the function is $f(x' - dx')$, and we have $\frac{dy'}{dx'} = \frac{f(x' - dx') - f(x')}{(x' - dx') - x'}$. By hypothesis $f(x' - dx') - f(x')$ is $-$*, and as $(x' - dx') - x'$ is evidently $-$, we have $\frac{dy'}{dx'} +$. Again, the immediately succeeding state to $y' = f(x')$ is $f(x' + dx')$; hence we have $\frac{dy'}{dx'} = \frac{f(x' + dx') - f(x')}{(x' + dx') - x'}$. By hypothesis $f(x' + dx') - f(x')$ is $-$*, and as $(x' + dx') - x'$ is evidently $+$, we have $\frac{dy'}{dx'} -$. Therefore where $y' = f(x')$ is a maximum $\frac{dy'}{dx'}$ changes sign from $+$ to $-$.

Second. If $y' = f(x')$ is at a minimum $\frac{dy'}{dx'} = \frac{f(x' - dx') - f(x')}{(x' - dx') - x'}$ is $-$, since by hypothesis $f(x' - dx') - f(x')$ is $+$, and $(x' - dx') - x'$ is evidently $-$. Again $\frac{dy'}{dx'} = \frac{f(x' + dx') - f(x')}{(x' + dx') - x'}$ is $+$, as by hypothesis $f(x' + dx') - f(x')$ is $+$, and $(x' + dx') - x'$ is evidently $+$.

Finally, since when a varying function changes sign it passes through 0 or ∞, we have $\frac{dy}{dx} = 0$ or ∞ for maxima and minima values of the function. Q. E. D.

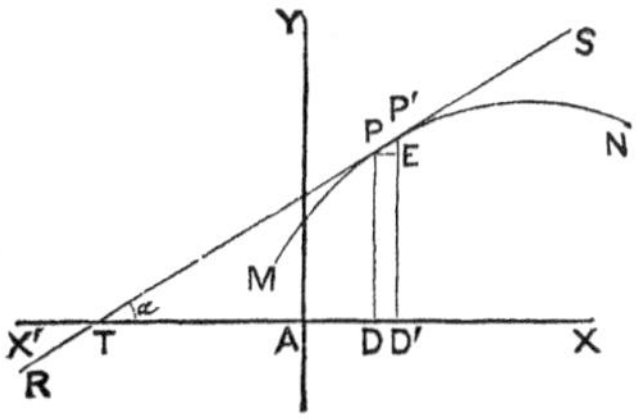

Geometrical Illustration.—**TS** being tangent to the curve **MN** at **P**, **P′** being a consecutive point so that **PE** represents dx, and **P′E** dy, we observe that the angle $\mathrm{P'PE} = \alpha$, the angle which the line makes with the axis of abscissas. Hence $\tan\alpha = \tan \mathrm{P'PE} = \frac{\mathrm{P'E}}{\mathrm{PE}} = \frac{dy}{dx}$; *i. e.* the first differential coefficient of the ordinate regarded as a function of the abscissa, represents the tangent of the angle which a tangent to a plane curve makes with the axis of abscissas.

Fig. 23.

Now, observing *Fig.* 22 we see that as x is increasing, and y approaching a maximum value as **PD**, the tangent to the curve makes an acute angle; hence approaching **P** from the left $\frac{dy}{dx}$ is $+$. *At* **P** the tangent becomes parallel to the axis of x; $\tan\alpha = \frac{dy}{dx} = 0$. Immediately upon passing **P**, α becomes obtuse, and consequently $\tan\alpha = \frac{dy}{dx}$ is $-$.

So also in approaching a minimum value as **P′D′** from the left it appears that α is obtuse, and hence $\frac{dy}{dx} -$; at this point, **P′**, $\alpha = 0$, and $\frac{dy}{dx} = 0$; and after passing **P′**, α becomes acute and $\frac{dy}{dx} +$.

* The hypothesis is that $y' = f(x')$ is a maximum, *i. e.* is greater than either the immediately preceding and the immediately succeeding states of the function. But $f(x' - dx')$ is the immediately preceding state, and $f(x' + dx)$ is the immediately succeeding state. Hence $f(x' - dx') < f(x')$, and $f(x' + dx') < f(x')$.

To illustrate the case in which $\frac{dy}{dx}$ changes sign by passing through ∞, consider $y = f(x)$ as the equation of **MN**, *Fig.* 24. **PD** is evidently a maximum ordinate. But in approaching **PD** from the left, α is an acute angle, and $\frac{dy}{dx}$, $+$. At **P**, $\alpha = 90°$, and $\frac{dy}{dx} = \infty$. After passing **PD**, α is obtuse and $\frac{dy}{dx}$, $-$. A similar illustration may be given of the case in which $\frac{dy}{dx}$ passes through ∞ at a minimum.

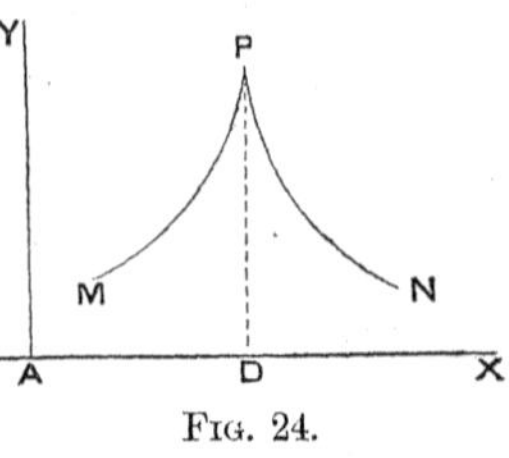

FIG. 24.

SCH.—The student needs to guard against the error of supposing that all values of the variable which render the first differential coefficient 0 or ∞, necessarily render the function a maximum or minimum. These values of the variable correspond to the maxima and minima values of the function if it has any maxima or minima values, since if the first differential coefficient changes sign, it must pass through 0 or ∞; *but a quantity may pass through 0 or ∞ without changing sign*, so that the values of the variable which render the first differential coefficient 0 or ∞ are simply *critical* values, *i. e.* values to be examined.

153. Prop.—*In an explicit function of a single variable,* $y = f(x)$, *the second differential coefficient,* $\frac{d^2y}{dx^2}$, *if not* 0, *is* $-$ *where the function is at a maximum, and* $+$ *where it is at a minimum.*

DEM.—Let $y = f(x)$ be the function. We have seen that when the function passes through a maximum $\frac{dy}{dx}$ changes sign from $+$ to $-$ for continuously increasing values of x, *i. e.* $\frac{dy}{dx}$ is *decreasing;* and when the function passes through a minimum $\frac{dy}{dx}$ changes sign from $-$ to $+$, *i. e.* $\frac{dy}{dx}$ is *increasing.* Now $\frac{d^2y}{dx^2} = \frac{d\frac{dy}{dx}}{dx} = \frac{df'(x)}{dx} = \frac{f'(x+dx) - f'(x)}{dx}$, which is $-$ when the numerator is $-$, and $+$ when the numerator is $+$, since dx is $+$ by hypothesis. But at a maximum $\frac{dy}{dx}$ is decreasing for increasing values of x, and $f'(x+dx) - f'(x)$ is $-$; and at a minimum $\frac{dy}{dx}$ is increasing for increasing values of x, and $f'(x+dx) - f'(x)$ is $+$. Therefore $\frac{d^2y}{dx^2}$ is $-$ at a maximum value of the function and $+$ at a minimum, unless it is 0, a case which is not yet provided for.

154. SCH. 1.—The ordinary method of examining an explicit function of a single variable for maxima and minima values is to form the first differential coefficient, put it equal to 0, and solve the resulting equation. Some or all of the values of the variable thus found may correspond to maxima and minima values of the function. They are then to be examined separately. To do this, form the second differential coefficient of the function and substituting in it the value of the variable to be examined, if it gives a $-$ result, this value of the variable corresponds to a maximum value of the function; but if it gives a $+$ result, it corresponds to a minimum value of the function. Thus all the values of the variable arising from equating the first differential coefficient with 0, are to be examined. If, however, any one of these critical values renders the second differential coefficient 0, it is best to examine the first differential coefficient for this value and see if it actually does change sign in passing from a value of the variable infinitesimally less to a value infinitesimally greater than that being examined.

155. Sch. 2.—The following axiomatic principles often facilitate the examination of a function for maxima and minima values:

1st. Whatever value of x renders $u = f(x)$ a maximum or minimum, renders $u' = af(x)$ or $u'' = \frac{f(x)}{a}$ a maximum or minimum. Hence constant factors or divisors may be dropped from the function.

2nd. Whatever value of x renders $u = f(x)$ positive and a maximum or minimum, renders $u' = [f(x)]^n$ a maximum or minimum, n being a positive integer; but if $u = f(x)$ is rendered negative for the particular value of x, $u' = [f(x)]^{2n}$ is a minimum when $u = f(x)$ is a maximum, and a maximum when $u = f(x)$ is a minimum. Hence the function may be involved to any power.

3rd. Whatever value of x renders $u = \log [f(x)]$ a maximum or minimum renders $u' = f(x)$ a maximum or minimum. Hence to examine the logarithm of a function we have to examine simply the function itself dropping the symbol log.

Ex. 1. What values of x render $y = \sqrt{4a^2x^2 - 2ax^3}$ a maximum or minimum; and what are the maxima and minima values of y?

Solution.*—Whatever value of x renders $y = \sqrt{4a^2x^2 - 2ax^3}$ a maximum or minimum renders y^2 or $y' = 4a^2x^2 - 2ax^3$ a maximum or minimum (***155***). And for a similar reason we may drop the constant factor $2a$, and examine $y'' = 2ax^2 - x^3$, since any value of x which renders the original function a maximum or minimum will also render this a maximum or minimum. Differentiating we have $\frac{dy}{dx} = 4ax - 3x^2$. Now whatever value of x renders the function a maximum or minimum renders $4ax - 3x^2 = 0$. From this $x = 0$, $x = \frac{4a}{3}$. If, therefore, there are any maxima or minima values of the function, they are those which correspond to

* This solution may seem needlessly prolix, but the author finds that comparatively few students really follow the argument through unless required to give it thus in detail.

one or the other or both of these values of x. Differentiating again, $\frac{d^2y}{dx^2}=4a-6x$. For $x=0$, $\frac{d^2y}{dx^2}=4a$; hence $x=0$ corresponds to a minimum value of the function. For $x=\frac{4a}{3}$, $\frac{d^2y}{dx^2}=4a-8a=-4a$; hence $x=\frac{4a}{3}$ corresponds to a maximum value of the function.

Substituting these values of x we find $y=\sqrt{4a^2x^2-2ax^3}=0$, a minimum value; and $y=\sqrt{4a^2x^2-2ax^3}=\sqrt{\frac{64a^4}{9}-\frac{128a^4}{27}}=\frac{8a^2}{3\sqrt{3}}$, a maximum.

Ex. 2. What values of x render $y=x^3-9x^2+24x-16$ a maximum or a minimum, and what are the maxima values of y?

Results, $x=2$ corresponds to a maximum, and $x=4$ to a minimum. The maximum value is $y=4$, and the minimum $y=0$.

Ex. 3. Examine $y=x^3-3x^2-24x+85$ for maxima and minima.

Results, For $x=4$, $y=5$, a minimum;
For $x=-2$, $y=113$, a maximum.

Ex. 4. Examine $y=5(x-x^2)$ for maxima and minima.

SUG.—Drop the 5. $x=\frac{1}{2}$, gives $y=\frac{5}{4}$, a maximum.

Ex. 5. Examine $y=(2ax-x^2)^{\frac{1}{2}}$ for maxima and minima.

SUG.—Use $y'=2ax-x^2$. $x=a$, gives $y=a$, a maximum, and $-a$, a minimum.

Ex. 6. Examine $y=x^4-8x^3+22x^2-24x+12$ for maxima and minima.

SUG'S. $\frac{dy}{dx}=4x^3-24x^2+44x-24=0$, or $x^3-6x^2+11x-6=0$. To find the roots of this equation, observe that the factors of the absolute term with its sign changed are 1, 2, and 3 (COMPLETE SCHOOL ALGEBRA, ***111***). By trial these are found to be the values of x, $x=1$ gives $y=3$, a minimum; $x=2$ gives $y=4$, a maximum; $x=3$ gives $y=3$, a minimum (see ILL. *Fig.* 21).

Ex. 7. Examine $y=x^5-5x^4+5x^3+1$ for maxima and minima.

Results, The critical values of x are 0, 0, 1, 3. For $x=1$, $y=2$, a maximum; for $x=3$, $y=-26$, a minimum. $x=0$ does not correspond to either a maximum or minimum value of y.

SUG.—That $x=0$ does not correspond to either a maximum or a minimum is determined as follows:

Having $\frac{dy}{dx}=5x^4-20x^3+15x^2$, substitute $0-h$ and $0+h$ for x, and evaluate the expression for h infinitesimal, thus determining whether $\frac{dy}{dx}$ changes sign or not.

in passing through $x = 0$. Thus $\frac{dy}{dx} = 5(0 - h)^4 - 20(0 - h)^3 + 15(0 - h)^2 = 5h^4 + 20h^3 + 15h^2 = 15h^2$, when h is infinitesimal. Again $\frac{dy}{dx} = 5h^4 - 20h^3 + 15h^2 = 15h^2$, when h is infinitesimal. Therefore, as $\frac{dy}{dx}$ has like signs on both sides of $x = 0$, and consecutive with it, it does not change sign in passing through $x = 0$. Hence $x = 0$ does not correspond to either a maximum or a minimum.

Ex. 8. Examine $y = b + (x - a)^3$ for maxima and minima.

SUG'S. $\frac{dy}{dx} = 3(x - a)^2 = 0$, gives $x = a$. Hence if there is any maximum or minimum it must be $y = b$, as no other value of x than $x = a$ will render $\frac{dy}{dx} = 0$. Again, since this value renders $\frac{d^2y}{dx^2} = 0$, we examine it by ascertaining whether $\frac{dy}{dx}$ changes sign at $x = a$. $\frac{dy}{dx} = 3(a - h - a)^2 = 3h^2$ is the value of $\frac{dy}{dx}$ next preceding $x = a$; and $\frac{dy}{dx} = 3(a + h - a)^2 = 3h^2$ is the next succeeding value. Therefore, as $\frac{dy}{dx}$ does not change sign at $x = a$, the function has no maximum nor minimum value.

Ex. 9. Examine $y = a(x - b)^4 + c$ for maxima and minima values.

SUG'S. $\frac{dy'}{dx} = 4(x - b)^3 = 0$. $\therefore x = b$. $\frac{dy'}{dx} = 4(b - h - b)^3 = -4h^3$, and $\frac{dy'}{dx} = 4(b + h - b)^3 = 4h^3$, are the values of $\frac{dy'}{dx}$ immediately preceding and succeeding $x = b$; hence, as $\frac{dy'}{dx}$ changes sign from $-$ to $+$ at this point, $x = b$ corresponds to a minimum. $\therefore y = a(b - b)^4 + c = c$ is a minimum.

Ex. 10. Examine $y = (x - 1)^4(x + 2)^3$ for maxima and minima.

SUG'S. $\frac{dy}{dx} = 4(x - 1)^3(x + 2)^3 + 3(x - 1)^4(x + 2)^2 = \{(x - 1)^3(x + 2)^2\}\{4(x+2)+3(x-1)\} = (x-1)^3(x+2)^2(7x+5) = 0$. $\therefore x - 1 = 0$, $x + 2 = 0$, $7x + 5 = 0$, give $x = 1$, $x = -2$, $x = -\frac{5}{7}$ as the critical values of x.

$\frac{d^2y}{dx^2} = 3(x-1)^2(x+2)^2(7x+5) + 2(x-1)^3(x+2)(7x+5) + 7(x-1)^3(x+2)^2 = 0$ for $x = 1$, and $x = -2$, but is $-\frac{(12^3) \cdot 9^2}{7^4}$ for $x = -\frac{5}{7}$. The latter value, therefore, corresponds to a maximum, and gives $y = (-\frac{5}{7} - 1)^4(-\frac{5}{7} + 2)^3 = \frac{\overline{12}^4 \cdot 9^3}{7^7}$, a maximum.

To ascertain whether $x = 1$ corresponds to a maximum or minimum, notice that $\frac{dy}{dx} = (1 - h - 1)^3(1 - h + 2)^2(7 - 7h + 5) = - h^3(3 - h)^2(12 - 7h)$ is $-$,

and $\frac{dy}{dx} = (1 + h - 1)^3(1 + h + 2)^2(7 + 7h + 5) = h^3(3 + h)^2(12 + 7h)$ is $+$.

Hence at $x = 1$, $\frac{dy}{dx}$ changes sign from $-$ to $+$, and there is a minimum at this value. This minimum is $y = 0$.

Finally, to test $x = -2$, $\frac{dy}{dx} = (-2 - h - 1)^3(-2 - h + 2)^2(-14 - 7h + 5) = (-3 - h)^3(-h)^2(-9 - 7h)$, which is $+$. Again, $\frac{dy}{dx} = (-2 + h - 1)^3 (-2 + h + 2)^2(-14 + 7h + 5) = (-3 + h)^3(+h)^2(-9 + 7h)$ is also $+$. Therefore $x = -2$ does not correspond either to a maximum or a minimum.

156. Sch.—It is usually easy to see, without going through with the details of the substitution, whether $\frac{dy}{dx}$ changes sign with h in such cases as the above; that is, whether if $x = a$ is the critical value we are testing, $\frac{dy}{dx}$ will have a different sign when we substitute $a + h$ for x, from what it will when we substitute $a - h$ for x.

Ex. 11. Examine $y = \frac{(x + 2)^3}{(x - 3)^2}$ for maxima and minima.

Sug's. $\frac{dy}{dx} = \frac{(x+2)^2(x - 13)}{(x - 3)^3} = 0$, gives for the critical values $x = -2$, $x = 13$.
$\frac{dy}{dx} = \frac{(x + 2)^2(x - 13)}{(x - 3)^3} = \infty$, gives $(x - 3)^3 = 0$, whence $x = 3$.

In this case it is better not to form $\frac{d^2y}{dx^2}$ as it is complicated, but test the critical values by noticing whether $\frac{dy}{dx}$ changes sign or not for these points. $x = -2$ does not correspond to either a maximum or a minimum. $x = 13$, gives $y = 33\frac{3}{4}$, a minimum. $x = 3$, gives $y = \infty$, a maximum.

Sch.—The first 10 examples give $x = \infty$ for $\frac{dy}{dx} = \infty$, and hence give rise to no critical values, as $x = \infty$ cannot correspond to a maximum or minimum, there being no succeeding value of the function.

Ex. 12. Examine $y = \frac{(x - 1)^2}{(x + 1)^3}$ for maxima and minima.

Sug's.—Putting $\frac{dy}{dx} = 0$, gives $x = 1$, and 5, as the critical values. Putting $\frac{dy}{dx} = \infty$, gives $x = -1$. When $x = 1$, $y = 0$, a minimum. When $x = 5$, $y = \frac{2}{27}$, a maximum. When $x = -1$, y is neither a maximum nor a minimum.

Ex. 13. Examine $y = b + (x - a)^{\frac{3}{2}}$ for maxima and minima.

Result, The critical value of x is $x = a$. But this does not correspond to either a maximum or a minimum, since $\frac{dy}{dx}$ does not change sign at this value.

Sug.—In this example $\frac{d^2y}{dx^2} = \pm \infty$ for $x = a$, and hence cannot be used to discriminate between maxima and minima.

Ex. 14. Examine $y = b + (x - a)^{\frac{4}{3}}$ for maxima and minima.

Result, $y = b$ is a minimum.

Ex. 15. Examine $y = b - (x - a)^{\frac{8}{5}}$ for maxima and minima.

Result, $y = b$ is a maximum.

Ex. 16. Show that $y = x^3 - 3x^2 + 6x + 7$ has neither a maximum nor a minimum value.

Ex. 17. Show that $y = \frac{x}{1 + x \tan x}$ is a maximum when $x = \cos x$.*

Sug. $\frac{dy}{dx} = \frac{1 - \frac{x^2}{\cos^2 x}}{(1 + x \tan x)^2}$. When $x < \cos x$, $\frac{dy}{dx}$ is $+$; but when $x > \cos x$, $\frac{dy}{dx}$ is $-$.

Ex. 18. Show that $y = \sin^3 x \cos x$ is a maximum when $x = 60°$.

Ex. 19. Show that $y = \frac{\sin x}{1 + \tan x}$ is a maximum when $x = 45°$.

GEOMETRICAL PROBLEMS.

Ex. 1. Required the altitude of the maximum cylinder which can be inscribed in a given right cone with a circular base.

Solution.—Let **SO** $= a$ be the altitude, and **AO** $= b$ the radius of the base of the given cone. Let $ac = x$ be the altitude, and c**O** $= af = y$ be the radius of the base of the required cylinder. The *function which is to be a maximum is the volume of the cylinder*. Calling this V, we have $V = \pi y^2 x$. In this form V is a function of *two* variables x and y. But these variables being dependent upon each other, we can find the value of one in terms of the other. Thus, **S**f : **SO** :: af : **AO** ; or, in the notation, $a - x : a :: y : b$; whence $y = \frac{b}{a}(a - x)$. Substi-

* When $x = \cos x$, $x = 42°21'$ nearly.

tuting this value of y, we have $V = \frac{\pi b^2}{a^2}(a - x)^2 x$, which is to be a maximum. Dropping the constant factor $\frac{\pi b^2}{a^2}$ (**153**, 1st), we have $V' = (a - x)^2 x = a^2 x - 2ax^2 + x^3$. $\therefore \frac{dV'}{dx} = a^2 - 4ax + 3x^2 = 0$; whence $x = \frac{1}{3}a$; that is, the axis of the cylinder is $\frac{1}{3}$ of the axis of the cone. From this we readily find y, the radius of the base of the cylinder $= \frac{2}{3}b$. $\therefore$ volume of cylinder $= \frac{4}{27}\pi ab^2$. But volume of cone $= \frac{1}{3}\pi ab^2$; whence volume of cylinder $= \frac{4}{9}$ volume of cone.

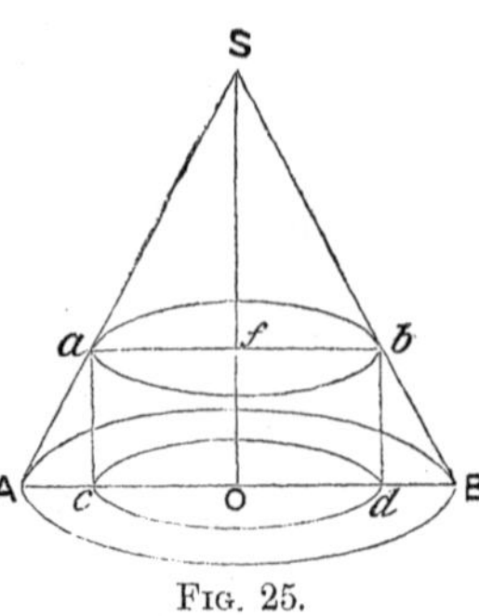

FIG. 25.

Ex. 2. To find the axis of the maximum cone which can be inscribed in a given sphere.

SUG'S.—Let **A**m**B** be the semicircle which, revolved upon **AB**, generates the sphere, and **A**ab the triangle which generates the cone. Let **AO** $= r$, **A**$b = x$, and $ab = y$. Then $V = \frac{1}{3}\pi y^2 x = \frac{1}{3}\pi x^2(2r - x)$, since $\overline{ab}^2 = y^2 =$ **A**$b \times b$**B** $= x(2r - x)$. $\therefore x = \frac{4}{3}r$, or the altitude of the cone is $\frac{2}{3}$ of the diameter of the sphere. Volume of sphere $= \frac{4}{3}\pi r^3$, volume of maximum cone $= \frac{8}{27} \times \frac{4}{3}\pi r^3$; or the cone $= \frac{8}{27}$ of the sphere.

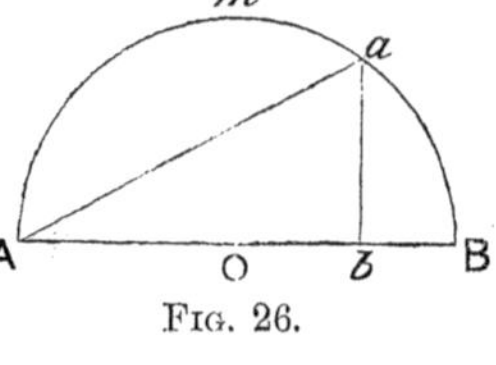

FIG. 26.

SCH.—In attempting the solution of such problems, first notice *what the function is* which is to be a maximum. Thus, in *Ex.* 1, it is *the volume of a cylinder;* in *Ex.* 2, it is *the volume of a cone.* Having obtained the equation expressing the function in terms of the variable or variables on which it depends, if there are two dependent variables involved, find from the conditions of the problem the relation between these variables, and substitute for one of them its value in terms of the other. Finally, we have a function of a single variable, which can be examined for maxima and minima values in the usual way.

Ex. 3. Required the cylinder of greatest convex surface which can be inscribed in a given right cone with a circular base.

SUG.—The function is *the convex surface of a cylinder.* Using the same notation as in *Ex.* 1, and letting **S** represent the function, we have **S** $= 2\pi yx = \frac{2\pi b}{a}(a - x)x$. $\therefore x = \frac{1}{2}a$, and **S** $= \frac{\pi ab}{2}$; that is, the altitude of the cylinder is $\frac{1}{2}$ that of the cone; and the convex surface of the cylinder is to the convex surface of the cone as $\frac{a}{2} : \sqrt{a^2 + b^2}$, or as $\frac{1}{2}$ the altitude of the cone is to its slant height.

Ex. 4. Required the maximum cylinder which can be cut from a given sphere. The axis of the cylinder $= \frac{2}{3}\sqrt{3}$ times radius of sphere. The cylinder is to the sphere as $1 : \sqrt{3}$.

Ex. 5. Required the area of the greatest rectangle which can be inscribed in a given circle.

The rectangle is a square, and its area $= 2r^2$.

Ex. 6. What is the altitude of the maximum rectangle which can be inscribed in a given parabola?

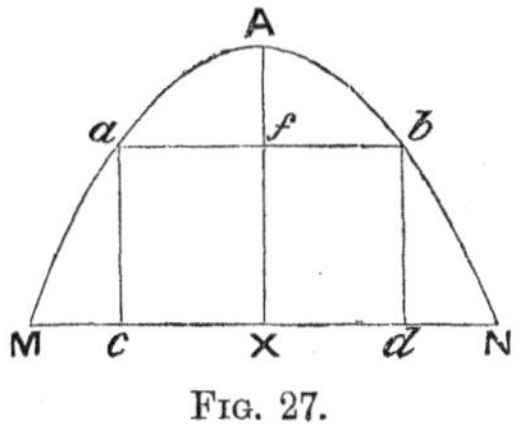

FIG. 27.

SUG'S.—Let $ac = x$, $af = y$, and $\mathbf{AX} = a$. Let A be the function, the area of the rectangle. Then $A = 2xy$. From the equation of the parabola $\overline{af}^2 = 2p \times \mathbf{A}f$, or $y^2 = 2p(a - x)$; whence $A = 2x\sqrt{2p(a - x)}$. $A' = ax^2 - x^3$, and $x = \frac{2}{3}a$.

Ex. 7. Required the axis of the cone of maximum convex surface which can be inscribed in a given sphere.

The axis $= \frac{4}{3}$ the radius of the sphere.

Ex. 8. Required the altitude of the maximum cone which can be inscribed in a given paraboloid, the vertex of the cone being at the intersection of the axis of the paraboloid with the base.

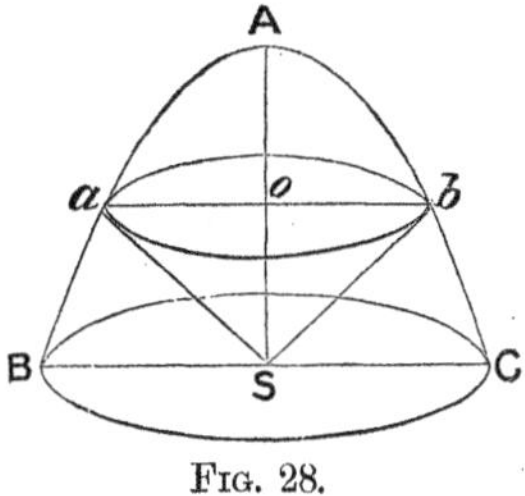

FIG. 28.

SUG'S.—Let $\mathbf{ABC}$ be the parabola whose revolution about $\mathbf{AS}$ as an axis generates the paraboloid. Let $\mathbf{AS} = a$ the axis of the paraboloid, $o\mathbf{S} = x$, the altitude of the cone, and $ao = y$ the radius of the base of the cone. The result is $x = \frac{1}{2}a$.

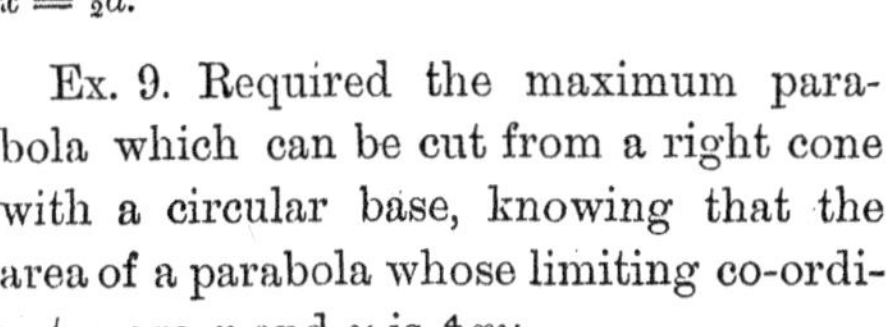
Ex. 9. Required the maximum parabola which can be cut from a right cone with a circular base, knowing that the area of a parabola whose limiting co-ordinates are x and y is $\frac{4}{3}xy$.

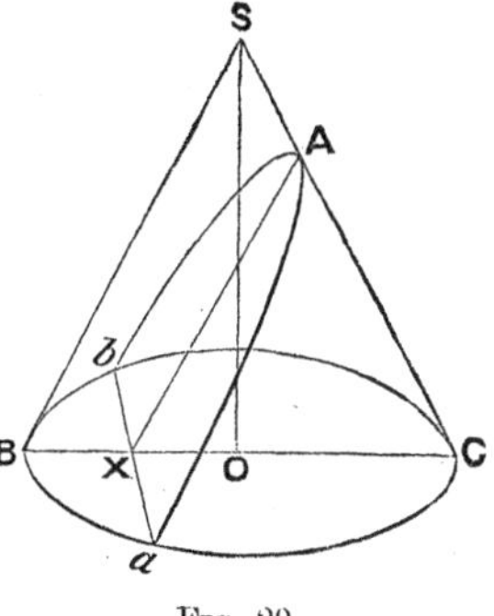

FIG. 29.

SUG'S.—Let $\mathbf{SO} = a$, $\mathbf{BO} = b$, $\mathbf{AX} = x$, and $a\mathbf{X} = y$. The function is A (the area) $= \frac{4}{3}xy$. But $a\mathbf{X} = y = \sqrt{\mathbf{BX} \times \mathbf{XC}}$; and $\mathbf{CX} : \mathbf{CB} : \mathbf{AX} : \mathbf{SB}$, or $\mathbf{CX} : 2b :: x : \sqrt{a^2 + b^2}$; whence $\mathbf{CX} = \frac{2bx}{S}$, letting $S = \sqrt{a^2 + b^2}$ for brevity. Then $\mathbf{BX} = \mathbf{CB} - \mathbf{CX} = 2b - \frac{2bx}{S}$

$= \frac{2b}{S}(S - x)$. Finally, $A = \frac{4}{3}x\sqrt{\frac{4b^2}{S^2}x(S - x)} = \frac{8b}{3S}\sqrt{x^3(S - x)}$, and $A' = Sx^3 - x^4$. The result is $x = \frac{3}{4}S$, that is, the axis of the parabola is $\frac{3}{4}$ the slant height of the cone. The area of the parabola $= \frac{1}{2}bS\sqrt{3}$. Notice that **CX** $= \frac{3}{4}$**CB**.

Ex. 10. From a given quantity of material a cylindrical vessel with circular base and open top is to be made, so as to contain the greatest amount. What must be its proportions?

Sug's.—Let x = the altitude, y the radius of the base, and V the volume. Then $V = \pi y^2 x$ is to be a maximum. Hence $\frac{dV}{dx} = 2yx\frac{dy}{dx} + y^2 = 0$, or $y = -2x\frac{dy}{dx}$. But $2\pi yx + \pi y^2 = s$, the surface. Differentiating $2\pi x dy + 2\pi y dx + 2\pi y dy = 0$; whence $\frac{dy}{dx} = -\frac{y}{x + y}$. Substituting, $y = \frac{2xy}{x + y}$. $\therefore\ y = x$, that is, the altitude = the radius of the base. The altitude $= \sqrt{\frac{s}{3\pi}}$.

Ex. 11. Of all right cones of a given convex surface to determine that whose solidity is the greatest.

The altitude $= \sqrt{2}$ into the radius of the base.

Ex. 12. To find the maximum rectangle inscribed in a given ellipse.

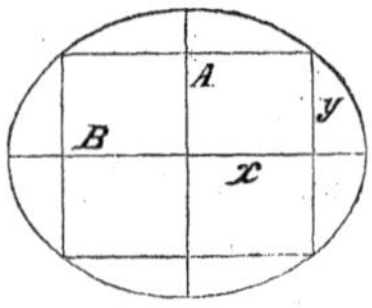

Sug's. $A = 4xy$. $A' = xy$. $\frac{dA'}{dx} = y + x\frac{dy}{dx} = 0$. $A^2y^2 + B^2x^2 = A^2B^2$. $\therefore\ \frac{dy}{dx} = -\frac{B^2x}{A^2y}$. $y = -x\frac{dy}{dx} = \frac{B^2x^2}{A^2y}$. $x : y :: A : B$. That is, the sides of the rectangle are to each other as the axes of the ellipse. The sides of the rectangle are $A\sqrt{2}$, and $B\sqrt{2}$.

Fig. 30.

Ex. 13. To find the maximum cylinder which can be inscribed in a given ellipsoid, generated by the revolution of an ellipse about its transverse axis.

The axis of the cylinder $= \frac{2}{\sqrt{3}}A$.

Ex. 14. A person being in a boat 3 miles from the nearest point of the beach, wishes to reach in the shortest time a place 5 miles from that point along the shore; supposing he can walk 5 miles an hour, but pull only at the rate of 4 miles an hour, required the place where he must land.

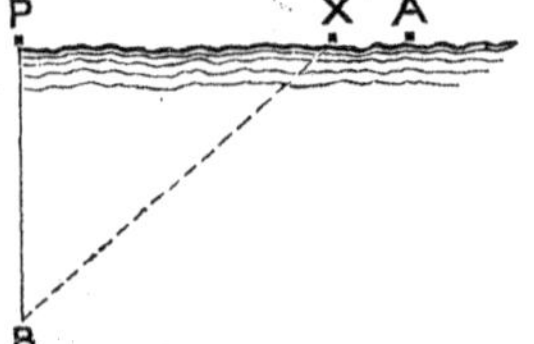

Fig. 31.

Sug's.—Let **AX** $= x$, and t = the time required to reach **A** by rowing from **B** to **X**, and walking from **X** to **A**. $t = \frac{x}{5} + \frac{\sqrt{(5 - x)^2 + 9}}{4}$ is to be a minimum. He must land at **X**, 1 mile from **A**.

Ex. 15. Divide a into two factors the sum of which shall be a minimum. *Result*, The factors are equal.

Ex. 16. The difference between two numbers is a; required that the square of the greater divided by the less shall be a minimum.
Result, The greater = twice the less.

Ex. 17. To find the number of equal parts into which a must be divided, so that their continued product shall be a maximum.

SUG'S.—The function is $u = \left(\frac{a}{x}\right)^x$. $\log u = x(\log a - \log x)$. $u' = x\log a - x\log x$. $\frac{du'}{dx} = \log a - \log x - 1 = 0$. $x = \frac{a}{e}$. Arithmetically the problem is possible only when $\frac{a}{e}$ is integral.

Ex. 18. Find a number x such that its xth root shall be a maximum. $x = e$.

Ex. 19. A privateer wishes to get to sea unobserved, but has to pass between two lights, A and B, on opposite headlands, the distance between which is a. The intensity, at a unit's distance, of A is b, and of B, c. At what point must the privateer cross the line joining the lights, so as to be as little in the light as possible; it being understood that the intensity of a light at any point equals its intensity at a unit's distance divided by the square of the distance from the light.

SUG.—Letting x = the distance from A, the function is $u = \frac{b}{x^2} + \frac{c}{(a - x)^2}$.

$$x = \frac{ab^{\frac{1}{3}}}{b^{\frac{1}{3}} + c^{\frac{1}{3}}}.$$

Ex. 20. The intensity of illumination from a given light varies as the sine of the angle under which the light strikes the illuminated surface, divided by the square of its distance from the surface. Required the height of a light directly over the centre of a given circle, so that it shall illuminate the circumference as much as possible.

SUG'S.—Let I represent the illumination at P, which is to be a maximum; $\mathsf{PO} = R$; and $\mathsf{LO} = x$. $I = \frac{\sin \mathsf{LPO}}{\overline{\mathsf{LP}}^2}$. But $\sin \mathsf{LPO} = \frac{\mathsf{LO}}{\mathsf{LP}} = \frac{x}{\mathsf{LP}}$. $\therefore I = \frac{x}{\overline{\mathsf{LP}}^3} = \frac{x}{(R^2 + x^2)^{\frac{3}{2}}}$; whence

$$\frac{dI}{dx} = \frac{(R^2 + x^2)^{\frac{3}{2}} - 3x^2(R^2 + x^2)^{\frac{1}{2}}}{(R^2 + x^2)^3} = 0. \quad (R^2 + x^2)^{\frac{3}{2}} - 3x^2(R^2 + x^2)^{\frac{1}{2}} = 0.$$

$R^2 + x^2 - 3x^2 = 0$, and $x = R\sqrt{\tfrac{1}{2}}$.

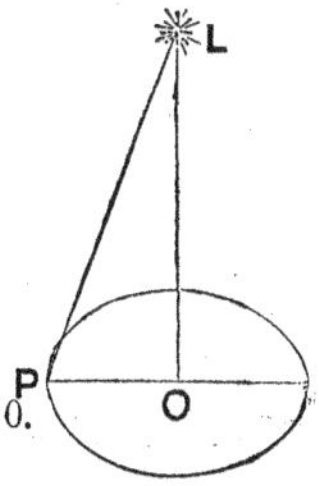

FIG. 32.

Ex. 21. To find in a line joining the centres of two spheres, the point from which the greatest portion of spherical surface is visible.

SUG'S.—The function is the sum of the two zones whose altitudes are **MD** and md; hence we must obtain an expression for the areas of these zones. Let **CO** $= R$, $co = r$, **O**$o = a$, **PO** $= x$, and **P**$o = x' = a - x$. From the right angled triangle **PCO**, $R^2 =$ **DO** $\times x$; whence **MD** $= R -$ **DO** $= \dfrac{Rx - R^2}{x}$, the altitude of the zone seen on this sphere. In like manner $md = \dfrac{rx' - r^2}{x'}$. Now the area of a zone being to the area of the surface of its sphere as the altitude of the zone is to the diameter of the sphere, letting Z and z be the zones, $Z : 4\pi R^2 :: \dfrac{Rx - R^2}{x} : 2R$, $\therefore Z = 2\pi R\dfrac{Rx - R^2}{x}$. And in like manner $z = 2\pi r\dfrac{rx' - r^2}{x'} = 2\pi r^2\dfrac{a - x - r}{a - x}$.

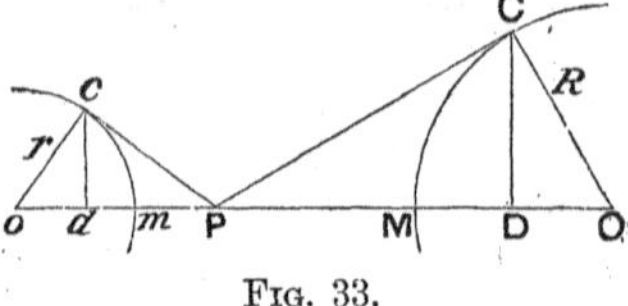

FIG. 33.

Hence, letting S represent the function, we have, $S = 2\pi R^2\dfrac{x - R}{x} + 2\pi r^2\dfrac{a - x - r}{a - x}$, $S' = R^2 - \dfrac{R^3}{x} + r^2 - \dfrac{r^3}{a - x}$. $\dfrac{dS'}{dx} = \dfrac{R^3}{x^2} - \dfrac{r^3}{(a - x)^2} = 0$; whence $x = \dfrac{aR^{\frac{3}{2}}}{r^{\frac{3}{2}} + R^{\frac{3}{2}}}$; and the entire surface $= 2\pi\left[r^2 + R^2 - \dfrac{(r^{\frac{3}{2}} + R^{\frac{3}{2}})^2}{a}\right]$.

Since $2\pi r^2 + 2\pi R^2$ is the sum of the hemispheres, S is always less than this sum except when $a = \infty$.

GENERAL SCHOLIUM.

The student should now resume the study of General Geometry at Chapter IV.

CHAPTER III.

THE INTEGRAL CALCULUS.

SECTION I.

Definitions and Elementary Forms.

157. ***The Integral Calculus*** is that branch of the Infinitesimal Calculus which treats of the methods of deducing the relations between finite values of variables, from given relations between the contemporaneous infinitesimal elements of those variables. It is the inverse of the Differential Calculus.

158. ***The Integral*** of a differential function is another function which being differentiated produces the differential.

159. ***Integration*** is the process of deducing the integral function from its differential.

160. ***The Sign of Integration*** is $\int$, which is a form derived from the old, or long *s*. It is the initial of the word *sum*, and came into use from the conception that integration is a process of summing an infinite series of infinitesimals.

Ill's.—Suppose we have given $dy = \frac{4x dx}{(1 - x^2)^2}$. This is a differential function, and we have given in the equation the relation between dy and dx. The Integral Calculus proposes to find the relation between y and x from such a relation between their differentials; or, in other words, to find the function which being differentiated produces the given differential. The function in this case is $y = \frac{2x^2}{1 - x^2}$, as will be proved by differentiating. The latter is therefore called the integral of the former. Using the sign of integration, we may write $\int dy = \int \frac{4x dx}{(1 - x^2)^2} = \frac{2x^2}{1 - x^2}$; and read, "the integral of dy equals the integral of $\frac{4x dx}{(1 - x^2)^2}$, which equals $\frac{2x^2}{1 - x^2}$."

The conception of integration as a process having for its object the summation of an infinite series of infinitesimals may be illustrated by considering the area of an ellipse as composed of an infinite number of infinitesimal segments, as represented in the figure. Let **A** represent the area of the ellipse; whence d**A** represents one of the infinitesimal segments, or elements of the area.

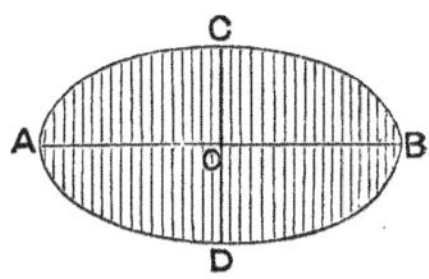

Fig. 34.

Now it is found that $dA = \frac{b}{a}(a^2 - x^2)^{\frac{1}{2}}dx$. By integration it is found that the entire area is πab, a and b being the semi-axes. But, as the entire area is the sum of the infinitesimal segments, the process of integration may be considered as having for its object the summing, or adding together of all the infinitesimals which go to make up the entire area.

161. Important General Statement.—Strictly speaking, there is no such thing as a *Process of Integration.* Whenever a differential is proposed for integration, the first question is, *Is this a Known Form?* that is, *Can we see by inspection what function, being differentiated, produces this?* If we cannot thus discern the integral by a simple inspection, the only question remaining is, *Can we transform the differential into an equivalent expression the integral of which we can recognize?* Thus, in any case, we pass from the differential to its integral by a simple inspection ; and the sufficient reason always is, *This expression is the integral of that, because, being differentiated, it produces it.*

THREE ELEMENTARY PROPOSITIONS.

162. Prop. 1.—*Constant factors or divisors appear in the integral the same as in the differential, and hence may be written before or after the sign of integration at pleasure.*

Dem.—This is a direct consequence of the fact that constant factors or divisors appear in the differential the same as in the function (***48***).

163. Prop. 2.—*To integrate the algebraic sum of several differentials, integrate each term separately, and connect the integrals by the same signs as their differentials were connected.*

Dem.—This is a direct consequence of the rule for differentiating the algebraic sum of several variables (***51***).

164. Prop. 3.—*An indeterminate constant must always be added to the integral of a function.*

Dem.—Since, in differentiating, *constant terms* disappear, in returning from the differential to the integral we have to represent any possible constant terms by an indeterminate constant.

Sch.—The method of disposing of this constant term, which we usually represent by C, will be presented hereafter.* The fact that there may be such a term is all that the student is expected to see at this point. To illustrate, suppose $y = 3ax^2 + 12b$, $dy = 6ax\,dx$. Now, if the latter alone were given, we might see that $y = 3ax^2$ was its integral, since being differentiated it would produce $dy = 6ax\,dx$. But so will $y = 3ax^2 +$ *any constant,* as $12b$, or, as we represent it, $y = 3ax^2 + C$.

---

* Section VII., closing illustration.

TWO ELEMENTARY RULES.

165. *RULE* 1.—Whenever a differential can be separated or transformed into three factors; viz., 1st. *Its constant factors;* 2nd. *A variable factor affected with any exponent except* -1; and 3rd. *A differential factor which is the differential of the 2nd factor without its exponent,* its integral is

THE PRODUCT OF THE SECOND FACTOR WITH ITS EXPONENT INCREASED BY 1, INTO THE 1ST OR CONSTANT FACTOR DIVIDED BY THE NEW EXPONENT.*

DEM.—This rule is evident from (**162**), and the rule for differentiating a variable affected with an exponent (**56**). Thus, if $y = m[f(x)]^n$, $dy = mn[f(x)]^{n-1} d.f(x)$, or $mn \times [f(x)]^{n-1} \times d.f(x)$; whence to pass from the latter to the former, we have to suppress the differential factor, $d.f(x)$, increase the exponent $n-1$ by 1 making it n, and divide the constant factor mn by this n.

In the exceptional case the exponent by which we would be required to divide according to the rule would be $1 - 1 = 0$, whence the result would be ∞.

Ex. 1. Integrate $dy = 3ax^2dx$.

SOLUTION. $dy = 3a \times x^2 \times dx$; whence $y = \int 3ax^2dx = \frac{3a}{3}x^3 + C = ax^3 + C$.

Ex. 2. Integrate $dy = ax^3dx$.

SOLUTION. $dy = a \times x^3 \times dx$. $\therefore$ $y = \int ax^3dx = \frac{a}{4}x^4 + C$.

Ex. 3. Integrate $dy = (a + 3x^2)^2 6xdx$.

SOLUTION. $dy = 1 \times (a + 3x^2)^2 \times 6xdx$, which corresponds to the requirements of the rule, since $d(a + 3x^2) = 6xdx$. $\therefore$ $y = \int (a+3x^2)^2 6xdx = \frac{1}{3}(a+3x^2)^3 + C$.

Ex. 4. Integrate $dy = (a + 3x^2)^3 xdx$.

SOLUTION.—The differential of the quantity within the parenthesis being $6xdx$, we write $dy = \frac{1}{6}(a + 3x^2)^3 \times 6xdx$, which conforms to the requirements of the rule. $\therefore$ $y = \int \frac{1}{6}(a + 3x^2)^3 6xdx = \frac{1}{24}(a + 3x^2)^4 + C$.

Ex. 5. Integrate $dy = a(ax + bx^2)^{\frac{1}{3}}dx + 2b(ax + bx^2)^{\frac{1}{3}}xdx$.

SUG'S. $y = \int [a(ax + bx^2)^{\frac{1}{3}}dx + 2b(ax+bx^2)^{\frac{1}{3}}xdx] = \int [(ax+bx^2)^{\frac{1}{3}}(a+2bx)dx]$ $= \int [1 \times (ax + bx^2)^{\frac{1}{3}} \times (a + 2bx)dx] = \frac{3}{4}(ax + bx^2)^{\frac{4}{3}} + C$.

166. *RULE* 2.—WHENEVER A DIFFERENTIAL CAN BE WRITTEN IN, OR TRANSFORMED INTO A FRACTION WHOSE NUMERATOR IS THE EXACT DIFFERENTIAL OF ITS DENOMINATOR, THE INTEGRAL IS THE NAPIERIAN LOGARITHM OF THE DENOMINATOR.*

---

* In giving such rules the constant term of the integral is not mentioned, as its addition is always implied.

DEM.—This is a direct consequence of the rule that the differential of the Napierian logarithm of a number is the differential of the number divided by the number. [This will be seen to be the exceptional case under the preceding rule.]

---

167. ELEMENTARY FORMS.

1. $y = \int x^n dx = \frac{1}{n+1} x^{n+1} + C.$ Same as *Rule* 1.

2. $y = \int \frac{dx}{x} = \log x + C.$ Same as *Rule* 2.

3. $y = \int a^x dx = \frac{1}{\log a} a^x + C.$ Converse of (60).

3_1. $y = \int e^x dx = e^x + C.$ " (61).

---

4. $y = \int \cos x \, dx = \sin x + C.$ " (66).
5. $y = \int - \sin x \, dx = \cos x + C.$ " (67).
6. $y = \int \frac{dx}{\cos^2 x}$, or $\int \sec^2 x \, dx = \tan x + C.$ " (69).
7. $y = \int - \frac{dx}{\sin^2 x}$, or $\int - \operatorname{cosec}^2 x \, dx = \cot x + C.$ " (70).
8. $y = \int \tan x \sec x \, dx = \sec x + C.$ " (71).
9. $y = \int - \cot x \operatorname{cosec} x \, dx = \operatorname{cosec} x + C.$ " (72).
10. $y = \int \sin x \, dx = \operatorname{vers} x + C.$ " (73).
11. $y = \int - \cos x \, dx = \operatorname{covers} x + C.$ " (74).

---

12. $y = \int \frac{dx}{\sqrt{1 - x^2}} = \sin^{-1} x + C.$ " (75).
13. $y = \int - \frac{dx}{\sqrt{1 - x^2}} = \cos^{-1} x + C.$ " (77).
14. $y = \int \frac{dx}{1 + x^2} = \tan^{-1} x + C.$ " (79).
15. $y = \int - \frac{dx}{1 + x^2} = \cot^{-1} x + C.$ " (80).
16. $y = \int \frac{dx}{x\sqrt{x^2 - 1}} = \sec^{-1} x + C.$ " (81).
17. $y = \int - \frac{dx}{x\sqrt{x^2 - 1}} = \operatorname{cosec}^{-1} x + C.$ " (82).
18. $y = \int \frac{dx}{\sqrt{2x - x^2}} = \operatorname{vers}^{-1} x + C.$ " (83).
19. $y = \int - \frac{dx}{\sqrt{2x - x^2}} = \operatorname{covers}^{-1} x + C.$ " (84).

168. SUBORDINATE CIRCULAR FORMS.

1. $y = \int \frac{dx}{\sqrt{a^2 - b^2x^2}} = \frac{1}{b} \sin^{-1} \frac{bx}{a} + C,$

$\text{or} = \int \frac{dx}{\sqrt{a^2 - x^2}} = \sin^{-1} \frac{x}{a} + C, \text{ when } b = 1.$

2. $y = \int - \frac{dx}{\sqrt{a^2 - b^2x^2}} = \frac{1}{b} \cos^{-1} \frac{bx}{a} + C,$

$\text{or} = \int - \frac{dx}{\sqrt{a^2 - x^2}} = \cos^{-1} \frac{x}{a} + C, \text{ when } b = 1.$

3. $y = \int \frac{dx}{a^2 + b^2x^2} = \frac{1}{ab} \tan^{-1} \frac{bx}{a} + C,$

$\text{or} = \int \frac{dx}{a^2 + x^2} = \frac{1}{a} \tan^{-1} \frac{x}{a} + C, \text{ when } b = 1.$

4. $y = \int - \frac{dx}{a^2 + b^2x^2} = \frac{1}{ab} \cot^{-1} \frac{bx}{a} + C,$

$\text{or} = \int - \frac{dx}{a^2 + x^2} = \frac{1}{a} \cot^{-1} \frac{x}{a} + C, \text{ when } b = 1.$

5. $y = \int \frac{dx}{x\sqrt{b^2x^2 - a^2}} = \frac{1}{a} \sec^{-1} \frac{bx}{a} + C,$

$\text{or} = \int \frac{dx}{x\sqrt{x^2 - a^2}} = \frac{1}{a} \sec^{-1} \frac{x}{a} + C, \text{ when } b = 1.$

6. $y = \int - \frac{dx}{x\sqrt{b^2x^2 - a^2}} = \frac{1}{a} \operatorname{cosec}^{-1} \frac{bx}{a} + C,$

$\text{or} = \int - \frac{dx}{x\sqrt{x^2 - a^2}} = \frac{1}{a} \operatorname{cosec}^{-1} \frac{x}{a} + C, \text{ when } b = 1.$

7. $y = \int \frac{dx}{\sqrt{2abx - b^2x^2}} = \frac{1}{b} \operatorname{vers}^{-1} \frac{bx}{a} + C,$

$\text{or} = \int \frac{dx}{\sqrt{2ax - x^2}} = \operatorname{vers}^{-1} \frac{x}{a} + C, \text{ when } b = 1.$

8. $y = \int - \frac{dx}{\sqrt{2abx - b^2x^2}} = \frac{1}{b} \operatorname{covers}^{-1} \frac{bx}{a} + C,$

$\text{or} = \int - \frac{dx}{\sqrt{2ax - x^2}} = \operatorname{covers}^{-1} \frac{x}{a}, \text{ when } b = 1.$

DEM.—These forms may be considered as the converse of *Ex's.* 1, 2, pages 38, 39. They may also be established by differentiating the result and showing that its differential is the given differential function. Thus, $d\left(\frac{1}{b} \sin^{-1} \frac{bx}{a}\right) =$

$$\frac{1}{b}\frac{d\left(\frac{bx}{a}\right)}{\sqrt{1-\frac{b^2x^2}{a^2}}}=\frac{1}{b}\frac{\frac{b}{a}dx}{\sqrt{\frac{a^2-b^2x^2}{a^2}}}=\frac{1}{b}\frac{bdx}{\sqrt{a^2-b^2x^2}}=\frac{dx}{\sqrt{a^2-b^2x^2}}.$$ [The student should verify all of them in this way.]

A direct way of obtaining these integrals, and one with which the student should not fail to become familiar, is the following :

To integrate $y=\int\frac{dx}{\sqrt{a^2-b^2x^2}}$, we observe that it has the general form of the differential of an arc in terms of its sine, which is $\frac{dx'}{\sqrt{1-x'^2}}$. To transform our expression into this form, we have first to make the first term under the radical 1. This can be readily done thus, $\int\frac{dx}{\sqrt{a^2-bx^2}}=\int\frac{dx}{a\sqrt{1-\frac{b^2x^2}{a^2}}}=\frac{1}{a}\int\frac{dx}{\sqrt{1-\frac{b^2x^2}{a^2}}}$, since the constant divisor a appears in the same form in the integral as in the differential (***162***). Now to make the quantity under the sign of integration the differential of an arc in terms of its sine, the numerator ought to be the differential of the square root of the second term in the denominator, which is the square of the sine. But $d\left(\frac{bx}{a}\right)=\frac{b}{a}dx$. We, therefore, need to introduce $\frac{b}{a}$ into the numerator. This can be done by putting $\frac{a}{b}$ outside the sign of integration as they will neutralize each other (***162***). Hence $\frac{1}{a}\int\frac{dx}{\sqrt{1-\frac{b^2x^2}{a^2}}}=\frac{1}{a}\cdot\frac{a}{b}\int\frac{\frac{b}{a}dx}{\sqrt{1-\frac{b^2x^2}{a^2}}}=\frac{1}{b}\int\frac{\frac{b}{a}dx}{\sqrt{1-\frac{b^2x^2}{a^2}}}$. The quantity now under the sign of integration is the exact differential of $\sin^{-1}\frac{bx}{a}$, since it is the differential of the sine, $\frac{bx}{a}$, divided by the square root of 1 — the square of the sine, $\frac{bx}{a}$(***75***). Hence, finally, as $\int dy=y$, we have $y=\int\frac{dx}{\sqrt{a^2-b^2x^2}}=\frac{1}{b}\sin^{-1}\frac{bx}{a}+C$.

[The student should produce all these subordinate integrals in this way, for the benefit of the exercise. We give the outline of two more, which should be explained at length as above.]

$$y=\int\frac{dx}{a^2+b^2x^2}=\int\frac{dx}{a^2\left(1+\frac{b^2x^2}{a^2}\right)}=\frac{1}{a^2}\int\frac{dx}{1+\frac{b^2x^2}{a^2}}=\frac{1}{a^2}\cdot\frac{a}{b}\int\frac{\frac{b}{a}dx}{1+\frac{b^2x^2}{a^2}}=$$

$$\frac{1}{ab}\int\frac{\frac{b}{a}dx}{1+\frac{b^2x^2}{a^2}}=\frac{1}{ab}\tan^{-1}\frac{bx}{a}+C.$$

$$y = \int -\frac{dx}{\sqrt{2ax - x^2}} = \int -\frac{dx}{a\sqrt{2\left(\frac{x}{a}\right) - \frac{x^2}{a^2}}} = \frac{1}{a}\int -\frac{dx}{\sqrt{2\left(\frac{x}{a}\right) - \frac{x^2}{a^2}}} =$$

$$\frac{1}{a}\cdot a\int -\frac{\frac{dx}{a}}{\sqrt{2\left(\frac{x}{a}\right) - \frac{x^2}{a^2}}} = \int -\frac{\frac{dx}{a}}{\sqrt{2\left(\frac{x}{a}\right) - \frac{x^2}{a^2}}} = \text{covers}^{-1}\frac{x}{a} + C.$$

---

169. LOGARITHMIC TRIGONOMETRICAL FORMS.

1. $y = \int \frac{dx}{\sin x}$ or $\int \operatorname{cosec} x dx = \int \frac{dx}{2\sin(\frac{1}{2}x)\cos(\frac{1}{2}x)} = \int \frac{\frac{dx}{2\cos^2(\frac{1}{2}x)}}{\frac{\sin(\frac{1}{2}x)}{\cos(\frac{1}{2}x)}} =$

$\int \frac{\frac{1}{2}\sec^2(\frac{1}{2}x)dx}{\tan(\frac{1}{2}x)} = \int \frac{d[\tan(\frac{1}{2}x)]}{\tan(\frac{1}{2}x)} = \log\tan(\frac{1}{2}x) + C.$

2. $y = \int \frac{dx}{\cos x}$ or $\int \sec x dx = \int \frac{dx}{\sin(\frac{1}{2}\pi - x)} = -\int \frac{d(\frac{1}{2}\pi - x)}{\sin(\frac{1}{2}\pi - x)} =$ [by (1)]
$-\log\tan(\frac{1}{4}\pi - \frac{1}{2}x) + C.$

3. $y = \int \frac{dx}{\tan x}$ or $\int \cot x dx = \int \frac{\cos x dx}{\sin x} = \int \frac{d(\sin x)}{\sin x} = \log\sin x + C.$

4. $y = \int \frac{dx}{\cot x}$ or $\int \tan x dx = \int \frac{\sin x dx}{\cos x} = -\int \frac{d(\cos x)}{\cos x} = -\log\cos x =$
$\log\frac{1}{\cos x} = \log\sec x + C.$

5. $y = \int \frac{dx}{\sin x\cos x} = \int \frac{2dx}{\sin(2x)} = \int \frac{d(2x)}{\sin(2x)} =$ [by (1)] $\log\tan x + C.$

SCH —The above 32 forms must be so thoroughly memorized as to be instantly recognized. There is no doing anything in the integral calculus without this. These forms are to integration what the multiplication table is to arithmetical operations. Thus we say, 7 goes into 56 8 times, *because 8 times* 7 = 56. In like manner we say that $\cot^{-1}x + C$ is the integral of $-\frac{dx}{1+x^2}$, *because* $\cot^{-1}x + C$ *differentiated* $= -\frac{dx}{1+x^2}$.

Ex. 1. Integrate $dy = 3ax^3dx$.

SOLUTION.—The integral of dy is y, since y differentiated $= dy$. To integrate $3ax^3dx$, notice that $3a \times x^3 \times dx$, conforms to (***165***). $\therefore\ y = \int 3ax^3dx = \frac{3a}{4}x^4 + C.$

Ex. 2. Integrate $dy = (2a + 3bx)^3dx$.

SUG'S. $y = \int (2a + 3bx)^3dx = \int (8a^3 + 36a^2bx + 54ab^2x^2 + 27b^3x^3)dx =$

$\int 8a^3dx + \int 36a^2bxdx + \int 54ab^2x^2dx + \int 27b^3x^3dx = 8a^3x + 18a^2bx^2 + 18ab^2x^3 + \frac{27}{4}b^3x^4 + C$ (**163**).

This may also be integrated by (**165**). Thus $y = \int \frac{1}{3b} \times (2a + 3bx)^3 \times 3bdx = \frac{1}{12b}(2a + 3bx)^4 + C$; which is the same as the preceding.

Ex. 3. Integrate $dy = \frac{xdx}{\sqrt{a^2 + x^2}}$.

SUG'S. $y = \int \frac{xdx}{\sqrt{a^2 + x^2}} = \int (a^2 + x^2)^{-\frac{1}{2}}xdx = \int \frac{1}{2} \times (a^2 + x^2)^{-\frac{1}{2}} \times 2xdx = (a + x^2)^{\frac{1}{2}} + C.$

Ex. 4. Integrate $dy = bx^{\frac{1}{2}}dx$. $\quad y = \frac{2}{3}bx^{\frac{3}{2}} + C.$

Ex. 5. Integrate $dy = 3x^{-4}dx$. $\quad y = -x^{-3} + C.$

Ex. 6. Integrate $dy = 2x^{\frac{2}{3}}dx$. $\quad y = \frac{6}{5}x^{\frac{5}{3}} + C.$

Ex. 7. Integrate $dy = -5mx^{-\frac{1}{3}}dx$. $\quad y = -\frac{15}{2}mx^{\frac{2}{3}} + C.$

Ex. 8. Integrate $dy = \frac{dx}{x^4}$. $\quad y = -\frac{1}{3x^3} + C.$

Ex. 9. Integrate $dy = \frac{x^2dx}{(a^2 + x^3)^{\frac{1}{2}}}$. $\quad y = \frac{2}{3}(a^2 + x^3)^{\frac{1}{2}} + C.$

Ex. 10. Integrate $dy = -\frac{2ax - x^2}{(3ax^2 - x^3)^{\frac{1}{3}}}dx$.

SUG'S. $-\frac{2ax - x^2}{(3ax^2 - x^3)^{\frac{1}{3}}}dx = -(3ax^2 - x^3)^{-\frac{1}{3}}(2ax - x^2)dx = -\frac{1}{3} \times (3ax^2 - x^3)^{-\frac{1}{3}} \times (6ax - 3x^2)dx.$ $\therefore y = \int -\frac{2ax - x^2}{(3ax^2 - x^3)^{\frac{1}{3}}}dx = -\frac{1}{2}(3ax^2 - x^3)^{\frac{2}{3}} + C.$

Ex. 11. Integrate $dy = 12bx(4bx^2 - 2cx^3)^{\frac{1}{3}}dx - 9cx^2(4bx^2 - 2cx^3)^{\frac{1}{3}}dx$.

SUG'S. $12bx(4bx^2 - 2cx^3)^{\frac{1}{3}}dx - 9cx^2(4bx^2 - 2cx^3)^{\frac{1}{3}}dx = (4bx^2 - 2cx^3)^{\frac{1}{3}}(12bx - 9cx^2)dx$. Now in order that the factor $(12bx - 9cx^2)dx$ should be the differential of $4bx^2 - 2cx^3$ we should have 8 instead of 12 and 6 instead of 9. Hence we write $\frac{3}{2}(4bx^2 - 2cx^3)^{\frac{1}{3}}(8bx - 6cx^2)dx$. $\therefore y = \frac{9}{8}(4bx^2 - 2cx^3)^{\frac{4}{3}} + C.$

170. SCH.—It is not always easy to determine just what constant factor is required in order to make the differential factor the differential of the quantity within the parenthesis; nor can such a factor always be found. To determine whether there is such a factor or not; and, if there is, to find it, we may proceed as in the following examples.

Ex. 12. Integrate $dy = \frac{2ax - 5x^2}{(2b + 3ax^2 - 5x^3)^{\frac{1}{3}}} dx.$

SUG'S. $\frac{2ax - 5x^2}{(2b + 3ax^2 - 5x^3)^{\frac{1}{3}}} dx = (2b + 3ax^2 - 5x^3)^{-\frac{1}{3}}(2ax - 5x^2)dx.$ Suppose A to be the constant factor sought, so that $y = \int \frac{1}{A}(2b + 3ax^2 - 5x^3)^{-\frac{1}{3}}(2aAx - 5Ax^2)dx.$ It is required that A should fulfill the condition $d(2b + 3ax^2 - 5x^3) = (2aAx - 5Ax^2)dx$, or $6ax - 15x^2 = 2aAx - 5Ax^2$. Now, as this is to be true for all values of x, we have $6a = 2aA$, or $A = 3$; and also $15 = 5A$, or $A = 3$. Hence 3 is the factor sought, and we have

$y = \int (2b + 3ax^2 - 5x^3)^{-\frac{1}{3}}(2ax - 5x^2)dx = \int \frac{1}{3}(2b + 3ax^2 - 5x^3)^{-\frac{1}{3}}(6ax - 15x^2)dx =$
$\frac{1}{2}(2b + 3ax^2 - 5x^3)^{\frac{2}{3}} + C.$

Ex. 13. Integrate $dy = x(1 + x^2 - 2x^5)^{\frac{3}{5}}dx - 3x^4(1 + x^2 - 2x^5)^{\frac{3}{5}}dx.$

SUG'S. $dy = x(1 + x^2 - 2x^5)^{\frac{3}{5}}dx - 3x^4(1 + x^2 - 2x^5)^{\frac{3}{5}}dx = (1 + x^2 - 2x^5)^{\frac{3}{5}}(x - 3x^4)dx.$ We are to seek a factor A, which fulfills the condition $d(1 + x^2 - 2x^5) = (Ax - 3Ax^4)dx$, or performing the differentiation, and dropping dx, $2x - 10x^4 = Ax - 3Ax^4$; whence the first condition required is $A = 2$, and the second is $3A = 10$, or $A = \frac{10}{3}$. These conditions being incompatible with each other, there is no factor which meets the conditions, and the integration cannot be performed in the manner now under consideration.

Ex. 14. Integrate $dy = \left(\frac{12}{x^2} - \frac{3}{x^4}\right)dx.$ $\qquad y = -\frac{12}{x} + \frac{1}{x^3} + C.$

Ex. 15. Integrate $dy = (\frac{5}{2}ax^{\frac{3}{2}} - \frac{3}{2}bx^{\frac{1}{2}})dx.$ $\qquad y = ax^{\frac{5}{2}} - bx^{\frac{3}{2}} + C.$

Ex. 16. Integrate $dy = ax^2dx + \frac{dx}{2\sqrt{x}}.$ $\qquad y = \frac{ax^3}{3} + x^{\frac{1}{2}} + C.$

Ex. 17. Integrate $dy = (2x^4 - 3x^2 + 1)^{\frac{1}{2}}(x^3 - \frac{3}{4}x)dx.$

$$y = \tfrac{1}{12}(2x^4 - 3x^2 + 1)^{\frac{3}{2}} + C.$$

Ex. 18. Which of the following can be integrated by the method used in the preceding examples; viz., $dy = (5x^2 - 2x)^{\frac{1}{3}}(3x - 5)dx$; $dy = 5(5x^2 - 2x)^{\frac{1}{3}}xdx - (5x^2 - 2x)^{\frac{1}{3}}dx$; $dy = \frac{2x - 1}{(1 - x + x^3)^2}dx$; $dy = \frac{2x^2 - 1}{(1 - x + x^3)^2}dx$?

SCH.—***Caution.*** The student must not fail to observe that it is only *constant* factors which he can introduce in the manner illustrated in the preceding examples. When the differential factor is not of the right form as to the *variable*, nothing can be done with it by this process. Were we to attempt to introduce a variable factor into the differential factor, its reciprocal would have to be introduced into the 1st, or constant factor, and this would destroy the condition that the first factor is constant. *It is always well, if there is the least doubt whether the integral found is correct, to differentiate it and see if it gives the proposed differential.*

---

171. Prop.—*It is sometimes possible to bring a differential to the form required in (**165**) by transposing one or more factors of the variable from the factor in the parenthesis to the differential factor,* or vice versa.

Ex. 19. Integrate $dy = \dfrac{axdx}{(2bx + x^2)^{\frac{3}{2}}}$.

SOLUTION. $\dfrac{axdx}{(2bx + x^2)^{\frac{3}{2}}} = a(2bx + x^2)^{-\frac{3}{2}}xdx = a(2bx^{-1} + 1)^{-\frac{3}{2}}x^{-2}dx = -\dfrac{a}{2b} \times (2bx^{-1} + 1)^{-\frac{3}{2}} \times (-2bx^{-2}dx)$, in which $-2bx^{-2}dx$ is the differential of $(2bx^{-1} + 1)$. $\therefore y = \int \dfrac{axdx}{(2bx+x^2)^{\frac{3}{2}}} = \int -\dfrac{a}{2b} \times (2bx^{-1}+1)^{-\frac{3}{2}} \times (-2bx^{-2}dx) = \dfrac{a}{b}(2bx^{-1}+1)^{-\frac{1}{2}} + C = \dfrac{ax}{b\sqrt{2bx + x^2}} + C.$

Ex. 20. Integrate $dy = \dfrac{adx}{x\sqrt{3bx + 4c^2x^2}}$.

SUG. $\dfrac{adx}{x\sqrt{3bx + 4c^2x^2}} = \dfrac{a}{3b}(3bx^{-1} + 4c^2)^{-\frac{1}{2}}3bx^{-2}dx,\ y = -\dfrac{2a(3bx + 4c^2x^2)^{\frac{1}{2}}}{3bx} + C.$

Ex. 21. Integrate $dy = \dfrac{2xdx}{1 + x^2}$(**166**). $\quad y = \log(1 + x^2) + C.$

SCH.—When the integral is a logarithm, it is customary to write the constant also as a logarithm. Thus, in the above example if we let $\log \boldsymbol{c} = C$, *i. e.* call the constant term $\log \boldsymbol{c}$, instead of C, we have $y = \log(1 + x^2) + \log \boldsymbol{c} = \log[\boldsymbol{c}(1 + x^2)]$, or $\log(\boldsymbol{c} + \boldsymbol{c}x^2)$.

Ex. 22. Integrate $dy = \dfrac{x^2 - 2x}{2x^3 - 6x^2 + 1}dx$.

SOLUTION. $\dfrac{x^2 - 2x}{2x^3 - 6x^2 + 1}dx = \dfrac{1}{6} \times \dfrac{6x^2 - 12x}{2x^3 - 6x^2 + 1}dx$, in which the numerator is the exact differential of the denominator. $\therefore y = \int \dfrac{x^2 - 2x}{2x^3 - 6x^2 + 1}dx = \dfrac{1}{6}\int \dfrac{6x^2 - 12x}{2x^3 - 6x^2 + 1}dx = \dfrac{1}{6}\log(2x^3 - 6x^2 + 1) + \log \boldsymbol{c} = \log[\boldsymbol{c}(2x^3 - 6x^2 + 1)^{\frac{1}{6}}].$

Ex. 23. Integrate $dy = \frac{5x^3dx}{15x^4 + 21}$. $y = \log[c^*(15x^4 + 21)^{\frac{1}{12}}]$

Ex. 24. Integrate $dy = \frac{2x^{\frac{1}{2}}dx}{1 - x^{\frac{3}{2}}}$.

$$y = -\tfrac{4}{3}\log(1 - x^{\frac{3}{2}}) + \log c = \log[c(1 - x^{\frac{3}{2}})^{-\frac{4}{3}}] = \log\frac{c}{(1 - x^{\frac{3}{2}})^{\frac{4}{3}}}.$$

Ex. 25. Integrate $dy = \frac{2bxdx}{a + bx^2}$. $y = \log c(a + bx^2)$.

Ex. 26. Integrate $dy = \frac{7xdx}{8a - 3x^2}$. $y = \log\frac{c}{(8a - 3x^2)^{\frac{7}{6}}}$.

Ex. 27. Which of the following can be integrated by the method used in the last 6 examples; viz., $dy = \frac{x - \frac{1}{3}}{3x^2 - 2x + 1}dx$; $dy = \frac{2x^2}{1 - x^3}dx$; $dy = \frac{x - x^2}{3 - x^3}dx$; $dy = \frac{3x^2dx}{2x^3 - 5}$; $dy = \frac{2a - 10x^2}{2ax - 5x^3}dx$?

Ex. 28. Integrate $dy = \frac{b(3x - a^2)^4dx}{nx^3}$.

Sug's. $\frac{b(3x - a^2)^4dx}{nx^3} = \frac{b}{n}\left(\frac{81x^4dx - 108a^2x^3dx + 54a^4x^2dx - 12a^6xdx + a^8dx}{x^3}\right) =$

$$\frac{b}{n}\left(81xdx - 108a^2dx + \frac{54a^4dx}{x} - \frac{12a^6dx}{x^2} + \frac{a^8dx}{x^3}\right).$$

$$\therefore\ y = \frac{b}{n}\left(\frac{81}{2}x^2 - 108a^2x + 54a^4\log x + \frac{12a^6}{x} - \frac{a^8}{2x^2}\right) + C.$$

Ex. 29. Integrate $dy = (b - x^2)^3x^{\frac{1}{2}}dx$.

$$y = \tfrac{2}{3}b^3x^{\frac{3}{2}} - \tfrac{6}{7}b^2x^{\frac{7}{2}} + \tfrac{6}{11}bx^{\frac{11}{2}} - \tfrac{2}{15}x^{\frac{15}{2}} + C.$$

Ex. 30. Integrate $dy = \frac{5(2a - x^2)^3}{x^5}dx$.

$$y = 5\left[-\frac{2a^3}{x^4} + \frac{6a^2}{x^2} + 6a\log x - \tfrac{1}{2}x^2\right] + C.$$

Ex. 31. Integrate $dy = 3\log^2 x\frac{dx}{x}$.

Sug's. $y = \int 3 \times (\log x)^2 \times \frac{dx}{x} = \log^3 x + C$, since $\frac{dx}{x} = d.\log x$.

Ex. 32. Integrate $dy = 2\log^3 x\frac{dx}{x}$. $y = \frac{1}{2}\log^4 x + C$.

* In all these examples c represents the constant of integration.

Ex. 33. Integrate $dy = m \log^n x \frac{dx}{x}$. $y = \frac{m}{n+1} \log^{n+1} x + C.$

Ex. 34. Integrate $dy = a^{2x} \log a dx$.

SUG'S.—In order to make this conform to (**167**, 3), we should have $d(2x) = 2dx$ as a factor. Hence we write $y = \int a^{2x} \log a dx = \int \frac{1}{2} a^{2x} \log a \cdot 2dx = \frac{1}{2} a^{2x} + C.$ [The pupil should differentiate, verify, and so fully consider the case as to see the reason for the introduction of the constant factor.]

Ex. 35. Which of the following can be integrated by (**167**, 3); $dy = a^{x^2} \log a \, 2dx$, or $dy = 3a^{x^2} \log a \, x \, dx$?

Ans., The latter, $y = \frac{3}{2} a^{x^2} + C.$

Ex. 36. Integrate $dy = e^{\frac{x}{a}} dx$. $y = ae^{\frac{x}{a}} + C.$

Ex. 37. Integrate $dy = 3e^x dx$.

Ex. 38. Integrate $dy = ba^{3x} dx$.

SUG'S. $y = b \int a^{3x} dx = \frac{b}{3 \log a} \int a^{3x} \log a \, 3dx = \frac{b}{3 \log a} a^{3x} + C.$

Ex. 39. Integrate $dy = me^{nx} dx$. $y = \frac{m}{n} e^{nx} + C.$

Ex. 40. Integrate $dy = \cos(2x) dx$.

SUG'S.—In order to make this conform to (**167**, 4), we should have $2dx$, *i. e.* the differential of the arc $2x$, instead of dx. Hence $y = \int \cos 2x dx = \frac{1}{2} \int \cos 2x \cdot 2dx = \frac{1}{2} \sin 2x + C.$

Ex. 41. Which of the following forms can be integrated by (**167**, 4); $dy = \cos x^2 2dx$, or $dy = \cos x^2 x dx$?

Ans., The latter, $y = \frac{1}{2} \sin x^2 + C.$

Ex. 42. Integrate $dy = \sin^3 x \cos x dx$.

SUG'S. $y = \int \sin^3 x \cos x dx = \int 1 \times (\sin x)^3 \times \cos x dx = \frac{1}{4} \sin^4 x + C$, according to (**165** and **167**, 4).

Ex. 43. Integrate $dy = \sin(3x) dx$.

$y = -\frac{1}{3} \cos(3x) + C$, or $\frac{1}{3}$ vers $(3x) + C'$.

SCH. $\frac{1}{3}$ vers $(3x) + C = \frac{1}{3}[1 - \cos(3x)] + C = \frac{1}{3} - \frac{1}{3} \cos(3x) + C.$ $\therefore C' = C + \frac{1}{3}.$

Ex. 44. Integrate $dy = \sin^2(2x) \cos(2x) dx$.

$y = \frac{1}{6} \sin^3(2x) + C.$

Ex. 45. Integrate $dy = \cos^2(3x) \sin(3x) dx$.

$y = -\frac{1}{9} \cos^3(3x) + C.$

Ex. 46. Integrate $dy = \sec^2 x^2\, x\, dx$. $y = \frac{1}{2}\tan x^2 + C$.

Ex. 47. Integrate $dy = 5\sec^2 x^3 \cdot x^2 dx$. $y = \frac{5}{3}\tan x^3 + C$.

Ex. 48. Integrate $dy = 6\sec(4x)\tan(4x)dx$.

$$y = \tfrac{3}{2}\sec(4x) + C.$$

Ex. 49. Integrate $dy = 2\sin(a + 3x)dx$.

$$y = -\tfrac{2}{3}\cos(a + 3x) + C.$$

Ex. 50. Integrate $dy = \frac{3}{2}\operatorname{cosec}^2\sqrt{2x} \cdot x^{-\frac{1}{2}}dx$.

SUG'S. $y = \int \frac{3}{2}\operatorname{cosec}^2\sqrt{2x} \cdot x^{-\frac{1}{2}}dx = \frac{3}{\sqrt{2}}\int \operatorname{cosec}^2\sqrt{2x} \cdot \frac{1}{2}\sqrt{2} \cdot x^{-\frac{1}{2}}dx = -\frac{3}{\sqrt{2}}\cot\sqrt{2x} + C.$

Ex. 51. Integrate $dy = 2\operatorname{cosec}(nx) \cdot \cot(nx)dx$.

$$y = -\frac{2}{n}\operatorname{cosec}(nx) + C.$$

Ex. 52. Integrate $dy = e^{\sin x}\cos x dx$. $y = e^{\sin x} + C$.

Ex. 53. Integrate $dy = -e^{\cos x}\sin x dx$. $y = e^{\cos x} + C$.

Ex. 54. Integrate $dy = \dfrac{dx}{\cos^2(\frac{1}{2}x)}$. $y = 2\tan(\frac{1}{2}x) + C$.

Ex. 55. Integrate $dy = -\dfrac{x dx}{\sin^2(3x^2)}$. $y = \frac{1}{6}\cot(3x^2) + C$.

Ex. 56. Integrate $dy = \sin(ax)dx$.

$$y = \frac{1}{a}\operatorname{versin}(ax) + C, \text{ or } -\frac{1}{a}\cos(ax) + C'.$$

Ex. 57. Integrate $dy = -\cos(\frac{1}{2}x^2)x dx$.

$$y = \operatorname{covers}(\tfrac{1}{2}x^2) + C, \text{ or } -\sin(\tfrac{1}{2}x^2) + C'.$$

SCH.—In the last, $C' = C + 1$. In the 56th *Ex.*, $C' = C + \dfrac{1}{a}$.

Ex. 58. Integrate $dy = \dfrac{2dx}{\sqrt{1 - 4x^2}}$.

SUG'S.—*The form of the denominator* suggests at once that this may be the differential of some arc in terms of its sine. Observing that the numerator is the differential of the square root of $4x^2$, we are enabled to conclude that $y = \int \dfrac{2dx}{\sqrt{1 - 4x^2}} = \sin^{-1} 2x + C.$

Ex. 59. Integrate $dy = \dfrac{x dx}{\sqrt{1 - x^4}}$. $y = \frac{1}{2}\sin^{-1}(x^2) + C$.

Ex. 60. Integrate $dy = \dfrac{dx}{\sqrt{2 - 9x^2}}$.

SUG'S. $\dfrac{dx}{\sqrt{2-9x^2}} = \dfrac{dx}{\sqrt{2}\sqrt{1-\frac{9}{2}x^2}} = \dfrac{\sqrt{2}}{3} \cdot \dfrac{\frac{3}{\sqrt{2}}dx}{\sqrt{2}\sqrt{1-\frac{9}{2}x^2}}$. $\therefore\ y = \displaystyle\int \frac{dx}{\sqrt{2-9x^2}} =$
$\dfrac{1}{3}\displaystyle\int \frac{\frac{3}{\sqrt{2}}dx}{\sqrt{1-\frac{9}{2}x^2}} = \frac{1}{3}\sin^{-1}\frac{3x}{\sqrt{2}} + C.$

Ex. 61. Integrate $dy = \dfrac{-xdx}{\sqrt{2 - 5x^4}}$.

SUG'S. $y = \displaystyle\int \frac{-xdx}{\sqrt{2-5x^4}} = \int \frac{-xdx}{\sqrt{2}\sqrt{1-\frac{5}{2}x^4}} = \int \frac{-2\sqrt{\frac{5}{2}}xdx}{2\sqrt{\frac{5}{2}}\sqrt{2}\sqrt{1-\frac{5}{2}x^4}} =$
$\dfrac{1}{2\sqrt{5}}\displaystyle\int \frac{-2\sqrt{\frac{5}{2}}xdx}{\sqrt{1-\frac{5}{2}x^4}} = \frac{1}{2\sqrt{5}}\cos^{-1}\left[\left(\frac{5}{2}\right)^{\frac{1}{2}}x^2\right] + C.$

Ex. 62. Integrate $dy = \dfrac{3dx}{4 + 9x^2}$. $\qquad y = \frac{1}{2}\tan^{-1}\dfrac{3x}{2} + C.$

Ex. 63. Integrate $dy = \dfrac{x^{\frac{1}{2}}dx}{\sqrt{2 - 4x^3}}$. $\qquad y = \frac{1}{3}\sin^{-1}(2^{\frac{1}{2}}x^{\frac{3}{2}}) + C.$

Ex. 64. Integrate $dy = \dfrac{xdx}{\sqrt{a^2 - b^2x^4}}$. $\qquad y = \dfrac{1}{2b}\sin^{-1}\dfrac{bx^2}{a} + C.$

Ex. 65. Integrate $dy = \dfrac{x^2dx}{1 + x^6}$. $\qquad y = \frac{1}{3}\tan^{-1}x^3 + C.$

Ex. 66. Integrate $dy = \dfrac{8x^{-\frac{2}{3}}dx}{\sqrt{2x^{\frac{1}{3}} - 6x^{\frac{2}{3}}}}$.

SUG'S.—As far as the variable is concerned this conforms to the differential of an arc in terms of its versed sine. Thus, the numerator $= d\sqrt{6x^{\frac{4}{3}}}$, *as far as the variable* is concerned; and $x^{\frac{2}{3}} = (x^{\frac{1}{3}})^2$, which is the relation between the functions of the variable in the denominator of the form referred to. Hence, if we can adjust the *constant* factors to this form, the integral will be apparent. To effect the latter, we proceed as follows:

$$\frac{8x^{-\frac{2}{3}}dx}{\sqrt{2x^{\frac{1}{3}} - 6x^{\frac{2}{3}}}} = \sqrt{6}\,\frac{8x^{-\frac{2}{3}}dx}{\sqrt{2\cdot 6x^{\frac{1}{3}} - (6x^{\frac{1}{3}})^2}} = 4\sqrt{6}\,\frac{2x^{-\frac{2}{3}}dx}{\sqrt{2\cdot 6x^{\frac{1}{3}} - (6x^{\frac{1}{3}})^2}},$$

in which $6x^{\frac{1}{3}}$ being regarded as the variable, the expression has the desired form. $\therefore\ y = 4\sqrt{6}\ \text{vers}^{-1}(6x^{\frac{1}{3}}) + C.$

Ex. 67. Integrate $dy = \frac{2dx}{x\sqrt{3x^2 - 5}}$. $\qquad y = \frac{2}{\sqrt{5}} \sec^{-1} \frac{3^{\frac{1}{2}}x}{5^{\frac{1}{2}}} + C.$

Ex. 68. Integrate $dy = \frac{-2x^{-1}dx}{\sqrt{14x^2 - 3}}$.

$$y = \frac{2}{\sqrt{3}} \operatorname{cosec}^{-1} \left[\left(\frac{14}{3}\right)^{\frac{1}{2}} x\right] + C.$$

MISCELLANEOUS EXERCISES UPON THE ELEMENTARY FORMS.

[NOTE.—The following exercises are given without the integrals, as it is of first importance in the Integral Calculus, that the pupil be able to discover in the differential the probable form of the integral.]

Ex. 1. Integrate $dy = \frac{(1 - \sin x)dx}{x + \cos x}$.

Ex. 2. Integrate $dy = (2x^{\frac{2}{3}} + x^{-1})dx$.

Ex. 3. Integrate $dy = \frac{xdx}{1 + x^4}$; also $dy = \frac{x^2dx}{1 + x^4}$.

Ex. 4. Integrate $dy = -\frac{xdx}{\sqrt{1 - x^2}}$.

Ex. 5. Integrate $dy = \frac{x^2 - 4x + 3}{x^3 - 6x^2 + 9x}dx$.

Ex. 6. Integrate $dy = \frac{3dx}{2 + 5x^2}$; also $dy = \frac{xdx}{2 + 5x^2}$.

Ex. 7. Integrate $dy = (1 + \cos x)dx$.

Ex. 8. Integrate $dy = \frac{(a + \sqrt{x})^2dx}{\sqrt{x}}$.

Ex. 9. Integrate $dy = \frac{(2a^2 + 4x^2)dx}{\sqrt{a^2 + x^2}}$.

Ex. 10. Integrate $dy = -\frac{3dx}{\sqrt{5x^4 - 2x^2}}$.

Ex. 11. Integrate $dy = \cos^2 x \sin x dx$.

Ex. 12. Integrate $dy = \tan^2 x \sec^2 x \, dx$, or $(\tan^2 x + \tan^4 x)dx$.

Ex. 13. Integrate $dy = (a - \frac{b}{x^3} + cx^{\frac{3}{2}})dx$.

Ex. 14. Integrate $dy = (1 + x)(1 - x^2)xdx$.

Ex. 15. Integrate $dy = \frac{x^4dx}{x^2+1}$.

Ex. 16. Integrate $dy = \frac{x^{n-1}dx}{(a+bx^n)^m}$.

Ex. 17. Integrate $dy = \frac{(a-x)dx}{(2ax-x^2)^{\frac{1}{2}}}$.

Ex. 18. Integrate $dy = \frac{5x^3dx}{3x^4+7}$.

Ex. 19. Integrate $dy = \frac{dx}{1+x+x^2}$.

SOLUTION. $\frac{dx}{1+x+x^2} = \frac{4dx}{4+4x+4x^2} = \frac{4dx}{3+1+4x+4x^2} = \frac{4dx}{3+(1+2x)^2} = \frac{\frac{4}{3}dx}{1+\left(\frac{1+2x}{3^{\frac{1}{2}}}\right)^2}$, which is the form for the differential of an arc in terms of the tangent, except that the numerator should be $d.\frac{1+2x}{3^{\frac{1}{2}}} = \frac{2dx}{3^{\frac{1}{2}}}$. $\therefore y = \int\frac{dx}{1+x+x^2}$

$$= \int\frac{\frac{4}{3}dx}{1+\left(\frac{1+2x}{3^{\frac{1}{2}}}\right)^2} = \int\frac{\frac{2}{3^{\frac{1}{2}}}\cdot\frac{2dx}{3^{\frac{1}{2}}}}{1+\left(\frac{1+2x}{3^{\frac{1}{2}}}\right)^2} = \frac{2}{3^{\frac{1}{2}}}\tan^{-1}\frac{1+2x}{3^{\frac{1}{2}}} + C.$$

Ex. 20. Integrate $dy = \frac{m+nx}{a^2+x^2}dx$.

SECTION II.

Rational Fractions.

SEPARATION INTO PARTS BY INDETERMINATE COEFFICIENTS.

[NOTE.—The body of what is called the Integral Calculus is made up of *Special Expedients* by means of which differentials of various forms can be reduced to equivalent known or elementary forms. A few of the more important and fundamental of these processes are given in this and the two succeeding sections.]

172. DEF.—***A Rational Fraction,*** as the term is used here, is a fraction in which the variable is affected with none but positive, integral exponents. The general form is, therefore,

$$\frac{ax^m + bx^{m-1} + cx^{m-2} - - - - - - - lx + k}{mx^n + nx^{n-1} + px^{n-2} - - - - - - - rx + s}.$$

173. Prop. 1.—*If the highest exponent of the variable in the numerator of a rational fraction is greater than that in the denominator, the fraction can always be converted into an equivalent expression consisting of a series of monomial terms with or without a rational fraction (as the case may be), in which fraction, when it occurs, the highest exponent of the variable in the numerator shall be at least* 1 *less than the highest exponent in the denominator.*

ILL'S. $\frac{x^3 - ax^2 + bx + c}{x^2 - 1} = x - a + \frac{(b+1)x - (a-c)}{x^2 - 1}$. Again $\frac{x^3 - a^3}{x - a} = x^2 + ax + a^2$; etc.

---

174. Prop. 2.—*Whenever the denominator of a rational fraction, as* $\frac{f(x)dx}{\varphi(x)}$, *whose numerator is of lower dimensions than its denominator, is real and resolvable into* n REAL *and* UNEQUAL *factors of the first degree, the fraction can be decomposed into* n *partial fractions of the form*

$$\frac{Adx}{x+a} + \frac{Bdx}{x+b} + \frac{Cdx}{x+c} - - - - - - - \frac{Ndx}{x+n},$$

and these fractions integrated separately.

DEM.—Assume $\frac{f(x)}{\varphi(x)} = \frac{A}{x+a} + \frac{B}{x+b} + \frac{C}{x+c} - - - - - \frac{N}{x+n}$. Bringing the second member to a common denominator, each numerator will be multiplied by $n - 1$ factors of the form $x + m$, and hence will be of the $(n-1)$th degree. Collecting the coefficients of x^{n-1}, x^{n-2}, etc. in the numerator of the second member, it will take the form $Mx^{n-1} + Nx^{n-2} + - - - Px + Qx^0$, in which $M, N, - - - P,$ and Q are functions of A, B, C, etc.

Now $f(x) = Mx^{n-1} + Nx^{n-2} + - - - Px + Qx^0$, since the denominators of both members are equal. Then as $f(x)$ is not above the $(n-1)$th degree, it can be treated as a complete polynomial of this degree; and the coefficients of the like powers of x being equated by the principle of indeterminate coefficients, there will result n simple equations between A, B, C, etc., from which these coefficients can be determined. The values of A, B, C, etc., thus determined, being substituted in the assumed series and the factor dx introduced, the decomposition is effected.

Finally, as these numerators are independent of the variable, we have $\int \frac{Adx}{x+a} = A \log(x+a)$, etc.

---

175. Prop. 3.—*Whenever the denominator of a rational fraction, as* $\frac{f(x)dx}{\varphi(x)}$, *whose numerator is of lower dimensions than its denominator, is real and resolvable into* n REAL *and* EQUAL *factors of the first degree, the fraction can be decomposed into* n *partial fractions of the form*

$$\frac{Adx}{(x+a)^n} + \frac{Bdx}{(x+a)^{n-1}} + \frac{Cdx}{(x+a)^{n-2}} \cdots\cdots\cdots \frac{Ndx}{x+a},$$

and these fractions integrated separately.

Dem.—Assume $\frac{f(x)}{\varphi(x)} = \frac{A}{(x+a)^n} + \frac{B}{(x+a)^{n-1}} + \frac{C}{(x+a)^{n-2}} \cdots \frac{N}{x+a}$. To bring the terms of the second member to a common denominator, we have to multiply B by $x+a$, C by $(x+a)^2$ - - - - - and N by $(x+a)^{n-1}$; hence the numerators, when added together, will make a polynomial of the $(n-1)$th degree. Equating this numerator with $f(x)$, the coefficients of the corresponding powers of x being placed equal to each other will, as in the preceding demonstration, give rise to n simple equations between $A, B, C,$ - - - - - N, from which the latter can be determined.

Having separated the fraction into the partial fractions as proposed, it remains to be shown that these can be integrated. As the numerators are constant, and the general form of the denominator is $(x+a)^n$, we are to show that the form $\frac{dx}{(x+a)^n}$ can always be integrated. If $n=1$, $\int\frac{dx}{x+a} = \log(x+a)$. If n is anything other than 1, $\int\frac{dx}{(x+a)^n} = \int(x+a)^{-n}dx = -\frac{1}{n-1}(x+a)^{-n+1}$, or $-\frac{1}{(n-1)(x+a)^{n-1}}$. Hence the integration can always be effected.

Sch.—The last two propositions are equally true whether the factors of $\varphi(x)$ are real or imaginary; but as, in the latter case the integration by those methods would give logarithms of imaginary numbers, the method given in the next proposition is preferable. We will still farther premise that as $\varphi(x)$ is real, if it contains imaginary factors, they must enter by pairs of the form $x \pm a + b\sqrt{-1}$ and $x \pm a - b\sqrt{-1}$,* for only thus can a real product arise from imaginary factors. If therefore, there are imaginary factors in $\varphi(x)$, we shall have

$\varphi(x) = \psi(x) \cdot \{(x \pm a + b\sqrt{-1}) \times (x \pm a - b\sqrt{-1})\}^n = \psi(x) \cdot \{(x \pm a)^2 + b^2\}^n$,

in which $\psi(x)$ represents the product of the real factors.

---

176. Prop. 4.—*Whenever the denominator of a rational fraction, as* $\frac{f(x)dx}{\varphi(x)}$, *whose numerator is of lower dimensions than its denominator, is real and resolvable into* n *real and equal* QUADRATIC *factors, the fraction can be decomposed into* n *partial fractions of the form*

$$\frac{(Ax+B)dx}{[(x\pm a)^2+b^2]^n} + \frac{(Cx+D)dx}{[(x\pm a)^2+b^2]^{n-1}} + \frac{(Ex+F)dx}{[(x\pm a)^2+b^2]^{n-2}} \cdots \frac{(Mx+N)dx}{(x\pm a)^2+b^2},$$

and these fractions integrated separately.

Dem.—[The first part of the demonstration, showing that the fraction can be

---

* Called conjugate imaginary factors.

separated into this form, is identical with that of the last proposition, and the student can supply it.]

Having separated the fraction into partial fractions as proposed, it remains to be shown that these partial fractions can be integrated. The general form is $\frac{(Ax+B)dx}{[(x \pm a)^2 + b^2]^n}$, in which n is an integer. To reduce this to known forms put $x \pm a = z$, whence $x = z \mp a$, $dx = dz$, and $(x \pm a)^2 = z^2$. Substituting these values, we have

$$\int \frac{(Ax+B)dx}{[(x \pm a)^2 + b^2]^n} = \int \frac{(Az \mp Aa + B)dz}{(z^2+b^2)^n} = \int \frac{Azdz}{(z^2+b^2)^n} + \int \frac{(B \mp Aa)dz}{(z^2+b^2)^n} =$$

$$\int A(z^2+b^2)^{-n}zdz + \int \frac{A'dz}{(z^2+b^2)^n}, \text{ in which } A' = B \mp Aa.$$

By (**165**) we have $\int A(z^2+b^2)^{-n}z\,dz = -\frac{A}{2(n-1)(z^2+b^2)^{n-1}}$.

*In a subsequent article (**192,** formula D) it will be shown that the $\int \frac{A_1 dz}{(z^2+b^2)^n}$ may be made to depend upon $\int \frac{A_2 dz}{(z^2+b^2)^{n-1}}$, which in turn may be made to depend upon $\int \frac{A_3 dz}{(z^2+b^2)^{n-2}}$, thus, in the end giving either $\int \frac{A_{n+1}dz}{(z^2+b^2)^{n-n}}$ $= \int A_{n+1}dz$, or $\int \frac{A_n dz}{z^2+b^2}$. But $\int \frac{A_n dz}{z^2+b^2} = \frac{A_n}{b}\tan^{-1}\frac{z}{b}$ (**168,** 3).

177. SCH.—In case the factors of the denominator are not readily seen, put it equal to 0, and solve the equation for the variable. According to the theory of the composition of equations, as developed in Higher Algebra, the variable minus each of the several roots in turn will be the factors.

Ex. 1. Integrate $dy = \frac{(x^2+1)dx}{x^3+6x^2+11x+6}$.

SOLUTION.—Putting $x^3+6x^2+11x+6=0$, we find $x = -1, -2$, and -3.† $\therefore x^3+6x^2+11x+6 = (x+1)(x+2)(x+3)$, and we assume

$$\frac{x^2+1}{x^3+6x^2+11x+6} = \frac{A}{x+1} + \frac{B}{x+2} + \frac{C}{x+3} =$$

$$\frac{A(x+2)(x+3)}{(x+1)(x+2)(x+3)} + \frac{B(x+1)(x+3)}{(x+1)(x+2)(x+3)} + \frac{C(x+1)(x+2)}{(x+1)(x+2)(x+3)} =$$

$$\frac{Ax^2+5Ax+6A+Bx^2+4Bx+3B+Cx^2+3Cx+2C}{x^3+6x^2+11x+6}.$$

Whence $x^2+1 = (A+B+C)x^2 + (5A+4B+3C)x + 6A+3B+2C$.

These members being identical,

$A+B+C=1$ (1); $5A+4B+3C=0$ (2); and $6A+3B+2C=1$ (3).

From (1), (2), and (3) we find $A=1$, $B=-5$, and $C=5$.

Hence we have

---

* This reduction might be exhibited here, but as the formula referred to is better for practical purposes, it is thought best to give the process but once.

† COMPLETE SCHOOL ALGEBRA, Part II., (**111**).

$$y = \int \frac{(x^2+1)dx}{x^3+6x^2+11x+6} = \int \frac{dx}{x+1} - 5\int \frac{dx}{x+2} + 5\int \frac{dx}{x+3} = \log(x+1) -$$
$$5\log(x+2) + 5\log(x+3) + \log c = \log\left\{\frac{c(x+1)(x+3)^5}{(x+2)^5}\right\}.$$

Ex. 2. Integrate $dy = \dfrac{adx}{x^2 - a^2}$.

$$y = \int \frac{adx}{x^2-a^2} = \frac{1}{2}\int \frac{dx}{x-a} - \frac{1}{2}\int \frac{dx}{x+a} = \log\sqrt{\frac{c^2(x-a)}{x+a}}.$$

Ex. 3. Integrate $dy = \dfrac{(3x-5)dx}{x^2-6x+8}$.

$$y = \tfrac{7}{2}\log(x-4) - \tfrac{1}{2}\log(x-2) + \log c = \log\left\{\frac{c(x-4)^{\frac{7}{2}}}{(x-2)^{\frac{1}{2}}}\right\}.$$

Ex. 4. Integrate $dy = \dfrac{adx}{x^2+bx}$.

SUG'S. $\dfrac{a}{x^2+bx} = \dfrac{A}{x^*} + \dfrac{B}{x+b}$ gives $a = Ab$, and $0 = A + B$; whence $A = \dfrac{a}{b}$, and $B = -\dfrac{a}{b}$. $y = \log\left\{c\left(\dfrac{x}{x+b}\right)^{\frac{a}{b}}\right\}$.

Ex. 5. Integrate $dy = \dfrac{(2+3x-4x^2)dx}{4x-x^3}$.

$$y = \tfrac{1}{2}\log x + \tfrac{5}{2}\log(2+x) + \log(2-x) + C.$$

Ex. 6. Integrate $dy = \dfrac{(2x+3)dx}{x^3+x^2-2x}$. $\qquad y = \log\left\{\dfrac{c(x-1)^{\frac{5}{3}}}{x^{\frac{3}{2}}(x+2)^{\frac{1}{6}}}\right\}$.

Ex. 7. Integrate $dy = \dfrac{(5x+1)dx}{x^2+x-2}$.

Ex. 8. Integrate $dy = \dfrac{2-3x^2}{(x+2)^3}dx$.

SUG'S.—Assume $\dfrac{2-3x^2}{(x+2)^3} = \dfrac{A}{(x+2)^3} + \dfrac{B}{(x+2)^2} + \dfrac{C}{x+2} = \dfrac{A + B(x+2) + C(x+2)^2}{(x+2)^3}$; whence $2 - 3x^2 = A + Bx + 2B + Cx^2 + 4Cx + 4C$. $\therefore$ $A + 2B + 4C = 2$ (1); $B + 4C = 0$ (2); $C = -3$ (3); from which $B = 12$, $A = -10$, and $C = -3$.

$$y = \int \frac{2-3x^2}{(x+2)^3}dx = -10\int \frac{dx}{(x+2)^3} + 12\int \frac{dx}{(x+2)^2} - 3\int \frac{dx}{x+2} = \frac{5}{(x+2)^2} -$$
$$\frac{12}{x+2} - 3\log(x+2) + C.$$

---

* This factor is equivalent to $x + 0$, so that $x^2 + bx = (x+0)(x+b)$.

Ex. 9. Integrate $dy = \dfrac{3x - 1}{(x - 3)^2}dx.$

$$y = -\frac{8}{x - 3} + 3 \log (x - 3) + C.$$

Ex. 10. Integrate $dy = \dfrac{x^2 dx}{(x - a)^2(x + a)}.$

SUG.—When some equal and some unequal factors occur in the denominator, the assumed forms must be combined. Thus $\dfrac{x^2}{(x - a)^2(x + a)}$ must be put $= \dfrac{A}{(x - a)^2} + \dfrac{B}{x - a} + \dfrac{C}{x + a}$. $A = \frac{1}{2}a$, $B = \frac{3}{4}$, and $C = \frac{1}{4}$. $y = -\dfrac{a}{2(x - a)} + \frac{3}{4} \log (x - a) + \frac{1}{4} \log (x + a) + C.$

Ex. 11. Integrate $dy = \dfrac{x^2 - 4x + 3}{x^3 - 6x^2 + 9x}dx.$

SUG. $x^3 - 6x^2 + 9x = x(x^2 - 6x + 9) = x(x - 3)^2.$

$$y = \tfrac{1}{3} \log x + \tfrac{1}{3} \log (x - 3)^2 + \tfrac{1}{3} \log c = \log [cx(x - 3)^2]^{\frac{1}{3}}.$$

Ex. 12. Integrate $dy = \dfrac{dx}{(x - 2)^2(x + 3)^2}.$

SUG.—Assume $\dfrac{1}{(x - 2)^2(x + 3)^2} = \dfrac{A}{(x - 2)^2} + \dfrac{B}{x - 2} + \dfrac{C}{(x + 3)^2} + \dfrac{D}{x + 3}$; whence $A = \frac{1}{25}$, $B = -\frac{2}{125}$, $C = \frac{1}{25}$, and $D = \frac{2}{125}$.

$$y = -\frac{1}{25(x - 2)} - \tfrac{2}{125} \log (x - 2) - \frac{1}{25(x + 3)} + \tfrac{2}{125} \log (x + 3) + C.$$

Ex. 13. Integrate $dy = \dfrac{x^3 + x - 1}{(x^2 + 2)^2}dx.$

SUG.—The *simple* binomial factors of the denominator being imaginary, $(x + \sqrt{-2})^2(x - \sqrt{-2})^2$, we assume $\dfrac{x^3 + x - 1}{(x^2 + 2)^2} = \dfrac{Ax + B}{(x^2 + 2)^2} + \dfrac{Cx + D}{x^2 + 2}$; whence $A = -1$, $B = -1$, $C = 1$, and $D = 0$.

$$\therefore \int \frac{x^3 + x - 1}{(x^2 + 2)^2}dx = \int \frac{-x - 1}{(x^2 + 2)^2}dx + \int \frac{xdx}{x^2 + 2} = \int \frac{xdx}{x^2 + 2} - \int x(x^2 + 2)^{-2}dx$$
$$- \int \frac{dx}{(x^2 + 2)^2} = \tfrac{1}{2} \log (x^2 + 2) + \frac{1}{2(x^2 + 2)} - \int \frac{dx}{(x^2 + 2)^2}.$$

The last term can be integrated by formula $\mathfrak{D}$ (***192***, *Ex.* 10).

Ex. 14. Integrate $dy = \dfrac{dx}{x^3 - x^2 + 2x - 2}.$

SUG'S —The simple factors are $x + \sqrt{-2}$, $x - \sqrt{-2}$, and $x - 1$; and the form of the partial fractions is $\frac{Ax+B}{x^2+2}, \frac{C}{x-1}$. $A=-\frac{1}{3}$, $B=-\frac{1}{3}$, and $C=\frac{1}{3}$.

$$\therefore\ y=\int\frac{dx}{x^3-x^2+2x-2}=-\tfrac{1}{3}\int\frac{xdx}{x^2+2}-\tfrac{1}{3}\int\frac{dx}{x^2+2}+\tfrac{1}{3}\int\frac{dx}{x-1}=-\tfrac{1}{6}\log(x^2+2)$$

$$+\tfrac{1}{3}\log(x-1)-\frac{1}{3\sqrt{2}}\tan^{-1}\frac{x}{\sqrt{2}}+C;\text{ since }\int\frac{dx}{x^2+2}=\frac{1}{\sqrt{2}}\tan^{-1}\frac{x}{\sqrt{2}}.$$

Ex. 15. Integrate $dy=\frac{xdx}{x^3+x^2+x+1}$.

$$y=\tfrac{1}{2}\tan^{-1}x+\tfrac{1}{2}\log\left\{\frac{c(x^2+1)^{\frac{1}{2}}}{x+1}\right\}.$$

Ex. 16. Integrate $dy=\frac{x^2dx}{x^4+x^2-2}$.

SUG'S.—Assume $\frac{x^2}{x^4+x^2-2}=\frac{A}{x+1}+\frac{B}{x-1}+\frac{Cx+D}{x^2+2}$; whence $A=-\frac{1}{6}$, $B=\frac{1}{6}$, $C=0$, and $D=\frac{2}{3}$. $y=\log\left(\frac{x-1}{x+1}\right)^{\frac{1}{6}}+\frac{\sqrt{2}}{3}\tan^{-1}\frac{x}{\sqrt{2}}+C$.

Ex. 17. Integrate $dy=\frac{x^2+x}{x^3-2x^2+x-2}dx$.

Ex. 18. Integrate $dy=\frac{x^2-x+1}{x^3+x^2+x+1}dx$.

Ex. 19. Integrate $dy=\frac{9x^2+9x-128}{x^3-5x^2+3x+9}dx$.

Ex. 20. Integrate $dy=\frac{x^3-1}{x^2-4}dx$.

$$y=\frac{x^2}{2}+\tfrac{9}{4}\log(x+2)+\tfrac{7}{4}\log(x-2)+C.$$

178. SCH.—It will be observed that the foregoing processes of separating rational fractions into partial fractions make their integration depend on one or more of the following forms:

$$\int x^m dx,\ \int\frac{dx}{x\pm a},\ \int\frac{dx}{x^2+a^2},\ \int\frac{xdx}{x^2+a^2},\ \int\frac{xdx}{(x^2+a^2)^m},\ \int\frac{dx}{(x^2+a^2)^m}.$$

All of these forms except the last are integrable by the elementary processes. The integration of the last is effected by formula 𝔇 (***192***).

SECTION III.

Rationalization.

179. When polynomial radicals occur in a differential which we desire to integrate, it is *sometimes* possible and expedient to rationalize the expression by the substitution of a new variable which is some definite function of the variable in the given differential. A few of the more important cases are given in this section.

BINOMIAL DIFFERENTIALS.

180. Prop. 1.—*Every binomial differential can be reduced to the form* $x^m(a+bx^n)^p dx$, *in which* m *and* n *are integral, and* n *positive.*

DEM.—1st. If x occurs in both terms of the binomial, and the form is $x^r(ax^s+bx^t)^p dx$, we can remove from the parenthesis the factor x^s, or x^t, which has the less exponent. Thus suppose $s<t$, we can write $x^r(ax^s+bx^t)^p\,dx = x^r\cdot x^{ps}\left(a+b\frac{x^t}{x^s}\right)^p dx = x^{r+ps}(a+bx^{t-s})^p dx$. In this form $t-s$ is positive, since $t>s$, but it may be fractional. $r+ps$ may be either positive or negative, integral or fractional. Now let $r+ps=\pm\frac{c}{h}$, and $t-s=+\frac{e}{f}$; whence we have $x^{\pm\frac{c}{h}}(a+bx^{+\frac{e}{f}})^p dx$.

2nd. In the latter form put $x=z^{hf}$; whence $x^{\pm\frac{c}{h}}=z^{\pm cf}$, $x^{+\frac{e}{f}}=z^{+eh}$, and $dx = hf\,z^{hf-1}dz$. Substituting these values, we have $x^{\pm\frac{c}{h}}(a+bx^{+\frac{e}{f}})^p dx = z^{\pm cf}(a+bz^{+eh})^p hf\,z^{hf-1}dz = hf\,z^{\pm cf+hf-1}(a+bz^{+eh})^p dz$, in which the exponents of z are integral, since c, e, f, and h are integers, and eh is positive. Therefore putting $\pm cf+hf-1=m$, and $eh=n$, we have $hf\,z^m(a+bz^n)^p dz$. Q. E. D.

---

181. Prop. 2.—*A binomial differential of the form* $x^m(a+bx^n)^{\frac{p}{q}}dx$ (*any or all the exponents being fractions*) *may be rendered rational by putting* $a+bx^n=z^q$, *when* $\frac{m+1}{n}$ *is integral.*

DEM.—Putting $\quad a+bx^n = z^q, \qquad (1)$

we have $\quad (a+bx^n)^{\frac{p}{q}} = z^p, \qquad (2)$

Differentiating (1), $nb\,x^{n-1}dx = qz^{q-1}dz. \qquad (3)$

Also from (1), $\quad x^{m-n+1} = \left(\frac{z^q-a}{b}\right)^{\frac{m-n+1}{n}}. \qquad (4)$

Multiplying (2), (3), and (4), $nb\,x^m(a+bx^n)^p dx = qz^{pq+q-1}\left(\frac{z^q-a}{b}\right)^{\frac{m-n+1}{n}}dz,$

or $$x^m(a+bx^n)^{\frac{p}{q}}dx = \frac{q}{nb}z^{p+q-1}\left(\frac{z^q-a}{b}\right)^{\frac{m+1}{n}-1}dz.$$

Now by hypothesis $p+q-1$ is integral; hence, if $\frac{m+1}{n}$ is integral, the expression is rational. Q. E. D.

---

182. Prop. 3.—*A binomial differential of the form* $x^m(a+bx^n)^{\frac{p}{q}}dx$ *(any or all the exponents being fractions) may be rendered rational by putting* $a+bx^n=z^qx^n$, *when* $\frac{m+1}{n}+\frac{p}{q}$ *is integral.*

DEM.—Putting $\quad a+bx^n=z^qx^n, \qquad (1)$

we have $\quad x^n=\frac{a}{z^q-b}, \qquad (2)$

$$x=\frac{a^{\frac{1}{n}}}{(z^q-b)^{\frac{1}{n}}}=a^{\frac{1}{n}}(z^q-b)^{-\frac{1}{n}}, \qquad (3)$$

and $\quad x^m=a^{\frac{m}{n}}(z^q-b)^{-\frac{m}{n}}. \qquad (4)$

Multiplying (2) by b and adding a, we have

$$a+bx^n=\frac{ab}{z^q-b}+a=\frac{az^q}{z^q-b}, \qquad (5)$$

whence $\quad (a+bx^n)^{\frac{p}{q}}=a^{\frac{p}{q}}(z^q-b)^{-\frac{p}{q}}z^p. \qquad (6)$

Differentiating (3), $dx=-\frac{q}{n}a^{\frac{1}{n}}(z^q-b)^{-\frac{1}{n}-1}z^{q-1}dz. \qquad (7)$

Multiplying together (4), (6), and (7) and putting A for the constant factor, there results

$$x^m(a+bx^n)^{\frac{p}{q}}dx=A(z^q-b)^{-\left(\frac{m+1}{n}+\frac{p}{q}+1\right)}z^{p+q-1}dz.$$

Now by hypothesis $p+q-1$ is integral; hence, if $\frac{m+1}{n}+\frac{p}{q}$ is integral, the expression is rational. Q. E. D.

183. SCH.—Although the rationalization can always be effected as stated in the last two propositions, it does not always facilitate the integration. When in the former case $\frac{m+1}{n}-1$ is a positive integer or 0, or in the latter $\frac{m+1}{n}+\frac{p}{q}+1$ is a negative integer or 0, the binomial z^q-b will have a positive integral exponent and can be expanded into a series of a finite number of terms, or a 0 exponent and will be equal to 1. Hence in any such case the rationalization will lead directly to the integration. But if $\frac{m+1}{n}-1$ is a negative integer in the former, or $\frac{m+1}{n}+\frac{p}{q}+1$ is a positive integer in the latter, the exponent of z^q-b will be negative, and the rationalization will not generally lead to the integration; and in fact it is not usually expedient to rationalize in such cases.

184. COR.—*Every differential of the form* $dy = Ax^m(a+bx)^p dx$ *can be rationalized and integrated when either* m *or* p *is a positive integer.*

DEM.—If p is a positive integer, $a+bx$ can be expanded into a series of a finite number of terms, which multiplied by $x^m dx$ will give a series of monomials; and an algebraic monomial can always be integrated by (***165*** or ***166***).

If p is fractional or negative and m integral and positive, $\frac{m+1}{n} - 1 = \frac{m+1}{1} - 1$, will be a positive integer or 0, and *Prop.* **2**, will effect a rationalization which will lead directly to the integral.

Ex. 1. Integrate $dy = x^5(2 + 3x^2)^{\frac{1}{2}}dx$.

SUG'S.—Since $m = 5$, and $n = 2$, $\frac{m+1}{n} - 1 = 2$, a positive integer, and *Prop.* 2, will lead to the integration.

To rationalize, put $2 + 3x^2 = z^2$; (1)

whence $(2 + 3x^2)^{\frac{1}{2}} = z$, (2)

Differentiating (1) $xdx = \frac{1}{3}zdz$, (3)

Also from (1) $x^4 = \left(\frac{z^2 - 2}{3}\right)^2$. (4)

Multiplying (2), (3), and (4) together

$$dy = x^5(2 + 3x^2)^{\frac{1}{2}}dx = \tfrac{1}{27}(z^2 - 2)^2z^2dz = \tfrac{1}{27}(z^6dz - 4z^4dz + 4z^2dz).$$

$\therefore\ y = \frac{1}{27}\int(z^6dz - 4z^4dz + 4z^2dz) = \frac{1}{27}\left(\frac{z^7}{7} - \frac{4z^5}{5} + \frac{4z^3}{3}\right) + C$; or, restoring the value of z, $y = \frac{1}{27}\left\{\frac{(2 + 3x^2)^{\frac{7}{2}}}{7} - \frac{4(2 + 3x^2)^{\frac{5}{2}}}{5} + \frac{4(2 + 3x^2)^{\frac{3}{2}}}{3}\right\} + C$. [This result may be expanded and reduced, if desired.]

Ex. 2. Integrate $dy = x^3(a + bx^2)^{\frac{3}{2}}dx$.

$$y = (a + bx^2)^{\frac{5}{2}}\left(\frac{5bx^2 - 2a}{35b^2}\right) + C.$$

Ex. 3. Integrate $dy = x^3(a - x^2)^{-\frac{1}{2}}dx$.

$$y = -\tfrac{1}{3}(a - x^2)^{\frac{1}{2}}(2a + x^2) + C.$$

Ex. 4. Integrate $dy = x^2(a + x)^{\frac{1}{2}}dx$.

$$y = \tfrac{2}{35}(a + x)^{\frac{3}{2}}(5x^2 - 4ax + \tfrac{8}{3}a^2) + C.$$

Ex. 5. Integrate $dy = \dfrac{dx}{x^4(1 + x^2)^{\frac{1}{2}}}$.

SUG'S. $y = \int\frac{dx}{x^4(1 + x^2)^{\frac{1}{2}}} = \int x^{-4}(1 + x^2)^{-\frac{1}{2}}dx$. Here $m = -4$, $n = 2$, and $p = -\frac{1}{2}$; whence $\frac{m+1}{n} + \frac{p}{q} + 1 = \frac{-4+1}{2} - \frac{1}{2} + 1 = -1$, and *Prop.* 3, will lead to the integration.

Putting $1+x^2=z^2x^2$ (1); we have $x^2=\frac{1}{z^2-1}$ (2); $x=\frac{1}{(z^2-1)^{\frac{1}{2}}}$ (3);

$x^{-4}=(z^2-1)^2$ (4); $1+x^2=1+\frac{1}{z^2-1}=\frac{z^2}{z^2-1}$ (5); $(1+x^2)^{-\frac{1}{2}}=z^{-1}(z^2-1)^{\frac{1}{2}}$ (6);

and differentiating (3), $dx=-(z^2-1)^{-\frac{3}{2}}z\,dz$ (7).

Multiplying together (4), (6), and (7), there results

$$y=\int x^{-4}(1+x^2)^{-\frac{1}{2}}dx=-\int(z^2-1)dz=z-\tfrac{1}{3}z^3+C=\frac{(2x^2-1)(1+x^2)^{\frac{1}{2}}}{3x^3}+C.$$

Ex. 6. Integrate $dy=\frac{dx}{x^2(1+x^2)^{\frac{3}{2}}}$.

$$y=-\frac{1}{(1+x^2)^{\frac{1}{2}}}\left(\frac{1}{x}+2x\right)+C.$$

Ex. 7. Integrate $dy=a(1+x^2)^{-\frac{3}{2}}dx$. $\qquad y=\frac{ax}{(1+x^2)^{\frac{1}{2}}}+C.$

Ex. 8. Integrate $dy=x^{-4}(1-2x^2)^{-\frac{1}{2}}dx$.

$$y=-\frac{1+4x^2}{3x^3}(1-2x^2)^{\frac{1}{2}}+C.$$

IRRATIONAL FRACTIONS.

185. Prop. 1.—*When a fraction contains none but monomial surds, it can be rationalized by substituting a new variable with an exponent which is a common multiple of all the denominators of the fractional indices in the given expression.*

DEM.—The general form of such a fraction is

$$\frac{ax^{\frac{m}{n}}+bx^{\frac{p}{q}}+\text{etc.}}{a'x^{\frac{r}{s}}+b'x^{\frac{t}{u}}+\text{etc.}}dx.$$

In this put $x=z^{nqsu,\text{ etc.}}$; whence $x^{\frac{m}{n}}=z^{mqsu,\text{ etc.}}$, $x^{\frac{p}{q}}=z^{npsu,\text{ etc.}}$, $x^{\frac{r}{s}}=z^{nqru,\text{ etc.}}$, $x^{\frac{t}{u}}=z^{nqrt,\text{ etc.}}$, and $dx=(nqsu,\text{ etc.})\,z^{nqsu,\text{ etc.}-1}dz$. These values substituted in the given fraction, evidently render it rational.

186. COR.—*This method is equally applicable when the fraction involves no surd except one of the form* $(a+bx)^{\frac{m}{n}}$, *by treating* $a+bx$ *as the variable.*

---

187. Prop. 2.—*When a fraction contains no surd but one of the form* $\sqrt{a+bx\pm x^2}$, *it can be rationalized by putting* $\sqrt{a+bx+x^2}=z-x$

when x^2 *is* $+$; *and when* x^2 *is* $-$, $\sqrt{a+bx-x^2}=\sqrt{(x-\alpha)(\beta-x)}=(x-\alpha)z$, *in which* α *and* β *are the roots of the equation* $a+bx-x^2=0$.

DEM.—1st. *When* x^2 *is* $+$. Putting $\sqrt{a+bx+x^2}=z-x$, $a+bx+x^2=z^2-2zx+x^2$; whence $x=\dfrac{z^2-a}{2z+b}$, $dx=\dfrac{2(z^2+bz+a)}{(2z+b)^2}dz$, and $\sqrt{a+bx+x^2}=z-\dfrac{z^2-a}{2z+b}=\dfrac{z^2+bz+a}{2z+b}$. Hence as x, $\sqrt{a+bx+x^2}$, and dx are expressed in rational terms of z, the transformed fraction will be rational.

2nd. *When* x^2 *is* $-$. Assuming $\sqrt{a+bx-x^2}=\sqrt{(x-\alpha)(\beta-x)}=(x-\alpha)z$, and squaring we have $(x-\alpha)(\beta-x)=(x-\alpha)^2z^2$, or $\beta-x=(x-\alpha)z^2$; whence $x=\dfrac{\alpha z^2+\beta}{z^2+1}$, $dx=\dfrac{2(\alpha-\beta)z\,dz}{(z^2+1)^2}$, and $\sqrt{a+bx-x^2}=\left\{\dfrac{\alpha z^2+\beta}{z^2+1}-\alpha\right\}z=\dfrac{z(\beta-\alpha)}{z^2+1}$. Hence as x, $\sqrt{a+bx-x^2}$, and dx are expressed in rational terms of z, the transformed fraction will be rational. Q. E. D.

Ex. 1. Integrate $dy=\dfrac{2x^{\frac{1}{2}}-3x^{\frac{2}{3}}}{5x^{\frac{1}{6}}}dx$.

SUG'S.—Put $x=z^6$; whence $dy=\frac{12}{5}z^7\,dz-\frac{18}{5}z^8\,dz$. $y=\frac{3}{10}x^{\frac{4}{3}}-\frac{2}{5}x^{\frac{3}{2}}+C$.

Ex. 2. Integrate $dy=\dfrac{3x^{\frac{1}{2}}dx}{2x^{\frac{1}{2}}-x^{\frac{2}{3}}}$.

$$y=-18\left\{\frac{x^{\frac{5}{6}}}{5}+\frac{x^{\frac{2}{3}}}{2}+\frac{4x^{\frac{1}{2}}}{3}+4x^{\frac{1}{3}}+16x^{\frac{1}{6}}-32\log\left(2-x^{\frac{1}{6}}\right)\right\}+C.$$

Ex. 3. Integrate $dy=\dfrac{dx}{(1+x)^{\frac{3}{2}}+(1+x)^{\frac{1}{2}}}$.

SUG'S.—Putting $1+x=z^2$, $dy=\dfrac{2zdz}{z^3+z}=\dfrac{2dz}{z^2+1}$; whence $y=2\tan^{-1}(1+x)^{\frac{1}{2}}+C$.

Ex. 4. Integrate $dy=\dfrac{x^3dx}{(1+4x)^{\frac{5}{2}}}$.

$$y=\tfrac{1}{128}\left[\frac{(1+4x)^{\frac{3}{2}}}{3}-3(1+4x)^{\frac{1}{2}}-\frac{3}{(1+4x)^{\frac{1}{2}}}+\frac{1}{3(1+4x)^{\frac{3}{2}}}\right]+C.$$

Ex. 5. Integrate $dy=\dfrac{dx}{x\sqrt{1+x}}$. $\qquad y=\log\dfrac{\sqrt{1+x}-1}{\sqrt{1+x}+1}+C.$

SUG.—*Ex's* 4 and 5 can be performed by (***184***) or (***181***). In fact these methods are essentially identical when there is but *one* surd of the form $(a+bx)^{\frac{m}{n}}$.

Ex. 6. Integrate $dy = \frac{dx}{x\sqrt{1+x+x^2}}$.

SOLUTION.—Put $\sqrt{1+x+x^2} = z - x$; whence $z = x + \sqrt{1+x+x^2}$, $x = \frac{z^2-1}{2z+1}$, $dx = \frac{2(z^2+z+1)dz}{(2z+1)^2}$, and $\sqrt{1+x+x^2} = \frac{z^2+z+1}{2z+1}$. Substituting these values

$$y = \int \frac{dx}{x\sqrt{1+x+x^2}} = \int \frac{2(z^2+z+1)}{(2z+1)^2} \times \frac{2z+1}{z^2-1} \times \frac{2z+1}{z^2+z+1} dz = \int \frac{2dz}{z^2-1} =$$

$$\int \frac{dz}{z-1} - \int \frac{dz}{z+1} = \log \frac{z-1}{z+1} + C = \log \frac{x-1+\sqrt{1+x+x^2}}{x+1+\sqrt{1+x+x^2}} + C =$$

$$\log \frac{3x}{2+x+2\sqrt{1+x+x^2}} + C.$$

Ex. 7. Integrate $dy = \frac{dx}{\sqrt{x^2-x-1}}$.

$$y = \log[c(2x-1+2\sqrt{x^2-x-1})].$$

Ex. 8. Integrate $dy = \frac{dx\sqrt{2x+x^2}}{x^2}$.

SUG'S.—Putting $\sqrt{2x+x^2} = z - x$, there results, in the usual way $dy = \frac{(z+2)^2dz}{z^2(z+1)}$ $= \frac{z^2+4z+4}{z^2(z+1)}dz = \frac{z^2dz}{z^2(z+1)} + \frac{4(z+1)dz}{z^2(z+1)} = \frac{dz}{z+1} + \frac{4dz}{z^2}$. $\therefore y = \log(z+1) - \frac{4}{z} + C = \log(x+1+\sqrt{2x+x^2}) - \frac{4}{x+\sqrt{2x+x^2}} + C.$

Ex. 9. Integrate $dy = \frac{dx}{\sqrt{2-x-x^2}}$.

SOLUTION.—Put $\sqrt{2-x-x^2} = \sqrt{(x+2)(1-x)} = (x+2)z$; whence $x = \frac{1-2z^2}{z^2+1}$, $dx = -\frac{6zdz}{(z^2+1)^2}$, and $\sqrt{2-x-x^2} = \frac{3z}{z^2+1}$. $\therefore y = \int \frac{dx}{\sqrt{2-x-x^2}} = -2\int \frac{dz}{1+z^2} = -2\tan^{-1}z + C = -2\tan^{-1}\sqrt{\frac{1-x}{x+2}} + C.$

Ex. 10. Integrate $dy = \frac{dx}{\sqrt{1+x-x^2}}$.

SUG'S.—From $1 + x - x^2 = 0$, we learn that the factors are $x - (\frac{1}{2} + \frac{1}{2}\sqrt{5})$ and $(\frac{1}{2} - \frac{1}{2}\sqrt{5}) - x$. As these roots are so cumbrous it will be economy to take $x - \alpha$ and $\beta - x$ as the factors, as in the general demonstration. The differential in terms of z is $dy = -\frac{2dz}{1+z^2}$. $\therefore y = -2\tan^{-1}z + C = 2\tan^{-1}\sqrt{\frac{\frac{1}{2}-\frac{1}{2}\sqrt{5}-x}{x-\frac{1}{2}-\frac{1}{2}\sqrt{5}}} + C.$

SECTION IV.

Integration by Parts.

188. ***The Formula for Integration by Parts*** is

$$\int u\,dv = uv - \int v\,du.$$

This formula is deduced directly from the differential of a product. Thus $d(uv) = v\,du + u\,dv$; whence $uv = \int v\,du + \int u\,dv$, and $\int u\,dv = uv - \int v\,du$.

FORMULÆ OF REDUCTION 𝔄, 𝔅, ℭ, AND 𝔇.

189. ***Prob.***—*To produce a formula for reducing the exponent of* x *without the parenthesis by the exponent of* x *within, in the form* $dy = x^m(a + bx^n)^p dx$; i. e. *to make the integration of this form depend upon the form* $\int x^{m-n}(a + bx^n)^p dx$.

Solution.—The solution of this problem is effected by applying the formula for integration by parts to the form $dy = x^m(a + bx^n)^p dx$. To make the application, put $dv = (a + bx^n)^p x^{n-1} dx$, whence $u = x^{m-n+1}$.

Integrating the former and differentiating the latter, we have $v = \dfrac{(a + bx^n)^{p+1}}{nb(p + 1)}$, and $du = (m - n + 1)x^{m-n}dx$.

Substituting in the formula $\int u\,dv = uv - \int v\,du$, we have $y = \int x^m(a + bx^n)^p dx$

$$= \frac{x^{m-n+1}(a + bx^n)^{p+1}}{nb(p + 1)} - \frac{m - n + 1}{nb(p + 1)}\int x^{m-n}(a + bx^n)^{p+1}dx.^* \quad (1).$$

Now $\int x^{m-n}(a + bx^n)^{p+1}dx = \int x^{m-n}(a + bx^n)^p dx \times (a + bx^n) =$

$$a\int x^{m-n}(a + bx^n)^p dx + b\int x^m(a + bx^n)^p dx.$$

Introducing this value of $\int x^{m-n}(a + bx^n)^{p+1}dx$ in (1), we have

$$y = \int x^m(a + bx^n)^p dx = \frac{x^{m-n+1}(a + bx^n)^{p+1}}{nb(p + 1)} - \frac{m - n + 1}{nb(p + 1)}a\int x^{m-n}(a + bx^n)^p dx - \frac{m - n + 1}{n(p + 1)}\int x^m(a + bx^n)^p dx.$$

Transposing the last term, we have $\left\{1 + \dfrac{m - n + 1}{n(p + 1)}\right\}\int x^m(a + bx^n)^p dx =$

$\dfrac{x^{m-n+1}(a + bx^n)^{p+1} - (m - n + 1)a\int x^{m-n}(a + bx^n)^p dx}{nb(p + 1)}$ or, $(np + m + 1)\int x^m(a + bx^n)^p dx$

$= \dfrac{x^{m-n+1}(a + bx^n)^{p+1} - a(m - n + 1)\int x^{m-n}(a + bx^n)^p dx}{b}$; and, finally,

$$y = \int x^m(a + bx^n)^p dx = \frac{x^{m-n+1}(a + bx^n)^{p+1} - a(m - n + 1)\int x^{m-n}(a + bx^n)^p dx}{b(np + m + 1)}. \quad (𝔄).$$

---

* This operation diminishes m by n, but it also increases p by 1. The latter *may* be a disadvantage. It will evidently be more likely to prove advantageous if we can diminish or increase either m or p without affecting the other.

190. Prob.—*To produce a formula for increasing the exponent of* x *without the parenthesis by the exponent of* x *within, in the form* $dy = x^m(a + bx^n)^p dx$; i. e. *to make the integration of this form depend upon the form* $\int x^{m+n}(a + bx^n)^p dx$.

SOLUTION.—Clearing (A) of fractions, transposing the first member and the last term of the second member, and dividing by $a(m - n + 1)$, we have

$$\int x^{m-n}(a + bx^n)^p dx = \frac{x^{m-n+1}(a + bx^n)^{p+1} - b(np + m + 1)\int x^m(a + bx^n)^p dx}{a(m - n + 1)}.$$

Putting $m - n = m'$, whence $m = m' + n$, this becomes

$$y = \int x^{m'}(a + bx^n)^p dx = \frac{x^{m'+1}(a + bx^n)^{p+1} - b(np + m' + n + 1)\int x^{m'+n}(a + bx^n)^p dx}{a(m' + 1)};$$

or, dropping the accents,

$$y = \int x^m(a + bx^n)^p dx = \frac{x^{m+1}(a + bx^n)^{p+1} - b(np + m + n + 1)\int x^{m+n}(a + bx^n)^p dx}{a(m + 1)}. \quad (\text{B}.)$$

---

191. Prob.—*To produce a formula for diminishing the exponent of the parenthesis by* 1, *in the form* $dy = x^m(a + bx^n)^p dx$; i. e. *to make the integration of this form depend upon the form* $\int x^m(a + bx^n)^{p-1} dx$.

SOLUTION.—We may write

$$y = \int x^m(a + bx^n)^p dx = \int x^m(a + bx^n)^{p-1} dx \times (a + bx^n) =$$
$$a\int x^m(a + bx^n)^{p-1} dx + b\int x^{m+n}(a + bx^n)^{p-1} dx. \quad (1).$$

By applying formula (A) to the last integral, it becomes

$$\int x^{m+n}(a + bx^n)^{p-1} dx = \frac{x^{m+1}(a + bx^n)^p - a(m + 1)\int x^m(a + bx^n)^{p-1} dx}{b(np + m + 1)}.$$

Substituting this in (1) it becomes $y = \int x^m(a + bx^n)^p dx = a\int x^m(a + bx^n)^{p-1} dx +$

$$\frac{x^{m+1}(a + bx^n)^p - a(m + 1)\int x^m(a + bx^n)^{p-1} dx}{np + m + 1};$$

or, uniting terms,

$$y = \int x^m(a + bx^n)^p dx = \frac{x^{m+1}(a + bx^n)^p + anp\int x^m(a + bx^n)^{p-1} dx}{np + m + 1}. \quad (\text{C}).$$

---

192. Prob.—*To produce a formula for increasing the exponent of the parenthesis by* 1, *in the form* $dy = x^m(a + bx^n)^p dx$; i. e. *to make the integration of this form depend upon* $\int x^m(a + bx^n)^{p+1} dx$.

SOLUTION.—Transposing and reducing (C) in the same manner as we did (A) in producing (B), we have

$$\int x^m(a + bx^n)^{p-1} dx = \frac{x^{m+1}(a + bx^n)^p - (np + m + 1)\int x^m(a + bx^n)^p dx}{-anp}.$$

Putting $p - 1 = p'$, whence $p = p' + 1$, this becomes

$$y = \int x^m(a+bx^n)^{p'}dx = \frac{x^{m+1}(a+bx^n)^{p'+1} - (np'+m+n+1)\int x^m(a+bx^n)^{p'+1}dx}{-an(p'+1)};$$

or, dropping the accents,

$$y = \int x^m(a+bx^n)^p dx = \frac{x^{m+1}(a+bx^n)^{p+1} - (np+m+n+1)\int x^m(a+bx^n)^{p+1}dx}{-an(p+1)}. \quad (\mathfrak{D}).$$

SCH.—Binomial differentials of this form, or such as may be readily reduced to it, are of such frequent occurrence, and the formulæ $\mathfrak{A}$, $\mathfrak{B}$, $\mathfrak{C}$, and $\mathfrak{D}$, called *Formulæ of Reduction*, are so frequently efficient in reducing them to known forms, that these *formulæ* should be carefully memorized.

Ex. 1. Integrate $dy = \dfrac{x^2dx}{(a^2-x^2)^{\frac{1}{2}}}$.

SOLUTION. $y = \displaystyle\int \frac{x^2dx}{(a^2-x^2)^{\frac{1}{2}}} = \int x^2(a^2-x^2)^{-\frac{1}{2}}dx$, a form which corresponds to $\int x^m(a+bx^n)^p dx$, by considering $m = 2$, $n = 2$, $a = a^2$, $b = -1$, and $p = -\frac{1}{2}$. We now observe that if the exponent of x outside the parenthesis, or in the numerator in the given form, were 0, so that $x^0 = 1$, the integral would be a circular function. Now formula ($\mathfrak{A}$) will so reduce this exponent; hence we apply it, and have

$$y = \int \frac{x^2dx}{(a^2-x^2)^{\frac{1}{2}}} = \int x^2(a^2-x^2)^{-\frac{1}{2}}dx$$

$$= \frac{x^{2-2+1}(a^2-x^2)^{-\frac{1}{2}+1} - a^2(2-2+1)\int x^{2-2}(a^2-x^2)^{-\frac{1}{2}}dx}{-1[2(-\frac{1}{2})+2+1]}$$

$$= \frac{x(a^2-x^2)^{\frac{1}{2}}}{-2} + \frac{a^2}{2}\int (a^2-x^2)^{-\frac{1}{2}}dx$$

$$= -\tfrac{1}{2}x(a^2-x^2)^{\frac{1}{2}} + \frac{a^2}{2}\int \frac{dx}{(a^2-x^2)^{\frac{1}{2}}}$$

$$= -\tfrac{1}{2}x(a^2-x^2)^{\frac{1}{2}} + \frac{a^2}{2}\sin^{-1}\frac{x}{a} + C.$$

Ex. 2. Integrate $dy = \dfrac{x^3dx}{(1-x^2)^{\frac{1}{2}}}$.

$$y = -\tfrac{1}{3}x^2(1-x^2)^{\frac{1}{2}} - \tfrac{2}{3}(1-x^2)^{\frac{1}{2}} + C.$$

$$= -\tfrac{1}{3}(x^2+2)(1-x^2)^{\frac{1}{2}} + C.$$

Ex. 3. Integrate $dy = \dfrac{x^3dx}{(2+x^2)^{\frac{3}{2}}}$.

$$y = x^2(2+x^2)^{-\frac{1}{2}} + 4(2+x^2)^{-\frac{1}{2}} + C = \frac{x^2+4}{(2+x^2)^{\frac{1}{2}}} + C.$$

Ex. 4. Integrate $dy = \dfrac{x^5dx}{(1-x^2)^{\frac{1}{2}}}$.

SUG.—Apply **A** twice in succession, and we have

$$y = -\left(\frac{1}{5}x^4 + \frac{1\cdot 4}{3\cdot 5}x^2 + \frac{1\cdot 2\cdot 4}{1\cdot 3\cdot 5}\right)\sqrt{1-x^2} + C.$$

Ex. 5. Integrate $dy = \dfrac{x^6dx}{(1-x^2)^{\frac{1}{2}}}$.

$$y = -\left(\frac{1}{6}x^5 + \frac{1\cdot 5}{4\cdot 6}x^3 + \frac{1\cdot 3\cdot 5}{2\cdot 4\cdot 6}x\right)\sqrt{1-x^2} + \frac{1\cdot 3\cdot 5}{2\cdot 4\cdot 6}\sin^{-1}x + C.$$

Ex. 6. Integrate $dy = \dfrac{dx}{x^4(1-x^2)^{\frac{1}{2}}}$.

SUG.—Apply **B** twice. $y = -\dfrac{2x^2+1}{3x^3}(1-x^2)^{\frac{1}{2}} + C.$

Ex. 7. Integrate $dy = \dfrac{dx}{x^3(x^2-a^2)^{\frac{1}{2}}}$.

SOLUTION. $y = \displaystyle\int\frac{dx}{x^3(x^2-a^2)^{\frac{1}{2}}} = \int x^{-3}(-a^2+x^2)^{-\frac{1}{2}}dx.$ Now by increasing the exponent of x without the parenthesis by that within the form becomes known. Hence we apply (**B**). In this case $m = -3$, $n = 2$, $a = -a^2$, $b = 1$, and $p = -\frac{1}{2}$. Hence substituting in the formula

$$y = \int\frac{dx}{x^3(x^2-a^2)^{\frac{1}{2}}} = \int x^{-3}(-a^2+x^2)^{-\frac{1}{2}}dx$$

$$= \frac{x^{-3+1}(-a^2+x^2)^{-\frac{1}{2}+1} - [2(-\frac{1}{2}) - 3 + 2 + 1]\int x^{-3+2}(-a^2+x^2)^{-\frac{1}{2}}dx}{-a^2(-3+1)}$$

$$= \frac{x^{-2}(x^2-a^2)^{\frac{1}{2}}}{2a^2} + \frac{1}{2a^2}\int\frac{dx}{x(x^2-a^2)^{\frac{1}{2}}}$$

$$= \frac{(x^2-a^2)^{\frac{1}{2}}}{2a^2x^2} + \frac{1}{2a^3}\sec^{-1}\frac{x}{a} + C.$$

Ex. 8. Integrate $dy = (a^2-x^2)^{\frac{1}{2}}x^2dx$.

SUG'S.—Applying **A**, we have

$$y = \int x^2(a^2-x^2)^{\frac{1}{2}}dx = -\frac{x(a^2-x^2)^{\frac{3}{2}}}{4} + \frac{a^2}{4}\int(a^2-x^2)^{\frac{1}{2}}dx.$$

Applying **C** to $\int(a^2-x^2)^{\frac{1}{2}}dx$, we have

$$\int(a^2-x^2)^{\frac{1}{2}}dx = \frac{x(a^2-x^2)^{\frac{1}{2}}}{2} + \frac{a^2}{2}\int(a^2-x^2)^{-\frac{1}{2}}dx = \frac{x(a^2-x^2)^{\frac{1}{2}}}{2} + \frac{a^2}{2}\sin^{-1}\frac{x}{a} + C.$$

$$\therefore\ y = \int(a^2-x^2)^{\frac{1}{2}}x^2\,dx = -\frac{x(a^2-x^2)^{\frac{3}{2}}}{4} + \frac{a^2x(a^2-x^2)^{\frac{1}{2}}}{8} + \frac{a^4}{8}\sin^{-1}\frac{x}{a} + C.$$

Ex. 9. Integrate $dy = (1-x^2)^{\frac{3}{2}}dx$.

Sug.—Apply $\mathfrak{C}$ twice. $y = \frac{1}{4}x(1-x^2)^{\frac{3}{2}} + \frac{3}{8}x(1-x^2)^{\frac{1}{2}} + \frac{3}{8}\sin^{-1}x + C.$

Ex. 10. Integrate $dy = \dfrac{dx}{(a^2+x^2)^2}$.

Sug.—Apply $\mathfrak{D}$. $y = \dfrac{x}{2a^2(a^2+x^2)} + \dfrac{1}{2a^2}\int(a^2+x^2)^{-1}dx = \dfrac{x}{2a^2(a^2+x^2)} + \dfrac{1}{2a^2}\int\dfrac{dx}{a^2+x^2} = \dfrac{x}{2a^2(a^2+x^2)} + \dfrac{1}{2a^3}\tan^{-1}\dfrac{x}{a} + C.$

Ex. 11. Integrate $dy = \dfrac{dx}{(1+x^2)^3}$.

$$y = \frac{x}{4(1+x^2)^2} + \frac{3}{8}\frac{x}{(1+x^2)} + \frac{3}{8}\tan^{-1}x + C.$$

Sch.—These *formulæ* often fail, in consequence of reducing the expression to ∞, by making the denominator 0; or by making the expression indeterminate. Thus it would seem at first glance that formula $\mathfrak{D}$ would reduce $y = \int\dfrac{dx}{(1-x^2)^{\frac{3}{2}}}$; but it will be found to fail. Nevertheless the *formulæ* are of great practical value; and that, not only in such examples as the above, which they reduce directly to the elementary forms, but in many more complicated cases where they reduce the given expression to a form which can be integrated by methods yet to be given.

LOGARITHMIC DIFFERENTIALS.

193. Prob.*—To integrate the form* $dy = X\cdot\log^n x\,dx$, *in which* X *is an algebraic function of* x.

Solution.—Put $X\,dx = dv$, whence $\log^n x = u$, and substitute in $\int u\,dv = uv - \int v\,du$. Thus $y = \int X\cdot\log^n x\,dx = \log^n x\cdot\int X\,dx - \int\left[n\log^{n-1}x\int(X\,dx)\cdot\dfrac{dx}{x}\right]$. If now $\int X\,dx$ is a known form, we have made the integration to depend upon a form in which the exponent of $\log x$ is diminished. Thus, if $\int X\,dx = X'$, the form of the expression to be integrated becomes $\dfrac{X'}{x}\log^{n-1}x\,dx$. To this the formula may be applied again, and n diminished still farther if the algebraic function $\dfrac{X'}{x}dx$ can be integrated.

Ex. 1. Integrate $dy = \dfrac{\log x\, dx}{(1+x)^2}$.

SUG.—Put $\dfrac{dx}{(1+x)^2} = dv$; whence $\log x = u$, $v = -\dfrac{1}{1+x}$, and $du = \dfrac{dx}{x}$.

$\therefore\ y = \displaystyle\int \frac{\log x\, dx}{(1+x)^2} = -\frac{\log x}{1+x} + \int \frac{dx}{x(1+x)}$.

Separating $\dfrac{dx}{x(1+x)}$ into partial fractions (***174***), and integrating, we have

$\displaystyle\int \frac{dx}{x(1+x)} = \log x - \log(1+x) + C.$

Finally, $y = -\dfrac{\log x}{1+x} + \log x - \log(1+x) + C = \dfrac{x}{1+x}\log x - \log(1+x) + C.$

Ex. 2. Integrate $dy = \log x\, dx$. $\qquad y = x(\log x - 1) + C.$

Ex. 3. Integrate $dy = x^2 \log^2 x\, dx$.

$$y = \tfrac{1}{3}x^3(\log^2 x - \tfrac{2}{3}\log x + \tfrac{2}{9}) + C.$$

Ex. 4. Integrate $dy = \dfrac{dx}{x \log^2 x}$.

SOLUTION.—Put $dv = \dfrac{dx}{x}$; whence $u = \log^{-2} x$, $v = \log x$, and $du = -2\log^{-3} x\dfrac{dx}{x}$.

Substituting in the formula for integration by parts $y = \displaystyle\int \frac{dx}{x(\log^2 x)} = \frac{1}{\log x} + 2\int \frac{dx}{x \log^2 x}$. Transposing the last term $-\displaystyle\int \frac{dx}{x\log^2 x} = \frac{1}{\log x}$, or $y = -\dfrac{1}{\log x} + C.$

Ex. 5. Integrate $dy = \dfrac{\log^2 x}{x^{\frac{3}{2}}} dx$.

$$y = -\frac{2}{x^{\frac{1}{2}}}(\log^2 x + 4\log x + 8) + C.$$

Ex. 6. Integrate $dy = \dfrac{x \log x\, dx}{(a^2 + x^2)^{\frac{1}{2}}}$.

$$y = (a^2 + x^2)^{\frac{1}{2}}\log x + a\log\frac{a + (a^2 + x^2)^{\frac{1}{2}}}{x} - (a^2 + x^2)^{\frac{1}{2}} + C.$$

EXPONENTIAL DIFFERENTIALS.

194. Prob.—*To integrate* $dy = x^n a^{cx} dx$, *when* n *is a positive integer.*

SOL.—Put $dv = a^{cx}dx$; whence $u = x^n$, $v = \dfrac{1}{c\log a}a^{cx}$, $du = nx^{n-1}dx$. Substituting in the formula $\int u dv = uv - \int v du$, we have $y = \int x^n a^{cx} dx = \dfrac{1}{c\log a}a^{cx}x^n - \dfrac{n}{c\log a}\int x^{n-1}a^{cx}dx$. Applying the formula to $\int x^{n-1}a^{cx}dx$, the integration is made to depend upon the form $A\int x^{n-2}a^{cx}dx$. Thus, the exponent of x can be finally reduced to 0, and the integration made to depend upon the form $A'\int a^{cx}dx$, which $= \frac{1}{c\log a}a^{cx} + C$.

Ex. 1. Integrate $dy = x^3e^{ax}dx$.

Solution.—Put $dv = e^{ax}dx$; whence $u = x^3$, $v = \frac{1}{a}e^{ax}$, and $du = 3x^2dx$.

$$\therefore\ y = \int x^3e^{ax}dx = \frac{x^3}{a}e^{ax} - \frac{3}{a}\int x^2e^{ax}dx, \quad \text{(repeating the process)}$$

$$= \frac{x^3}{a}e^{ax} - \frac{3x^2}{a^2}e^{ax} + \frac{6}{a^2}\int x\,e^{ax}dx \quad \text{" " "}$$

$$= \frac{x^3}{a}e^{ax} - \frac{3x^2}{a^2}e^{ax} + \frac{6x}{a^3}e^{ax} - \frac{6}{a^3}\int e^{ax}dx$$

$$= \frac{x^3}{a}e^{ax} - \frac{3x^2}{a^2}e^{ax} + \frac{6x}{a^3}e^{ax} - \frac{6}{a^4}e^{ax} + C$$

$$= e^{ax}\left(\frac{x^3}{a} - \frac{3x^2}{a^2} + \frac{6x}{a^3} - \frac{6}{a^4}\right) + C.$$

Ex. 2. Integrate $dy = x^3a^xdx$.

$$y = \frac{a^x}{\log a}\left\{x^3 - \frac{3x^2}{\log a} + \frac{6x}{\log^2 a} - \frac{6}{\log^3 a}\right\} + C.$$

Ex. 3. Integrate $dy = e^xx^4dx$.

$$y = e^x(x^4 - 4x^3 + 12x^2 - 24x + 24) + C.$$

Ex. 4. Integrate $dy = \frac{x^2dx}{e^x}$.

Sug.—Put $dv = e^{-x}dx$, as before. $y = -e^{-x}(x^2 + 2x + 2) + C$.

SPECIAL FORMS OF EXPONENTIALS.

Ex. 5. Integrate $dy = \frac{e^{2x} - 1}{e^{2x} + 1}dx$.

Solution. $y = \int\frac{e^{2x} - 1}{e^{2x} + 1}dx = \int\frac{e^x - e^{-x}}{e^x + e^{-x}}dx = \log\left[c(e^x + e^{-x})\right]$, as $(e^x - e^{-x})dx$ $= d(e^x + e^{-x})$.

Ex. 6. Integrate $dy = e^{e^x}e^xdx$.

Sug.—Put $e^x = z$, whence $dy = e^zdz$, and $y = e^z + C = e^{e^x} + C$.

Ex. 7. Integrate $dy = \frac{1 + x^2}{(1 + x)^2}e^xdx$.

Solution.—Put $1 + x = z$; whence $x = z - 1$, and $dx = dz$.

Substituting, $y = \int\frac{1 + x^2}{(1 + x)^2}e^xdx = \int\frac{z^2 + 2 - 2z}{z^2}e^{z-1}dz =$

$$\int e^{z-1}dz + 2\int\frac{e^{z-1}dz}{z^2} - 2\int\frac{e^{z-1}dz}{z} =$$

$$\frac{1}{e}\left[e^z + 2\int\frac{e^z dz}{z^2} - 2\int\frac{e^z dz}{z}\right].$$

Applying the formula for integrating by parts to $\int\frac{e^z dz}{z^2}$, by putting $dv = z^{-2}dz$, and $u = e^z$, we have $\int\frac{e^z dz}{z^2} = -\frac{e^z}{z} + \int\frac{e^z dz}{z}$.

Substituting this value,

$$y = \frac{1}{e}\left[e^z - \frac{2e^z}{z} + 2\int\frac{e^z dz}{z} - 2\int\frac{e^z dz}{z}\right] = \frac{1}{e}\left(e^z - \frac{2e^z}{z}\right) + C = e^{z-1}\left(1 - \frac{2}{z}\right) + C =$$

$$e^x\left(1 - \frac{2}{1+x}\right) + C = e^x\left(\frac{x-1}{1+x}\right) + C.$$

Ex. 8. Integrate $dy = \frac{e^x x dx}{(1+x)^2}$. $\qquad y = \frac{e^x}{1+x} + C.$

TRIGONOMETRICAL DIFFERENTIALS.

195. Prob.—*To integrate the forms* $dy = \sin^m x dx$, $dy = \cos^n x dx$, $dy = \sin^m x \cos^n x dx$.

Solution.—To integrate $dy = \sin^m x\, dx$, put $\sin x = z$; whence $\cos x = (1-z^2)^{\frac{1}{2}}$, $dx = \frac{dz}{\cos x} = (1-z^2)^{-\frac{1}{2}}dz$; and we obtain $dy = z^m(1-z^2)^{-\frac{1}{2}}dz$, which may be rationalized by (***181, 182***), or reduced by one or more of formulæ 𝔄, 𝔅, ℭ, and 𝔇.

In like manner putting $\cos x = z$, $dy = \cos^n x\, dx$ becomes $dy = -z^n(1-z^2)^{-\frac{1}{2}}dz$, and can be disposed of as before.

Again, putting $\cos x = z$; whence $\sin^m x = (1-z^2)^{\frac{m}{2}}$, $\cos^n x = z^n$, and $dx = -\frac{dz}{(1-z^2)^{\frac{1}{2}}} = -(1-z^2)^{-\frac{1}{2}}dz$, we have

$$dy = \sin^m x \cos^n x\, dx = -z^n(1-z^2)^{\frac{m-1}{2}}dz;$$

or we may put $\sin x = z$, and have

$$dy = \sin^m x \cos^n x dx = z^m(1-z^2)^{\frac{n-1}{2}}dz.$$

Either of these forms may be treated as the first.

196. Sch.—It will be seen that this process will always effect the integration when m and n are either positive or negative integers, and frequently when they are fractions. Thus, when m and n are positive *even* integers, successive applications of 𝔄 will reduce the final integral to the form $\pm\int(1-z^2)^{-\frac{1}{2}}dz = \pm\int\frac{dz}{(1-z^2)^{\frac{1}{2}}} = \sin^{-1}z$, or $\cos^{-1}z = x$; and, when positive *odd* integers, the final form will be $\pm \int z(1-z^2)^{-\frac{1}{2}}dz = \mp (1-z^2)^{\frac{1}{2}} = \mp \cos x$, or $\mp \sin x$.

When m and n are negative integers, formula 𝔅 will reduce the first two cases.

The third case may require any or all of the four *formulæ*, but can always be integrated when m and n are integers.

In many cases it will not require the application of the *formulæ*, as will be seen in the following examples.

Ex. 1. Integrate $dy = \sin^3 x\,dx$.

SOLUTION.—Putting $\sin x = z$, and applying 𝔄, we have

$$y = \int \sin^3 x dx = \int z^3(1-z^2)^{-\frac{1}{2}}dz = \frac{z^2(1-z^2)^{\frac{1}{2}} - 2\int z(1-z^2)^{-\frac{1}{2}}dz}{-3}$$

$$= -\tfrac{1}{3}z^2(1-z^2)^{\frac{1}{2}} + \tfrac{2}{3}\int z(1-z^2)^{-\frac{1}{2}}dz$$

$$= -\tfrac{1}{3}z^2(1-z^2)^{\frac{1}{2}} - \tfrac{2}{3}(1-z^2)^{\frac{1}{2}} + C.$$

$$= -\tfrac{1}{3}\sin^2 x \cos x - \tfrac{2}{3}\cos x + C.$$

Ex. 2. Integrate $dy = \sin^4 x\,dx$.

$$y = -\frac{\cos x}{4}(\sin^3 x + \tfrac{3}{2}\sin x) + \tfrac{3}{8}x + C.$$

Ex. 3. Integrate $dy = \sin^5 x\,dx$.

$$y = -\frac{\cos x}{5}(\sin^4 x + \tfrac{4}{3}\sin^2 x + \tfrac{8}{3}) + C.$$

Ex. 4. Integrate $dy = \sin^6 x\,dx$.

$$y = -\frac{\cos x}{6}(\sin^5 x + \tfrac{5}{4}\sin^3 x + \tfrac{15}{8}\sin x) + \tfrac{5}{16}x + C.$$

Ex. 5. Integrate $dy = \cos^2 x\,dx$.

$$y = \tfrac{1}{2}(\sin x \cos x + x) + C = \tfrac{1}{2}(\tfrac{1}{2}\sin 2x^* + x) + C.$$

Ex. 6. Integrate $dy = \cos^3 x\,dx$.

$$y = \tfrac{1}{3}\sin x \cos^2 x + \tfrac{2}{3}\sin x + C = \tfrac{1}{12}\sin 3x + \tfrac{3}{4}\sin x + C.\dagger$$

Ex. 7. Integrate $dy = \cos^4 x\,dx$.

$$y = \tfrac{1}{32}\sin 4x + \tfrac{1}{4}\sin 2x + \tfrac{3}{8}x + C.$$

Ex. 8. Integrate $dy = \cos^5 x\,dx$.

$$y = \frac{\sin x}{5}(\cos^4 x + \tfrac{4}{3}\cos^2 x + \tfrac{8}{3}x) + C = \tfrac{1}{8}\left(\frac{\sin 5x}{10} + \frac{5\sin 3x}{6} + 5\sin x\right) + C.$$

---

* Trigonometry (*56*) $\sin x \cos x = \frac{1}{2}\sin 2x$.

† To effect the reduction substitute $1 - \sin^2 x$ for $\cos^2 x$; and then for $\sin^3 x$ substitute $\frac{1}{4}(3\sin x - \sin 3x)$. (See Trigonometry, page 28, *Ex.* 12.)

Ex. 9. Integrate $dy = \cos^5 x \sin^5 x\, dx$.

SUG'S.—Putting $\sin x = z$, there results $y = \frac{\sin^6 x}{2}\left(\frac{1}{3} - \frac{\sin^2 x}{2} + \frac{\sin^4 x}{5}\right) + C$.

Putting $\cos x = z$, we have $y = -\frac{\cos^6 x}{2}\left(\frac{1}{3} - \frac{\cos^2 x}{2} + \frac{\cos^4 x}{5}\right) + C'$.

QUERY.—What is the relation between C and C'?

Ex. 10. Integrate $dy = \sin^6 x \cos^3 x\, dx$.

$$y = \sin^7 x\left(\frac{1}{7} - \frac{\sin^2 x}{9}\right) + C.$$

SUG.—If one factor has an even and the other an odd exponent, it will be found expedient to put the function which has the even exponent $= z$.

Ex. 11. Integrate $dy = \sin^4 x \cos^4 x\, dx$.

SOLUTION.—Put $\sin x = z$, and apply **C**.

$$y = \int z^4(1-z^2)^{\frac{3}{2}}dz = \frac{z^5(1-z^2)^{\frac{3}{2}}}{8} + \tfrac{3}{8}\int z^4(1-z^2)^{\frac{1}{2}}dz$$

$$= \quad \text{"} \quad + \frac{z^5(1-z^2)^{\frac{1}{2}}}{16} + \frac{1}{16}\int z^4(1-z^2)^{-\frac{1}{2}}dz.$$

Now apply **A** to the last integral.

$$\int z^4(1-z^2)^{-\frac{1}{2}}dz = \frac{z^3(1-z^2)^{\frac{1}{2}}}{-4} + \tfrac{3}{4}\int z^2(1-z^2)^{-\frac{1}{2}}dz$$

$$= \quad \text{"} \quad -\frac{3}{4}\,\frac{z(1-z^2)^{\frac{1}{2}}}{2} + \frac{3}{4}\,\frac{}{2}\int (1-z^2)^{-\frac{1}{2}}dz.$$

$$\therefore\ y = \frac{z^5}{8}\left\{(1-z^2)^{\frac{3}{2}} + \tfrac{1}{2}(1-z^2)^{\frac{1}{2}}\right\} - \frac{(1-z^2)^{\frac{1}{2}}}{64}\left\{z^3 + \tfrac{3}{2}z\right\} - \tfrac{3}{128}\sin^{-1}z + C =$$

$$\frac{\sin^5 x}{8}\left\{\cos^3 x + \tfrac{1}{2}\cos x\right\} - \frac{\cos x}{64}\left\{\sin^3 x + \tfrac{3}{2}\sin x\right\} + \tfrac{3}{128}x + C.$$

Ex. 12. Integrate $dy = \frac{\sin^5 x}{\cos^2 x}dx$.

SUG'S.—Putting $\sin x = z$, we have $dy = z^5(1-z^2)^{-\frac{3}{2}}dz$.

Applying **A** twice $y = \frac{-1}{3\cos x}(\sin^4 x + 4\sin^2 x - 8) + C$.

Ex. 13. Integrate $dy = \frac{dx}{\sin^5 x}$.

SUG'S.—Put $\sin x = z$, and we have $dy = z^{-5}(1-z^2)^{-\frac{1}{2}}dz$.

Apply **B** twice and we have

$$y = -\frac{(1-z^2)^{\frac{1}{2}}}{4z^4} - \tfrac{3}{4}\cdot\frac{(1-z^2)^{\frac{1}{2}}}{2z^2} + \tfrac{3}{8}\int z^{-1}(1-z^2)^{-\frac{1}{2}}dz, \text{ or restoring } x,$$

$$= -\frac{\cos x}{4}\left(\frac{1}{\sin^4 x} + \frac{3}{2\sin^2 x}\right) + \tfrac{3}{8}\int\frac{dx}{\sin x},$$

$$= -\frac{\cos x}{4}\left(\frac{1}{\sin^4 x} + \frac{3}{2\sin^2 x}\right) + \tfrac{3}{8}\log\tan\left(\tfrac{1}{2}x\right) + C.$$ (See ***169,*** 1.)

Ex. 14. Integrate $dy = \frac{dx}{\cos^6 x}$.

$$y = \frac{\sin x}{5}\left(\frac{1}{\cos^5 x} + \frac{4}{3\cos^3 x} + \frac{8}{3\cos x}\right) + C.$$

SUG.—Applying **B** three times the last term reduces to 0, and we have the result without integrating.

Ex. 15. Integrate $dy = \frac{dx}{\sin^4 x \cos^2 x}$.

SOLUTION.—Putting $\cos x = z$, whence $\sin^{-4} x = (1 - z^2)^{-2}$, $\cos^{-2} x = z^{-2}$, and $dx = -(1 - z^2)^{-\frac{1}{2}} dz$, we have

$$y = \int \frac{dx}{\sin^4 x \cos^2 x} = \int \cos^{-2} x \sin^{-4} x\, dx = -\int z^{-2}(1 - z^2)^{-\frac{5}{2}} dz.$$

Applying **B**, - - - - $= -\left\{\frac{1}{-z(1-z^2)^{\frac{3}{2}}} + 4\int (1 - z^2)^{-\frac{5}{2}} dz\right\}$

" **D**, - - - - $= \frac{1}{z(1-z^2)^{\frac{3}{2}}} - 4\left\{\frac{z}{3(1-z^2)^{\frac{3}{2}}} + \frac{2}{3}\int (1 - z^2)^{-\frac{3}{2}} dz\right\}$

" " again, - $= \frac{1}{z(1-z^2)^{\frac{3}{2}}} - \frac{4z}{3(1-z^2)^{\frac{3}{2}}} - \frac{8}{3}\left\{\frac{z}{(1-z^2)^{\frac{1}{2}}} - 0\right\} + C$

Restoring x, - - - - $= \frac{1}{\cos x \sin^3 x} - \frac{4\cos x}{3\sin^3 x} - \frac{8\cos x}{3\sin x} + C.$

ANOTHER SOLUTION.—When the exponents are even, an elegant solution is obtained by means of a special expedient, as follows:

Introducing the factor $\sin^2 x + \cos^2 x$, which being equal to 1 does not change the value of the differential, we have

$$y = \int \frac{dx}{\sin^4 x \cos^2 x} = \int \frac{(\sin^2 x + \cos^2 x)dx}{\sin^4 x \cos^2 x} = \int \frac{dx}{\sin^2 x \cos^2 x} + \int \frac{dx}{\sin^4 x} =$$

$$\int \frac{(\sin^2 x + \cos^2 x)dx}{\sin^2 x \cos^2 x} + \int \frac{dx}{\sin^4 x} = \int \frac{dx}{\cos^2 x} + \int \frac{dx}{\sin^2 x} + \int \frac{dx}{\sin^4 x} =$$

$$= \tan x - \cot x - \frac{\cos x}{3}\left[\frac{1}{\sin^3 x} + \frac{2}{\sin x}\right] + C.$$

[It will afford the student a good exercise in trigonometrical reduction, to transform the former expression into the latter.]

Ex. 16. Integrate $dy = \frac{dx}{\sin x \cos^3 x}$.

$$y = \frac{1}{2\cos^2 x} + \log \tan x + C.$$

Ex. 17. Integrate $dy = \frac{\sin^4 x}{\cos^4 x} dx = \tan^4 x\, dx$.

SUG.—Put $\tan x = z$, whence $dx = \frac{dz}{\sec^2 x} = \frac{dz}{1 + z^2}$. $\therefore y = \int \frac{\sin^4 x}{\cos^4 x} dx = \int \tan^4 x\, dx = \int \frac{z^4 dz}{1 + z^2} = \int z^2 dz - \int dz + \int \frac{dz}{1 + z^2} = \frac{1}{3}z^3 - z + \tan^{-1} z + C = \frac{1}{3}\tan^3 x - \tan x + x + C.$

Ex. 18. Integrate $dy = \tan^5 x dx$.

$$y = \tfrac{1}{4}\tan^4 x - \tfrac{1}{2}\tan^2 x + \log \sec x + C.$$

---

197. Prob.—*To integrate the forms* $dy = x^n \sin x\, dx$, *and* $dy = x^n \cos x\, dx$.

SOLUTION.—To integrate $dy = x^n \sin x\, dx$, put $dv = \sin x\, dx$, whence $u = x^n$, $v = -\cos x$, and $du = nx^{n-1}dx$. $\therefore y = -x^n \cos x + n\int x^{n-1}\cos x\, dx$. By repeating the process the integration will finally depend upon the form $A\int \cos x\, dx$, or $A'\int \sin x\, dx$.

In like manner $y = \int x^n \cos x\, dx = x^n \sin x - n\int x^{n-1}\sin x\, dx$, and the final forms are the same as before.

Ex. 1. Integrate $dy = x^3 \cos x dx$.

$$y = x^3 \sin x + 3x^2 \cos x - 6x \sin x - 6 \cos x + C.$$

Ex. 2. Integrate $dy = x^4 \sin x\, dx$.

$$y = -x^4 \cos x + 4x^3 \sin x + 12x^2 \cos x - 24x \sin x - 24 \cos x + C.$$

---

198. Prop.—*When* m *and* n *are integers, the form* $dy = \sin^m x \cos^n x\, dx$ *may be integrated in simple terms of the sines and cosines of the multiple arcs.*

DEM.—The form $\sin^m x \cos^n x$ may be expressed in simple terms of multiple arcs by the use of the following formulæ :

(1)* $\sin x \sin y = \frac{1}{2}\cos(x - y) - \frac{1}{2}\cos(x + y)$,

(2) $\cos x \cos y = \frac{1}{2}\cos(x - y) + \frac{1}{2}\cos(x + y)$,

(3) $\sin x \cos y = \frac{1}{2}\sin(x + y) + \frac{1}{2}\sin(x - y)$.

The truth of this statement and the manner of applying the formulæ, may be seen most readily from the solution of a few examples.

Ex. 1. Integrate in simple terms of the sines or cosines of multiple arcs $dy = \sin^3 x \cos^2 x\, dx$.

SOLUTION. $\sin^3 x \cos^2 x = \sin x (\sin x \cos x)^2$

$= \sin x[\frac{1}{2}\sin 2x]^2$ [From (3), making $y = x$.]

$= \frac{1}{4}\sin x[\sin^2 2x]$

$= \frac{1}{4}\sin x[\frac{1}{2}\cos 0 - \frac{1}{2}\cos 4x]$ [From (1), making $x = y = 2x$.]

$= \frac{1}{4}\sin x[\frac{1}{2} - \frac{1}{2}\cos 4x]$

$= \frac{1}{8}\sin x - \frac{1}{8}\sin x \cos 4x$

$= \frac{1}{8}\sin x - \frac{1}{8}[\frac{1}{2}\sin 5x - \frac{1}{2}\sin 3x]$ [From (3), making $x = x$, and $y = 4x$.]

$= \frac{1}{8}\sin x - \frac{1}{16}\sin 5x + \frac{1}{16}\sin 3x$.

Hence $y = \int \sin^3 x \cos^2 x\, dx = \frac{1}{8}\int \sin x\, dx - \frac{1}{16}\int \sin 5x\, dx + \frac{1}{16}\int \sin 3x\, dx = -\frac{1}{8}\cos x - \frac{1}{48}\cos 3x + \frac{1}{80}\cos 5x + C$.

---

* These *formulæ* are essentially those of ART. 59, Plane Trigonometry. To put the *formulæ* of that article into this form simply change x into $\frac{1}{2}(x + y)$ and y into $\frac{1}{2}(x - y)$.

Ex. 2. Integrate in simple terms of the sines or cosines of multiple arcs, $dy = \sin^3 x \cos^3 x\, dx$.

SUG'S. $\sin^3 x \cos^3 x = \frac{1}{8}\sin^3 2x = \frac{1}{8}\sin 2x \sin^2 2x = \frac{1}{8}\sin 2x(\frac{1}{2} - \frac{1}{2}\cos 4x) = \frac{1}{16}\sin 2x - \frac{1}{16}\sin 2x \cos 4x = \frac{1}{16}\sin 2x - \frac{1}{16}(\frac{1}{2}\sin 6x - \frac{1}{2}\sin 2x) = \frac{1}{16}\sin 2x - \frac{1}{32}\sin 6x + \frac{1}{32}\sin 2x = \frac{3}{16}\sin 2x - \frac{1}{32}\sin 6x$.

$\therefore\ y = \int \sin^3 x \cos^3 x\, dx = \frac{3}{16}\int \sin 2x\, dx - \frac{1}{32}\int \sin 6x\, dx = -\frac{3}{32}\cos 2x + \frac{1}{192}\cos 6x + C.$

Ex. 3. Integrate in simple terms of the sines or cosines of multiple arcs $dy = \sin^6 x\, dx$.

SUG'S. $\sin^6 x = (\sin^2 x)^3 = \frac{1}{8}(1 - \cos 2x)^3 = \frac{1}{8} - \frac{3}{8}\cos 2x + \frac{3}{8}\cos^2 2x - \frac{1}{8}\cos^3 2x = \frac{1}{8} - \frac{3}{8}\cos 2x + \frac{3}{8}(\frac{1}{2} + \frac{1}{2}\cos 4x) - \frac{1}{8}\cos 2x(\frac{1}{2} + \frac{1}{2}\cos 4x) = \frac{5}{16} - \frac{7}{16}\cos 2x + \frac{3}{16}\cos 4x - \frac{1}{16}\cos 2x \cos 4x = \frac{5}{16} - \frac{7}{16}\cos 2x + \frac{3}{16}\cos 4x - \frac{1}{16}(\frac{1}{2}\cos 2x + \frac{1}{2}\cos 6x) = \frac{5}{16} - \frac{15}{32}\cos 2x + \frac{3}{16}\cos 4x - \frac{1}{32}\cos 6x$.

$\therefore\ y = \frac{1}{32}(-\frac{1}{6}\sin 6x + \frac{3}{2}\sin 4x - \frac{15}{2}\sin 2x + 10x) + C.$ [The student should be careful to observe that the three formulæ given under the proposition are sufficient to effect the required reductions.]

Ex. 4. Integrate as above $dy = \cos^3 x\, dx$.

$$y = \frac{1}{4}\left(\frac{\sin 3x}{3} + 3\sin x\right) + C.$$

CIRCULAR DIFFERENTIALS.

199. Prob.—*To integrate the forms* $dy = f(x)\sin^{-1} x\, dx$, $dy = f(x)\cos^{-1} x\, dx$, $dy = f(x)\tan^{-1} x\, dx$, *etc.*

METHOD OF SOLUTION.—Put $f(x)dx = dv$, and $\sin^{-1} x$, $\cos^{-1} x$, or $\tan^{-1} x$, as the case may be, $= u$, and substitute in the formula for integration by parts.

Ex. 1. Integrate $dy = x^2 \sin^{-1} x\, dx$.

SOLUTION.—Putting $x^2 dx = dv$, whence $\sin^{-1} x = u$, $v = \frac{1}{3}x^3$, and $du = \frac{dx}{\sqrt{1 - x^2}}$, we have $y = \frac{1}{3}x^3 \sin^{-1} x - \frac{1}{3}\int \frac{x^3 dx}{\sqrt{1 - x^2}}$. Applying formula $\mathfrak{A}$ to the last integral we have $y = \frac{1}{3}x^3 \sin^{-1} x - \frac{1}{15}(x^2 + 2)\sqrt{1 - x^2} + C.$

Ex. 2. Integrate $dy = \sqrt{1 - x^2}\cos^{-1} x\, dx$.

SOLUTION.—Putting $(1 - x^2)^{\frac{1}{2}} dx = dv$; whence $u = \cos^{-1} x$, $v = \frac{1}{2}x(1 - x^2)^{\frac{1}{2}} + \frac{1}{2}\int \frac{dx}{\sqrt{1 - x^2}}$ * $= \frac{1}{2}x(1 - x^2)^{\frac{1}{2}} - \frac{1}{2}\cos^{-1} x$, and $du = -\frac{dx}{\sqrt{1 - x^2}}$.

Substituting in the formula for integration by parts,

---

* By applying $\mathfrak{C}$.

$$y = \tfrac{1}{2}[x(1-x^2)^{\frac{1}{2}} - \cos^{-1}x]\cos^{-1}x - \int \tfrac{1}{2}[x(1-x^2)^{\frac{1}{2}} - \cos^{-1}x]\left(-\frac{dx}{\sqrt{1-x^2}}\right)$$

$$= \quad \text{"} \quad + \tfrac{1}{2}\int x\,dx + \tfrac{1}{2}\int \cos^{-1}x\left(-\frac{dx}{\sqrt{1-x^2}}\right)$$

$$= \quad \text{"} \quad + \tfrac{1}{4}x^2 + \tfrac{1}{2}\int \cos^{-1}x\,d(\cos^{-1}x)$$

$$= \tfrac{1}{2}[x(1-x^2)^{\frac{1}{2}} - \cos^{-1}x]\cos^{-1}x + \tfrac{1}{4}x^2 + \tfrac{1}{4}(\cos^{-1}x)^2 + C.$$

$$= \tfrac{1}{2}[x(1-x^2)^{\frac{1}{2}}]\cos^{-1}x - \tfrac{1}{4}(\cos^{-1}x)^2 + \tfrac{1}{4}x^2 + C.$$

Ex. 3. Integrate $dy = \dfrac{x^2 \tan^{-1}x}{1+x^2}dx.$

$$y = \tan^{-1}x\,(x - \tfrac{1}{2}\tan^{-1}x) - \log\sqrt{1+x^2} + C.$$

---

200. Prob.—*To integrate the forms* $dy = e^{ax}\sin^n x\,dx$, *and* $dy = e^{ax}\cos^n x\,dx$.

METHOD OF SOLUTION.—Put $e^{ax}dx = dv$; whence $\sin^n x = u$, $v = \frac{1}{a}e^{ax}$, and $du = n\sin^{n-1}x\cos x\,dx$.

$$\therefore\ y = \frac{1}{a}e^{ax}\sin^n x - \frac{n}{a}\int e^{ax}\sin^{n-1}x\cos x\,dx.$$

Applying the formula for integration by parts again, putting $dv = e^{ax}dx$, $u = \sin^{n-1}x\cos x$, $v = \frac{1}{a}e^{ax}$, $du = (n-1)\sin^{n-2}x\cos^2 x\,dx - \sin^n x\,dx$, and we have $\therefore\ y = \int e^{ax}\sin^n x\,dx = \frac{1}{a}e^{ax}\sin^n x -$

$$\frac{n}{a}\left\{\frac{1}{a}e^{ax}\sin^{n-1}x\cos x - \frac{n-1}{a}\int e^{ax}\sin^{n-2}\cos^2 x\,dx + \frac{1}{a}\int e^{ax}\sin^n x\,dx\right\}.$$

Transposing the last term, uniting it with the first, and dividing by $\frac{a^2+n}{a^2}$, we have

$$y = \frac{a}{a^2+n}e^{ax}\sin^n x - \frac{n}{a^2+n}e^{ax}\sin^{n-1}x\cos x + \frac{n(n-1)}{a^2+n}\int e^{ax}\sin^{n-2}\cos^2 dx$$

$$= \quad \text{"} \qquad \text{"} \quad + \frac{n(n-1)}{a^2+n}\int e^{ax}\sin^{n-2}dx - \frac{n(n-1)}{a^2+n}\int e^{xn}\sin^n x\,dx.$$

If now the last term be transposed and united with the first member and we divide by the coefficient, we shall have made the integration to depend upon a form in which n is diminished by 2.

By successive applications of the same process, the integration may be made to depend upon the form $A\int e^{ax}dx$ when n is even, or $A'\int e^{ax}\sin x\,dx$ when n is odd. The former is an elementary form; and the formula for integration by parts being applied to the latter, it becomes an integral without further process, since the coefficient of the unintegrated term contains a factor $n-1$, as will be seen above.

By a process altogether similar the form $dy = e^{ax}\cos^x x\,dx$, can be integrated.

Ex. 1. Integrate $dy = e^{ax}\cos x\,dx.$

SOLUTION.—Put $dv = e^{ax}dx$; whence $u = \cos x$, $v = \frac{1}{a}e^{ax}$, and $du = -\sin x\,dx$.

$$y = \int e^{ax}\cos x\,dx = \frac{1}{a}e^{ax}\cos x + \frac{1}{a}\int e^{ax}\sin x\,dx$$

$$= \frac{1}{a}e^{ax}\cos x + \frac{1}{a^2}e^{ax}\sin x - \frac{1}{a^2}\int e^{ax}\cos x\,dx.$$

$$\therefore\ y = \frac{a}{a^2+1}e^{ax}\cos x + \frac{1}{(a^2+1)}e^{ax}\sin x + C.$$

$$= \frac{e^{ax}}{a^2+1}(a\cos x + \sin x) + C.$$

Ex. 2. Integrate $dy = e^x \sin^3 x\,dx$.

$$y = \tfrac{1}{10}e^x(\sin^3 x + 3\cos^3 x + 3\sin x - 6\cos x) + C.$$

Ex. 3. Integrate $dy = e^{-ax}\sin kx\,dx$.

$$y = -\frac{a\sin kx + k\cos kx}{(a^2+k^2)e^{ax}} + C.$$

---

201. Prob.—*To integrate the form* $dy = \frac{dx}{(a + b\cos x)^n}$.

SOLUTION. $y = \int \frac{dx}{(a+b\cos x)^n} = \int \frac{(a+b\cos x)dx}{(a+b\cos x)^{n+1}} = a\int \frac{dx}{(a+b\cos x)^{n+1}} +$ $b\int \frac{\cos x\,dx}{(a+b\cos x)^{n+1}}$. Applying the formula for integration by parts to the last integral by putting $\cos x\,dx = dv$, we have $\int \frac{\cos x\,dx}{(a+b\cos x)^{n+1}} = \frac{\sin x}{(a+b\cos x)^{n+1}} -$

$$(n+1)\int \frac{b\sin^2 x\,dx}{(a+b\cos x)^{n+2}} = \frac{\sin x}{(a+b\cos x)^{n+1}} - (n+1)\int \frac{(b - b\cos^2 x)dx}{(a+b\cos x)^{n+2}}.$$

Substituting this value in the preceding we have

$$\int \frac{dx}{(a+b\cos x)^n} = a\int \frac{dx}{(a+b\cos x)^{n+1}} + \frac{b\sin x}{(a+b\cos x)^{n+1}} - (n+1)\int \frac{(b^2 - b^2\cos^2 x)dx}{(a+b\cos x)^{n+2}}$$

$$= \frac{b\sin x}{(a+b\cos x)^{n+1}} + a\int \frac{dx}{(a+b\cos x)^{n+1}} -$$

$$(n+1)\int \frac{b^2 - a^2 + 2a(a+b\cos x) - (a+b\cos x)^{2*}}{(a+b\cos x)^{n+2}}dx = \frac{b\sin x}{(a+b\cos x)^{n+1}} +$$

$$a\int \frac{dx}{(a+b\cos x)^{n+1}} - (n+1)(b^2-a^2)\int \frac{dx}{(a+b\cos x)^{n+2}} - 2a(n+1)\int \frac{dx}{(a+b\cos x)^{n+1}}$$

$$+ (n+1)\int \frac{dx}{(a+b\cos x)^n}.$$

Transposing and uniting similar integrals,

$$(n+1)(b^2-a^2)\int \frac{dx}{(a+b\cos x)^{n+2}} = \frac{b\sin x}{(a+b\cos x)^{n+1}} - a(2n+1)\int \frac{dx}{(a+b\cos x)^{n+1}}$$

$$+ n\int \frac{dx}{(a+b\cos x)^n}.$$

Dividing by $(n+1)(b^2-a^2)$, and writing $n-2$ for n, we obtain

---

* By adding and subtracting $2a^2 + 2ab\cos x$, $b^2 - b^2\cos^2 x = b^2 + 2a^2 + 2ab\cos x - 2a^2 - 2ab\cos x - b^2\cos^2 x = b^2 - a^2 + 2a(a+b\cos x) - (a+b\cos x)^2$.

$$y=\int\frac{dx}{(a+b\cos x)^n}=\frac{b\sin x}{(n-1)(b^2-a^2)(a+b\cos x)^{n-1}}-\frac{a(2n-3)}{(n-1)(b^2-a^2)}\int\frac{dx}{(a+b\cos x)^{n-1}}$$
$$+\frac{n-2}{(n-1)(b^2-a^2)}\int\frac{dx}{(a+b\cos x)^{n-2}}.$$

By means of repeated applications of this formula, or the process by which it was produced, the integration may be made to depend upon the form $A\int\frac{dx}{a+b\cos x}$.

To integrate $dy=\frac{dx}{a+b\cos x}$, we remember that $\cos x=\cos^2\frac{x}{2}-\sin^2\frac{x}{2}$, and $\cos^2\frac{x}{2}+\sin^2\frac{x}{2}=1$. Hence

$$y=\int\frac{dx}{a+b\cos x}=\int\frac{dx}{a\left(\cos^2\frac{x}{2}+\sin^2\frac{x}{2}\right)+b\left(\cos^2\frac{x}{2}-\sin^2\frac{x}{2}\right)}=$$

$$\int\frac{dx}{(a+b)\cos^2\frac{x}{2}+(a-b)\sin^2\frac{x}{2}}=\int\frac{\frac{dx}{\cos^2\frac{x}{2}}}{(a+b)+(a-b)\frac{\sin^2\frac{x}{2}}{\cos^2\frac{x}{2}}}=$$

$$\frac{1}{a+b}\int\frac{\sec^2\frac{x}{2}\,dx}{1+\frac{a-b}{a+b}\tan^2\frac{x}{2}}=\frac{1}{a+b}\int\frac{2d\left(\tan\frac{x}{2}\right)}{1+\frac{a-b}{a+b}\tan^2\frac{x}{2}}=\frac{2}{a+b}\times$$

$$\frac{(a+b)^{\frac{1}{2}}}{(a-b)^{\frac{1}{2}}}\int\frac{\frac{(a-b)^{\frac{1}{2}}}{(a+b)^{\frac{1}{2}}}d\left(\tan\frac{x}{2}\right)}{1+\frac{a-b}{a+b}\tan^2\frac{x}{2}}=\frac{2}{(a^2-b^2)^{\frac{1}{2}}}\tan^{-1}\left\{\frac{(a-b)^{\frac{1}{2}}}{(a+b)^{\frac{1}{2}}}\tan\frac{x}{2}\right\}+C,\text{ when } a>b.$$

When $a<b$, we have

$$y=\int\frac{dx}{a+b\cos x}=\int\frac{\sec^2\frac{x}{2}\,dx}{(b+a)-(b-a)\tan^2\frac{x}{2}}=\int\frac{2d\left(\tan\frac{x}{2}\right)}{(b+a)-(b-a)\tan^2\frac{x}{2}}$$

$$=\int\frac{\frac{1}{(b+a)^{\frac{1}{2}}}d\left(\tan\frac{x}{2}\right)}{(b+a)^{\frac{1}{2}}-(b-a)^{\frac{1}{2}}\tan\frac{x}{2}}+\int\frac{\frac{1}{(b+a)^{\frac{1}{2}}}d\left(\tan\frac{x}{2}\right)}{(b+a)^{\frac{1}{2}}+(b-a)^{\frac{1}{2}}\tan\frac{x}{2}}$$

$$=\frac{-1}{(b^2-a^2)^{\frac{1}{2}}}\int\frac{-(b-a)^{\frac{1}{2}}d\left(\tan\frac{x}{2}\right)}{(b+a)^{\frac{1}{2}}-(b-a)^{\frac{1}{2}}\tan\frac{x}{2}}+\frac{1}{(b^2-a^2)^{\frac{1}{2}}}\int\frac{(b-a)^{\frac{1}{2}}d\left(\tan\frac{x}{2}\right)}{(b+a)^{\frac{1}{2}}+(b-a)^{\frac{1}{2}}\tan\frac{x}{2}}$$

$$=\frac{1}{(b^2-a^2)^{\frac{1}{2}}}\log\left\{\frac{(b+a)^{\frac{1}{2}}+(b-a)^{\frac{1}{2}}\tan\frac{x}{2}}{(b+a)^{\frac{1}{2}}-(b-a)^{\frac{1}{2}}\tan\frac{x}{2}}\right\}+C.$$

SCH.—The four preceding sections comprise the greater part of what is known concerning abstract methods of passing from the differential to the *exact* integral function of a single variable; but there are many other methods of great practical importance for determining the *approximate* value of the integral of a differential which cannot be integrated by these methods. One of the most useful and simple is given in the next section.

SECTION V.

Integration by Infinite Series.

202. It often occurs that a differential can be expanded into an infinite series, the terms of which can be integrated separately. If the result is a converging series, the value of the integral may be determined with sufficient accuracy for practical purposes by summing a finite number of terms. It may also sometimes happen that the law of the series is such that its exact sum may be found, although the series itself is infinite.

This method is not only a last resort when the methods of exact integration fail, but it is sometimes serviceable by being more simple than they, even when they are applicable; moreover, it affords a method of developing such a function.

Ex. 1. Integrate $dy = x^{\frac{1}{2}}(1 - x^2)^{\frac{1}{2}}dx$.

SOLUTION.—Expanding $(1 - x^2)^{\frac{1}{2}}$ by the binomial or Maclaurin's theorem, we have

$$(1 - x^2)^{\frac{1}{2}} = 1 - \tfrac{1}{2}x^2 - \tfrac{1}{8}x^4 - \tfrac{1}{16}x^6 - \tfrac{5}{128}x^8 - \text{, etc.}$$

$$\therefore y = \int x^{\frac{1}{2}}(1-x^2)^{\frac{1}{2}}dx = \int x^{\frac{1}{2}}dx - \tfrac{1}{2}\int x^{\frac{5}{2}}dx - \tfrac{1}{8}\int x^{\frac{9}{2}}dx - \tfrac{1}{16}\int x^{\frac{13}{2}}dx - \tfrac{5}{128}\int x^{\frac{17}{2}}dx - \text{etc}$$

$$= \tfrac{2}{3}x^{\frac{3}{2}} - \tfrac{1}{7}x^{\frac{7}{2}} - \tfrac{1}{44}x^{\frac{11}{2}} - \tfrac{1}{120}x^{\frac{15}{2}} - \tfrac{5}{1216}x^{\frac{19}{2}} - \text{etc.} + C.$$

This series is converging for $x < 1$ and more rapidly converging as x is less.

Ex. 2. Integrate $dy = \dfrac{dx}{(1 + x^4)^{\frac{1}{2}}}$.

$$y = x - \tfrac{1}{2}\frac{x^5}{5} + \frac{1 \cdot 3}{2 \cdot 4}\frac{x^9}{9} - \frac{1 \cdot 3 \cdot 5}{2 \cdot 4 \cdot 6}\frac{x^{13}}{13} + \text{etc} + C.$$

Ex. 3. Integrate $dy = \dfrac{(x - 1)^{\frac{2}{3}}dx}{x^{\frac{1}{2}}}$.

$$y = \frac{6}{7}x^{\frac{7}{6}} + 4x^{\frac{1}{6}} + \frac{2}{15}x^{-\frac{5}{6}} - \frac{8}{297}x^{-\frac{11}{6}} - \text{etc.} + C.$$

Ex. 4. Integrate $dy = \frac{dx}{1+x^2}$ in an infinite series, and thus obtain a development of $y = \tan^{-1}x$.

$$y = \int \frac{dx}{1+x^2} = \tan^{-1}x^* + C = x - \frac{x^3}{3} + \frac{x^5}{5} - \frac{x^7}{7} + \frac{x^9}{9} - \text{etc.} + C.$$

Ex. 5. Integrate $dy = \frac{dx}{\sqrt{1-x^4}}$ in an infinite series and thus obtain a development of $y = \sin^{-1}x$.

$$y = \int \frac{dx}{\sqrt{1-x^2}} = \sin^{-1}x + C = x + \frac{x^3}{6} + \frac{3x^5}{40} + \frac{5x^7}{112} + \text{etc.} + C.$$

Ex. 6. Integrate $dy = \frac{dx}{1+x}$ in an infinite series and thus obtain a development of $y = \log(1+x)$.

$$y = \int \frac{dx}{1+x} = \log(1+x) + C = x - \frac{x^2}{2} + \frac{x^3}{3} - \frac{x^4}{4} + \text{etc.} + C.$$

SECTION VI.

Successive Integration.

203. Prob.—*To integrate a second, third, or* n*th differential function of a single equicrescent variable.*

Solution.—Since dx is constant we may write $\int d^2y = \int f(x)dx^2 = dx\int f(x)dx$, and integrate $f(x)dx$ as before, thus reducing the degree of the element (differential) by unity. Putting $\int f(x)dx = f_1(x) + C_1$, we have $\int d^2y = \int f(x)dx^2 = dx\int f(x)dx = f_1(x)dx + C_1dx$. But $\int d^2y = dy$; hence $dy = f_1(x)dx + C_1dx$.

Integrating again we have $y = f_2(x) + C_1x + C_2$.

In like manner from $d^3y = f(x)dx^3$, we have $\int d^3y = d^2y = dx^2\int f(x)dx = dx^2[f_1(x) + C_1] = f_1(x)dx^2 + C_1dx^2$.

Integrating again,

$$\int d^2y = dy = dx\int f_1(x)dx + dx\int C_1dx = dx[f_2(x) + C_1x + C_2] = f_2(x)dx + C_1x\,dx + C_2dx.$$

Integrating a third time, we have

$$\int dy = y = \int f_2(x)dx + C_1\int x\,dx + C_2\int dx = f_3(x) + \tfrac{1}{2}C_1x^2 + C_2x + C_3.$$

From these processes we deduce

$$\int^{n\dagger} d^ny = y = f_n(x) + \frac{C_1x^{n-1}}{1\cdot2\cdot3\cdots(n-1)} + \frac{C_2x^{n-2}}{1\cdot2\cdot3\cdots(n-2)} + \frac{C_3x^{n-3}}{1\cdot2\cdot3\cdots(n-3)} + \cdots + C_{n-2}\frac{x^2}{1\cdot2} + C_{n-1}x + C_n.$$

* By preceding methods.

† This notation signifies the nth integral.

Ex. 1. Integrate $d^3y = 6a\,dx^3$.

SOLUTION. $\int d^3y = d^2y = \int 6a\,dx^3 = 6a\,dx^2 \int dx = 6ax\,dx^2 + C_1{}^*dx^2$.
Again, $\int d^2y = dy = \int [6ax\,dx^2 + C_1\,dx^2] = 6a\,dx \int x\,dx + C_1 dx \int dx =$
$3a\,x^2dx + C_1 x\,dx + C_2 dx$.
Finally, $y = ax^3 + \frac{1}{2}C_1x^2 + C_2x + C_3$.

Ex. 2. Integrate $d^4y = \cos x\,dx^4$.

PROCESS.† $d^3y = \sin x\,dx^3 + C_1dx^3$,
$d^2y = -\cos x\,dx^2 + C_1x\,dx^2 + C_2dx^2$,
$dy = -\sin x\,dx + \frac{1}{2}C_1x^2dx + C_2x\,dx + C_3dx$,
$y = \cos x + \frac{1}{6}C_1x^3 + \frac{1}{2}C_2x^2 + C_3x + C_4$.

SCH. 1.—It will be observed that whatever value we assign the constants, we get the original differential by differentiating the integral as many times as we integrated. The reason for this, if not seen at once, will appear upon performing the operation.

Ex. 3. Given $\frac{d^3y}{dx^3} = 0$, to find the integral y.

SOLUTION.—Since the third differential coefficient is the differential of the second differential coefficient, divided by the differential of the variable, *i. e.* $\frac{d^3y}{dx^3} = \frac{d\left(\frac{d^2y}{dx^2}\right)}{dx}$, when $\frac{d^3y}{dx^3} = 0$, $\frac{d^2y}{dx^2}$ must be a *constant*. Hence in this example, $\frac{d^2y}{dx^2} = C_1$, or $d^2y = C_1dx^2$. From this we obtain $y = \frac{1}{2}C_1x^2 + C_2x + C_3$.

Ex. 4. Given $\frac{d^3y}{dx^3} = \frac{2}{x^3}$ to find the integral y.

$$y = \log x + \frac{1}{2}C_1x^2 + C_2x + C_3.$$

Ex. 5. Integrate $d^2y = \sin x \cos^2 x\,dx^2$.

SOLUTION.—Put $\sin x = v$; whence $dv = \cos x\,dx$, $\cos^2 x\,dx^2 = dv^2$, and $d^2y = \sin x \cos^2 x\,dx^2 = v\,dv^2$. From $d^2y = v\,dv^2$, we obtain as before $y = \frac{1}{6}v^3 + C_1v + C_2$.
$\therefore\ y = \frac{1}{6}\sin^3 x + C_1 \sin x + C_2$.

Ex. 6. Integrate $d^2y = \cos x \sin^2 x\,dx^2$.

$$y = \frac{1}{6}\cos^3 x + C_1 \cos x + C_2.$$

SCH. 2.—In order to integrate successively $d^ny = f(x)dx^n$, according to the foregoing process, it is necessary that we be able to integrate exactly $f(x)dx$, $f_1(x)dx$, $f_2(x)dx$, etc. It is evident that this will be frequently impossible. A method of approximation which is often serviceable in such cases, is readily obtained by means of Maclaurin's Formula.

* We might write $6aC_1$ as the coefficient of this term; but as C_1 represents any constant, it is unnecessary to retain the $6a$.

† This is simply a convenient form in which to write the operation; the student should understand the process and be able to explain it, as in the preceding solution.

204. Prop.—*The* n*th integral of* f(x)dx^n *may be developed into a series by the following formula:*

$$y = \int^n d^n y = \int^n f(x)dx^n = C_n + C_{n-1}\frac{x}{1} + C_{n-2}\frac{x^2}{1\cdot 2} + C_{n-3}\frac{x^3}{1\cdot 2\cdot 3} + \text{- - - -}$$

$$C_1\frac{x^{n-1}}{1\cdot 2\cdot 3\text{ - - - }(n-1)} + [f(x)]\frac{x^n}{1\cdot 2\cdot 3\text{ - - - }n} + \left[\frac{df(x)}{dx}\right]\frac{x^{n+1}}{1\cdot 2\cdot 3\text{ - - - }(n+1)}$$

$$+ \left[\frac{d^2f(x)}{dx^2}\right]\frac{x^{n+2}}{1\cdot 2\cdot 3\text{ - - - }(n+2)} + \left[\frac{d^3f(x)}{dx^3}\right]\frac{x^{n+3}}{1\cdot 2\cdot 3\text{ - - - }(n+3)} +, \text{ etc.}$$

Dem.—1st. Developing $y = \int^n f(x)dx^n$ by Maclaurin's formula, we have

$$y = \int^n f(x)dx^n = [\int^n f(x)dx^n] + [\int^{n-1} f(x)dx^{n-1}]^*\frac{x}{1} + [\int^{n-2} f(x)dx^{n-2}]\frac{x^2}{1\cdot 2},$$

$$+ [\int^{n-3} f(x)dx^{n-3}]\frac{x^3}{1\cdot 2\cdot 3} + \text{- - - - - - } [\int f(x)dx]\frac{x^{n-1}}{1\cdot 2\cdot 3\text{ - - - }(n-1)},$$

$$+ [f(x)]\frac{x^n}{1\cdot 2\cdot 3\text{ - - - }n} + \left[\frac{df(x)}{dx}\right]\frac{x^{n+1}}{1\cdot 2\cdot 3\text{ - - - }(n+1)} + \left[\frac{d^2f(x)}{dx^2}\right]\frac{x^{n+2}}{1\cdot 2\cdot 3\text{ - - - }(n+2)} + \text{etc.}$$

2nd. Comparing this development with the development of $\int^n f(x)dx^n$ as made in the solution Art. **203,** remembering that $x = 0$ in the bracketed factors of the present series, and that these factors are therefore constant, we see that the present series is that of Art. **203** reversed, extended and generalized; that $[\int^n f(x)dx^n]$ is C_n, $[\int^{n-1} f(x)dx^{n-1}]$ is C_{n-1}, etc., and that the former series begins with the term $[f(x)]\frac{x^n}{1\cdot 2\cdot 3\text{ - - - }n}$ of the latter. Hence using C_n, C_{n-1}, etc. for these constant factors the above development becomes

$$y = \int^n d^n y = \int^n f(x)dx^n = C_n + C_{n-1}\frac{x}{1} + C_{n-2}\frac{x^2}{1\cdot 2} + C_{n-3}\frac{x^3}{1\cdot 2\cdot 3} + \text{- - - - - - -}$$

$$\text{- - - } C_1\frac{x^{n-1}}{1\cdot 2\cdot 3\text{ - - - }(n-1)} + [f(x)]\frac{x^n}{1\cdot 2\cdot 3\text{ - - - }n} + \left[\frac{df(x)}{dx}\right]\frac{x^{n+1}}{1\cdot 2\cdot 3\text{ - - - }(n+1)}$$

$$+ \left[\frac{d^2f(x)}{dx^2}\right]\frac{x^{n+2}}{1\cdot 2\cdot 3\text{ - - - }(n+2)} + \left[\frac{d^3f(x)}{dx^3}\right]\frac{x^{n+3}}{1\cdot 2\cdot 3\text{ - - - }(n+3)} + \text{etc.} \quad \text{Q. E. D.}$$

Sch.—This formula is readily remembered and applied by noticing the law of the first part of this series (that containing the constants C_n, C_{n-1}, etc.); and then observing that the second part of the series {that from and including $[f(x)]\frac{x^n}{1\cdot 2\cdot 3\text{ - - - }n}$}, is the development of $f(x)$ by Maclaurin's Formula, each term being multiplied by x^n, and the successive terms divided by $1\cdot 2\text{ - - - }n$, $2\cdot 3\text{ - - - }(n+1)$, $3\cdot 4\text{ - - - }(n+2)$ respectively.

Ex. Develop $y = \int^4 d^4y = \int^4 \frac{dx^4}{\sqrt{1-x^2}}$.

Solution.—Here we have $n = 4$, and $f(x) = \frac{1}{\sqrt{1-x^2}} = (1-x^2)^{-\frac{1}{2}}$

Developing $(1-x^2)^{-\frac{1}{2}}$ by Maclaurin's Formula we have

* To differentiate an integral is to depress its order, as will be seen from the nature of the processes.

$$(1-x^2)^{-\frac{1}{2}} = 1 + \tfrac{1}{2}x^2 + \frac{1\cdot 3}{2\cdot 4}x^4 + \frac{1\cdot 3\cdot 5}{2\cdot 4\cdot 6}x^6 +, \text{ etc.}$$

Hence by the above scholium, we obtain

$$y = \int^4 d^4y = \int^4 \frac{dx^4}{\sqrt{1-x^2}} = C_4 + C_3\frac{x}{1} + C_2\frac{x^2}{1\cdot 2} + C_1\frac{x^3}{1\cdot 2\cdot 3} + \frac{x^4}{1\cdot 2\cdot 3\cdot 4} + \frac{x^6}{1\cdot 2\cdot 3\cdot 4\cdot 5\cdot 6}$$
$$+ \frac{1\cdot 3x^8}{1\cdot 2\cdot 4\cdot 5\cdot 6\cdot 7\cdot 8} + \frac{1\cdot 3\cdot 5x^{10}}{1\cdot 2\cdot 4\cdot 6\cdot 7\cdot 8\cdot 9\cdot 10} +, \text{ etc.}$$

SECTION VII.

Definite Integration and the Constants of Integration.

205. Def.—***An Indefinite Integral*** is one in which the constant or constants of integration remain undetermined, and which has not been satisfied by any particular value of the variable.

Ill.—All the integrals hitherto produced are indefinite integrals.

206. Def.—***A Corrected Integral*** is one in which the value of the constant (or constants) of integration has been determined and substituted for the general symbol C.

207. Def.—An integral is said to be taken between ***Limits,*** when the indefinite integral has been satisfied for two different values of the variable, and the difference between these results taken.

208. Def.—***A Definite Integral*** is an integral taken between limits.

A definite integral and the limits between which it is taken is symbolized thus: $y = \int_a^b f(x)dx$. This signifies that the indefinite integral of $f(x)dx$ is to be obtained, and first satisfied by substituting b for x, then by substituting a for x; and that, finally, the latter is to be subtracted from the former. $x = a$ and $x = b$ are called the limits of the integral, the former being called the *inferior* and the latter the *superior* limit.*

Ex. 1. Find the value of $y = \int_1^3 5x^2dx$.

Sug's.—The indefinite integral is $y' = \frac{5}{3}x^3 + C$.

Now this is true for all values of x, hence for $x = 3$. For this value of x, we have

---

* It is assumed that x can have such values as we assign it, and that the function is continuous between these values, *i. e.* that it does not become imaginary or infinite for any intermediate value of x.

we have $y'' = 45 + C$. In like manner for $x = 1$
$y''' = \frac{5}{3} + C$. Subtracting the latter

from the former, $y'' - y''' = 43\frac{1}{3}$. $\therefore y = y'' - y''' = \int_1^3 5x^2 dx = 43\frac{1}{3}$.

Ex. 2. Find the definite integral of $dy = nx\,dx$ between the limits a and b. $y = \int_a^b nx\,dx = \frac{n(b^2 - a^2)}{2}$.

Ex. 3. Find the value of $y = \int_0^b (x^3dx - b^2x\,dx)$. *Ans.*, $-\frac{1}{4}b^4$.

---

209. *Disposing of the Constant of Integration.* There are two principal methods of disposing of the constant of integration, C:

1st. By integrating between limits, the constant is eliminated. This is illustrated in the preceding examples.

2nd. When there is anything in the nature of the problem under discussion, from which we can know the value of the function for some particular value of the variable, by substituting these values in the indefinite integral, the value of the constant C can be found. And, as C is a constant, if we find its value for any particular value of the variable, it has the same value for *all* values of the variable.

Ex. 1. Find the corrected integral of the function $dz = (dx^2 + \frac{9}{4}ax\,dx^2)^{\frac{1}{2}}$, on the hypothesis that $z = 0$ when $x = 0$.

SUG'S.—The indefinite integral is $z = \frac{8}{27a}(1 + \frac{9}{4}ax)^{\frac{3}{2}} + C$. Now if $z = 0$ when $x = 0$, we have $0 = \frac{8}{27a} + C$. $\therefore C = -\frac{8}{27a}$. Substituting this value of C in the indefinite integral, we have, as the corrected integral, $z = \frac{8}{27a}(1 + \frac{9}{4}ax)^{\frac{3}{2}} - \frac{8}{27a}$.

Ex. 2. What is the value of C, when $du = \frac{2\pi y}{p}(p^2 + y^2)^{\frac{1}{2}}dy$, if $u = 0$ when $y = 0$? *Ans.*, $C = -\frac{2}{3}\pi p^3$.

Ex. 3. Given $du = (2r)^{\frac{1}{2}}(2r - y)^{-\frac{1}{2}}dy$, what is the value of C, if $u = 0$ when $y = 0$? What if $u = 0$ when $y = 2r$?

Answers, $4r$, 0.

ILL.—A differential is one of the infinitesimal elements of which a quantity is conceived as composed. Thus, let A represent the *area* of the surface lying between **AM** and **AX**, *Fig.* 35; **PD** being *any* ordinate, and ab the consecutive

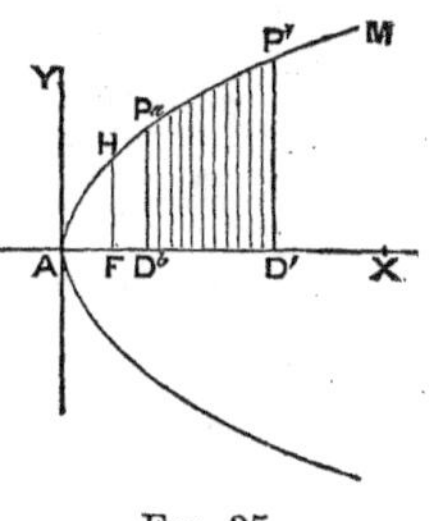

FIG. 35.

ordinate, **PD**ab may be considered as representing an element of the area, or dA. This element in the case of the parabola is found to be $\sqrt{2px}\,dx$; hence $dA = \sqrt{2px}\,dx$. This expression therefore represents *any* one of the infinitesimal elements of which the quantity A is composed.

The *Indefinite Integral* is $A = \frac{2}{3}(2p)^{\frac{1}{2}}x^{\frac{3}{2}} + C$.

This is indefinite in two respects. 1st, it is true for *any* value of x; 2nd, it is indefinite, and in fact indeterminate, as regards C, the value of which may be anything.

AS REGARDS THE CONSTANT, *if we choose to estimate the area from* **A**, so that $A = 0$ when $x = 0$, we have $0 = 0 + C$. $\therefore$ $C = 0$. Hence the corrected integral is $A = \frac{2}{3}(2p)^{\frac{1}{2}}x^{\frac{3}{2}}$. This represents the area, *estimated from* **A** to any value we may choose to give x. If $x =$ **AD**$'$, $A =$ area **AP**$'$**D**$'$. Again, *if we choose to estimate the area from the focal ordinate* **HF**, calling $A = 0$, when $x = \frac{1}{2}p$, we have $0 = \frac{2}{3}(2p)^{\frac{1}{2}}(\frac{1}{2}p)^{\frac{3}{2}} + C$, or $\frac{1}{3}p^2 + C$. $\therefore$ $C = -\frac{1}{3}p^2$, and the corrected integral is $A = \frac{2}{3}(2p)^{\frac{1}{2}}x^{\frac{3}{2}} - \frac{1}{3}p^2$. This represents the area estimated from the focal ordinate **HF** to any value we choose to give x. Thus if $x =$ **AD**$'$, this corrected integral represents the area **HFP**$'$**D**$'$. Finally, suppose we ask, Where must the area be conceived as beginning in order that $C = -m$? To meet this case we have $0 = \frac{2}{3}(2p)^{\frac{1}{2}}x^{\frac{3}{2}} - m$, whence $x = \sqrt[3]{\frac{m^2}{2p}}$. If therefore the area is conceived as commencing at the ordinate corresponding to $x = \sqrt[3]{\frac{9m^2}{8p}}$, $C = -m$. Thus we perceive the indeterminate character of C. We also observe the limits of its possible values. In this case C may be *any* negative quantity, but *no* positive quantity.

INTEGRATION BETWEEN LIMITS is illustrated by considering the area as estimated from some possible place (no matter where) and extending 1st to $x = a$, whence $A' = \frac{2}{3}(2p)^{\frac{1}{2}}a^{\frac{3}{2}} + C$; and 2nd to $x = b$, whence $A'' = \frac{2}{3}(2p)^{\frac{1}{2}}b^{\frac{3}{2}} + C$. Now the difference between A' and A'' will represent the area between the ordinates corresponding to $x = a$ and $x = b$. Call this A'''; and $A''' = \frac{2}{3}(2p)^{\frac{1}{2}}(b^{\frac{3}{2}} - a^{\frac{3}{2}})$, if $a < b$. Letting **AD** $= a$ and **AD**$' = b$, **A**$''' = \frac{2}{3}(2p)^{\frac{1}{2}}(b^{\frac{3}{2}} - a^{\frac{3}{2}}) =$ area **PDP**$'$**D**$'$.

This subject will have more ample illustration in *Sections IX.—XIII.* inclusive, which the student is now prepared to read.

THE END.

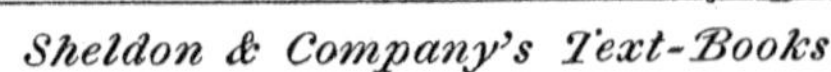

www.ingramcontent.com/pod-product-compliance
Lightning Source LLC
LaVergne TN
LVHW011200110826
845150LV00006B/1270

* 9 7 8 1 4 2 5 5 7 1 9 7 9 *